IoT for Smart Grid

IEEE Press
445 Hoes Lane
Piscataway, NJ 08854

IoT for Smart Grid

Revolutionizing Electrical Engineering

Edited by

Rahiman Zahira
Senior Member IEEE, B.S. Abdur Rahman Crescent Institute of Science and Technology
Chennai, Tamil Nadu, India

Palanisamy Sivaraman
Senior Member IEEE, Anna University, Chennai, Tamil Nadu, India

Chenniappan Sharmeela
Senior Member IEEE, Anna University, Chennai, Tamil Nadu, India

Sanjeevikumar Padmanaban
Senior Member IEEE, University of South-Eastern Norway, Norway

IEEE Press
WILEY

Published by John Wiley & Sons, Inc., Hoboken, New Jersey.
Published simultaneously in Canada.

Library of Congress Cataloging-in-Publication Data Applied for:

Hardback ISBN: 9781394279371

Cover Design: Wiley
Cover Image: © jamesteohart/Adobe Stock Photos

Set in 9.5/12.5pt STIXTwoText by Straive, Chennai, India

Contents

About the Editors

Dr. Rahiman Zahira is an IEEE Senior Member. She received her BE in Electrical and Electronics Engineering from the University of Madras in 2004, her ME in Power Systems Engineering from B.S. Abdur Crescent Engineering College, Anna University, Chennai, in 2006, and earned her Doctoral degree from Anna University in February 2018. With 18 years of teaching experience, she began her academic career as a lecturer in 2006 and currently serves as an associate professor at B.S. Abdur Rahman Crescent Institute of Science and Technology, Chennai. She has published over 70 publications in national and international journals and conference proceedings, contributed to 11 book chapters, edited 3 books. She also holds one international patent (granted) and three patents (published). Dr. Zahira is actively involved in the IEEE Standards Association. She serves as a working group member of IEEE P1729 – Recommended Practice for Electric Power Distribution System Analysis, IEEE P2882 WG-STDS-P2882, WG-SG2 Steady-state, Harmonic and Dynamic Stability, and as secretary and working group member of the IEEE PES Task Force session on Demand Flex metrics standardization for grid-interactive buildings and customer systems.

She has received awards from Guinness World Records (UK), World Book of Records (London) for participating in the longest 150-hour conference, Asian Book, and Indian Book of Records for being an editor and author in a record book titled “Most authors contributing for a single book: COVID-19 and its impact.” She has also been honored with the Best Innovation in Teaching Award 2021, the Young Educator & Scholar Award in the 10th National Teachers’ Day Awards 2019, the Women Researcher Award, the Outstanding Scientist Award, the Innovative Technological Researcher & Dedicated Academician Award (Electrical Engineering), and the Best Academics award from the BSA Crescent Alumni Association during Hangout 2019. Dr. Zahira has guided over 20 undergraduates and postgraduate students and 2 research scholars. She is a life member in 7 Professional Bodies. She serves as a reviewer for reputable journals, an editorial board member, and an advisory member for numerous conferences. Her areas of interest include power quality, harmonic suppression, active filter control techniques, renewable energy systems, microgrids, smart grids, and electric vehicle charging systems.

ORCID ID: https://orcid.org/my-orcid?orcid=0000-0002-5492-9048
LinkedIn: https://www.linkedin.com/in/drrzahira/
Google Scholar ID: https://scholar.google.com/citations?user=NcKB9_UAAAAJ&hl=en

Mr. Palanisamy Sivaraman (Member'20, Senior Member'21, IEEE) was born in Vellalur, Madurai district, Tamil Nadu, India. He completed schooling in Government Higher Secondary School, Vellalur and earned a BE in Electrical and Electronics Engineering and an ME in Power Systems Engineering from Anna University, Chennai, India, in 2012 and 2014, respectively. With over 10 years of industrial experience, he specializes in the field of power system studies and grid code compliance for renewable power plants, including solar and wind power plants and battery energy storage systems. Currently, he is an industry working professional and also a Research Scholar, Department of EEE, Anna University, Chennai, India. He has trained over 500 personnel on renewable energy and power quality. A proficient user of power system simulation software like ETAP, PSCAD, DIGSILENT POWER FACTORY, PSSE, and MATLAB, he actively participates in the IEEE Standards Association. Mr. Sivaraman is a working group member of IEEE standard, including P2800.2 (Recommended Practice for Test and Verification Procedures for Inverter-based Resources (IBRs) Interconnecting with Bulk Power Systems), P1729 (IEEE Recommended Practice for Electric Power Distribution System Analysis), P2418.5 (Standard for Blockchain in Power and Energy), P1854 (Guide for Smart Distribution Systems), P2688 (Recommended Practice for Energy Storage Management Systems in Energy Storage Applications), and IEEE 3001.9-2023 (Design of Power Systems Supplying Lighting Systems in Commercial and Industrial Facilities).

He is also a member of the IEEE PES task force on Energy Storage. He had authored/co-authored/edited ten books in the field of electrical engineering with Elsevier and Wiley-IEEE Press and published several papers at national and international conferences. Mr. Sivaraman is a Senior Member of the Institute of Electrical and Electronics Engineers (IEEE), a member of the International Council on Large Electric Systems (CIGRE), and an Associate Member of the Institution of Engineers (India). He holds a Professional Engineer (PEng) certification from the Institution of Engineers, India. He is a recognized speaker well versed in both National and International Standards.

Google Scholar link: https://scholar.google.co.in/citations?user=XLdd0mgAAAAJ&hl=en&authuser=1

Dr. Chenniappan Sharmeela holds a BE in Electrical and Electronics Engineering, an ME in Power Systems Engineering from Annamalai University, Chidambaram, and a PhD in Electrical Engineering from the College of Engineering, Guindy, Anna University, Chennai. She currently serves as Professor and Professor-In-Charge of Power Engineering and Management in the Department of Electrical and Electronics Engineering at C.E.G., Anna University, Chennai. She is also actively involved in research as a professor at the Centre for E-Vehicle Technologies and the Centre for Energy Storage Technology, Anna University, Chennai. From 2015 to 2018, she served as Assistant Director of the Centre for Entrepreneurship Development at C.E.G., Anna University, Chennai. Dr. Sharmeela has undertaken numerous consultancies on Renewable Energy Systems, including Solar Photovoltaic (SPV) Power Systems, power quality measurements, and the design of compensators for industries. She has coordinated and organized several short-term courses on power quality for Tamil Nadu State Electricity Board Engineers. She has delivered several invited talks and trained over 1000 engineers on the importance of Power Quality, Power Quality Standards, and the design of SPV power systems for more than 12 years in leading organizations such as CII, FICCI, CPRI, MSME, GE (Alsthom), and APQI. In 2011, she received a grant from CTDT, Anna University, for a two-year project on “Energy Efficient Solar-Based Lighting System for Domestic Application.” In 2020, she received a research grant from AICTE – RPS, New Delhi, India, on “Smart EV Charging Station.” Dr. Sharmeela has authored over 30 journal papers in refereed international journals and more than 60 papers in international and national conferences. She has authored/co-authored/edited 12 book chapters, edited 10 books, and authored 2 books. Her areas of interest include power quality, power electronics applications to power systems, smart grid, energy storage systems, renewable energy systems, electric vehicles, battery management systems, and electric vehicle supply equipment. She is a Senior Member of IEEE; a Member of the IEEE – Power and Energy Society; a Fellow of the Institution of Engineers (India); a Life Member of ISTE; a member of the Central Board of Irrigation and Power (CBIP), New Delhi, India; and a member of SSI, India. With over 24 years of experience in teaching, research, and consultancy in the areas of power quality and power systems, Dr. Sharmeela is an active participant in the IEEE Standards Association. She is a working group member of IEEE standards P2800.2 (Recommended Practice for Test and Verification Procedures for Inverter-based Resources Interconnecting with Bulk Power Systems) and P1729 (Recommended Practice for Distribution System Analysis). She is also a working group member of the IEEE PES task force on Energy Storage. She has authored/co-authored/edited nine books in the field of electrical engineering with Elsevier and Wiley-IEEE Press, notably “Fast charging infrastructure for electric and hybrid electric vehicles” by Wiley-IEEE Press in 2023 and “Power system operation with 100% renewable energy resources” by Elsevier in 2023.

ORCID ID: https://orcid.org/0000-0001-6706-4779

Dr. Sanjeevikumar Padmanaban (Member’12–Senior Member’15, IEEE) received a Ph.D. degree in electrical engineering from the University of Bologna, Bologna, Italy 2012. He is a Full Professor in Electrical Power Engineering at the Department of Electrical Engineering, Information Technology, and Cybernetics, University of South-Eastern Norway, Norway. S. Padmanaban has authored over 750+ scientific papers and received the Best Paper cum Most Excellence Research Paper Award from IET-SEISCON’13, IET-CEAT’16, IEEE-EECSI’19, IEEE-CENCON’19, and five best paper awards from ETAEERE’16 sponsored Lecture Notes in Electrical Engineering, Springer book. He is a Fellow of the Institution of Engineers, India, the Institution of Electronics and Telecommunication Engineers, India, and the Institution of Engineering and Technology, U.K. He received a lifetime achievement award from Marquis Who’s Who – USA 2017 for contributing to power electronics and renewable energy research. He is listed among the world’s top 2 scientists (from 2019) by Stanford University USA.

He served an Editor/Associate Editor/Editorial Board for refereed journals, in particular the *IEEE Systems Journal*, *IEEE Transaction on Industry Applications*, *IEEE Access*, *IET Power Electronics*, *IET Electronics Letters*, and Wiley-*International Transactions on Electrical Energy Systems*, Subject Editorial Board Member—*Energy Sources—Energies Journal*, MDPI, and the Subject Editor for the *IET Renewable Power Generation*, *IET Generation, Transmission and Distribution*, and *FACETS Journal* (Canada).

List of Contributors

Sukumaran Aasha Nandhini
Department of Electronics and Communication Engineering
Sri Sivasubramaniya Nadar College of Engineering
Chennai
Tamil Nadu
India

Naser S. Almutairi
Department of Computer System and Technology
Faculty of Computer Science and Information Technology
Universiti Malaya
Kuala Lumpur
Malaysia

Veerasamy Balaji
Department of Electrical and Electronics Engineering
PSG College of Technology
Coimbatore
Tamil Nadu
India

Nalini P. Behera
School of Computer Engineering
KIIT Deemed to be University
Bhubaneswar
Odisha
India

Debashish Bhowmik
Electrical Engineering Department
Mizoram University
Aizawl
Mizoram
India

Zain Buksh
Department of Computer Science and Mathematics
The University of Fiji
Lautoka
Fiji

Rishal Chand
Department of Computer Science and Mathematics
The University of Fiji
Lautoka
Fiji

Jayachandran Divya Navamani
Department of Electrical and Electronics Engineering
SRM Institute of Science and Technology
Kattankulathur
Tamil Nadu
India

Pijush K. Dutta Pramanik
School of Computer Applications and Technology
Galgotias University
Greater Noida
Uttar Pradesh
India

Gunasekaran Ezhilarasi
Department of EEE
Sri Sairam Institute of Technology
Chennai
Tamil Nadu
India

Khang W. Goh
School of Engineering
Shinawatra University
Pathum Thani
Thailand

and

Faculty of Data Science and Information Technology
INTI International University
Nilai
Malaysia

Balan Gunapriya
Department of Electrical and Electronics Engineering
New Horizon College of Engineering
Bangalore
Karnataka
India

G. Jagadish
School of Electrical Engineering
Vellore Institute of Technology
Chennai
Tamil Nadu
India

Naseer Ahamed Javed
Department of Computer Science and Engineering
Jerusalem College of Engineering
Anna University
Chennai
Tamil Nadu
India

and

Bharath Institute of Higher Education and Research
Chennai
Tamil Nadu
India

G. Kabilan
School of Electrical Engineering
Vellore Institute of Technology
Chennai
Tamil Nadu
India

Kallankurichy P. Kaliyamurthie
Department of Information Technology
Bharath Institute of Higher Education and Research
Chennai
Tamil Nadu
India

A. Kamalasegaran
School of Electrical Engineering
Vellore Institute of Technology
Chennai
Tamil Nadu
India

Ravichandran Karthick Manoj
Department of EEE
AMET Deemed to be University
Chennai
Tamil Nadu
India

Burhan Khan
Department of Computer System and Technology
Faculty of Computer Science and Information Technology
Universiti Malaya
Kuala Lumpur
Malaysia

Jashnil Kumar
Department of Computing Sciences and Information Systems
Fiji National University
Natabua
Lautoka
Fiji

Saurav Kumar
School of Computer Engineering
KIIT Deemed to be University
Bhubaneswar
Odisha
India

Ajay Kushwaha
Electrical and Instrumentation Engineering Department
Thapar Institute of Engineering and Technology
Patiala
Punjab
India

Dhandapani Lakshmi
Department of EEE
AMET Deemed to be University
Chennai
Tamil Nadu
India

Anbazhagan Lavanya
Department of Electrical and Electronics Engineering
SRM Institute of Science and Technology
Kattankulathur
Tamil Nadu
India

Shanmugasundaram Logeshkumar
Department of Electronics and Communication Engineering
Christ the King Engineering College
Anna University
Chennai
Tamil Nadu
India

Rajagopal Logesh Krishna
Department of Electrical and Electronics Engineering
PSG Institute of Technology and Applied Research
Coimbatore
Tamil Nadu
India

Aabid A. Mir
Malaysian Institute of Information Technology
Universiti Kuala Lumpur
Kuala Lumpur
Malaysia

Ravikumar Mithra
Department of Electrical and Electronics Engineering
PSG Institute of Technology and Applied Research
Coimbatore
Tamil Nadu
India

Mohamed Mustafa Mohamed Iqbal
Department of Electrical and Electronics Engineering
PSG Institute of Technology and Applied Research
Coimbatore
Tamil Nadu
India

Balasubramanian Nandhan
Department of Electrical and Electronics Engineering
PSG Institute of Technology and Applied Research
Coimbatore
Tamil Nadu
India

S. Nazrin Salma
Department of Electrical and Electronics Engineering
Thamirabharani Engineering College
Anna University
Chennai
Tamil Nadu
India

Niyas Ahamed
Department of Electronics and Communication Engineering
Thamirabharani Engineering College
Anna University
Chennai
Tamil Nadu
India

Rajesh K. Padmashini
Department of EEE
AMET Deemed to be University
Chennai
Tamil Nadu
India

Vallikanu Pramila
Department of Electrical and Electronics Engineering
B.S. Abdur Rahman Crescent Institute of Science and Technology
Chennai
Tamil Nadu
India

Kumaresa P. Punitha
Department of Electrical and Electronics Engineering
P S R Engineering College
Anna University
Chennai
Tamil Nadu
India

Rajasekharan Rajasree
Department of EEE
AMET Deemed to be University
Chennai
Tamil Nadu
India

Janarthanan N. Rajesh Kumar
Department of Computer science and Engineering
Sree Sastha Institute of Engineering and Technology
Chembarambakkam
Tamil Nadu
India

Yogesh Rajkumar
Department of Information Technology
Bharath Institute of Higher Education and Research
Chennai
Tamil Nadu
India

Nur F.L.M. Rosely
School of Computer Science
Faculty of Innovation and Technology
Taylor's University
Kuala Lumpur
Malaysia

Sanjeevikumar Padmanaban
Faculty of Technology, Natural Sciences and Maritime Sciences
Department of Electrical Engineering, Information Technology and Cybernetics
Campus Porsgrunn
University of South-Eastern Norway
Norway

Harsh Saran
School of Computer Engineering
KIIT Deemed to be University
Bhubaneswar
Odisha
India

Krishnakumar Shanmugasundaram
Capgemini US Corp
Charlotte
NC
USA

Neeraj A. Sharma
Department of Computer Science and Mathematics
The University of Fiji
Lautoka
Fiji

Chenniappan Sharmeela
Department of EEE
CEG
Anna University
Chennai
Tamil Nadu
India

A. B. M. Shawkat Ali
Bangladesh University of Business and Technology (BUBT)
Dhaka
Bangladesh

Palanisamy Sivaraman
Research Scholar
Department of EEE
Anna University
Chennai
Tamil Nadu
India

G. Srinivasan
Department of Electrical and Electronics Engineering
MVJ College of Engineering
Visvesvaraya Technological University
Bengaluru
Karnataka
India

P. Sriramalakshmi
School of Electrical Engineering
Vellore Institute of Technology
Chennai
Tamil Nadu
India

Sakthivel Sruthi
Department of Electrical and Electronics Engineering
PSG Institute of Technology and Applied Research
Coimbatore
Tamil Nadu
India

Aleena Swetapadma
School of Computer Engineering
KIIT Deemed to be University
Bhubaneswar
Odisha
India

Bijoy K. Upadhyaya
Department of Electronics and Communication Engineering
Tripura Institute of Technology
Narsingarh
Tripura
India

Rahiman Zahira
Department of Electrical and Electronics Engineering
B.S. Abdur Rahman Crescent Institute of Science and Technology
Chennai
Tamil Nadu
India

1

Introduction to the Internet of Things

Anbazhagan Lavanya[1], Jayachandran Divya Navamani[1], and Rahiman Zahira[2]

[1]*Department of Electrical and Electronics Engineering, SRM Institute of Science and Technology, Kattankulathur, Tamil Nadu, India*
[2]*Department of Electrical and Electronics Engineering, B.S. Abdur Rahman Crescent Institute of Science and Technology, Chennai, Tamil Nadu, India*

1.1 Introduction

The Internet of Things (IoT) is a cutting-edge technique that facilitates interface, communications, and data sharing among IoT devices. In the IoT, the information is transmitted since many sources are gathered for the purpose of creating up-to-date decisions and conducting analysis. Developing IoT applications encounters numerous problems, with security being a significant one. The IoT holds the possibility to enhance the overall quality of human existence by offering sophisticated applications that cater to the diverse requirements of humans across various domains, such as commercial, private, and industry. The IoT is constructed using the existing style of cyberspace and integrates mutually the Internet structure and developing engineering. The outcome of this amalgamation facilitates the seamless interconnection of an immense number of embedded systems, resulting in cost-effective service management, as well as enhanced scalability and flexibility.

The IoT has immense possibilities for creating value and is becoming recognized as the subsequent phase in the extensive digitization of the global economy. A rising phenomenon involves the interconnection of all devices that might benefit from a connection in a smart, energy-efficient, and profitable method. This revolutionizes the way individuals and companies connect to the physical world and with one other, to impact several aspects of everyday life, with its application domains spanning different sectors like health care, intelligent grids, transportation systems, automation in industry, and agriculture [1–6].

The IoT is a technique that involves interconnecting almost all items with intelligence, communication skills, and the ability to sense and act through Internet Protocol (IP) networks. The current state of the Internet has experienced a significant shift, moving away from being primarily driven by physical components such as computers, fibers, and Ethernet connections, to being driven by market forces and opportunities. This phenomenon has occurred as a consequence of the integration of seemingly separate intranets with robust software competences [7–9]. The IoT requires the use of open environments and a unified architecture consisting of interoperable platforms. Smart products and cyber-physical systems, commonly referred to as "things," are the most recent devices of the IoT. Such items are common products that have been improved using microcomputers, optoelectronic and/or radio transceivers, actuators and sensors, and interface stacks. They can gather data from their surroundings, act on that data, and interact with the physical world in environments with limited resources.

IoT for Smart Grid: Revolutionizing Electrical Engineering, First Edition.
Edited by Rahiman Zahira, Palanisamy Sivaraman, Chenniappan Sharmeela, and Sanjeevikumar Padmanaban.

The IoT, as a developing technology, is anticipated to provide innovative solutions for revolutionizing the function and purpose of several established industrial systems, including transportation and manufacturing systems. IoT devices are essential for the general advancement of IoT as they provide several applications in various domains [10–18]. Considering the growing demand and speedy advancement in sensors, particularly, have greatly impacted and transformed our daily lives, it is necessary to conduct an in-depth investigation of embedded platforms and boards.

1.2 Evolution of IoT

The Apollo Guidance Computer (AGC) was initially introduced by the National Aeronautics and Space Administration (NASA) in 1965. NASA purposefully designed this computer for the Apollo moon landing mission. NASA outfitted the Apollo Lunar Module and Command Module with a device that offered interfaces and processing capabilities for spacecraft control and navigation. Afterward, a sequence of important events occurred, influencing the direction of the IoT industry. These milestones encompass the founding of N M Electronics, presently Intel in 1968, the debut mobile phone call took place in 1973, the release of the initial personal computer in 1975, and the dedicated endeavors toward internet development in the 1980s. The evolution of the IoT technique is shown in Figure 1.1.

In the 1980s, a group of university students initially explored the idea of incorporating sensors and advanced capabilities into tangible items by enabling a Coca-Cola vending machine to remotely monitor its inventory. However, the cumbersome nature of the technology limited its advancements. The term "IoT" originated in 1999, credited to computer scientist Kevin Ashton. During Ashton's tenure at Procter & Gamble, he suggested the implementation of radio-frequency identification (RFID) chips on items as a means of monitoring their movement around the supply chain. Allegedly, he incorporated the popular term "internet" into his presentation in order to capture the executives' interest. The phrase became firmly established. In the following decade, there was a surge in public interest in IoT technology as an increasing number of interconnected products became available on the market. LG launched the first smart refrigerator in 2000, while Apple released the first iPhone in 2007. By 2008, the number of interconnected devices had surpassed

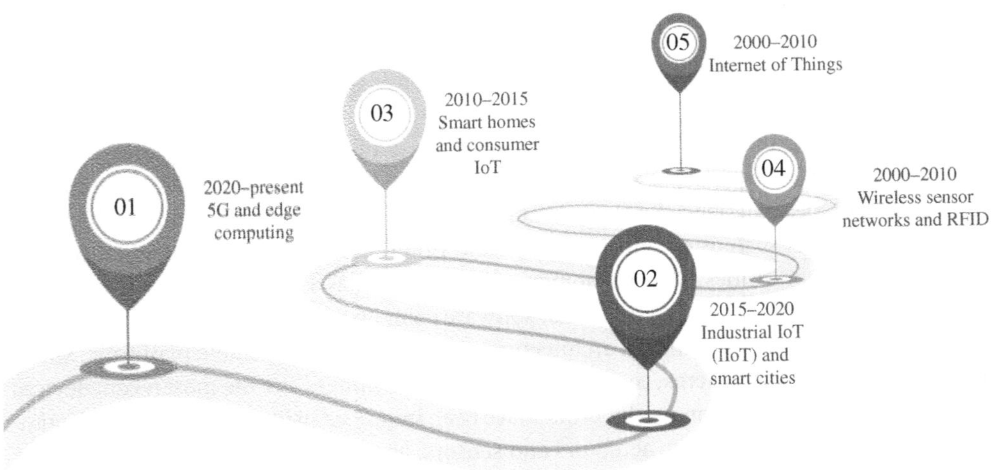

Figure 1.1 Evolution of IoT.

the global population. Google commenced trials of autonomous vehicles in 2009, and in 2011, Google introduced the Nest smart thermostat, enabling users to remotely regulate their central heating systems.

Developments in science and technology, along with the greater availability of the internet, have been the primary catalysts for the growth of the IoT in India. The IoT market in India has experienced growth since 2013 due to factors such as the growing recognition of cloud computing and data analytics, the expansion of data analytics, and a higher level of awareness. India has more than 100 proposed smart city initiatives that prioritize seamless communication and increased efficiency. By the end of 2019, the IoT will facilitate communication advancement, leading to a probable surge in business in India.

1.3 Need for IoT

The IoT empowers devices to autonomously perform everyday tasks without the need for human interaction. Companies can implement automation to streamline processes, reduce labor expenses, minimize waste, and improve service delivery efficiency. Figure 1.2 provides the requirements of the IoT technique. The IoT enables cost reduction in the manufacturing and delivery of goods, as well as visibility into customer transactions.

The IoT aims to establish connectivity between various items, enabling them to interact with each other over the internet. The connectivity enhances the security and convenience of human existence. The IoT enables a high level of interconnectedness in our world. In the present era, internet infrastructure is nearly ubiquitous, allowing us to access it at any time. Embedded computing equipment may be vulnerable to the internet's impact. Examples of embedded computer devices include MP3 players, MRI machines, signals, microwave ovens, washing machines, dishwashers, Global Positioning System (GPS) devices, heart monitoring implants, and biochips. The IoT intends toward building enhanced connectivity among various devices, systems, and services with the

Figure 1.2 Requirements for IoT.

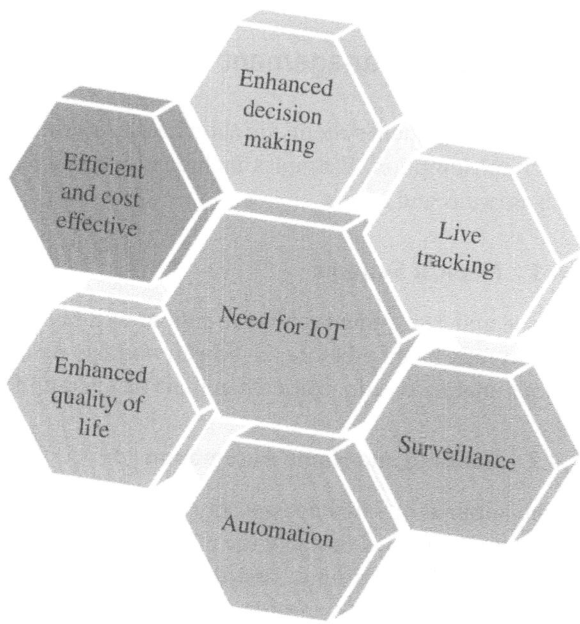

assistance of the internet. This gradual process enables automation in several domains. Imagine a scenario where various protocols interconnect all things and send information across different domains and applications. The IoT aims to establish connectivity between various items, enabling them to interact with each other over the internet. The goal of this connectivity is to improve human lives by offering security and convenience. Recent research indicates that by 2020, there will be more than 20 billion IoT devices in operation [19–25]. The IoT's ability to control devices and reduce radio expenses is due to its implementation. However, these vast areas present issues such as a scarcity of IP addresses and the need to design compatible and functional protocols and environments.

1.3.1 Environmental Monitoring

IoT is employed in the following process and monitors the status of the health of the system:

- Water, soil, or air measurement device
- Earthquake or tsunami warning systems
- Monitor wildlife habit

1.3.2 Infrastructure Management

- Infrastructure organization is a valuable tool for keeping track of potential issues in both urban and rural infrastructures, such as bridges and railways.
- Its purpose is to mitigate and minimize the risk of danger and any structural failures. It quickly assesses the infrastructure's strength and alerts for immediate repairs.

1.3.3 Industrial Applications

In real time, industrial applications analyze product quality to optimize marketing strategies. This involves identifying the target audience for each product and determining how little modification might enhance its marketability.

1.4 Energy Management

Energy management systems are classified as internet-connected systems that use sensors to reduce power consumption. Examples include cloud-based systems and remote control for appliances such as ovens and lamps.

1.4.1 Medical Systems

Medical and care systems enhance patient well-being by observing and regulating vital signs such as heart rate, blood pressure, and dietary intake. A smart tablet shows the exact dosage needed at various angles, assisting patients in their recuperation.

1.4.2 Building and Home Automation

Home automation encompasses any devices within a household that may be automatically controlled and governed, notably the air conditioning system, security locks, lighting, heating, ventilation, telephone systems, and televisions. Its purpose is to create a comfortable, secure, and energy-efficient living environment.

Table 1.1 Features of IoT.

S. no.	Parameters	Features
1	Efficiency	Processes is streamlined and eliminates manual control
2	Data insights	Real-time data is collected and utilized for decision-making
3	Cost savings	It maximizes resource efficiency and minimizes operational expenses
4	Automation	Remotely observing and regulating the devices
5	Improved security	Effective security protocols for interconnected devices
6	Environmental impact	Ensures sustainability through resource conservation
7	Innovation	It supports technological advancements and creates new business opportunities
8	Quality of life	Improves everyday life with intelligent home systems and healthcare solutions

1.4.3 Transport Systems

Transport systems implement various technologies to optimize urban and environmental transportation. These include automated traffic light systems, intelligent parking solutions, traffic cameras that identify congested roads and suggest alternate routes, and smart cameras that issue fines to speeding drivers.

1.4.4 Large-Scale Deployments

Smart cities are crowded with a wide range of IoT devices favored by wireless technology.

Connectivity is a crucial element that enables communication between devices. The IoT ecosystem incorporates a range of connectivity alternatives, such as Wi-Fi, cellular networks, Bluetooth, and others, to facilitate smooth data flow across different contexts. The IoT ecosystem flourishes through the collaborative synergy of its numerous components. The fundamental basis of IoT [26–28] is established by the integration of equipment, communication standards, and sophisticated data processing algorithms. By comprehending the complex interconnections among the various elements of the IoT, the knowledge about the extraordinary capacity of this technology to transform industries, improve productivity, and fundamentally change our everyday existence. Table 1.1 outlines the parameters used in this study to analyze the performance of IoT.

1.5 Main Components Used in IoT

Sensors and actuators serve as the sensory organs in the realm of the IoT. Temperature motion detectors and humidity sensors gather live data from their surroundings. Figure 1.3 shows the key components involved in the IoT technique.

On the other hand, actuators allow IoT equipment to trigger physical arrangements based on the data they have gathered. Sensors and actuators serve as a means of connecting the physical world with the digital world [29–31].

Connectivity modules are essential components of the IoT, as they enable devices to establish connections and transport data without any interruptions. Connectivity segments, such as Wi-Fi,

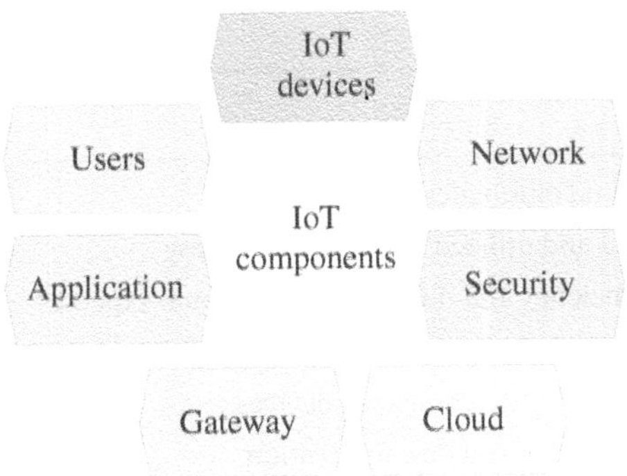

Figure 1.3 Components of IoT.

and Bluetooth, provide efficient inter-device communication, confirming the effective transmission and reception of collected data.

Data Processing Units: Gather the data, process it, and analyze it to extract significant insights. Data processing units, which encompass a variety of technologies such as edge computing devices and cloud servers, are responsible for doing the necessary computational activities to transform raw data into usable information.

Edge computing involves the computation of data near the vicinity of its origin, leading to decreased latency. In contrast, cloud servers are responsible for handling activities that require a higher number of resources.

Control Interfaces: The capability to govern and oversee IoT components is a notable benefit. Control gateways facilitate the interaction with IoT components, enabling operators to track and control their equipment, modify settings, and initiate activities. This architectural design highlights the primary elements of the IoT and their interconnectedness. The hardware elements of IoT, such as sensors, networking modules, data processing units, and control interfaces, combine to form a varying interconnected ecosystem.

The complex interplay of these components is crucial for the operation of IoT devices and enables the development of groundbreaking applications in several sectors, revolutionizing our interaction with technology and the world around us.

1.6 IoT Devices

Hardware components like sensors, actuators, gadgets, appliances, or machines, specifically configured for specific purposes and capable of data transfer via the internet or other networks, are known as IoT devices. Figure 1.4 provides various IoT devices in several applications that can integrate mobile devices, industrial equipment, etc. [32–35].

Nowadays, IoT devices incorporate artificial intelligence (AI) and machine learning to improve the intelligence and flexibility of many systems. This includes applications in automated e-vehicle, industry 4.0, medical devices, and residential automation.

Many of these devices are compact, reduce energy consumption, lower costs, and rely on microcontrollers as their core system. Increasing net bandwidth and evolving customer expectations about information security and functionality necessitate greater reliance on on-device processing. This involves processing data in the IoT endpoint compared to employing methods that are hosted in the cloud. Figure 1.5 illustrates the smart IoT revolution in various applications.

IoT devices are tangible gadgets that create wireless links to the web or within a neighborhood hub. These remote-managed gadgets can transmit and receive data from other devices.

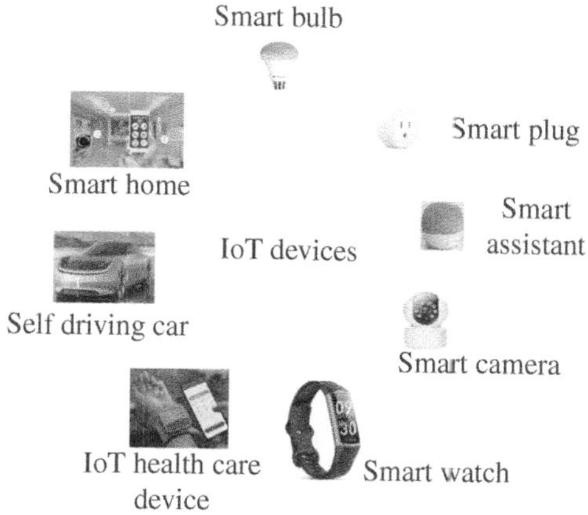

Figure 1.4 IoT devices.

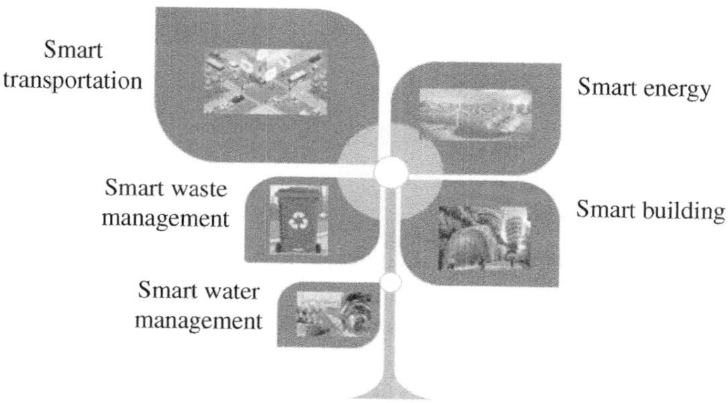

Figure 1.5 Smart device applications.

Everyday examples of smart devices include a smart automobile, a smart doorbell, and a smart refrigerator. IoT devices enable the transformation of everyday items into informational tools IoT devices gather data via their sensors and utilize software to interpret it, allowing machines to make choices based on the information. Usually, these devices make a link with a central server to obtain further information.

Additionally, these devices perform data comparisons and transmissions to community websites and services for the purpose of collecting data. They also establish connections with a message server that enables email, text, or call functionalities. In order to provide instructions, IoT devices have the capability to make links with other connected devices through the utilization of a common Wi-Fi network. IoT devices have advantages in improving productivity, facilitating simplicity, and growing task capacities. However, they also pose a potential threat to privacy and safety when they are targeted by hackers or compromised.

1.7 IoT Characteristics

The characteristics of IoT pertain to the distinctive qualities and aspects that differentiate IoT technology. These include connectivity, data gathering, live monitoring, scalability, and security. The Internet of Things, also referred to as IoT, is the central component of the ongoing technology revolution.

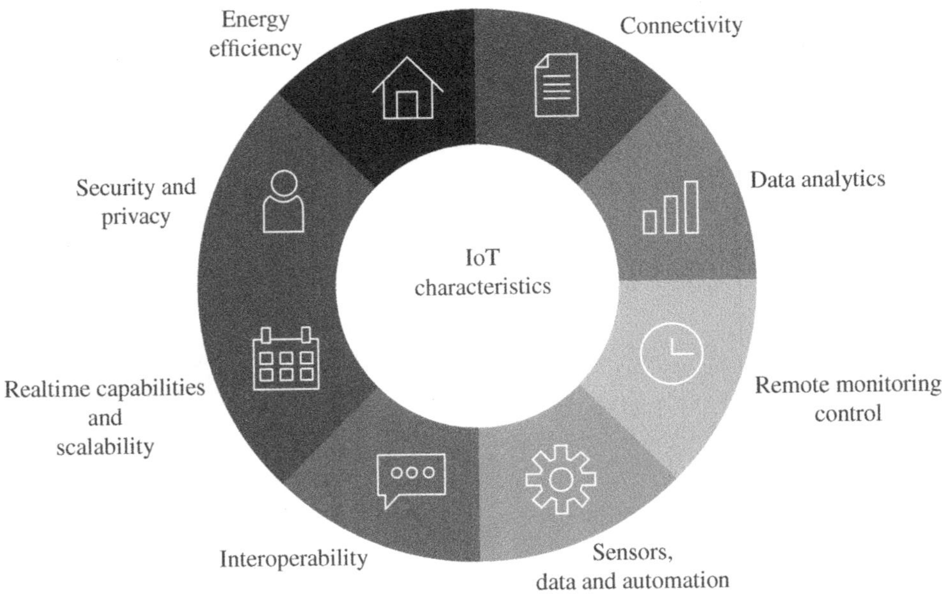

Figure 1.6 Characteristics of IoT.

The IoT is an extensive, interconnected network that includes tangible objects, household appliances, and cars. Figure 1.6 illustrates the various characteristics of the IoT. What distinguishes it is the incorporation of sensors, software, and networking into these items. This fusion enables them to go beyond mere existence, allowing them to gather, exchange, and interact with data and their environment [36–38].

The following are the main characteristics and properties of IoT:

1) Network connectivity is an important aspect of IoT technology because it enables the interconnection of different Internet devices, such as laptops and mobiles. Information on any subject is readily accessible to everybody, regardless of their location or time. The IoT enables the connection of various wireless devices, such as sensors, mobile phones, and trackers. Users can operate these devices even when they are not online thanks to this connectivity.
2) The concept of identifying objects is crucial. The process of deriving knowledge from the collected data is crucial. For instance, only a correct understanding of the data produced by a sensor qualifies it as genuine. Every IoT device possesses a distinct and individual identity. This identity facilitates the monitoring of the equipment and enables the retrieval of its current condition.
3) Ability to scale: the IoT is experiencing continuous growth, and ensuring scalability is of the highest priority for the system because it is an essential attribute of IoT. Scalability refers to the capacity of a system to expand its size or scope without causing any negative impact on its performance. Augmenting the existing design with extra computing power or programming layers might accelerate the process.
4) Dynamics: the ability to self-adapt is crucial for the IoT, as it needs to possess the capability to comprehend and respond to alterations in its surroundings. Consider a camera, initially designed to capture images but later enhanced with the ability to alter the image's quality. As a result, dynamism is critical for system development.

5) Self-improvement: AI enables autonomous self-improvement on the IoT, eliminating the need for human intervention. It enables the network to configure new IoT devices. Consequently, the technology can commence functioning promptly.
6) IoT architecture cannot be uniform. The IoT network should be hybrid, accommodating various manufacturers. The IoT becomes a reality when many areas converge.
7) Compatibility between different systems or devices allows them to work together seamlessly. Interoperability, a key attribute of the IoT, refers to the capacity of IoT devices and systems to communicate and share data, irrespective of the underlying technology or manufacturer. It facilitates the smooth operation of various devices and systems, ensuring an optimal user experience.

   IoT devices utilize standardized communication protocols and technologies to establish communication with one another and other systems, as well as data formats to ensure compatibility. These standards facilitate uniform and dependable data comprehension and processing across various devices, enabling seamless data transmission across devices and systems, irrespective of the technology employed. Lack of interoperability restricts IoT systems to isolated data and device repositories, impeding information sharing and the development of novel services and applications.
8) Intelligence is a crucial attribute of the IoT. The intelligence of IoT devices refers to the cognitive abilities of smart sensors and devices to perceive data, communicate with one another, and gather data for analysis. Digital personal assistants such as Alexa, Cortana, and Siri exemplify the cognitive capabilities of electronic devices. To ensure the intelligence of your IoT gadget, it is imperative that you are well informed on the most recent technological advancements.
9) The growing number of IoT devices has led to the emergence of security concerns, particularly regarding the protection of personal data. Data leakage is a potential risk that might arise when a large amount of data is collected, transmitted, and generated. The unauthorized transfer of personal data is a significant cause for concern. In order to address this obstacle, the IoT has developed networks, systems, and devices that effectively protect privacy.
10) Ensuring safety and security is a significant challenge for the IoT. Nevertheless, it manages to handle it seamlessly. Self-configuring IoT devices have the capability to update their software as needed with minimal user involvement. In addition, they have the capability to establish the network, enabling the incorporation of new devices into an existing network. This is a crucial attribute of the IoT.
11) Network: as the number of IoT devices in a network increases, it becomes challenging to ensure smooth connections for optimal operation. Cloud services and gateways are effective and adaptive approaches to address these difficulties, as they enable communication between IoT devices, surpassing the capabilities of other existing technologies. Often, one device can utilize the connectivity of another device to establish a network connection, even if another device is not directly linked to the network.
12) Diversity: the presence of heterogeneity is a crucial characteristic of the IoT. The gadgets utilize many hardware platforms and networks. Through various networks, they establish communication with other devices or service platforms. The IoT design enables direct network communication between diverse networks. The fundamental design criteria for heterogeneous entities are scalability, modularity, extension, and interoperability.
13) The system integrates sensors and actuators. The analog input from the external environment is known as sensory information. IoT technology provides a deep comprehension of our intricate surroundings, which is one of its notable characteristics. IoT sensors identify and quantify environmental changes, producing data for environmental reporting or engagement.

Sensing technologies enable the creation of capabilities that provide precise representations of knowledge about the physical environment and its inhabitants.

14) The data IoT function facilitates the collection of data for future predictions. For instance, what is the daily caloric expenditure of our body? This feature facilitates the regulation of daily calorie intake. Therefore, the data gathered is utilized from these gadgets in various ways, greatly enhancing the accessibility of our lives.

1.7.1 Technology Behind IoT

A digital twin is a dynamic virtual representation of an object that captures real-time data collected by sensors, providing a comprehensive view of its whole existence. These ecosystems utilize IoT, big data analytics, and AI to oversee and trace products. Figure 1.7 shows the several technologies behind the growth of the IoT.

Fog computer: This horizontally structured system design facilitates the efficient distribution of computer resources. It enables objects from the cloud to be allocated storage, control points, and network bandwidth. These solutions facilitate the connection between the cloud and the object, resulting in the formation of more tightly integrated smart systems.

Edge computing involves the direct allocation of resources to applications, bypassing the need to rely on local area networks (LANs) for application availability. Studies predict that by 2025, network edge computing platforms will transmit around 45% of data.

1.7.2 Hurdles of IoT Adoption

1) Insufficient infrastructure and limited connectivity: a major obstacle to the widespread adoption of IoT in India is the insufficient infrastructure and connections in most areas. Although urban areas may have reliable internet connectivity, rural regions continue to encounter difficulties in accessing stable networks. In order to fully utilize the potential of the IoT, it is crucial to have uninterrupted and smooth connectivity. Nevertheless, the Digital India program and continuous endeavors to enhance internet infrastructure have demonstrated encouraging outcomes, progressively narrowing the disparity in connectivity between urban and rural areas.

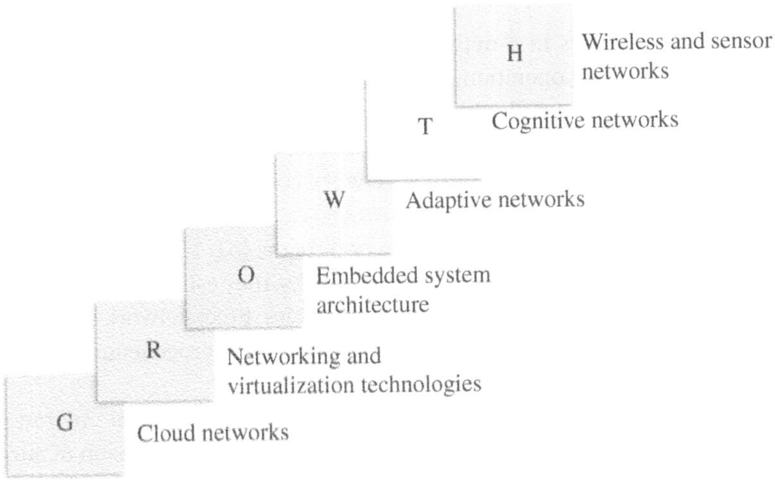

Figure 1.7 Technologies behind the growth of IoT.

2) Concerns about data privacy and security have emerged. The widespread use of IoT devices has led to significant concerns over data privacy and security, which are now essential priorities for both individuals and businesses. India's absence of comprehensive data protection legislation necessitates that IoT development companies emphasize the implementation of strong security measures. Users who address these concerns will gain confidence and be more likely to adopt IoT technologies.
3) The topic of discussion is the skill gap that exists in the field of IoT development. IoT development necessitates expertise in hardware and software engineering, data analytics, and cybersecurity. However, India's lack of IoT expertise hinders the advancement of innovative IoT solutions. To address this issue, it is imperative that the Indian government and business sector allocate resources toward upskilling programs and training projects. This will help cultivate a proficient workforce that can effectively contribute to the development of IoT services in the country.
4) Exorbitant implementation expenses: small and medium-sized organizations (SMEs) may find the initial expenditure needed to implement IoT solutions to be too expensive. Furthermore, the ongoing upkeep and operational expenses exacerbate the financial strain. In order to enhance the accessibility of IoT adoption, it is imperative for IoT development companies to prioritize the creation of cost-effective solutions and scalable models that can cater to the varied requirements of businesses in India.
5) A lack of uniformity: the absence of standardized norms and protocols in the IoT ecosystem has impeded smooth integration and compatibility. The lack of compatibility across IoT devices from various manufacturers might result in fragmentation and increased complexity. Promoting industry-wide standards adoption and advocating for open-source frameworks will facilitate collaboration and cultivate a flourishing IoT ecosystem in India.
6) The complexity entailed in implementing IoT solutions: deploying and overseeing IoT solutions can be complex and challenging. IoT ecosystems consist of several networked devices, sensors, networks, and software components that need to coordinate seamlessly. As deployment scale increases, so does the complexity of IoT projects. Therefore, it is essential for organizations to have the required technical experience and resources to effectively manage the intricacies of such projects.
7) Obstacles to realizing return on investment (ROI): quantifying the ROI of IoT technologies can present enterprises with intricate and demanding tasks. Although the IoT presents various advantages, such as increased efficiency, decreased expenses, and higher customer satisfaction, accurately measuring these advantages in financial terms can be challenging.

ROI assessments for IoT initiatives sometimes encompass numerous intangible aspects, hence rendering it more challenging to substantiate the initial expenditure. Moreover, the ROI for long-term IoT implementations may require a significant amount of time to become apparent, thus discouraging certain organizations from adopting IoT.

1.8 IoT Market Share

The IoT presents a lucrative market potential for manufacturers of equipment, distributors of Internet services, and developers of applications. By the end of 2020, projections indicate a global installation of 212 billion IoT smart objects [9]. By 2022, projections indicate that machine-to-machine (M2M) traffic will account for approximately 45% of the total Internet traffic. In addition to these forecasts, the McKinsey Global Institute has reported a 300% increase in the overall number of connected machines (units) during the past five years. Cellular network traffic monitoring in the United States also indicated a 250% surge in M2M traffic volume in 2011.

The economic growth resulting from the implementation of IoT-based services is significant for organizations. The healthcare and industrial sectors have the largest economic influence. By 2025, healthcare applications and IoT-based services like m-Health and telecare are projected to generate annual growth of $1.1–$2.5 trillion for the global economy. These services facilitate the efficient delivery of medical wellness, prevention, diagnosis, treatment, and monitoring through electronic media. By 2025, the anticipated yearly economic effect of the IoT is projected to reach between $2.7 trillion and $6.2 trillion.

However, Wiki Bon predicts that by 2020, the Industrial Internet will generate over $1279 billion in value, increasing the ROI from 13% in 2012 to 149%. Furthermore, Navigant recently projected a 60% growth in the Building Automation Systems (BAS) market, from $58.1 billion in 2013 to $100.8 billion by 2021. These figures indicate a potentially substantial and rapid expansion of the IoT in the near future, along with its associated industries and services. This development presents a distinct chance for conventional equipment and appliance makers to convert their products into "intelligent objects." To expand the reach of the IoT and its associated services worldwide, internet service providers (ISPs) must configure their networks to ensure quality of service (QoS) for a combination of M2M, person-to-machine (P2M), and person-to-person (P2P) traffic.

Three prominent technologies primarily control the global connectivity of the IoT: Wi-Fi, Bluetooth, and cellular IoT.

1) **Wireless Fidelity:** Wi-Fi accounts for 31% of all IoT connections. Over 50% of Wi-Fi-enabled devices deployed globally in 2022 were built using the most recent Wi-Fi 6 and Wi-Fi 6E technologies, ensuring enhanced wireless connectivity with improved speed and reliability. The implementation of these technologies has strengthened the efficiency of communication to IoT devices, contributing to improved client experiences and productivity in general. Wi-Fi technology is the dominant form of connectivity for the IoT in industries like smart homes, buildings, and healthcare.
2) **Bluetooth Technology:** Bluetooth is the basis for 27% of all IoT connections worldwide. Bluetooth low energy (BLE), commonly denoted as Bluetooth Smart, has been continuously developed to ensure reliable connectivity for IoT devices while lowering power consumption. As a result, BLE has become the preferred option for IoT devices that depend on batteries, including intelligent home sensors and surveillance devices. There is a growing interest in IO-Link Wireless technology within the industrial sector. This technology employs IEEE 802.15.1, which is the technical standard for Bluetooth, to establish wireless links among sensors, actuators, and an input–output master.
3) **Cellular IoT**: Cellular IoT, which includes recently 5G, LTE-M, and narrowband Internet of Things (NB-IoT), currently accounts for around 20% of IoT connections worldwide. Based on the worldwide cellular IoT connectivity tracker and forecast from IoT analytics, worldwide cellular IoT connections experienced a 27% year-on-year rise in 2022, significantly outpacing the growth rate of global IoT connections. The increase in growth can be attributed to the implementation of more advanced technologies while older technologies like 2G and 3G are being gradually replaced. Despite a YoY gain of over 100% in 2022, the rate at which 5G module shipments increased was not as fast as anticipated by many.

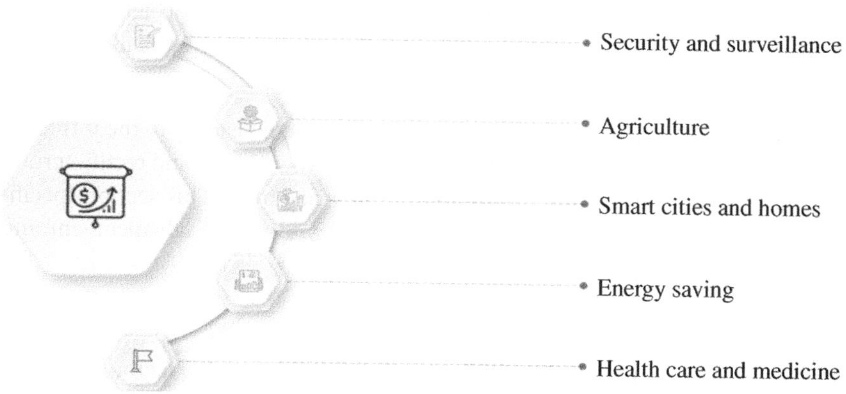

Figure 1.8 Applications of IoT.

By 2023, the five leading network operators like China Mobile, Telecom, Unicom, Vodafone, and AT&T collectively controlled 84% of the total worldwide cellular IoT links. The top five network operators dominate the IoT network operator market, accounting for 64% of the revenue. These operators are the key players in this industry.

The projections indicate a 14.5% increase in IoT expenditure, reaching $1.1 trillion by the end of 2025, in terms of market size. The Indian market reached a value of $9 billion by the end of 2020, exhibiting a growth rate of 31%. Among the global population of 7.6 billion individuals, approximately 3.7 billion people possess the ability to connect to the internet.

Notably, India is home to almost 24% of this internet-connected population. Therefore, it is evident that the IoT has the potential to enhance connections and revolutionize communication in India. India presently has more than 120 specialized IoT companies equipped with the technical expertise required to facilitate this transformative transformation. Nearly 70% of the IoT startups operating in India today have their origins in recent times. The healthcare and manufacturing industries attracted significant investment, leading to the establishment of numerous companies. Few important applications of IoT are presented in Figure 1.8.

Among Indian companies, WeMakeIOT, QBurst, Altizon, Happiest Minds, and Traxroot appear to be maintaining their positions and performing well. About 65% of Indian firms that manage IoT applications are start-ups with the intention of expanding, and the total number of IoT devices in India exceeds 200 million.

- With the increasing use of digitalization and automation in India's economy, there are significant prospects for both domestic and international enterprises in the field of IoT development services. IoT development organizations could establish themselves as trailblazers in the rapidly growing Indian market by understanding its unique obstacles and customizing solutions accordingly.
- IoT development firms should strategically link their solutions with government efforts, such as the Smart Cities Mission and Make in India, in order to leverage the extensive opportunities presented by public–private collaborations.
- Customized solutions for businesses of all sizes: creating affordable and easily expandable IoT solutions customized for all types of businesses can encourage wider acceptance and promote economic development.
- Emphasize data security and privacy: prioritizing strong data security measures and adhering to growing data protection rules can increase customer trust and confidence.

- Investing in research and development (R&D) is crucial for firms to develop advanced IoT technology. This investment will help organizations establish themselves as leaders in the fast-changing IoT industry.

Researchers have intensively investigated the service selection problem, leading to the introduction of many study techniques. However, it is critical to acknowledge the address and rectify certain outstanding issues in the academic and industrial sectors in the near future. This section specifically addresses the obstacles and areas for future investigation regarding the advancement and establishment of capable and effective service selection solutions.

Challenges

- **Battery Life and Energy Efficiency:** The capacity of the battery is a limiting factor for mobile devices. Prolonging battery life and improving energy efficiency remain major challenges. Efficiently identifying and choosing suitable services from nearby cloud data centers help decrease energy usage in a mobile setting. Furthermore, tackling this problem necessitates optimizing the service selection in order to minimize device resource utilization.
- **QoS:** QoS poses a number of significant issues. Ensuring efficient and dependable communication is of utmost importance, as it guarantees users a satisfactory and constant level of service. This is particularly critical in situations where network conditions, service providers, and user requirements may vary.

1.9 Conclusion

This paper provides a concise overview of the development of context-aware technologies and their growing significance in contemporary applications. Initially, a wide range of IoT products to identify several areas for future work is analyzed, including scalability and interoperability, real-time adaptability, and economic considerations, particularly in the context of utilizing blockchain technology. Additionally, integration with emerging technologies, such as AI-driven access control, is imperative. Lastly, it is crucial to prioritize real-world implementations. Real-world implementations and assessments of access control systems in a variety of IoT deployments. When it comes to context-aware computing, there are a substantial number of IoT solutions available in the industry marketplace is discussed. These assessments can offer pragmatic observations and detect unexpected obstacles to further encourage innovation.

The concept of the IoT is gaining momentum in our contemporary society, with the goal of enhancing the overall quality of life through the interconnection of various intelligent devices, technologies, and applications. Overall, the IoT will allow the full mechanization of every aspect of our environment. This paper provides a comprehensive examination of the fundamental concept and the technology that facilitates it, the protocols employed, the potential applications, and the ongoing research that delves into various facets of the IoT. This will serve as a solid basis for researchers and practitioners who wish to get a deep understanding of IoT technologies and protocols. It will enable them to comprehend the comprehensive structure and function of all the parts and protocols of the IoT. Additionally, this paper has addressed certain obstacles and concerns related to the design and implementation of IoT solutions. Furthermore, it discusses the interaction between the IoT, big data analytics, cloud computing, etc. This chapter showcased comprehensive application use cases to exemplify common protocol integration scenarios, ultimately providing the necessary IoT services.

References

1 Ziegeldorf, J.H., Morchon, O.G., and Wehrle, K. (2014). Privacy in the Internet of Things: threats and challenges. *Security and Communication Networks* 7 (12): 2728–2742.
2 Al-Fuqaha, A., Guizani, M., Mohammadi, M. et al. (2015). Internet of Things: a survey on enabling technologies, protocols, and applications. *IEEE Communications Surveys and Tutorials* 17 (4): 2347–2376.
3 Atzori, L., Iera, A., and Morabito, G. (2010). The Internet of Things: a survey. *Computer Networks* 54 (15): 2787–2805.
4 Perera, C., Liu, C.H., Jayawardena, S., and Chen, M. (2014). A survey on Internet of Things from industrial market perspective. *IEEE Access* 2: 1660–1679.
5 Da Xu, L., He, W., and Li, S. (2014). Internet of Things in industries: a survey. *IEEE Transactions on Industrial Information* 10 (4): 2233–2243.
6 Khan, R., Khan, S.U., Zaheer, R., and Khan, S. (2012). Future internet: the Internet of Things architecture, possible applications and key challenges. *Proc. 10th Int. Conf. FIT*. pp. 257–260.
7 Gubbi, J., Buyya, R., Marusic, S., and Palaniswami, M. (2013). Internet of Things (IoT): a vision, architectural elements, and future directions. *Future Generation Computer Systems* 29 (7): 1645–1660.
8 Lopez, P., Fernandez, D., Jara, A.J., and Skarmeta, A.F. (2013). Survey of Internet of Things technologies for clinical environments. *Proc. 27th Int. Conf. WAINA*. pp. 1349–1354.
9 Yang, D., Liu, F., and Liang, Y. (2010). A survey of the Internet of Things. *Proc. 1st ICEBI*. pp. 358–366.
10 Gluhak, A., Krco, S., Nati, M. et al. (2011). A survey on facilities for experimental Internet of Things research. *IEEE Communications Magazine* 49 (11): 58–67.
11 Gronbaek, I. (2008). Architecture for the Internet of Things (IoT): API and interconnect. *Proc. Int. Conf. Sensor Technol. Appl.* p. 802807.
12 Viswanathan, H., Lee, E.K., and Pompili, D. (2012). Mobile grid computing for data- and patient-centric ubiquitous healthcare. *Proc. 1st IEEE Workshop Enabling Technol. Smartphone Internet Things (ETSIoT)*. p. 3641.
13 Zhao, W., Chaowei, W., and Nakahira, Y. (2011). Medical application on Internet of Things. *Proc. IET Int. Conf. Commun. Technol. Appl. (ICCTA)*. p. 660665.
14 Yang, N., Zhao, X., and Zhang, H. (2012). A non-contact health monitoring model based on the Internet of Things. *Proc. 8th Int. Conf. Natural Comput. (ICNC)*. p. 506510.
15 Imadali, S., Karanasiou, A., Petrescu, A. et al. (2012). eHealth service support in IPv6 vehicular networks. *Proc. IEEE Int. Conf. Wireless Mobile Comput., Netw. Commun. (WiMob)*. p. 579585.
16 Iqal, Z.M., Selamat, A., and Krejcar, O. (2024). A comprehensive systematic review of access control in IoT: requirements, technologies, and evaluation metrics. *IEEE Access* 12: 12636.
17 Genç, D., Tomur, E., and Erten, Y.M. (2019). Context-aware operation-based access control for Internet of Things applications. *Proc. Int. Symp. Netw., Comput. Commun. (ISNCC)*. pp. 1–6.
18 Zhou, Q., Elbadry, M., Ye, F., and Yang, Y. (2021). Towards fine-grained access control in enterprise-scale Internet-of-Things. *IEEE Transactions on Mobile Computing* 20 (8): 2701–2714.
19 Dramé-Maigné, S., Laurent, M., and Castillo, L. (2019). Distributed access control solution for the IoT based on multi-endorsed attributes and smart contracts. *Proc. 15th Int. Wireless Commun. Mobile Comput. Conf. (IWCMC)*. pp. 1582–1587.
20 Moghe, R., Lambert, F.C., and Divan, D. (2012). Smart 'stick-on' sensors for the smart grid. *IEEE Transactions on Smart Grid* 3 (1): 241–252.

21 Wu, Y.C., Cheung, L.F., Lui, K.S., and Pong, P.W.T. (2012). Efficient communication of sensors monitoring overhead transmission lines. *IEEE Transactions on Smart Grid* 3 (3): 1130–1136.

22 Jiang, J. and Qian, Y. (2016). Distributed communication architecture for smart grid applications. *IEEE Communications Magazine* 54 (12): 60–67.

23 Fateh, B., Govindarasu, M., and Ajjarapu, V. (2013). Wireless network design for transmission line monitoring in smart grid. *IEEE Transactions on Smart Grid* 4 (2): 1076–1086.

24 Abdalzaher, M.S. and Muta, O. (2020). A game-theoretic approach for enhancing security and data trustworthiness in IoT applications. *IEEE Internet of Things Journal* 7 (11): 11250–11261. [82] C. Muralidharan.

25 Neiat, A.G., Bouguettaya, A., and Bahutair, M. (2022). A deep reinforcement learning approach for composing moving IoT services. *IEEE Transactions on Services Computing* 15 (5): 2538–2550.

26 Khelloufi, A., Ning, H., Naouri, A. et al. (2024). A multimodal latent-features-based service recommendation system for the social Internet of Things. *IEEE Transactions on Computational Social Systems* 11: 5388. https://doi.org/10.1109/TCSS.2024.3360518.

27 Zhang, N. (2021). Service discovery and selection based on dynamic QoS in the Internet of Things. *Complexity* 2021: 1–12.

28 Hamrouni, A., Ghazzai, H., and Massoud, Y. (2022). Service discovery in social Internet of Things using graph neural networks. *Proc. IEEE 65th Int. Midwest Symp. Circuits Syst. (MWSCAS)*. pp. 1–4.

29 Rad, M.M., Rahmani, A.M., Sahafi, A., and Qader, N.N. (2023). Community detection and service discovery on social Internet of Things. *International Journal of Communication Systems* 36 (11): e5501.

30 Muteba, K., Djouani, K., and Olwal, T. (2022). 5G NB-IoT: design, considerations, solutions and challenges. *Procedia Computer Science* 198: 86–93.

31 Shafique, K., Khawaja, B.A., Sabir, F. et al. (2020). Internet of Things (IoT) for next-generation smart systems: a review of current challenges, future trends and prospects for emerging 5G-IoT scenarios. *IEEE Access* 8: 23022–23040.

32 Zeng, X. and Bao, S. (2023). *Key Technologies of Internet of Things and Smart Grid*. Springer Science and Business Media LLC.

33 Venkat Narayana Rao, T., Raghavendra Rao, M., and Bhavana, S. (2024). *Integration of Data Science and Cloud-Based IoT Networks*, Chapter 19. IGI Global.

34 Al-Fuqaha, A., Guizani, M., Mohammadi, M. et al. (2015). Internet of Things: a survey on enabling technologies, protocols, and applications. *IEEE Communications Surveys & Tutorials* 17: 2347.

35 Pons, M., Valenzuela, E., Rodríguez, B. et al. (2023). Utilization of 5G technologies in IoT applications: current limitations by interference and network optimization difficulties—a review. *Sensors* 23: 3876.

36 Dede, G., Fragiadakis, G., Michalakelis, C., and Kamalakis, T. (2019). Brokering intelligence as a service for the Internet of Things. *International Journal of Technology Diffusion* 10: 18.

37 Vinora, A., Nancy, D.R., Sivakarthi, G. et al. (2024). *IoT Devices for Natural Disasters*, Chapter 7. IGI Global.

38 Peng, S.-L., Pal, S., and Huang, L. (2020). *Principles of Internet of Things (IoT) Ecosystem: Insight Paradigm*. Springer Science and Business Media LLC.

2

IoT Fundamentals: Platforms, Architectures, and Sensor Technologies

Naseer Ahamed Javed[1,2], Yogesh Rajkumar[3], and Kallankurichy P. Kaliyamurthie[3]

[1]Department of Computer Science and Engineering, Jerusalem College of Engineering, Anna University, Chennai, Tamil Nadu, India
[2]Research scholar, Bharath Institute of Higher Education and Research, Chennai, Tamil Nadu, India
[3]Department of Information Technology, Bharath Institute of Higher Education and Research, Chennai, Tamil Nadu, India

2.1 Introduction

In the ever-changing realm of technology, the Internet of Things (IoT) is a revolutionary power, changing industries and daily interactions. In this chapter, key components fueling this revolution are examined. Beginning with IoT system architecture and design principles, the chapter unveils the frameworks that underpin these systems and the principles guiding their development. Sensors and transducers form the core, converting physical data into digital signals for real-time monitoring and decision-making, essential across diverse applications. Key to optimizing low-power wide-area network (LPWAN) infrastructures, the LoWPAN network management protocol (LNMP) ensures robust performance and scalability. Diagnostic tools for wireless sensor networks (WSNs) are critical, enhancing reliability in dynamic environments. The chapter also delves into IoT communication technologies like Bluetooth, Zigbee, and cellular networks, crucial for selecting suitable frameworks tailored to specific operational needs. Practical applications highlight IoT's transformative impact, from enhancing efficiency to fostering innovation across industries.

2.2 Overview of IoT System Architectures and Design Principles

2.2.1 IoT System Architecture

IoT architectures can be classified into

1) Three-Layer Architecture
2) Four-Layer Architecture
3) Five-Layer Architecture
4) Service-Oriented Architecture
5) Fog Computing Architecture
6) Cloud-Based Architecture

A four or five-layered architecture is commonly used and it also gives us a complete overview of how it works in real life [1].

IoT for Smart Grid: Revolutionizing Electrical Engineering, First Edition.
Edited by Rahiman Zahira, Palanisamy Sivaraman, Chenniappan Sharmeela, and Sanjeevikumar Padmanaban.

2.2.1.1 Three-Layer Architecture

Perception Layer: This is the physical layer equipped with sensors that collect information from the surroundings. It detects certain physical attributes or recognizes other intelligent objects in the surroundings [2].

Network Layer: It is in charge of linking up with other smart devices, network equipment, and servers. Its characteristics are additionally utilized for sending and analyzing sensor information.

Application Layer: In charge of providing user with application-specific services. It outlines different uses for the IoT, such as smart homes, smart cities, and smart health (Figure 2.1).

2.2.1.2 Four-Layer Architecture

The functions of *Perception layer, Network layer* and *Application layer* is same as in Three-layer architecture [2].

Middleware Layer: Consists of middleware software, data processing services, storage systems, application programming interface (API) management. It acts as an intermediary to process, store, and analyze data before it reaches the application layer. It ensures interoperability among various IoT devices and platforms, manages data flow, and provides a unified interface for application developers (Figure 2.2).

2.2.1.3 Five-Layer Architecture

The functions of *Perception layer* and *Application layer* is same as in Three-layer Architecture [2].

Transport Layer: The sensor data is moved between the perception layer and the processing layer by the transport layer using various networks like wireless, 3G, local area network (LAN), Bluetooth, radio-frequency identification (RFID), and near field communication (NFC).

Processing Layer: The middleware layer is another term for the processing layer. It gathers, examines, and handles vast quantities of data originating from the transport layer. It has the ability to handle and deliver a variety of services to the layers below. It utilizes various technologies like databases, cloud computing, and modules for processing big data.

Business Layer: It oversees the entire IoT system, which includes applications, business and profit models, and users' privacy. The paper does not cover the business layer. Therefore, we will not delve into it any more (Figure 2.3).

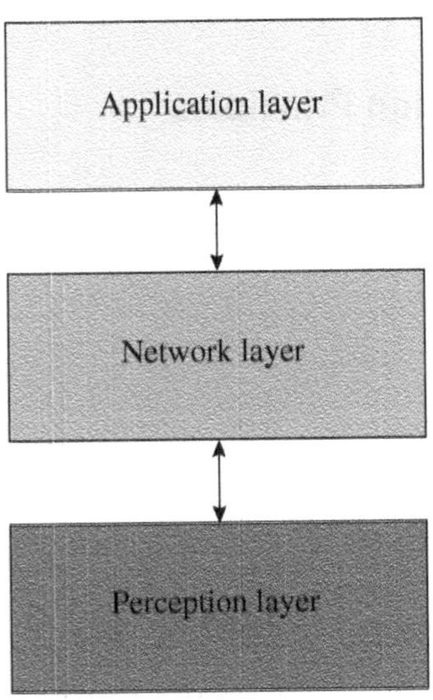

Figure 2.1 IoT three-layer architecture.

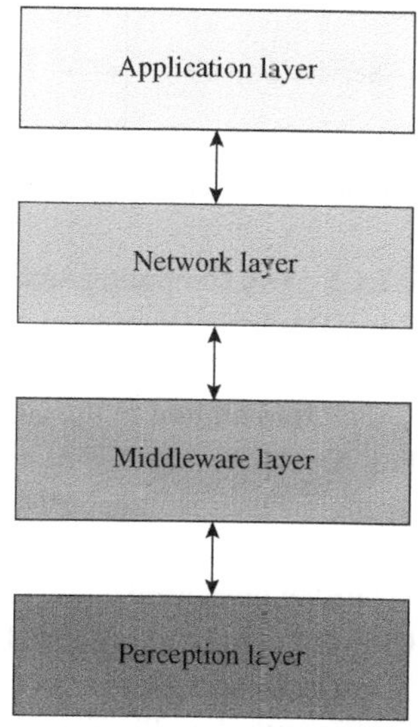

Figure 2.2 IoT four-layer architecture.

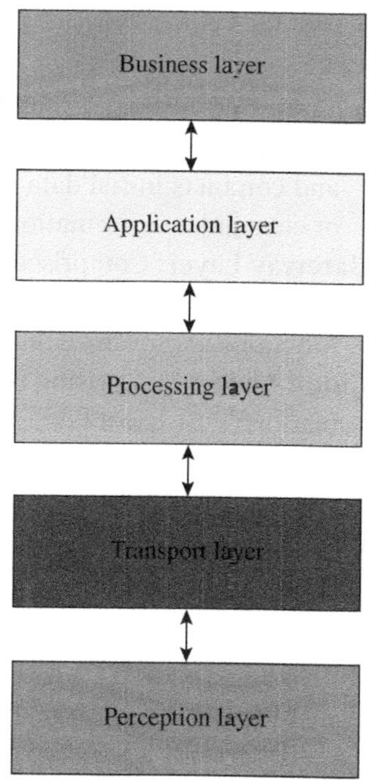

Figure 2.3 IoT five-layer architecture.

2.2.1.4 Service-Oriented Architecture

Service Provider: It consists of Web services, microservices. It offers various services over the network. Services are designed to be reusable, discoverable, and loosely coupled, allowing for flexible integration and interaction.

Service Broker: It consists of Service registry, service discovery mechanisms. It acts as an intermediary that facilitates the finding and accessing of services by clients. It maintains a registry of available services and provides information on how to use them.

Service Requester: It consists of client applications, user interfaces. It consumes services offered by the service provider. Service requesters can be end-user applications, other services, or automated systems that need to access functionality provided by the service provider (Figure 2.4).

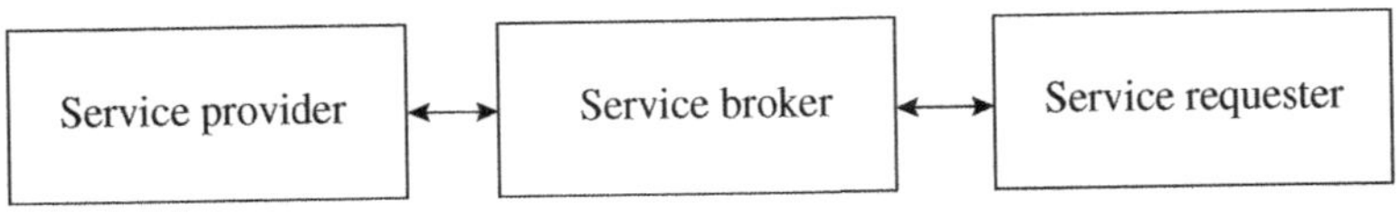

Figure 2.4 Service-oriented architecture.

2.2.1.5 Fog Computing Architecture

Edge Devices: Comprising of sensors, actuators, and smart devices (e.g., cameras, thermostats), known as IoT. Data is gathered and prepared at the periphery of the network. This decreases the data transmitted to the cloud and reduces latency by processing important tasks on-site.

Fog Nodes: Composed of intermediary devices including gateways, routers, and local servers. It offers closer storage, processing, and networking services to edge devices. Fog nodes carry out functions such as data filtering, consolidation, and real-time analysis, which helps alleviate the burden on central cloud servers.

Cloud: Encompasses centralized data centers, cloud storage, and computing services. It manages intricate and extensive data processing, storage, and analysis tasks that fog nodes cannot efficiently handle. The cloud offers strong computing power, machine learning abilities, and extended data storage (Figure 2.5) [1].

2.2.1.6 Cloud-Based Architecture

Edge Layer: Includes IoT sensors and actuators. It gathers information from the surroundings and conducts initial data analysis. Edge devices frequently conduct basic calculations to refine or consolidate information before transmitting it to the cloud.

Gateway Layer: Comprised of gateways, routers, and edge servers. It links edge devices with the cloud. Gateways perform protocol translation, data encryption, and occasionally edge computing functions to minimize the data sent to the cloud.

Cloud Layer: Comprising cloud infrastructure, data storage, computing resources, and analytics platforms. It consolidates the processing and storage of data. The layer of clouds provides

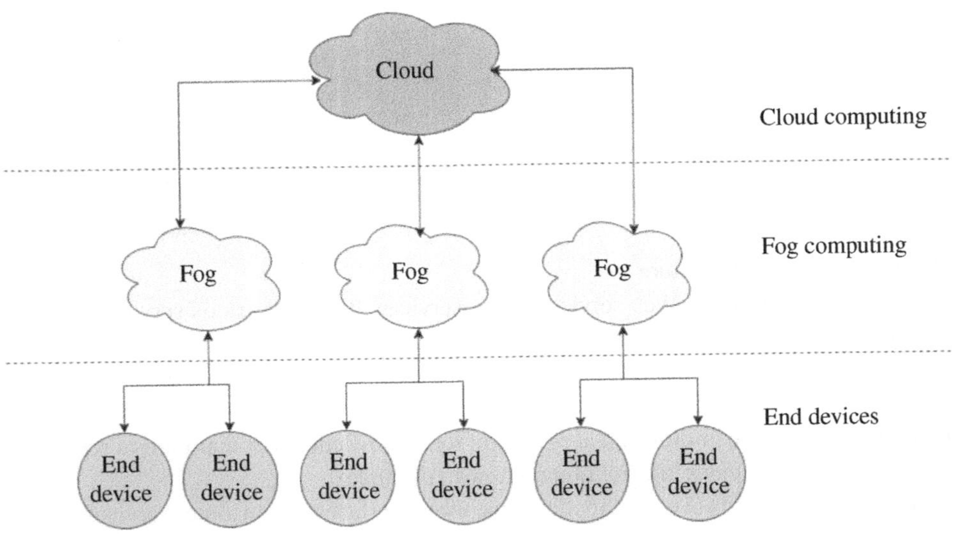

Figure 2.5 Fog computing architecture.

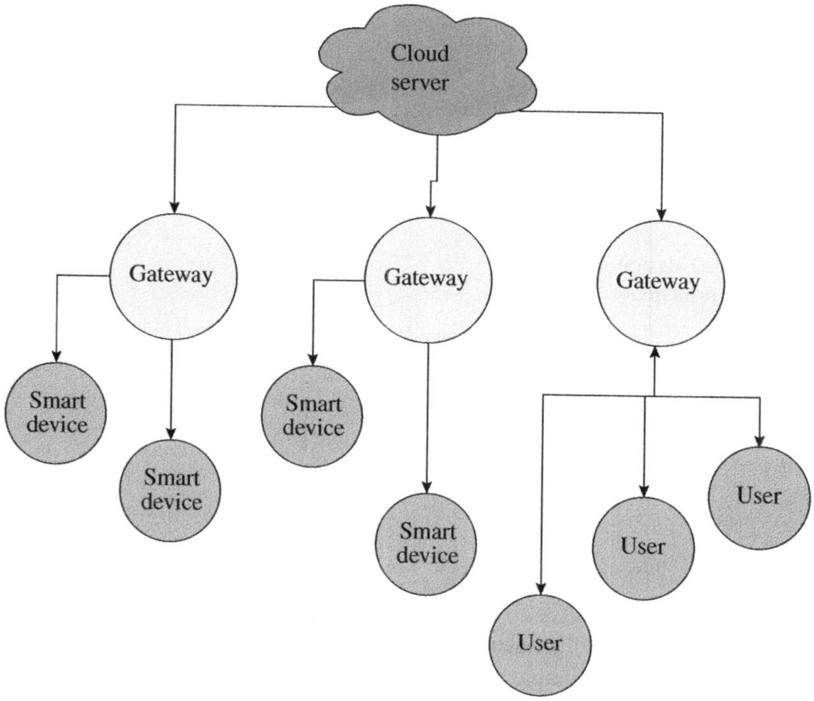

Figure 2.6 Cloud-based architecture.

resources that can be scaled and adjusted easily for analyzing complex data, implementing machine learning, and managing large amounts of data. It offers a medium for creating and launching IoT applications capable of managing large volumes of data from many devices (Figure 2.6) [3].

2.2.2 Design Principles

Product designers must address industrial product requirements, IT components, business needs, and user experience (UX) design simultaneously. Successfully incorporating these elements can lead to highly effective IoT devices and systems. Here are eight key design principles that guide the development of IoT systems (Figure 2.7) [4].

2.2.2.1 Research Before Building

Understand the Purpose and User Base: Research the device's purpose and its user base thoroughly. Think from the perspective of the end-users to identify pain points and how the IoT product can solve them.

Gather Insights: Speak to customer support and sales teams to understand user needs and expectations. Attend industry conferences to learn about current trends and potential customer interests.

2.2.2.2 Match Features to User Value

Identify User Problems: Use research to discover which end-user problems are worth solving with an IoT device.

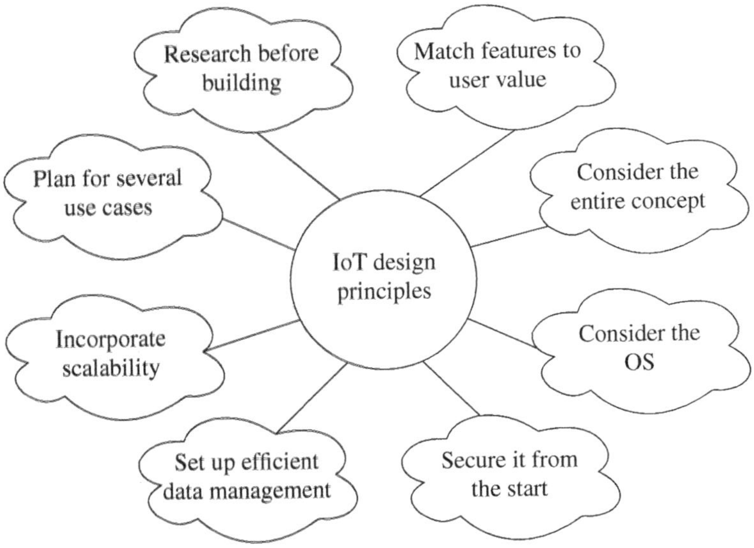

Figure 2.7 IoT design principles.

Provide Clear Value: Ensure the product's feature set addresses these problems or barriers effectively, making it easier for users to see the value and adopt the product [4].

2.2.2.3 Consider the Entire Concept

System Integration: Consider the entire IoT ecosystem, including how different devices, systems, and applications interact with each other.

Interoperability: Design devices that can connect, control, and communicate seamlessly within the network to support multiple use cases [4].

2.2.2.4 Consider the Operating Settings

Contextual Design: Develop features that are timely and purposeful for the intended use case.

Adaptability: Ensure the device can operate under varying conditions, such as low power environments or harsh weather.

2.2.2.5 Secure It from the Start

Early Security Integration: Address security from the beginning of the design process.

Comprehensive Protection: Include both hardware and software measures to protect against IoT security threats and ensure data privacy.

2.2.2.6 Set Up Effective Data Management

Data Workflow: Create effective data management processes to handle the massive volumes of data generated by IoT devices.

Reduce Latency: Improve data transmission and storage to reduce latency and provide timely access to information.

2.2.2.7 Incorporate Scalability

Design for Scale: Ensure the IoT system can handle the addition of numerous devices without performance degradation.

Fleet Management: Incorporate gateways and management software to handle large-scale deployments and data collection.

2.2.2.8 Plan for Several Use Cases

Flexibility: Design IoT devices to be adaptable for various applications, including unintended or innovative uses.

Track and Adapt: Monitor how users employ devices and incorporate these insights into future iterations to support new use cases.

2.3 Exploring IoT/M2M Systems and Their Role in Connectivity

2.3.1 What Exactly Is M2M?

Machine-to-machine (M2M) technology is a hybrid of hardware and software that allows machines to communicate with one another. Sensors that measure data are often used, along with network infrastructure for data transmission and machines or network entities that interact with the data.

2.3.2 Historical Context

M2M technology has existed since 1968, when Theodore Paraskevako devised a method for telephones to share caller information, marking the first example of M2M communication. Despite its 50-year history, M2M technology continues to drive cutting-edge applications, revolutionizing industries and enhancing daily lives [5].

2.3.3 M2M and IoT

M2M and IoT are frequently used interchangeably to refer to linked devices, although there are important differences:

M2M: This implies a direct point-to-point connection between equipment, which is frequently employed in particular sectors and localities.

IoT: A larger category of linked devices that primarily use cloud-based communication, but can also contain point-to-point connectivity.

Both terms are functionally equivalent, using both wired and wireless connections and not necessitating an internet connection but rather a network connection (Figure 2.8).

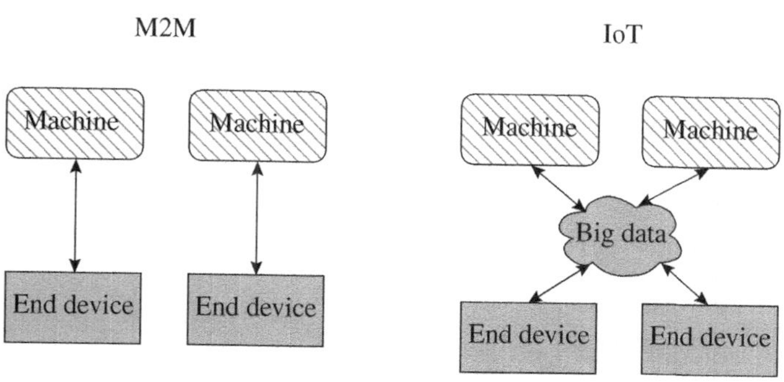

Figure 2.8 M2M and IoT.

2.3.4 Working

The primary objective of M2M communication is to gather sensor data and transmit it through networks. Unlike supervisory control and data acquisition (SCADA) or similar remote monitoring tools, M2M systems utilize open networks and connectivity methods such as cellular or Ethernet to ensure cost-effectiveness.

Key components of an M2M system include sensors, RFID, Wi-Fi or cellular communication modules, and autonomic computing software. These components enable network devices to interpret data and make informed decisions. M2M applications then translate this data to initiate predefined automated actions.

Telemetry, a widely recognized form of M2M communication, has been utilized for transmitting operational data since the early 1900s. Initially employing telephone lines and later radio waves, telemetry pioneers transmitted performance metrics gathered from remote monitoring instruments. Today, telemetry is integrated into everyday devices like heaters, electric meters, and internet-connected appliances, thanks to advancements in wireless technology and the Internet. Originally confined to sectors like manufacturing and engineering, the primary advantage of M2M is its ability to monitor and communicate with various devices and systems (Figure 2.9) [5].

2.3.5 Advantages

Visibility: M2M technology allows organizations and consumers to monitor data without being physically present. For example, utility firms may employ smart meters to automatically communicate use data for invoicing and optimization.

Automation: Data collected by M2M devices can trigger automated actions based on predefined conditions. This includes applications such as automated billing, security alerts, heating, ventilation, and air conditioning (HVAC) control based on temperature, and predictive maintenance.

Remote Access: M2M technology enables remote control and access to devices through applications. Businesses may diagnose issues and respond to crises without being present on-site, increasing operational efficiency.

Scalability: M2M technology enables the deployment and administration of huge numbers of devices, automating operations where feasible and concentrating efforts on key areas.

2.3.6 Applications

Smart Meters: Provide and receive data, enabling for more efficient resource allocation, automatic invoicing, and extensive insights into resource usage.

Smart Security: Modern security systems connect to apps, allowing for real-time warnings and remote monitoring of homes, companies, and assets.

Fleet Management: Global Positioning System (GPS) trackers and telematics devices provide real-time position and diagnostic data, allowing fleet managers to better coordinate assets and address concerns.

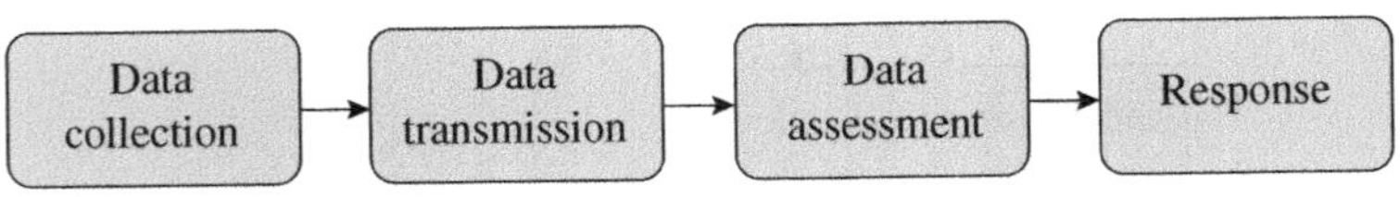

Figure 2.9 M2M working.

Industry 4.0: The fourth industrial revolution employs M2M technology to automate and collect large data throughout production, hence enhancing safety, efficiency, and cost-effectiveness in smart factories and power plants.

2.4 Introduction to Sensors and Transducers in IoT

2.4.1 Sensors

Sensors are devices that detect and measure physical attributes or changes in their surrounding. They convert these observations into signals that can be interpreted, displayed, or processed further. Sensors find application across various fields, including measuring temperature, humidity, speed, or motion. Essentially, sensors act as the sensory inputs of electronic systems, delivering essential information that supports decision-making processes [6].

2.4.1.1 Working

Sensors are miniature interpreters constantly monitoring specific environmental aspects [7]. The procedure may be divided into the following steps:

Physical Measurement: Sensors detect physical quantities like temperature or pressure.
Signal Conversion: The detected physical phenomenon is converted into an electrical signal.
Data Processing: The electrical signal is processed into valuable data, often transmitted wirelessly.
Action or Analysis: This data triggers actions (e.g., adjusting thermostat settings) or is analyzed for insights (e.g., tracking fitness metrics) (Figure 2.10).

2.4.1.2 Key Characteristics of Sensors

Sensitivity: Refers to the ability to detect tiny changes in the measured amount.
Resolution: The capacity to detect minor variations in measures, similar to finer gradations on a ruler.
Accuracy: Precision in reflecting actual physical quantities.
Linearity: Consistency in output relative to changes in the measured property.
Range: Minimum and maximum values measurable by the sensor.
Selectivity: Capacity to isolate specific quantities from environmental variables.
Response Time: Speed of sensor response to changes in measured parameters.

2.4.1.3 Classification of Sensors

By Operating Principle: Examples include piezoelectric pressure sensors, thermocouples for temperature, photoelectric light sensors, and electrochemical sensors for chemical processes.
By Application: Includes sensors for temperature, pressure, proximity, motion, and more, tailored to specific IoT functions.

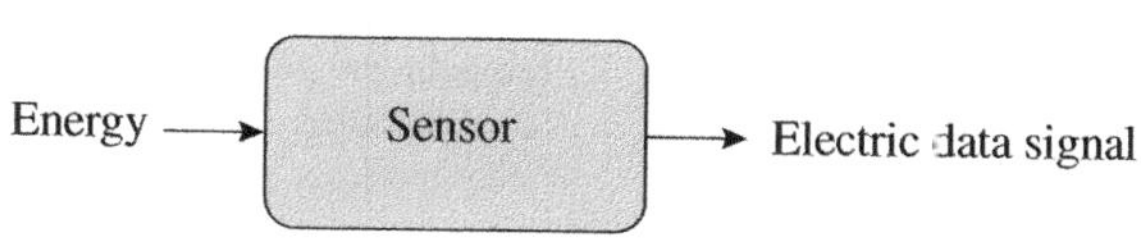

Figure 2.10 Sensor.

2.4.1.4 Role of Sensors in IoT Architecture

In IoT systems, sensors serve foundational roles,

Data Acquisition: Primary source of real-time environmental data.
Preprocessing: Basic data filtering and formatting for transmission.
Communication: Wireless data transmission to IoT network devices, gateways, or cloud platforms.
Automation: Use of sensor data to automate actions within IoT systems.
Real-time Monitoring: Continuous monitoring of environmental parameters.
Data-driven Insights: Analysis of sensor data for predictive maintenance and process optimization.

2.4.2 Transducers

Transducers are created to change energy from one form to another, specifically transforming non-electrical measurements into electrical ones in instrumentation. This transformation covers mechanical, optical, chemical, and electromagnetic energy, designed to meet the needs of various transducer types and uses. In IoT, transducers play a crucial role by combining data from the physical world into digital systems [8].

2.4.2.1 Working

A sensor and signal conditioning unit are utilized by a transducer for transduction. The sensor detects any variation in the physical quantity or energy and delivers a non-electric signal. It is then transformed into a proportional electrical signal through a signal conditioning unit (Figure 2.11).

2.4.2.2 Classification

Transducers are classified into five as,

1) On the basis of transduction form used
2) As primary and secondary transducers
3) As passive and active transducers
4) As analog and digital transducers
5) As transducers and inverse transducers

2.4.2.3 Factors to be Consider When Choosing a Transducer

There are a number of important factors that impact the choice of transducer.

Sensitivity: It is the capacity to recognize and react to slight alterations in input.
Precision: The extent to which the transducer's reading aligns with the genuine recorded quantity.
Operating Range: The range of input values within which the transducer functions efficiently.

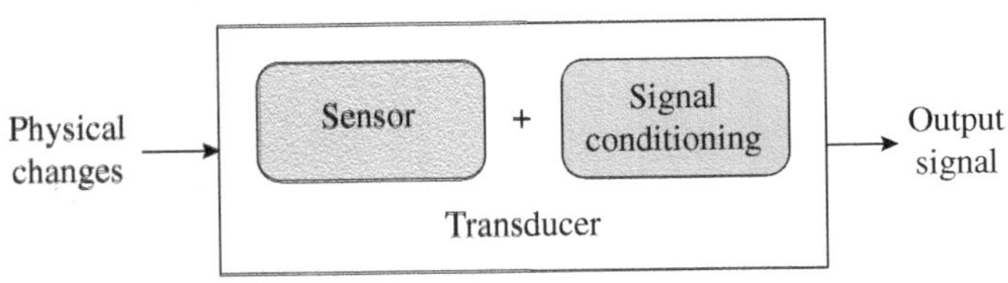

Figure 2.11 Transducer.

Ruggedness: Refers to the ability to withstand both environmental conditions and mechanical stress.

Linearity: Maintaining a consistent correspondence between input and output. Crucial for accurate measurements.

Consistency: Consistency in results when the identical input is used in the same conditions.

Technical Details: Covers signal-to-noise ratio, frequency response, and interoperability with other electronic systems.

2.5 LoWPAN Network Management Protocol (LNMP)

LNMP is a protocol created to manage LoWPANs, which are Low-Power Wireless Personal Area Networks. LoWPANs are networks usually made up of devices powered by batteries with restricted processing power and memory, communicating through low-power wireless connections, like IEEE 802.15.4.

2.5.1 Key Features and Functions of LNMP

2.5.1.1 Topology Management

Formation: LNMP assists in forming and maintaining the network topology of LoWPANs. It helps devices join the network, establish connections, and organize themselves into efficient communication structures.

Maintenance: It ensures that the network topology remains stable and optimized. This includes handling changes in device connectivity (e.g., device movement or failure) and reconfiguring the network as necessary [9].

2.5.1.2 Addressing and Routing

Address Assignment: LNMP manages the assignment of unique addresses to devices within the LoWPAN. This can involve dynamic address allocation and management to accommodate the dynamic nature of LoWPAN environments.

Routing: It facilitates efficient routing of data packets within the LoWPAN. This involves selecting optimal paths based on network conditions, device capabilities, and energy constraints [9].

2.5.1.3 Security Management

Key Management: LNMP facilitates the control and allocation of security keys employed for data encryption, authentication, and integrity protection in the LoWPAN.

Access Control: Makes sure that only approved devices can connect to the network and reach its resources, guarding against unauthorized access and potential security risks.

2.5.1.4 Monitoring and Optimizing Performance

Monitoring: LNMP provides mechanisms to monitor the performance and health of the LoWPAN. This includes monitoring traffic patterns, device status, and network conditions.

Optimization: Based on monitoring data, LNMP can optimize network parameters such as routing paths, transmission power levels, and scheduling to improve overall network performance and energy efficiency [9].

2.5.1.5 Interoperability and Standards Compliance

LNMP adheres to standards and protocols relevant to LoWPANs, ensuring interoperability between devices from different manufacturers and compatibility with existing networking infrastructures.

2.5.2 Implementation and Deployment

LNMP can be implemented as part of the network stack in LoWPAN-enabled devices or as a separate management entity within the network infrastructure It usually works in conjunction with other protocols like IEEE 802.15.4 for physical and medium access control (MAC) layer tasks, and IPv6 for network layer addressing.

LNMP is created to efficiently oversee LoWPANs by using both operational and informational architecture (Figure 2.12).

2.5.3 Operational Architecture of LNMP

2.5.3.1 Network Discovery and Device Detection

Automatic network discovery is crucial in LoWPANs due to the large-scale deployment of WSNs, making manual discovery impractical. LNMP facilitates automatic discovery distributed across the 6LoWPAN network. Coordinators, which are central devices in LoWPANs, are responsible for discovering and managing devices. They maintain lists of connected devices, monitor their statuses, and facilitate queries to ensure efficient network operation. This decentralized approach reduces communication overhead and enhances network reliability [10].

2.5.3.2 Device Categorization and Management

Devices in LNMP are categorized into:

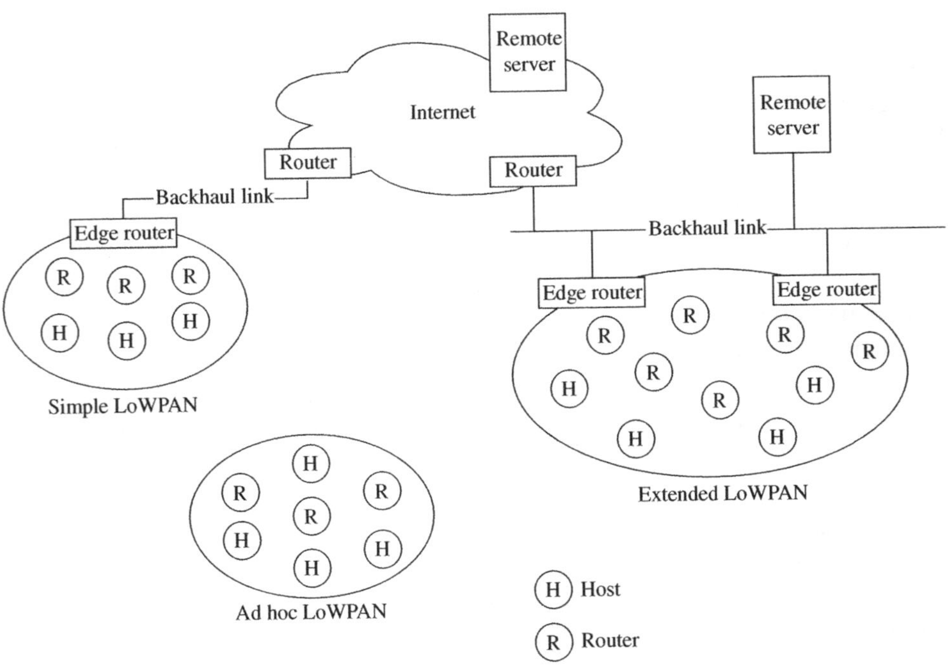

Figure 2.12 LNMP architecture.

End Devices (FFD and RFD): Full function devices (FFDs) handle comprehensive communication tasks and respond directly to coordinator queries. Reduced function devices (RFDs) have limited functionality, often acting as sensors that transmit data periodically or on-demand.

Coordinators: These devices play a pivotal role in managing the state information of all connected devices within the LoWPAN. They populate device lists, maintain state tables, and manage state transitions to optimize network performance and resource utilization.

Gateways: Act as interfaces between the 6LoWPAN network and external Internet Protocol (IP)-based networks. Gateways aggregate device status information from coordinators, translate them into IP addresses, and maintain a comprehensive management information base (MIB). The MIB contains critical network device data, facilitating network monitoring and control via standard protocols like simple network management protocol (SNMP).

2.5.3.3 SNMP and 6LoWPAN Integration

SNMP is supported on the IPv6 network side, and packets are converted to a simplified format for 6LoWPAN transport.

The gateway translates SNMP queries to user datagram protocol (UDP) requests containing an ID and converts responses back to SNMP.

This approach allows existing network management tools to be used without modification.

For instance, a network management system (NMS) might send an SNMP query to check the status of a sensor, with the 6LoWPAN gateway handling the necessary conversions and communications [10].

2.5.4 Informational Architecture of LNMP

2.5.4.1 Management Information Base (MIB) Standardization

LNMP emphasizes the standardization of MIBs across different layers of 6LoWPAN networks. Standardized MIB structures ensure uniform representation of network parameters such as device status, topology, and communication metrics. This uniformity enhances interoperability between devices and NMS, supporting scalable and reliable network operations.

2.5.4.2 Protocol Reuse for Efficiency

Efficient network management in LNMP is achieved through the reuse of established protocols such as SNMP. Despite the limited packet size (127 bytes) of 6LoWPAN networks, LNMP enables SNMP integration through gateway proxies. Gateway proxies translate SNMP queries and responses into simplified formats suitable for 6LoWPAN communication, ensuring compatibility with existing network management tools. This approach simplifies monitoring, troubleshooting, and configuration management across diverse IoT environments.

LNMP offers a robust framework for managing LoWPANs by combining advanced operational and informational architectures. By automating network discovery, categorizing devices, standardizing MIB structures, and promoting protocol reuse, LNMP enhances network efficiency, scalability, and interoperability. This comprehensive approach supports diverse IoT applications, ensuring reliable and adaptive network management across various deployment scenarios.

2.6 WSN Diagnostic Tools: Ensuring Reliability and Performance

WSNs are essential in various sectors like environmental monitoring and industrial automation, relying on multiple interconnected sensor nodes for operation. WSN diagnostic tools are crucial to guaranteeing the dependability, effectiveness, and functionality of WSNs. These tools consist of different software applications and methods utilized for mimicking, showcasing, monitoring, debugging, and analyzing WSNs from start to finish. These diagnostic tools provide information on network behavior, protocol effectiveness, energy consumption patterns, and current operational parameters. They assist researchers and developers in enhancing network design, communication protocols, node lifespan, and ensuring dependable and sustainable WSN deployments through controlled simulations and real-world scenarios [11].

2.6.1 Simulation Tools

Simulation utilities are essential for simulating and assessing the functionality of WSNs in controlled settings prior to implementation. They offer understanding of network performance, protocol effectiveness, and scalability across different scenarios.

OMNeT++: Famous for its modular structure, enabling researchers to simulate intricate network scenarios by representing specific elements like nodes, protocols, and environments. It is versatile for WSN research since it can be used for both discrete event simulation and real-time simulation.

Network Simulator 2 (ns-2): A commonly utilized tool in academic and research settings, enables in-depth simulation of network protocols and behaviors. It contains different wireless communication technology models and can be tailored for specific WSN uses.

***TinyOS Simulator (TOSSIM)*:** Designed specifically for use with TinyOS-based sensor networks, allowing researchers to simulate TinyOS applications and protocols effectively. It offers information on energy usage, node actions, and network efficiency, essential for enhancing applications in environments with limited resources [12].

2.6.2 Visualization Tools

Visualization tools are crucial in comprehending the spatial and temporal aspects of WSN function, helping with debugging, analyzing performance, and visualizing networks.

WSN Visualization (WSNVis): Provides graphic visualizations of sensor node arrangements, network structures, and information transmission. Assisting researchers in visualizing network dynamics, spotting communication bottlenecks, and improving node placement for better coverage and connectivity.

WSNLab: Created for the purpose of displaying sensor data and network performance metrics in real-time, allowing researchers to observe sensor nodes in field trials. It supports data aggregation, visualization of environmental changes, and debugging of sensor node interactions in real-world scenarios.

2.6.3 Debugging and Monitoring Tools

Identifying and solving issues in WSNs during development, deployment, and operation requires essential debugging and monitoring tools.

D-SimSPIN: This debugging tool provides a graphical interface to visualize the execution flow and behavior of sensor nodes during simulation. It helps in diagnosing protocol errors, optimizing code efficiency, and understanding node interactions in simulated environments.

SensorScope: A monitoring tool that gathers sensor data, network statistics, and energy consumption metrics in real-time from nodes in WSNs. It provides ongoing monitoring, identifies abnormalities, and analyzes performance to guarantee dependable function and effective resource handling.

2.6.4 Energy Profiling Tools

Energy profiling tools are essential for evaluating and enhancing energy usage in sensor nodes, which is vital for extending network longevity and eco-friendliness.

***PowerTOSSIM*:** It is an enhancement of TOSSIM that incorporates energy usage simulation into TinyOS simulations. It allows researchers to profile energy usage at the node level, evaluate the impact of protocols on battery life, and optimize power management strategies for energy-efficient WSN deployments.

***Energy Debugger*:** This tool provides real-time monitoring and analysis of energy consumption patterns in deployed sensor nodes. It helps researchers identify power-hungry components, optimize energy harvesting strategies, and implement energy-aware protocols to maximize network longevity in autonomous and remote environments [11, 12].

2.6.5 Network Analysis Tools

Network analysis tools concentrate on capturing, analyzing, and interpreting data traffic in WSNs to enhance communication efficiency and protocol performance.

***Sensor Network Interception Framework (SNIF)*:** A tool for sniffing packets in WSNs, capturing and analyzing network traffic in real-time. It helps researchers debug communication issues, assess protocol overhead, and validate data integrity across sensor nodes. SNIF supports protocol debugging, performance tuning, and network security analysis in diverse WSN deployments.

***MoteLab*:** Designed for long-term monitoring and management of sensor networks, MoteLab facilitates remote access and control of deployed nodes. It supports data logging, performance monitoring, and experimental reproducibility in field trials, enabling researchers to evaluate network reliability, scalability, and environmental adaptability over extended periods.

These diagnostic tools collectively contribute to advancing WSN research by providing insights into network dynamics, protocol efficiency, energy management, and reliability in diverse operational scenarios. They are essential for developing robust WSN applications, optimizing resource utilization, and addressing challenges associated with deployment in real-world environments.

2.7 Overview of IoT Communication Technologies

Networked ecosystems are built on IoT communication technologies, which provide seamless control and data flow between a variety of systems and devices. These technologies come in a range of forms, catering to different requirements including data speed, distance, energy consumption, and expandability. They include wireless, cellular, and cable choices. To fully benefit from a connected society, IoT communication technologies are essential. These comprise LTE for machines (LTE-M) and 5G new radio (5G NR) for high-speed cellular communications and Bluetooth low energy (BLE) and Zigbee for low-power, short-range applications. Ethernet is a networked technology that offers dependable, consistent connections, which makes it ideal for industrial automation and corporate applications. Safe connections between IoT devices across a range of industries, including as consumer electronics, smart cities, healthcare, and industrial IoT, are made possible by the quick data transfer from the message queuing telemetry transport (MQTT) and constrained application protocol (CoAP) protocols. A thorough grasp of these technologies is essential for creating and implementing scalable, reliable IoT solutions that foster productivity and international innovation in a variety of industries [12, 13].

2.7.1 Wireless Technologies

2.7.1.1 Bluetooth Low Energy (BLE)

An energy-efficient wireless technology that allows devices to run on very little power.

Due to its low energy consumption, BLE is a popular choice in the IoT. This makes it ideal for situations where devices must run for lengthy periods of time on a limited amount of battery life without regular recharge. Wearable health monitors, proximity-detecting sensors in retail settings, and smart home appliances like door locks and thermostats are a few examples of this.

2.7.1.2 Zigbee

Frequently utilized in smart home systems to enable wireless communication between devices that are close to one another. Utilizing the IEEE 802.15.4 standard, Zigbee is widely used to establish mesh networks in which devices connect with one another via intermediate nodes. This is perfect for situations that require dependable communication, such industrial settings for equipment supervision and control, smart lighting systems, and residential automation (smart plugs, sensors) [13].

2.7.1.3 LoRaWAN

IoT extends connectivity to devices across urban and rural areas spanning several kilometers using LoRaWAN, renowned for its extensive range. Asset tracking enhances smart supply chains, while smart cities benefit from features like parking management and environmental monitoring. Similarly, smart agriculture leverages IoT for monitoring crop health and soil moisture levels.

2.7.1.4 Narrowband Internet of Things (NB-IoT)

Narrowband Internet of Things (NB-IoT) utilizes licensed spectrum to ensure dependable connectivity, enabling reliable communication even in challenging environments. It supports critical applications such as smart metering for monitoring gas, water, and electricity usage, environmental monitoring using air quality sensors, and remote patient monitoring in healthcare.

2.7.2 Cellular Technologies

2.7.2.1 LTE for Machines (LTE-M)

More data rates are available with LTE-M than with NB-IoT, making it a good choice for applications with moderate to high data throughput needs including industrial automation (which monitors machines remotely), smart city infrastructure (which controls waste management and smart lighting), and real-time asset tracking [14].

2.7.2.2 5G New Radio (5G NR)

5G NR promises transformative capabilities for IoT with ultra-low latency, high reliability, and support for massive device connectivity. It is poised to revolutionize industries like autonomous vehicles (vehicle-to-everything communication), augmented reality (real-time interactions), smart manufacturing (robotics and automation), and healthcare (telemedicine and remote surgery).

2.7.3 Wired Technologies

2.7.3.1 Ethernet

In IoT applications where dependability, large bandwidth, and low latency are critical, Ethernet is still the backbone. Building management systems and campus networks in business contexts, smart grid applications (energy management, distribution automation), and industrial automation (process control, factory automation) all make substantial use of it.

2.7.4 IoT Protocols and Standards

2.7.4.1 Message Queuing Telemetry Transport (MQTT)

Due to its lightweight messaging protocol, MQTT is very effective in IoT contexts, allowing devices to publish and subscribe to data topics. It is used in telemetry and logistics tracking systems, as well as remote monitoring and control (smart home appliances, industrial sensors) (Figure 2.13).

2.7.4.2 Constrained Application Protocol (CoAP)

CoAP is a lightweight alternative to hyper text transfer protocol (HTTP) for communication across restricted networks, tailored for IoT devices with low resources. It is utilized in precision agriculture (crop monitoring, irrigation systems), smart city deployments (street lighting control, traffic management), and environmental monitoring (weather stations, pollutant sensors) (Figure 2.14).

Every type of wireless, cellular, and wired solution that meets certain IoT needs like scalability, power consumption, data rate, and range is included in the wide variety of IoT communication

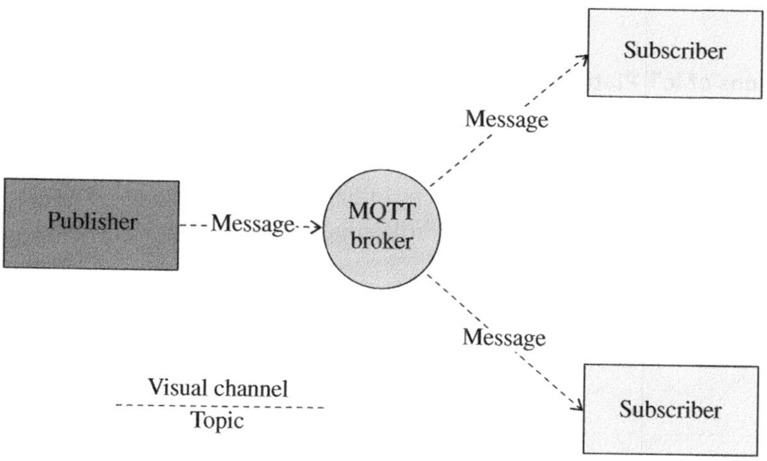

Figure 2.13 MQTT architecture.

Figure 2.14 CoAP architecture.

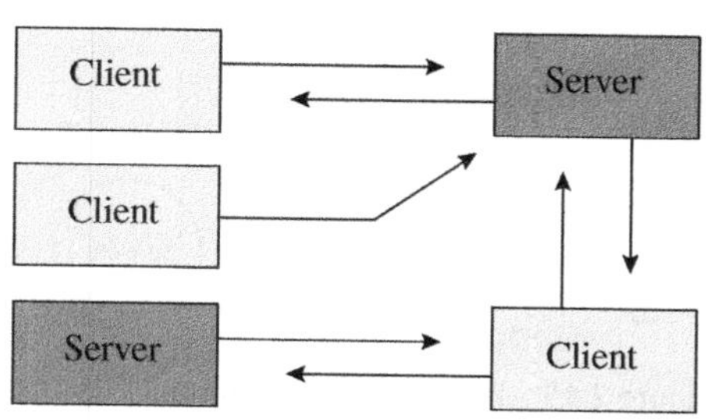

technologies. Optimizing IoT installations across a variety of sectors, increasing operational efficiency, opening up new applications, and boosting overall quality of service all depend on selecting the appropriate communication technology and protocol. As IoT develops, communication technology advancements will be crucial in accelerating digital transformation and building smarter, more linked ecosystems around the globe.

2.8 Practical Applications of IoT Platforms, Sensor Technologies and Communication Protocols

The integration of sensor technologies, IoT platforms, and communication protocols has significantly transformed the way we engage with and improve our surroundings. Through the collection, arrangement, and examination of information from various interconnected devices, IoT systems act as pivotal centers for smooth automation and management in multiple sectors. Real-time data recording from physical environments, such as industrial machinery performance, ambient variables, and medical diagnostics, is simultaneously facilitated by sensor technology. By providing industries with useful data, these sensors promote productivity, sustainability, and creativity [15]. Furthermore, communication protocols enable reliable operations across a range of networks, including smart homes and intricate industrial systems, and provide strong connectivity and data transmission between IoT devices in addition to these developments. Together, these technologies provide the foundation for contemporary digital ecosystems that are revolutionizing industries, boosting standards of living, and opening the door to a future in which communication will be much more widespread (Figure 2.15).

2.8.1 Practical Applications of IoT Platforms

2.8.1.1 Smart Home Automation

Homeowners have the ability to automate and oversee a diverse array of devices, such as lights, thermostats, security cameras, and entertainment systems, by utilizing IoT platforms like Amazon Alexa, Google Home, and Apple HomeKit. Through the utilization of smartphone applications,

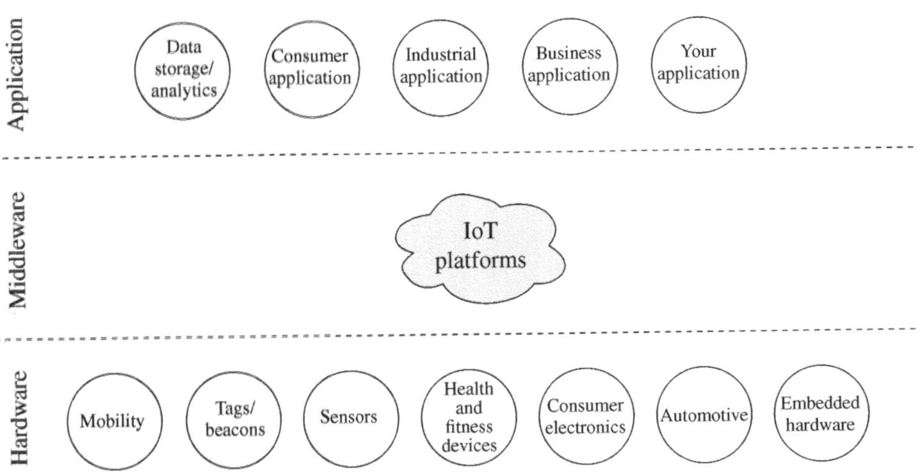

Figure 2.15 Practical application of IoT platforms.

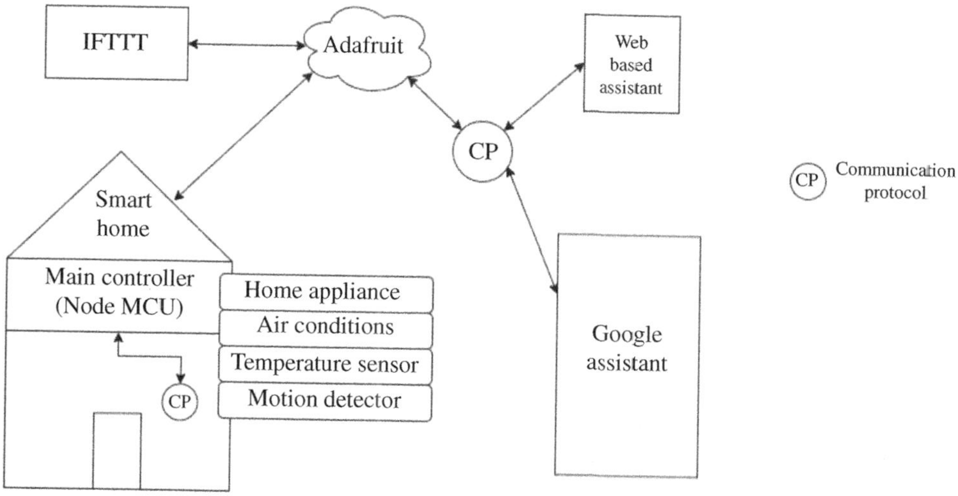

Figure 2.16 Practical application of IoT platforms – SMART HOME AUTOMATION.

clients can create personalized routines and remotely manage their households by leveraging the integration of these platforms with a wide range of smart devices (Figure 2.16).

2.8.1.2 Industrial Automation

The advancement of industrial automation is greatly facilitated by various IoT platforms like Siemens MindSphere, GE Predix, and Dassault Systèmes, which seamlessly connect machinery, sensors, and enterprise systems. These platforms enable users to monitor equipment performance in real-time, anticipate maintenance requirements, and efficiently optimize production processes [16]. Consequently, operational efficiency is heightened, downtime is reduced, and overall productivity is significantly enhanced in manufacturing and industrial settings (Figure 2.17).

2.8.1.3 Healthcare

IoT platforms support remote patient monitoring, telemedicine, and personalized medicine initiatives. Examples include Medtronic Sugar.IQ for diabetes management, Philips Healthcare's IntelliSite for digital pathology, and Fitbit's HealthKit integration for health tracking and analytics. Patient care is significantly improved through the utilization of these platforms, which allow for

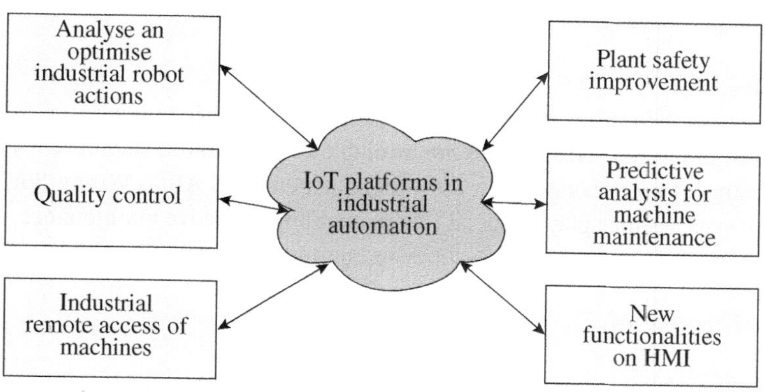

Figure 2.17 Practical application of IoT platforms – INDUSTRIAL AUTOMATION.

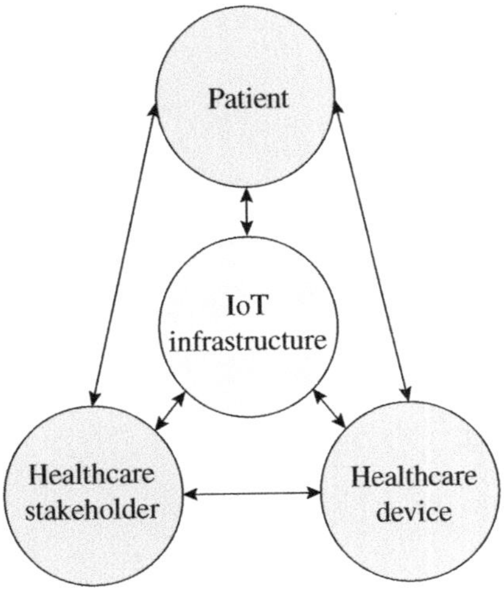

Figure 2.18 Practical application of IoT platforms – HEALTHCARE.

continuous health monitoring, prompt intervention, and customized treatment strategies. This ultimately leads to better patient results and a decrease in healthcare expenses (Figure 2.18).

2.8.1.4 Transportation

IoT platforms play a crucial role in enhancing transportation systems through the enhancement of traffic management, optimization of route planning, and tracking of vehicle fleets. Notable examples of such platforms are the high efficiency RFID-enabled (HERE) navigation software development kit (SDK), which offers precise mapping and navigation services, Google Maps, which provides real-time traffic updates and directions, and GPS tracking systems designed for fleet management purposes. The utilization of these platforms results in improved transportation efficiency, decreased congestion, and heightened safety levels on roadways.

2.8.2 Practical Applications of Sensor Technologies

2.8.2.1 Environmental Monitoring

Sensors monitor environmental parameters such as air quality, water quality, temperature, humidity, and weather patterns. Examples include AirVisual for air quality monitoring, Water.io for smart water management, and Weather Underground for weather data analytics. Environmental sensors provide valuable insights for pollution control, water resource management, and weather forecasting, contributing to environmental sustainability efforts (Figure 2.19).

2.8.2.2 Industrial Monitoring

Sensors track machine performance, energy consumption, and equipment health in industrial environments. Various examples of industrial sensor suites are available in the market, such as GE's Predix for predictive maintenance, Emerson's AMS for asset management, and ABB's Wrong-Bot for robotic automation. These sensor suites play a crucial role in enabling proactive maintenance, optimizing energy usage, and ensuring operational efficiency in manufacturing plants and industrial facilities.

2.8.2.3 Healthcare

Sensors monitor vital signs, diagnose diseases, and track patient conditions in healthcare settings. Examples include health monitoring wristbands like WearFit and Fitbit, which track heart rate,

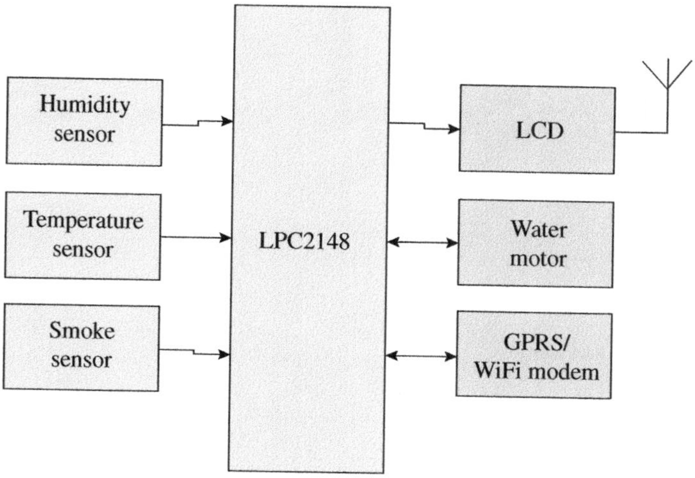

Figure 2.19 Practical application of sensor technologies – ENVIRONMENTAL MONITORING.

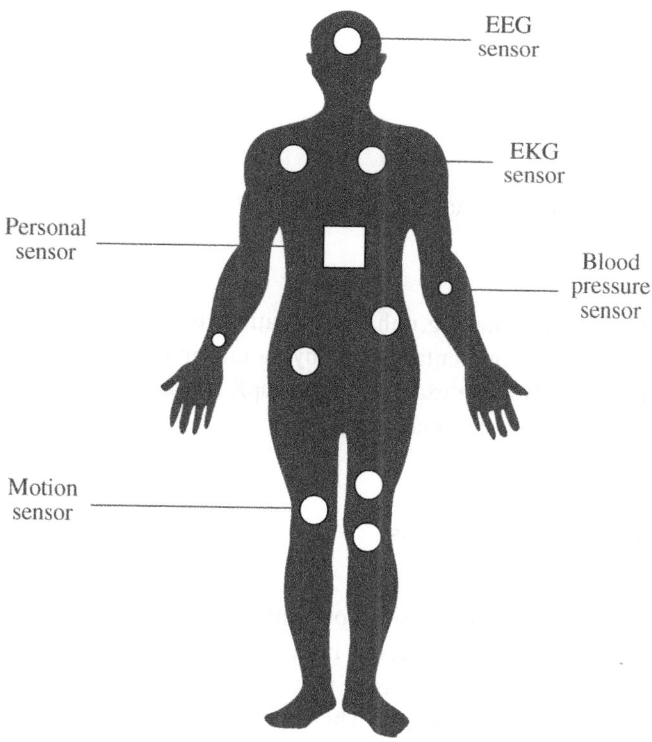

Figure 2.20 Practical application of sensor technologies – HEALTHCARE.

activity levels, and sleep patterns, and continuous glucose monitors (CGMs) for diabetes management [17]. Healthcare sensors enable remote patient monitoring, early detection of health issues, and personalized treatment approaches, improving patient outcomes and reducing hospital visits (Figure 2.20).

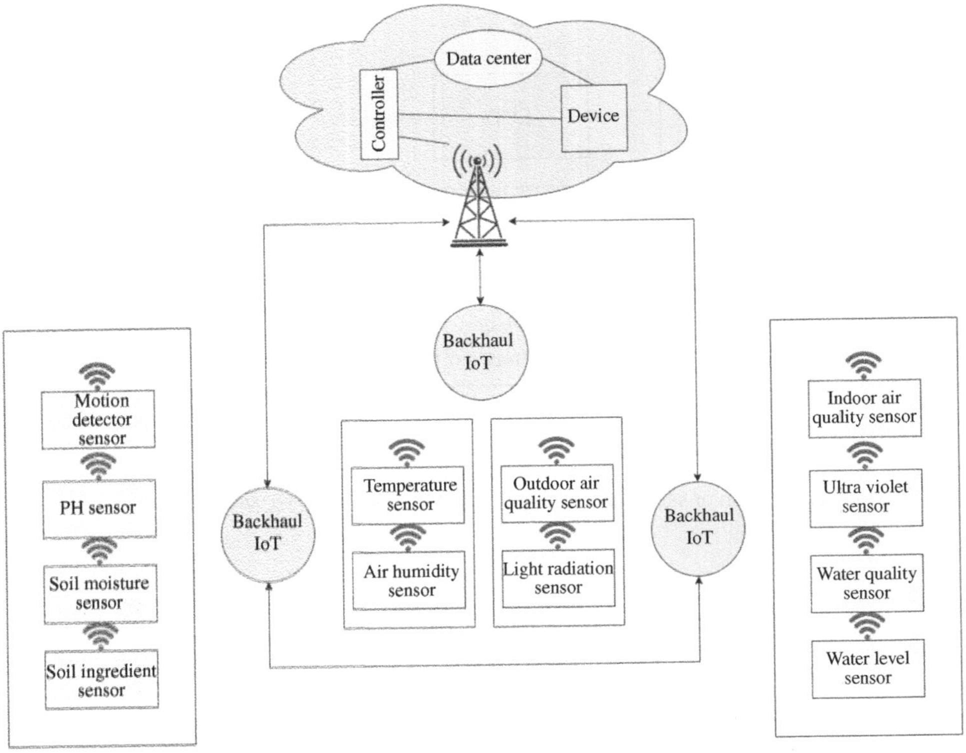

Figure 2.21 Practical application of sensor technologies – AGRICULTURE.

2.8.2.4 Agriculture

Agricultural sensors facilitate data-driven decision-making, enhance irrigation and fertilizer usage, and maximize crop yields while minimizing environmental impact by monitoring soil moisture, temperature, nutrient levels, and crop health. Notable examples are CropX for soil monitoring, FarmLogs for crop management, and Granular for farm analytics (Figure 2.21).

2.8.3 Practical Applications of Communication Protocols

2.8.3.1 Low-Power Wide-Area Networks (LPWANs)

LPWAN protocols provide cost-effective and energy-efficient connectivity solutions, enabling low-power, long-range communication for IoT devices located in remote areas or areas with unreliable internet access. For instance, Sigfox is used for asset tracking, LoRaWAN for smart city applications, and NB-IoT for cellular IoT deployments. These protocols support various applications such as asset tracking, smart metering, and environmental monitoring (Figure 2.22).

2.8.3.2 Wi-Fi and Bluetooth

They facilitate wireless communication over short distances among IoT devices within a LAN. For instance, Wi-Fi routers connect homes and offices, while Bluetooth speaker systems stream audio. These protocols are essential for smart home automation, wearable technology, and consumer electronics, ensuring rapid data transfer and effortless connectivity across devices.

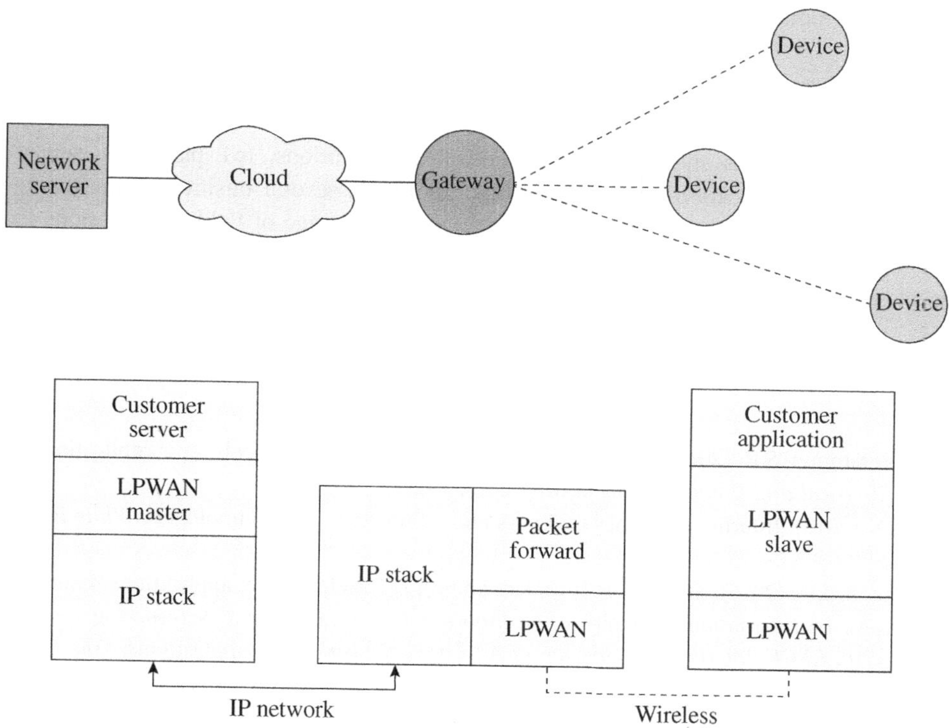

Figure 2.22 LPWAN network architecture.

2.8.3.3 Cellular Networks

For IoT devices to be able to connect everywhere, they need long-range connectivity and easier-to-use protocols. Cellular networks, 2G, 3G, 4G, 5G and others, have been developed to provide wide coverage to ensure effective communication across geographical areas. Each cellular network generation is excellent compared to its previous generation by having better internet network. Cellular IoT modules like the SIM800L used for M2M communication and connectivity providers, for example, from AT&T, Verizon, or T-Mobile. The cellular protocol is used for applications like vehicle tracking, remote monitoring, and industrial automation because it completes the transmission of data packets and the functionality of the network.

2.8.3.4 Mesh Networking

Mesh networking protocols enable decentralized communication among IoT devices by allowing devices to relay data through neighboring nodes. Examples include Zigbee for smart home networks, Thread for low-power IoT devices, and Bluetooth Mesh for lighting control and building automation. Mesh networking enhances reliability, extends network coverage, and supports scalable IoT deployments in environments where traditional communication methods may be impractical or costly.

2.8.4 Integration and Impact

The integration of IoT platforms, sensor technologies, and communication protocols drives significant impacts across industries:

i) Efficiency and Productivity Gains
ii) Cost Savings
iii) Sustainability
iv) Enhanced Quality of Life

These real-world examples show how mesh networking solutions, IoT platforms, sensor technologies, and communication protocols are revolutionizing several businesses, increasing productivity, sharpening judgment, and spurring creativity in a range of fields. Innovations in these technologies will speed up digital transformation as IoT develops, resulting in smarter, more interconnected ecosystems that benefit companies, governments, and society at large.

References

1 Sethi, P. and Sarangi, S.R. (2017). Internet of Things: architectures, protocols, and applications. *Journal of Electrical and Computer Engineering* 2017: 1–25.

2 El Hakim, A. (2018). Internet of Things (IoT) system architecture and technologies. *White Paper* 10: 1–5.

3 Soumyalatha, S.G.H. (2016). Study of IoT: understanding IoT architecture, applications, issues and challenges. *1st International Conference on Innovations in Computing & Net-working (ICICN16), CSE, RRCE. International Journal of Advanced Networking & Applications*. Vol. 478.

4 Vogel, B. and Gkouskos, D. (2017). An open architecture approach: towards common design principles for an IoT architecture. *Proceedings of the 11th European Conference on Software Architecture: Companion Proceedings*.

5 Putera, C.A.L. and Lin, F.J. (2015). Incorporating OMA lightweight M2M protocol in IoT/M2M standard architecture. *2015 IEEE 2nd World Forum on Internet of Things (WF-IoT)*, Milan, Italy. pp. 559–564.

6 Misra, S., Mukherjee, A., and Roy, A. (2021). IoT sensing and actuation. In: *Introduction to IoT*, 97–114. Cambridge University Press.

7 Sharma, A., Sharma, S., and Gupta, D. (2021). A review of sensors and their application in Internet of Things (IoT). *International Journal of Computer Applications* 174: 27–34.

8 Arpaia, P., Bonavolontà, F., Moccaldi, N., and Cioffi, A. (2021). Reproducibility enhancement by optimized power analysis attacks in vulnerability assessment of IoT transducers. *IEEE Transactions on Instrumentation and Measurement* 70: 3523608.

9 Mukhtar, H., Kang-Myo, K., Chaudhry, S.A. et al. LNMP-management architecture for IPv6 based low-power wireless personal area networks (6LoWPAN). *NOMS 2008 - 2008 IEEE Network Operations and Management Symposium*, Salvador, Brazil. pp. 417–424.

10 Lamaazi, H., Benamar, N., Jara, A., Ladid, L., and El Ouadghiri, D. (2013). Internet of thing and networks' management: LNMP, SNMP, COMAN protocols. *First Int. Work. Wirel. Networks Mob. Commun. (WINCOM 2013)*, 1–5.

11 Rodrigues, A., Camilo, T., Silva, J.S., and Boavida, F. (2012). Diagnostic tools for wireless sensor networks: a comparative survey. *Journal of Network and Systems Management* 21 (3): 408–452.

12 Al-Sarawi, S., Anbar, M., Alieyan, K., and Alzubaidi, M. (2017). Internet of Things (IoT) communication protocols: review. *2017 8th International Conference on Information Technology (ICIT)*, Amman, Jordan. pp. 685–690.

13 Machorro-Cano, I., Alor-Hernández, G., Cruz-Ramos, N.A. et al. (2018). A brief review of IoT platforms and applications in industry. In: *New Perspectives on Applied Industrial Tools*

and Techniques (ed. J.L. García-Alcaraz, G. Alor-Hernández, A.A. Maldonado-Macías, and C. Sánchez-Ramírez), 293–324. Cham: Springer.

14 Afzal, B., Umair, M., Shah, G.A., and Ahmed, E. (2019). Enabling IoT platforms for social IoT applications: vision, feature mapping, and challenges. *Future Generation Computer Systems* 92: 718–731.

15 Ullo, S.L. and Sinha, G.R. (2021). Advances in IoT and smart sensors for remote sensing and agriculture applications. *Remote Sensing* 13 (13): 2585.

16 Sidna, J., Amine, B., Abdallah, N. et al. (2020). Analysis and evaluation of communication protocols for IoT applications. *Proceedings of the 13th International Conference on Intelligent Systems: Theories and Applications*.

17 Kraijak, S. and Tuwanut, P. (2015). A survey on IoT architectures, protocols, applications, security, privacy, real-world implementation and future trends. *11th International Conference on Wireless Communications, Networking and Mobile Computing (WiCOM 2015)*. IET.

3

Communication Protocols for Transactive IoT

A. Kamalasegaran, G. Kabilan, and P. Sriramalakshmi

School of Electrical Engineering, Vellore Institute of Technology, Chennai, Tamil Nadu, India

3.1 Introduction

The fast-paced progress on the Internet of Things (IoT) has transformed several sectors by enabling real-time connectivity and smart automation. One of the most promising applications of IoT is in the domain of smart grids (SGs), where it enables a more sustainable, reliable, and decentralized method of power management called transactive IoT. This approach leverages IoT devices to a dynamic and self-governing energy ecology, which is enabled to act on real-time data and marketplace signals. Critical to the implementation of transactive IoT, there are communication protocols that enable seamless information flow between millions of devices, sensors, and control systems. This chapter explores the critical communication protocols that form the backbone of transactive IoT, with a specific focus on message queuing telemetry transport (MQTT), extensible messaging presence protocol (XMPP), advanced message queuing protocol (AMQP), and data distribution service (DDS). Each of these protocols offers distinct benefits that are critical to the impactful functioning of SGs. MQTT is known for its lightweight design and efficiency, making it ideal for low-bandwidth and high-latency conditions commonly found in IoT applications. XMPP offers robust real-time data exchange and device presence capabilities, while AMQP ensures reliable and secure message delivery, vital for the coordination of distributed energy resources. DDS excels in high overall performance and real-time data distribution, which is essential to the disturbing requirements of SG systems.

The integration of these communication protocols within the SG domain enables a host of benefits such as superior grid reliability, optimal energy consumption, decentralized power management, and dynamic pricing mechanisms. By understanding and implementing those protocols, stakeholders can unlock the whole potential of transactive IoT to build more resilient and efficient power infrastructures. In this bankruptcy, it is possible to go through every protocol in detail and scan their roles, functionalities, and packages in the context of transactive IoT in SGs. With this, the readers shall achieve a comprehensive understanding of how the combination of these protocols contributes to enhancing smart, responsive, and sustainable strength systems.

3.2 Transactive Systems in Smart Grids

Transactive systems in SGs are one of the innovative ways of managing energy distribution and consumption. These systems make decentralized and dynamic interactions among different

IoT for Smart Grid: Revolutionizing Electrical Engineering, First Edition.
Edited by Rahiman Zahira, Palanisamy Sivaraman, Chenniappan Sharmeela, and Sanjeevikumar Padmanaban.

stakeholders: energy producers, consumers, and prosumers – entities that produce and consume energy [1]. Making use of advanced communication protocols and IoT technologies, transactive systems have enabled real-time energy trading, grid stability, and promotion of the integration of renewable sources of energy. This chapter delves into a detailed explanation of transactive systems in SGs.

3.2.1 Key Components of Transactive Systems in Smart Grids

3.2.1.1 Prosumers

- **Role:** A prosumer generates energy. For example, solar panels generate electricity and consume it simultaneously. They take an active part in energy markets; they sell back excess energy to the grid or directly to other consumers.
- **Technologies:** Prosumers, with the help of smart meters and IoT devices, monitor their energy production and consumption in real time [2].

3.2.1.2 Decentralized Energy Markets

- **Functionality:** Decentralized energy markets facilitate peer-to-peer (P2P) energy trading. Consumers and prosumers have a chance to buy and sell energy directly to the market. It operates through an automated transaction system based on the real-time supply and demand in the market.
- **Benefits:** The markets save energy, reduce energy costs, and encourage renewable energy adoption.

3.2.1.3 Dynamic Pricing

- **Mechanism:** Transactive systems work with dynamic pricing models whereby energy prices vary depending on real-time market conditions and grid demands. This mechanism inspires consumers to regulate their usage of electricity through the responsive use of rate signals.
- **Impact:** Dynamic pricing actually balances power supply and demand; it reduces the need for peak electricity generation and gives grid balance.

3.2.1.4 Smart Contracts

- **Purpose:** Smart contracts are self-executable contracts whose terms are directly written into lines of code. They digitize, verify, and enforce transaction agreements between parties in the energy marketplace.
- **Benefits:** They bring about transparency, reduce transaction costs, and eliminate intermediaries.

3.2.1.5 Grid Edge Intelligence

- **Implementation:** Edge computing (EC) devices use various techniques that record locally, which permits real-time decision-making not reliant on a centralized statistics center.
- **Impact:** It enhances the responsiveness and resilience of the energy system [3].

3.2.2 Conceptual TE Model

Figure 3.1 is the conceptual model of the transactive energy (TE) system. It allows for the two-way flow of energy and information between producer and consumer. This helps to improve the efficiency and reliability of the grid.

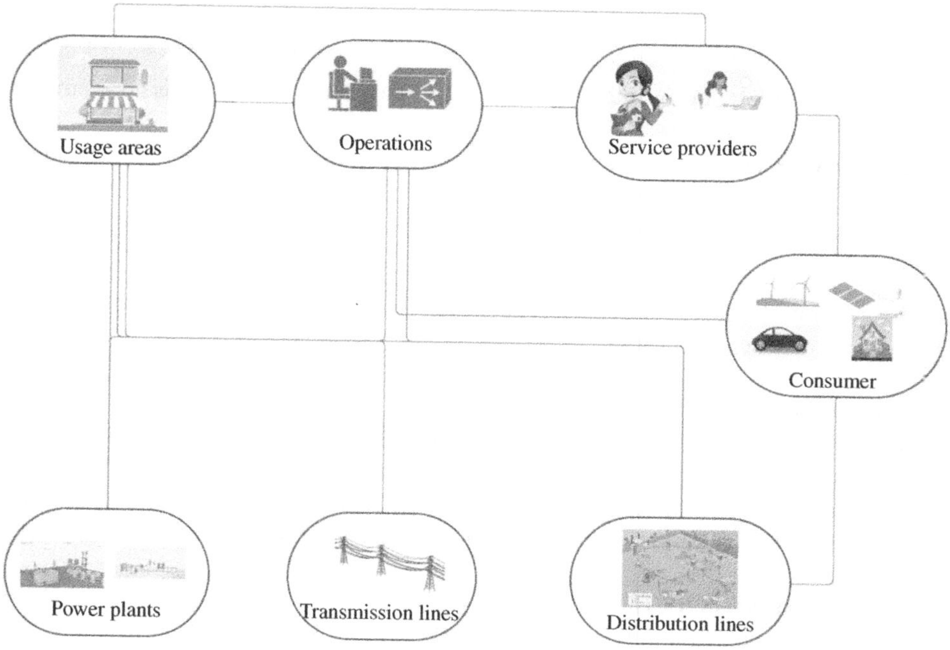

Figure 3.1 Conceptual TE model.

- **Usage Area:** This represents the usage area where the electricity is bought and sold.
- **Operations:** This represents the physical infrastructure of the grid, including transmission and distribution lines.
- **Service Provider:** This represents the companies that generate and sell electricity.
- **Customer:** The people and businesses that use electricity. This can improve the efficiency of the grid by allowing the consumers to feed the excess electricity back to the grid on a payment basis.

This helps to reduce the greenhouse gas emission by letting the consumers use renewable energy resources. This improves the reliability of the grid by allowing for a more distributed and flexible power supply.

3.3 MQTT, CoAP, and Other Protocols in Transactive Systems

3.3.1 Message Queuing Telemetry Transport (MQTT)

MQTT is a lightweight messaging protocol and it's designed for excessive-latency, low-bandwidth, and unreliable networks. MQTT is massively utilized in IoT applications because it is exclusively for inexperienced small systems and it can manipulate many communication styles. MQTT serves as a communication backbone facilitating the exchange of data, commands, and notifications between many sensors, gadgets, and applications applied in transactive operations [4].

3.3.2 Key Features of MQTT

3.3.2.1 Publish-Subscribe Model

This messaging model is used in MQTT and they are ideal for transactive structures wherein many messages communicate asynchronously.

3.3.2.2 Quality of Service (QoS)

MQTT helps three ranges of quality of service (QoS) to make sure dependable message delivery:

QoS 0 (At Maximum Once): The message is brought at the maximum once. There is no assurance of transport even once.

QoS 1 (At Least as Soon as): The message is delivered at least once. It may be brought a couple of times if vital to make certain delivery.

QoS 2 (Exactly Once): The message is delivered exactly once and as quickly as by the use of a step handshake mechanism.

In transactive systems, the choice of QoS degree depends on the significance of message delivery and the tolerance for message duplication.

3.3.2.3 Retained Messages

MQTT allows the supplier to keep the ultimate message posted on a subject. When a new subscriber connects to the dealer and subscribes to a topic with a retained message, the supplier sends the most recent retained message to the subscriber immediately. This function is useful in transactive systems for presenting the modern state or data to newly related devices or applications [8].

3.3.2.4 Last Will and Testament (LWT)

MQTT clients can specify a "remaining will" message that the broker will post on behalf of the consumer if the customer disconnects unexpectedly. In transactive structures, this selection may be used to inform extraordinary entities regarding the unexpected disconnection of a device or software-taking element in transactive operations.

3.3.2.5 Lightweight Protocol

One of the key benefits of MQTT is its lightweight nature. The protocol header is small, and the overhead for message transmission is minimal. This makes MQTT suitable for useful resource-confined gadgets usually determined in IoT deployments, making sure green communication in transactive structures without consuming immoderate bandwidth or computational resources.

3.3.2.6 Scalability

MQTT is designed to be scalable, permitting heaps or even thousands of gadgets to connect to a single provider. MQTT dealers can be deployed in an allotted manner to address expanded load and ensure immoderate availability in huge-scale transactive systems.

3.3.2.7 Security

While MQTT itself no longer provides incorporated protection abilities, it may be used in the aspect of transport layer security/secure socket layer (TSL/SSL) for encrypted communication and authentication mechanisms, which include username/password or client certificates. In transactive systems, handling sensitive data or crucial operations, making secure communication over MQTT is paramount [6].

3.3.2.8 Interoperability

MQTT is an open protocol with support for various programming languages and platforms. This enables interoperability between devices and applications from different vendors, facilitating seamless integration within transactive systems where diverse devices and components need to communicate with each other.

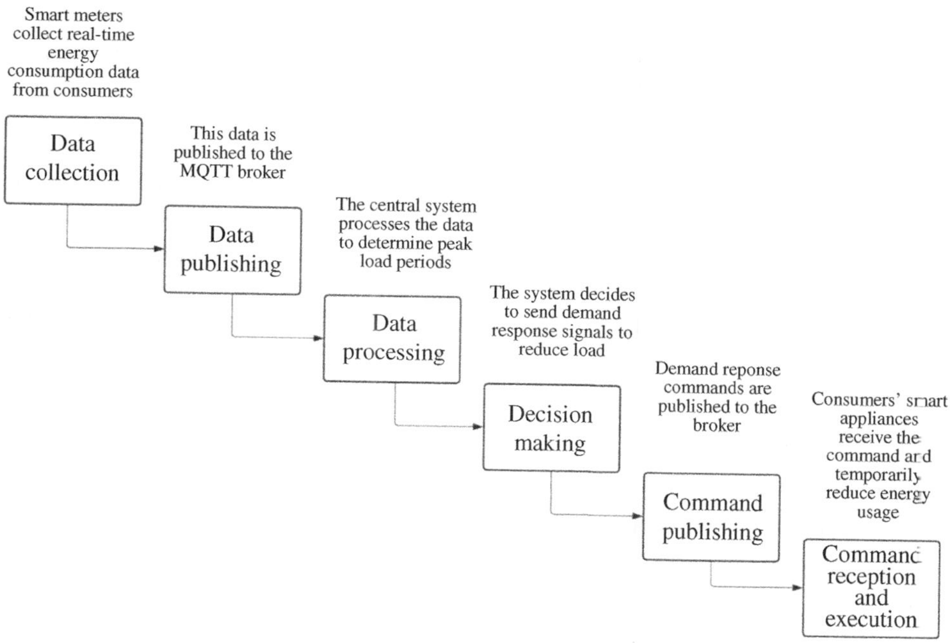

Figure 3.2 Demand response (sample use case of MQTT).

Figure 3.2 is the flowchart representation of demand response, a use case of MQTT protocol in transactive systems. From this flowchart, it is inferred that MQTT can be used for managing energy consumption in SGs. In this case, the SG constantly publishes data to a central hub using MQTT. This hub is the central system that preprocesses the data, followed by analyzing the data, identifying peak usage periods, and sending targeted commands through MQTT to consumer appliances. These appliances receive the command and temporarily respond by reducing energy use.

3.3.3 Constrained Application Protocol (CoAP)

The CoAP is a specialized web transfer protocol for constrained devices and constrained networks in the context of the IoT. It is designed to be so lightweight and efficient that it can be used within resource-constrained environments. Resources are usually characterized in terms of processing power, memory, and battery life [5].

3.3.4 Key Features of CoAP

3.3.4.1 Request/Response Model

CoAP supports a request–response model that is similar to hypertext transfer protocol (HTTP). In this type of version, a consumer CoAP sends a request to the server, and the server responds with the requested data or the acknowledgment of the stated request. CoAP can guide four sorts of messages: confirmable, non-confirmable, acknowledgment, and reset.

3.3.4.2 Protocol Stack

Because CoAP has to paint in relatively resource-limited surroundings, the CoAP protocol stack is designed to be as simple as possible. CoAP is designed for the use of the User Datagram Protocol (UDP) as a Transport Layer Protocol. This is a connectionless protocol with no guarantee of shipping. Thus, the use of UDP over Transmission Control Protocol (TCP) reduces overhead and conserves sources – a first-rate requirement for useful resource-restricted gadgets with low processing power and reminiscence [7].

3.3.4.3 Resource Model

CoAP introduces the idea of sources. Resources are identifiable by the usage of uniform resource identifiers (URIs) and are commonly hosted on CoAP servers. CoAP uses the identical request methods as HTTP, along with GET, PUT, POST, and DELETE, making it less difficult for people who are familiar with HTTP to study and use. These strategies correspond to moves inclusive of retrieving statistics (GET), updating statistics (PUT), creating new data (POST), and deleting data (DELETE) [9].

3.3.4.4 Message Format

CoAP messages are considerably smaller than HTTP messages, consequently minimizing the usage of bandwidth for higher optimization of communication in low-resources-equipped devices. They encompass a hard and fast-length header followed by nonobligatory tokens, options, and payload fields. The header consists of records along with message type, code, message ID, and token length.

3.3.4.5 Security

CoAP is able to secure the use of the Datagram TLS to offer a further layer of encryption and authentication of sensitive statistics at discretion. This will make the data continue to be personal and steady among gadgets, even on untrusted networks [7].

3.3.4.6 Scalability

Resource performance in CoAP designs makes it higher in terms of system overall performance for a better quantity of resource-restrained devices to communicate in IoT surroundings.

3.3.4.7 Optional Reliability

There exist in CoAP confirmable and non-confirmable messages. Confirmable messages offer a mechanism through which a consumer might get acknowledgment from the server for a message. The non-confirmable message might be true in conditions in which the aim is to lessen the overhead; it is uncertain that the messages will reach the server, so it is applicable in situations whilst the loss of statistics is bearable.

3.3.5 Extensible Messaging and Presence Protocol (XMPP)

XMPP is important to SGs because it allows for efficient, secure, and scalable communication among the many components in the grid. It makes it easy to monitor and control the grid through real-time messaging among smart meters, sensors, and other communicating appliances. XMPP also provides strong encryption with TLS and authentication with a simple authentication and security layer (SASL), making it greatly suited to critical energy infrastructure. The ability of the protocol to aggregate and distribute data in real-time ensures that the right stakeholders are provided with information in a timely fashion. Its scalability means that it is capable of large-scale SG networks, while its extensibility through XMPP extension protocols makes it guaranteed to interoperate with other systems and protocols in the SG ecosystems [11]. Applications of XMPP in SGs range from demand response – where utilities send real-time signals to adjust energy consumption – to the remote monitoring and management of grid components, communication with distributed energy resources such as solar panels and wind turbines, and even real-time updates to consumers on energy usage and outages. All of these features put together make XMPP a key protocol in efficient, reliable, and sustainable modern energy management.

3.4 Data Distribution Service (DDS)

DDS is a key protocol in SGs; it is a high-performance, scalable, and real-time data exchange framework important for managing complex distributed systems. A publish-subscribe model of communication allows DDS to provide efficient and low-latency dissemination of data between SG components, such as sensors, controllers, and monitoring systems. This model gives the assurance of reliable and predictable delivery of data in real-time applications within an SG: monitoring, control, and automation. DDS is designed to be scalable to keep up with the huge amounts of data produced by myriad connected devices in SGs, which continue to grow.

DDS is integrated with QoS policies that provide adaptive, dependable data delivery based on the QoS required for specific reliability, bandwidth, and latency. It is then certain that critical data reaches its destination on time, which is of importance for responsiveness and stability in the grid. DDS helps in smooth and effective communication, which thus increases the SG capability of adapting dynamically to changes, optimizing energy distribution, and integrating renewable energy sources; therefore, it continues to uphold the reliability and efficiency of the grid. This capability to manage real-time data flow and support distributed architectures makes DDS a vital protocol in the modernization and improvement of operations within the SG [10].

3.4.1 Advanced Message Queuing Protocol (AMQP)

The AMQP plays a dynamic role in SGs by offering a robust, flexible, and interoperable framework for messaging between various grid components. AMQP supports the reliable asynchronous messaging system, which provides safe, efficient communication to and from different devices or systems within the SG. This constitutes a central aspect of handling the multifield and complex interactions among smart meters, sensors, controllers, and other grid infrastructures.

AMQP supports complex routing, queuing, and delivery guarantees that ensure the messages are delivered just in order, without loss, and as they ought to be. This is vital in the retention of the integrity and reliability of the grid. Thus, the standardized nature of the protocol fosters interoperability between different vendors and technologies, easing the integration of diverse components in an SG. Interoperability is key to managing distributed energy resources, facilitating real-time demand response, and ensuring efficient energy distribution [12].

Further, AMQP allows point-to-point and publish-subscribe messaging patterns, which set up various modes of communication within the SG. These may include real-time monitoring and control, data analytics, and customer interaction. Hence, by facilitating secure, reliable, and scalable communication, AMQP enhances the SG's capacity for dynamic adjustment to changed conditions, optimization of energy performance, integration of renewable sources, and maintenance of overall grid stability and efficiency. In this manner, the comprehensive messaging capabilities of AMQP render it a very important protocol in the advancement of functionality and resilience of modern SGs.

In the Table 3.1, the comparison between different communication protocols is illustrated. From the table, it is inferred that both MQTT and AMQP are well suited for communication and applications of SG as they have a supporting QoS and data security. CoAP, on the other hand, is

Table 3.1 Comparison of IoT protocols.

IoT protocols	QoS	Data security	Message pattern	Complexity	Application in smart grid
AMQP	Yes	TLS SSL	Reg-Res Pub-Sub	Low	Smart meter AMI
CoAP	Yes	DTLS	Req-Res Pub-Sub	Low	Smart home
DDS	Yes	TCP UDP	Pub-Sub	High	EMS
MQTT	Yes	TCP	Pub-Sub	Low	Smart home, Smart meter
XMPP	No	TCP	Req-Res Pub-Sub Push–Pull	High	Grid management system

a lightweight protocol with lower complexity but may not be efficient enough for complex data changes. Last, XMPP is a complex protocol and doesn't guarantee message delivery, making it less suitable for some mission-critical applications in SGs.

3.5 Edge Computing and Real-Time Implementation

Enormous amounts of data are continuously generated within SGs every second. Sophisticated data analysis algorithms are essential to convert this data into actionable information and insights, which can then be leveraged for SG operations and services. Typically, these data analytics heavily rely on information and communication technologies (ICTs), playing a crucial role in collecting, transmitting, and processing the data [13]. Within cloud computing, devices and equipment spread across different geographic locations are linked to cloud data centers, allowing for centralized decision-making and control command issuance. However, this approach faces various drawbacks, including restricted bandwidth resources, diverse environments, and privacy issues. In response to these challenges, EC emerges as an alternative solution. EC shifts the boundary of computational applications away from centralized nodes toward the extremities of the communication network. By harnessing computing resources nearer to sensors and users, EC facilitates data analytics in a decentralized manner. It gains advantages for its ability to effectively reduce system delay, lighten the burden of cloud computing centers, improve system scalability and availability, and protect data security and privacy [15].

Figure 3.3 is the SG task computing queue. This shows how tasks are offloaded from a source to the edge node, cloud, and neighboring edge node.

- **Source:** In this, the SG task originates and it could be the user equipment (UE) or network element.
- **Edge Node:** The edge node is a network device that is located at the edge of the network, closer to the UE. It can process some of the SG tasks locally, reducing the load on the core network.
- **Cloud:** The cloud is the major location where the tasks can be operated. Because it has more resources than edge, it can handle more complex tasks. It gets the data from the edge and processes the task and sends it back to the receiver edge node.
- **Neighboring Edge Node:** This is used to process the SG tasks. This is used when the local edge is overloaded or if the task requires resources that are not available on the local edge node.

The specific decision of where to send a task likely depends on factors like:

- **Task Complexity:** Simpler tasks can be done by the local nodes and complex tasks can be done by the cloud node.

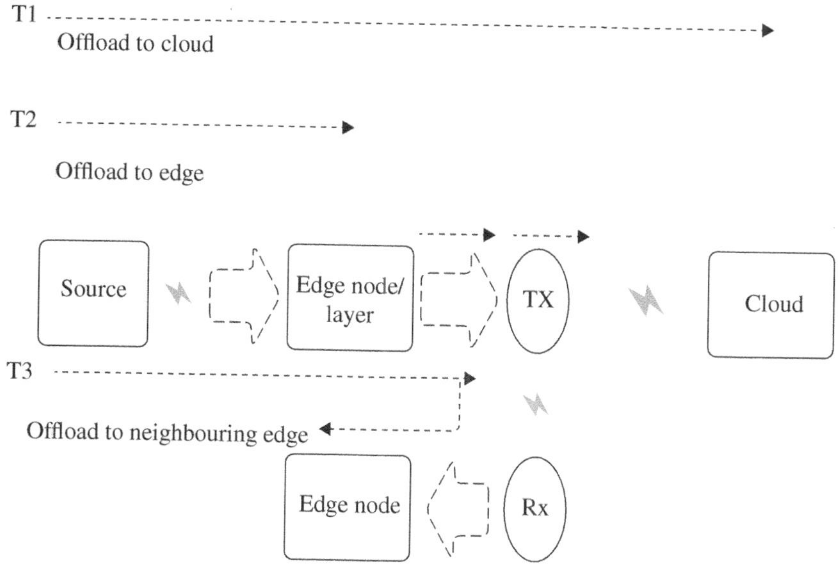

Figure 3.3 Smart grid task computation queue.

- **Resource Availability:** If the local edge node is overloaded, the tasks are offloaded to a neighbor node which will help to complete the task.
- **Latency Requirements:** Tasks that need an immediate response from the node can be sent for processing the tasks in the nodes.

Three possible offloading destinations for SG tasks:

- **T1:** Offload to Cloud – Cloud will handle complex tasks.
- **T2:** Offload to Local Edge – Local edge nodes handle tasks and send them to the cloud for storage.
- **T3:** Offload to Neighboring Edge – When local edge node is overloaded.

Table 3.2 identifies potential users of EC in SGs. Users in the table:

- **Grid User:** This could be a place/area that uses electricity.
- **Market Operator:** The organization that manages the buying and selling of electricity on the grid is market operator.
- **Energy Service Company:** This provides energy services to customers; they are electricity generation and retail sales.
- **System Operator:** This is responsible for the reliable operation of the transmission and distribution system.

Figure 3.4 depicts a three-tier architecture for implementing EC in SGs. This architecture consists of three layers, which are user layer, edge layer, and cloud layer. The user layer consists of the physical components of the SGs, which include power plants, wind plants, distribution lines, transformers, solar panels, and other energy sources. Followed by edge layer, which consists of devices that collect the data and then process the data at the grid's periphery.

Examples of these devices are smart meters, routers, Personal Computers (PCs), servers, and base stations. The last layer is the cloud layer, which includes cloud servers and a control center. The cloud layer receives the data from the edge layer and stores it. Its main purpose is to analyze and monitor the data received from the edge layer and provide overall monitoring to the SGs.

Table 3.3 shows us the main EC characteristics that benefit SGs.

Table 3.2 Potential users of edge computing in smart grids.

Data transmission	User interface	Operator	Energy company	Operating system
Data source	Applications Sensors Meter	Meter	Company server	Meter-electric, harmonic IED Sensor PMU Asset monitoring sensor
Data generated	Feedback from Apps Data from sensors Load	Injecting power Status of operation	Info on tariff Plan on usage	Frequency-voltage, current, power Quality of power Status of equipment and information on faults Geographical and meteorological data Data from phasors Data from videos, emissions, etc.
Data received	Control order Info tariff	Feedback from users Offers	Feedback profile of load	GIS Clearing outcomes

Source: Adapted from Shi et al. [15].

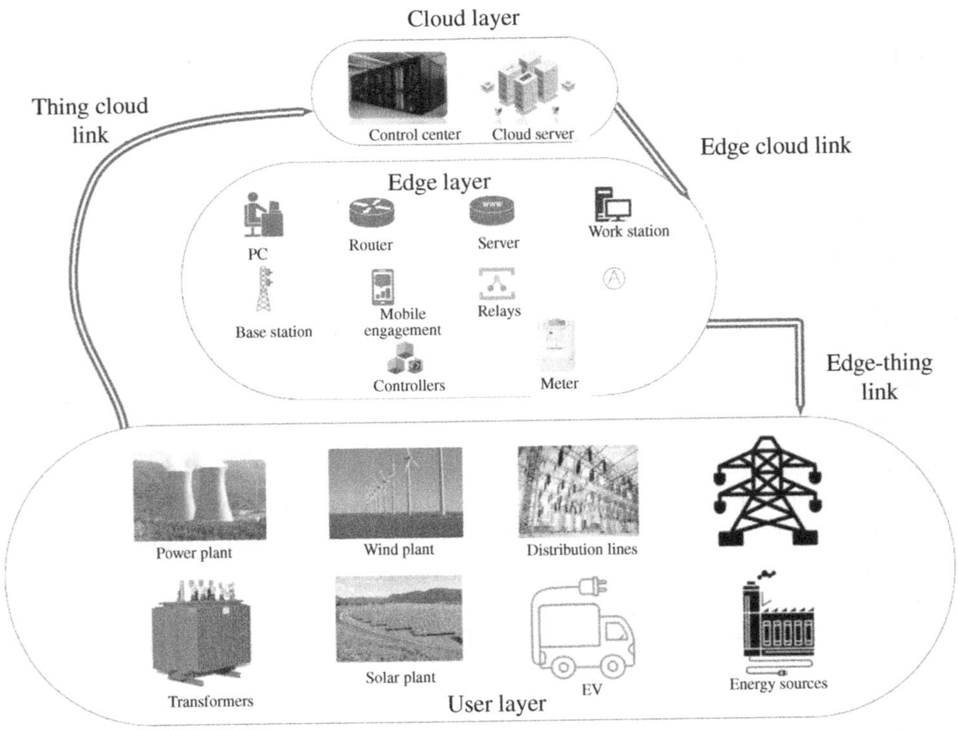

Figure 3.4 Architecture for implementing edge computing in smart grids. Source: Leaflet/Wikimedia Commons/CC BY SA 3.0.

Table 3.3 EC characteristics and strengths.

Characteristics	Detailed description	Application scenarios
Low-latency	Less hops; less middleware; isolation from other parts	Time-sensitive Applications
Cognition	Similar environment with SG agents; tailored with user-side knowledge	Cognitive Applications
Flexibility	Plug and play/on-demand; distribute tasks within the community	Community Applications
Reliability	Resilient to $N-1$ failure; cybersecurity and privacy	Security Applications

EC can offload tasks to different places to reduce latency. The low latency refers to fewer hops, which requires less middleware and isolation from other parts; it is used in time-sensitive applications. The cognition refers to the environment with SG agents and is tailored with user-side knowledge. Flexibility refers to the plug-and-play and distribution tasks within the community. Reliability is resilient to N-1 failure and even cyber security and privacy. EC can enhance security compared with clouds; however, it still faces several security challenges. The geo-distribution of EC resources increases the risk of physical attacks. The existing security mechanisms typically rely on the assumption that the majority of the data sources.

Figure 3.5 shows how SGs can help make EC more sustainable. From this figure, a few points are important to be noted. First, to have a sustainable energy ecosystem SGs and EC should have

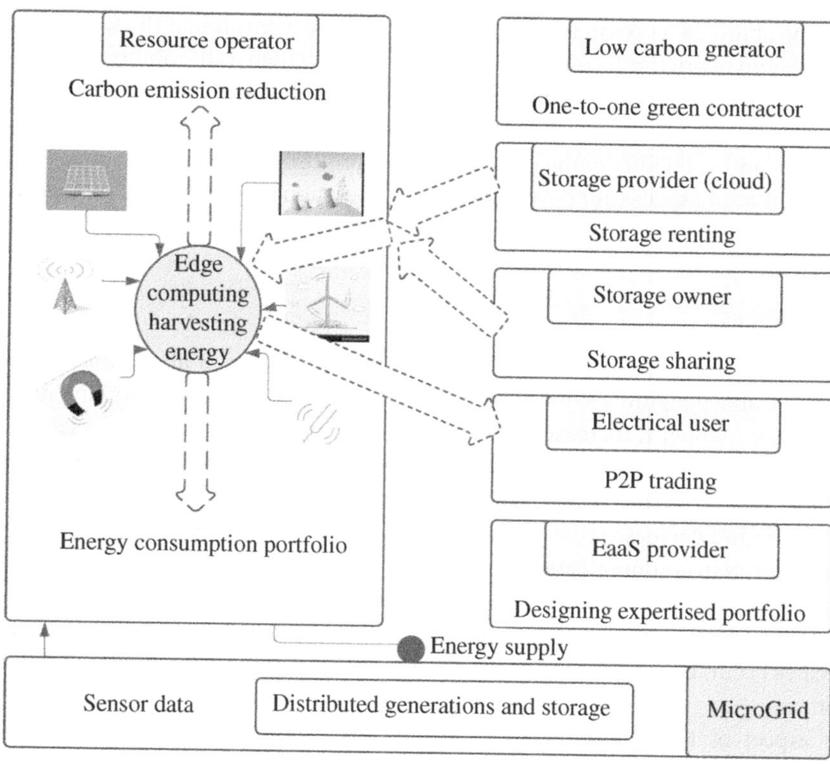

Figure 3.5 Various approaches for EC. Source: Bera et al. [14]/IEEE.

a complementary relationship. Second, EC analyzes data in real-time from SGs to make real-time decisions for efficient energy use. Then, renewable energy sources such as solar and wind power are essential in making the system sustainable. Distribution generation means using smaller, decentralized power generation units (SGs) for improved sustainability. Finally, the inclusion of elements such as P2P trading and green contractors improves the potential for customer participation in sustainable practices. EC can enhance security compared with clouds; however, it still faces several security challenges.

The geo-distribution of EC resources increases the risk of physical attacks. The existing security mechanisms typically rely on the assumption that the majority of the data sources or cybersecurity-relevant challenges can also be addressed with EC. Authentication, authorization, and accounting (AAA) are fundamental elements of cybersecurity. They control who is permitted to use network resources (through authentication), and what they are authorized to do (through authorization), and capture the actions performed while accessing the network (through accounting). Using the cloud to monitor a large number of geo-distributed devices incurs massive overhead. EC nodes can delegate the task of security monitoring and AAA management [14]. Edge-based AAA management is particularly significant for the power distribution system, where SG equipment communicates through a public communication network instead of the utilities' private network.

3.6 Reliability and Scalability

Scalability can be defined as the ability of a system to change its scale in order to meet growing volumes of demand [16]. A system is understood as a set of interacting elements with similar boundary conditions. The ability of a system to scale does not necessarily mean that the scaled-up system performs well. A more restrictive formulation defines scalability then as the ability of a system to maintain its performance (i.e., relative performance) and function, and retain all its desired properties when its scale is increased without having a corresponding increase in the system's complexity [17]. The reliability and scalability of the communication protocols used in transactive IoT systems are very important in the successful development and operation of SGs. IoT technologies in SGs help in the efficiency, scalability, reliability, and sustainability of electricity distribution.

3.6.1 Reliability Challenges

The reliability of the system is one of the most important factors of SGs, and it also comes with some challenges. From the above Figure 3.6, it can be seen that the uncertainty, diversity, and distribution of energy supplies – which is increased mainly due to environmental and sustainability concerns – drive grid congestion [18]. Real-time power flow patterns can significantly differ from those considered during design or offline analysis. More frequent and larger transfers over longer distances increase volatility and decrease reliability margins, which are exacerbated by the energy market. Achieving low latency in communication is very important but is cumbersome, especially in large-scale deployments with diverse devices. Wireless communication is prone to interference and signal degradation, which disrupts data transmission. As the connected devices increase in number, scalability issues occur, making it hard to maintain reliable communication without performance degradation.

Security is a core aspect of IoT communication, so protocols should be formed, keeping in view the encryption and authentication mechanisms that can protect sensitive data from

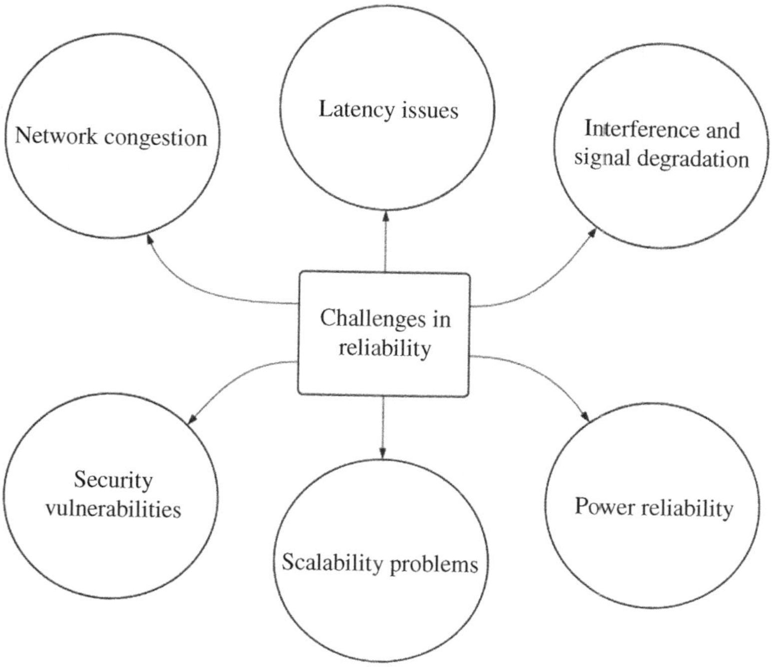

Figure 3.6 Challenges in reliability.

cyberattacks [20]. Continuous power supply to all IoT devices is crucial, as power outages will lead to communication breakdowns. Sometimes physical failure of the network components breaks communication; hence, there is a need for reliable failover mechanisms. Bugs in protocols and the need to update them may hamper the reliability of data communication and the system's functioning [19]. Environmental factors such as weather can affect communication reliability, especially for outdoor IoT devices.

3.6.2 Strengthening Reliability

Several techniques can be used to improve the communication protocol reliability. Redundancy is the means of deploying multiple systems for vital components, which include network paths, sensors, and controllers. This may be coupled with mechanisms for automated failover to replace the backup systems while failure is recognized. Strong encryption protocols for integrity and confidentiality are used to enhance protection, while steady authentication strategies are used to permit the handiest authorized devices to get the right of entry to the network. This, similarly, includes intrusion detection structures, which display and reply to protection threats in actual time. Optimizing communication protocols with integrated blunder correction and detection capability is critical to keeping records integrity and assembling actual-time needs; protocols with low latency and excessive throughput are essential [22].

Resilience is designed by the use of mesh network topology for making sure of continuous communication; it uses EC to manage records locally if you want to reduce dependency on vital servers. Scalable solutions, such as adaptive protocols that adjust to varying network conditions and elastic cloud resources for centralized data storage and analytics, support system scaling [21]. Regular upkeep and updates involve recurring fitness tests and software program updates to save unexpected failures and enhance overall performance. Implementing QoS mechanisms to prioritize critical communication traffic and employing bandwidth management techniques to allocate network resources efficiently ensures smooth data flow. These techniques applied can remarkably boost the reliability of IoT transactive exchange protocols for SGs to facilitate easy and uninterrupted data change for powerful grid control.

3.6.3 Scalability Challenges

The scalability issue in communication protocols used for transactive IoT in SGs arises because of several motives. First, as the number of interconnected devices increases, managing communication among them becomes more complex, potentially leading to congestion and decreased performance. Second, it is tough to accommodate several forms of devices and directive technologies in the community, as a result posing compatibility challenges to scale up [23]. The devices can also have exceptional capabilities and resource constraints. Moreover, ensuring consistent communication quality and reliability across a growing network can be challenging, especially when devices have varying capabilities and connectivity requirements. Additionally, with a large quantity of data required for shifting amongst devices, there exist scalability problems regarding information processing and storage; as a result, best aid allocation and management strategies are needed. Last, the model of communication protocols to deal with the changing desires and dynamics of SGs – such as modifications in community topology and calls for patterns – gives ongoing scalability challenges that require continuous optimization and model [24].

3.6.4 Enhancing Scalability

Enhancing the scalability of communication protocols in transactive IoT for SGs involves several strategies to manage the increased number of devices, diverse communication technologies, and large data volumes. As shown in Figure 3.7, the key approaches include employing lightweight protocols such as MQTT and CoAP, which are designed for efficient performance in resource-constrained environments and reduce overhead to handle many devices. Implementing adaptive communication protocols that dynamically adjust their parameters based on network conditions ensures efficient communication as the network scales. Shifting from centralized to decentralized control architectures, such as EC, distributes the communication load and reduces the burden on central servers [25]. Creating a hierarchical network structure that segments the grid into smaller, manageable subnetworks helps reduce complexity and improve scalability. Utilizing

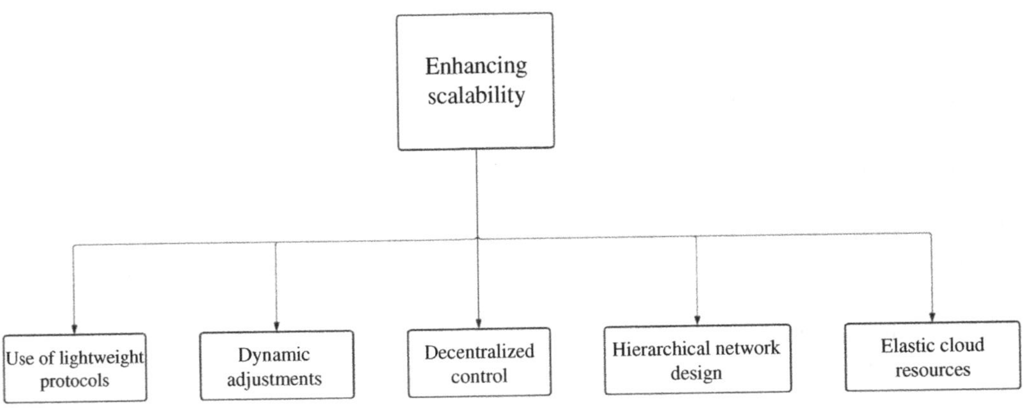

Figure 3.7 Scalability enhancement.

edge and fog computing for local data processing reduces data transmission to central servers, while data aggregation techniques combine multiple data points into a single message, lowering communication overhead.

Leveraging elastic cloud resources provides scalable storage, processing power, and services. Implementing QoS mechanisms prioritizes critical communication traffic and manages bandwidth efficiently. Designing IoT devices and systems to be modular and easily integrable facilitates network expansion without significant reconfiguration. Using mesh network topologies increases redundancy and improves scalability by allowing devices to communicate through multiple paths. Implementing software-defined networking (SDN) dynamically manages network traffic, improving routing paths and resource allocation [17]. Employing standardized communication protocols ensures interoperability between different devices and systems. Regularly monitoring network performance, identifying faults, and updating algorithms improve protocol efficiency and scalability [26]. By integrating these strategies, the scalability of communication protocols in transactive IoT for SGs can be significantly enhanced, ensuring the network can grow and adapt to increasing demands without compromising performance or reliability.

3.7 Case Studies and Real-Life Implementations

- Various transactive systems are applied globally, especially in Europe and America. In the Brooklyn microgrid, a project titled Open Access Technology International (OATI) microgrid center is a community-primarily based strength initiative that leverages the blockchain era to enable P2P power trading amongst citizens. This undertaking is a collaboration among (LO3) Energy and Siemens, specializing in developing a decentralized energy market inside a small city location. This venture makes use of IoT gadgets to display and control power production and consumption in actual time, making sure balanced and efficient power distribution. Moreover, this microgrid can perform independently from the primary grid in case of outages, enhancing power protection for the community [27].
- The project Power Matching City, it is one of the first huge-scale SGs within the globe. This undertaking entails imposing a Virtual Power Plant where the system aggregates allotted electricity resources to shape a digital energy plant, which could take part in power markets and provide ancillary offerings to the grid. The mission validated sizable improvements in power performance and decreased carbon emissions [28].
- In a venture titled SG Gotland, it's a Swedish primarily based undertaking that targets to integrate a high share of renewable power into the nearby grid while retaining stability and reliability. It engages purchasers in demand response applications to balance delivery and demand, particularly at some stage in high renewable power production intervals. It focuses on integrating wind and solar energy into the nearby grid, with an aim of reaching a high percent of renewable power penetration [29].
- Under the mission Piclo [30], community operators' software program systems for P2P energy buying and selling were mounted within the United Kingdom. Germany has mounted a decentralized device for electric vehicle (EV) charging, transactions, and information sharing through the share and charge mission.
- Pacific Northwest National Laboratory (PNNL) – Transactive control system (Washington, USA), PNNL's transactive management gadget pursuits to dynamically stabilize electricity delivery by integrating disbursed power assets (DERs) such as wind, solar, and battery garage. The gadget utilizes rate signals to talk with DERs and purchasers, incentivizing them to adjust their power

production and consumption in response to grid conditions. This improved stability between strength deliver and call for, decreasing the want for peaking strength plant life and improving grid balance [31].

- A mission titled Advanced Distributed Energy Platform Technology (ADEPT) specializes in creating a distributed electricity market with the usage of transactive power ideas. This mission makes use of IoT devices to provide actual-time records on power production and consumption, enabling dynamic pricing and green strength use. This enabled greater participation of distributed energy sources inside the electricity marketplace [32].
- The Buffalo Distributed System Platform (DSP) challenge is a collaborative initiative aimed at modernizing the electrical grid in Buffalo, New York, to enhance its reliability, performance, and capability to integrate DERs. The principal aim of the challenge changed into the improvement of microgrid structures, which could perform independently in case of outages, ensuring continuous strength delivery to critical centers. This challenge includes advanced metering infrastructure (AMI), sensors, and automated manipulation systems to permit real-time monitoring and control of the grid. It engages customers in calls for reaction projects, letting them alter their electricity usage in the course of peak times in reaction to price indicators or incentives [33].
- The Transactive Energy Service System (TESS) is a framework for dealing with and coordinating distributed power assets using marketplace-primarily based mechanisms to optimize strength use and distribution. It employs superior control structures to automate and optimize electricity flows based mostly on contemporary market situations and grid goals and also ensures interoperability between numerous devices and structures through standardized communication protocols. Another advantage of this mission is that it was designed to scale from small local systems to massive regional grids, accommodating numerous styles of DERs and energy offerings [34].
- The challenge named micro transactive grid is a localized electricity system that allows for the generation, garage, and trading of energy inside a described community or neighborhood, focusing on P2P interactions. It makes use of smart devices and IoT technology to display and manipulate power manufacturing, storage, and intake in real-time and deploys battery garage structures to control extra power and offer stability to the local grid. It reduced environmental effects by maximizing the use of renewable power sources [35].
- Grid exchange is a platform designed to facilitate the exchange of energy between manufacturers, clients, and the grid usage of a TE version. It was set up, as an internet marketplace in which energy producers and customers can exchange electricity primarily based on real-time supply and call for demand response. It employs IoT gadgets and advanced analytics to reveal electricity manufacturing and consumption, imparting data for optimizing transactions and imposing dynamic pricing mechanisms to mirror actual-time marketplace conditions and incentivize efficient power use. This increased participation in energy markets by providing a person-pleasant platform for buying and selling power [36].

3.8 Conclusion

This chapter explored the communication protocols that underpin transactive IoT systems within SGs. Furthermore, the discussion on EC and real-time implementation understood the importance of processing data closer to its source to provide more data security, reduce latency, and enhance decision-making capabilities. Reliability and scalability are paramount in a transactive IoT system, as the system must handle increasing volumes of data and devices while maintaining robust performance and fault tolerance. Reliable data delivery is also essential for ensuring the smooth operation of these systems. Last, the case studies and real-life implementations provided practical insights into how these frameworks and protocols are implemented and maintained in a real-world scenario. These examples provide the benefits and challenges of deploying transactive systems. In conclusion, the integration of suitable communication protocols is vital for the advancement and sustainability of transactive IoT systems. By selecting the right protocol based on factors such as reliability, scalability, and data complexity, we can create a robust and efficient communication infrastructure that facilitates optimized energy use, market participation, and a more sustainable energy ecosystem. As these systems continue to evolve, innovation and adaptation are essential for leveraging new opportunities that contribute to a more efficient and reliable infrastructure.

References

1 Huang, Q., Amin, W., Umer, K. et al. (2021). A review of transactive energy systems: concept and implementation. *Energy Reports* 7: 7804–7824.

2 Nunna, H.K. and Srinivasan, D. (2017). Multiagent-based transactive energy framework for distribution systems with smart microgrids. *IEEE Transactions on Industrial Informatics* 13 (5): 2241–2250.

3 Zamani, R., Moghaddam, M.P., and Haghifam, M.R. (2021). Evaluating the impact of connectivity on transactive energy in smart grid. *IEEE Transactions on Smart Grid* 13 (3): 2491–2494.

4 Çorak, B.H., Okay, F.Y., Güzel, M. et al. (2018). Comparative analysis of IoT communication protocols. *2018 International Symposium on Networks, Computers and Communications (ISNCC)*. pp. 1–6. IEEE.

5 Mahmood, A., Javaid, N., and Razzaq, S. (2015). A review of wireless communications for smart grid. *Renewable and Sustainable Energy Reviews* 41: 248–260.

6 Nguyen, K.T., Laurent, M., and Oualha, N. (2015). Survey on secure communication protocols for the Internet of Things. *Ad Hoc Networks* 32: 17–31.

7 Sharma, C. and Gondhi, N.K. (2018). Communication protocol stack for constrained IoT systems. *2018 3rd International Conference On Internet of Things: Smart Innovation and Usages (IoT-SIU)*. pp. 1–6. IEEE.

8 Soni, D. and Makwana, A. (2017). A survey on MQTT: a protocol of Internet of Things (IoT). *International Conference on Telecommunication, Power Analysis and Computing Techniques (ICTPACT-2017)*. Vol. 20, pp. 173–177.

9 Dave, M., Doshi, J., and Arolkar, H. (2020). MQTT-CoAP interconnector: IoT interoperability solution for application layer protocols. *2020 Fourth International Conference on I-SMAC (IoT in Social, Mobile, Analytics and Cloud) (I-SMAC)*. pp. 122–127. IEEE.

10 Laszka, A., Dubey, A., Walker, M., and Schmidt, D. (2017). Providing privacy, safety, and security in IoT-based transactive energy systems using distributed ledgers. *Proceedings of the Seventh International Conference on the Internet of Things*. pp. 1–8.

11 Nadeem, F., Aftab, M.A., Hussain, S.S. et al. (2019). Virtual power plant management in smart grids with XMPP based IEC 61850 communication. *Energies* 12 (12): 2398.

12 Jun, H.J. and Yang, H.S. (2021). Performance of the XMPP and the MQTT protocols on IEC 61850-based micro grid communication architecture. *Energies* 14 (16): 5024.

13 Fang, X., Misra, S., Xue, G., and Yang, D. (2011). Smart grid—The new and improved power grid: a survey. *IEEE Communications Surveys & Tutorials* 14 (4): 944–980.

14 Bera, S., Misra, S., and Rodrigues, J.J.P.C. (2015). Cloud computing applications for smart grid: a survey. *IEEE Transactions on Parallel and Distributed Systems* 26 (5): 1477–1494.

15 Shi, W., Cao, J., Zhang, Q. et al. (2016). Edge computing: vision and challenges. *IEEE Internet of Things Journal* 3 (5): 637–646.

16 Bonnefoy, P.A. (2008). Scalability of the air transportation system and development of multi-airport systems: a worldwide perspective. Doctoral dissertation. Massachusetts Institute of Technology.

17 Gupta, K. and Kumar, V. (2024). KMS–AMI: an efficient and scalable key management scheme for secure two-way communications in advanced metering infrastructure of smart grid. *The Journal of Supercomputing* 80 (7): 8668–8701.

18 Moslehi, K. and Kumar, R. (2010). Smart grid-a reliability perspective. *2010 Innovative Smart Grid Technologies (ISGT)*. pp. 1–8. IEEE.

19 Kabalci, E. and Kabalci, Y. (ed.) (2019). *Smart Grids and Their Communication Systems (No. 1)*. Singapore: Springer.

20 Yan, Y., Qian, Y., Sharif, H., and Tipper, D. (2012). A survey on cyber security for smart grid communications. *IEEE Communications Surveys & Tutorials* 14 (4): 998–1010.

21 Bergmann, J., Glomb, C., Götz, J. et al. (2010). Scalability of smart grid protocols: protocols and their simulative evaluation for massively distributed DERs. *2010 First IEEE International Conference on Smart Grid Communications*. pp. 131–136. IEEE.

22 Fateri, S., Ni, Q., Taylor, G.A. et al. (2012). Design and analysis of multicast-based publisher/subscriber models over wireless platforms for smart grid communications. *2012 IEEE 11th International Conference on Trust, Security and Privacy in Computing and Communications*. pp. 1617–1623. IEEE.

23 Ma, S., Zhang, H., and Xing, X. (2018). Scalability for smart infrastructure system in smart grid: a survey. *Wireless Personal Communications* 99: 161–184.

24 Kathuria, V., Mohanasundaram, G., and Das, S.R. (2013). A simulation study of routing protocols for smart meter networks. *2013 IEEE International Conference on Smart Grid Communications (SmartGridComm)*. pp. 384–389. IEEE.

25 Rajalingham, G., Ho, Q.D., and Le-Ngoc, T. (2013). Attainable throughput, delay and scalability for geographic routing on smart grid neighbor area networks. *2013 IEEE Wireless Communications and Networking Conference (WCNC)*. pp. 1121–1126. IEEE.

26 Appasani, B., Maddikara, J.B.R., and Mohanta, D.K. (2019). Standards and communication systems in smart grid. In: *Smart Grids and Their Communication Systems* (ed. E. Kabalci and Y. Kabalci), 283–327. Springer.

27 Mengelkamp, E., Gärttner, J., Rock, K., Kessler, S., Orsini, L., and Weinhardt, C. (2018). Designing microgrid energy markets: A case study: The Brooklyn Microgrid. *Applied energy* 210: 870–880.

28 Fan, Z., Cao, J., Jamal, T. et al. (2022). The role of 'living laboratories' in accelerating the energy system decarbonization. *Energy Reports* 8: 11858–11864.

29 Wallnerström, C.J., Tjernberg, L.B., Hilber, P., and Jürgensen, J.H. (2016). Framework for system analyses of smart grid solutions with examples from the Gotland case. *In 2016 International Conference on Probabilistic Methods Applied to Power Systems (PMAPS)*. pp. 1–9. IEEE.

30 Mujeeb, A., Hong, X., and Wang, P. (2019). Analysis of peer-to-peer (P2P) electricity market and piclo's local matching trading platform in UK. *In 2019 IEEE 3rd Conference on Energy Internet and Energy System Integration (EI2)*. pp. 619–624. IEEE.

31 Huang, P., Kalagnanam, J., Natarajan, R., Hammerstrom, D., Melton, R., Sharma, M., and Ambrosio, R. (2010). Analytics and transactive control design for the pacific northwest smart grid demonstration project. *In 2010 First IEEE international conference on smart grid communications*. pp. 449–454. IEEE.

32 Shah, R., Surinkaew, T., Islam, S., and Rutovitz, J. (2023). Neighbourhood batteries in Heyfield-technical analysis of impacts and benefits.

33 Alam, M. M., Haque, A., Hakami, J. et al. (2024). An optimal deep belief with buffalo optimization algorithm for fault detection and power loss in grid-connected system. *Soft Computing* 28 (3): 2577–2591.

34 Zia, M.F., Benbouzid, M., Elbouchikhi, E. et al. (2020). Microgrid transactive energy: review, architectures, distributed ledger technologies, and market analysis. *IEEE Access* 8: 19410–19432.

35 Zia, M.F., Benbouzid, M., Elbouchikhi, E., Muyeen, S.M., Techato, K., and Guerrero, J.M. (2020). Microgrid transactive energy: Review, architectures, distributed ledger technologies, and market analysis. *IEEE access* 8: 19410–19432.

36 Child, M., Kemfert, C., Bogdanov, D., and Breyer, C. (2019). Flexible electricity generation, grid exchange and storage for the transition to a 100% renewable energy system in Europe. *Renewable energy* 139: 80–101.

4

Transactive IoT: Merging Transactions and Connectivity

Burhan Khan[1], Aabid A. Mir[3], Naser S. Almutairi[1], and Khang W. Goh[2,4]

[1]*Department of Computer System and Technology, Faculty of Computer Science and Information Technology, Universiti Malaya, Kuala Lumpur, Malaysia*
[2]*School of Engineering, Shinawatra University, Pathum Thani, Thailand*
[3]*Malaysian Institute of Information Technology, Universiti Kuala Lumpur, Kuala Lumpur, Malaysia*
[4]*Faculty of Data Science and Information Technology, INTI International University, Nilai, Malaysia*

4.1 Introduction

4.1.1 IoT and Smart Grids

The Internet of Things (IoT) is a disruptive technology transforming the industry sector, exemplified by the interconnection of Internet-enabled devices for sending, receiving, and transmitting data. This concept refers to billions of physical objects equipped with sensors, connectivity, and software that can connect and exchange data from anywhere on the planet to anywhere else [1]. IoT technologies have an intense penetration in the energy sector, especially in developing smart grid infrastructures.

A smart grid is an advanced form of a classic electrical grid integrated with advanced information and communication technologies to enhance electricity distribution efficiency, reliability, and sustainability. The main goal of smart grids is to create a much more responsive and adaptive energy system that will meet the dynamic demands of modern society. Therefore, IoT's real-time data acquisition and communication capabilities are crucial in this evolutionary step toward realizing an effective method for monitoring and managing energy flows [2].

4.1.2 Significance

The reason IoT is relevant to smart grids is the capability of the technology to provide granular insights into energy usage patterns and system performances. For example, smart meters, using IoT devices, provide measurements of electricity usage that are sent back in real time, enabling utilities to distribute the load more effectively and to rapidly identify issues [3]. In addition, IoT sensors in the grid infrastructure can be applied to monitor physical asset health and make predictions of failure and timely maintenance to minimize downtime and improve system reliability indices in general [4]. Integrating IoT and smart grids enables more sophisticated grid management strategies, such as demand response (DR) programs, where consumers are incentivized to consume less during peak demands by reducing their electricity usage. Thus, IoT fosters unique energy

IoT for Smart Grid: Revolutionizing Electrical Engineering, First Edition.
Edited by Rahiman Zahira, Palanisamy Sivaraman, Chenniappan Sharmeela, and Sanjeevikumar Padmanaban.

management, quick and effective real-time feedback, and automatic control [5]. Data collection from IoT devices can identify trends and optimize and distribute energy to improve grid adaptiveness in different situations.

4.1.3 Chapter Aims

This chapter provides a complete insight into how implementing IoT utilizing transactive energy models revolutionizes smart grid technology. This chapter examines this convergence in detail to justify the vast enhancements and pragmatic benefits of IoT-enabled transactive systems for modern energy management. This chapter aims to achieve the following objectives.

- **Understanding Transactive Energy Systems (TESs):** This chapter aims to define TESs and provide a perspective on the concept and its goals. This will involve an in-depth examination of how such systems facilitate decentralized energy transactions and real-time demand–supply balances.
- **Role of IoT in Transactive Models:** To explain the necessity of IoT technology for enabling and improving TESs. This includes researching IoT devices that collect, communicate, and provide real-time data, which is essential for the efficient functioning of transactive models.
- **Benefits for Grid Management:** To discuss the benefits of implementing IoT in TESs related to improved grid management and its performance (higher efficiency, higher reliability, renewable energy integration), which can enable more sustainable and resilient energy systems.
- **Real-World Applications and Case Studies:** To comprehend real-world implementations of transactive IoT across several verticals, such as smart grids and smart cities. This chapter also explores the actual effects of such technologies on energy efficiency, reliability, and sustainability by presenting case studies and examples.
- **Economic and Market-Based Strategies:** This chapter explains the effectiveness of financial and market approaches in transactive IoT systems. It illustrates mechanisms such as real-time pricing (RTP) and DR that serve as a road map for understanding consumers' behavior toward energy market efficiency.
- **Technical and Architectural Components:** Describing physical and software elements that form transactive IoT systems. This comprises a detailed understanding of IoT device architectures, communication layers, data processing layers, and security mechanisms that should be in place to secure these systems.
- **Addressing Challenges:** List and describe the technical, regulatory, and operational challenges experienced in transactive IoT system implementation. This chapter highlights new interventions, describes the continued challenges in the field, and explains the research efforts dedicated to solving these problems to enable technological deployment.
- **Future Directions:** Provide an outlook on the development of smart grids and transactive IoT systems. Discuss recent advances, possible technological breakthroughs, and areas that demand further scientific research and innovation.

4.2 IoT Integration with Transactive Models

4.2.1 Transactive Energy Systems

TESs represent a highly advanced approach to managing electricity generation, distribution, and consumption. These systems use economic and control strategies to establish decentralized market-driven energy networks that balance supply and demand dynamically. The basic objective of TESs is to provide a platform where consumers, also referred to as prosumers, engage effectively and actively in the production, consumption, and trading of energy [6]. The operational principles of TESs are discussed below.

- **Market-Based Transactions:** A TES operates based on market-based transaction mechanisms. Electricity is traded similarly to commodity trading in financial markets, where prices vary according to real-time demand and supply.
- **Dynamic Pricing:** These systems implement dynamic pricing with electricity rates varying according to real-time market situations. Rates increase during peak demands so that consumption is lowered, and vice versa.
- **Distributed Energy Resources (DERs):** DERs include solar panels, wind turbines, and battery storage systems, thus making the grid more resilient through localized energy generation.
- **Consumer Participation:** The consumer is transformed into a prosumer, implying that he participates in energy production and consumption. This form ensures the best distribution method and reduced consumption through informed real-time decision-making processes [7].

4.2.2 Role of IoT in Transactive Models

IoT technology plays a significant role in enabling the functionality and efficiency of TESs. This technology is instrumental in delivering capabilities for acquiring real-time data, communication, and control through the functionality of IoT devices and facilitates such systems.

- **Real-Time Data Acquisition:** IoT sensors and smart meters constantly collect data on energy consumption, generation, and environmental conditions. These data are instrumental in making informed decisions regarding energy distribution and pricing [8].
- **Communication Networks:** Because robust communication networks link IoT devices, data are exchanged instantly throughout the grid. All stakeholders involved, including consumers, grid operators, and energy providers, will always have access to the most recent information owing to this interconnection [9].
- **Automated Control Mechanisms:** IoT enables the automation of grid operations. For instance, smart thermostats can adjust heating and cooling systems in response to real-time price signals. Such adjustments consider optimal energy use without the need for any human intervention. Automated DR systems restrict or shift loads during peak periods to conserve grid stability [10].
- **Predictive Analytics and Machine Learning (ML):** Insights from IoT data are analyzed using ML algorithms and advanced analytics. Such technologies forecast energy demand patterns, identify anomalies, and optimize grid operations by improving efficiency and reliability [11].

4.2.3 IoT and TES Integration for Grid Management

The IoT embedded with TESs has various advantages that improve the capabilities and operation of electrical grids.

4.2.3.1 Improved Efficiency

- **Load Balancing:** In IoT-enabled transactive systems, the energy cost is adjusted in real time to ensure adaptive balancing across the grid. It helps to avoid overloads and, therefore, costly infrastructure upgrades [5].

- **Resource Optimization**: This provides details about how and in what form energy is being used, which further helps in making better allocation of resources, thereby preventing unnecessary wastage or mismanagement and making grid operations more effective [12].

4.2.3.2 Enhanced Reliability

- **Real-Time Monitoring and Maintenance:** Real-time IoT-based monitoring and maintenance ensures the reliability improvement of grids by consistently monitoring the health of grid components, predicting their failures, and scheduling predictive maintenance, thereby reducing downtime [13].
- **Demand Response**: Automated DR allows the grid to respond rapidly to changes in demand, avoid blackouts during peak times, and maintain stability [14].

4.2.3.3 Facilitation of Renewable Energy Integration

- **Distributed Energy Resources (DERs):** IoT delivers real-time insights into renewable output to support the grid integration of DERs easily. Consequently, the energy system becomes more resilient and sustainable [15].
- **Energy Trading:** Users with renewable energy infrastructure can trade their surplus back to the grid or with nearby consumers through IoT-supported platforms, preferring renewable energy resources to conventional approaches [16].

4.3 Transactive IoT in Modern Applications

4.3.1 Smart Grids

The concept of transactive IoT (shown in Figure 4.1) adds state-of-the-art communication and decision automation capabilities to classic grid systems, turning them into interactive, responsive, efficient networks that can cope with modern energy demands [17]. This feature helps in the smarter and faster movement of electricity from the generation to the consumption side for improved grid operations. TESs using IoT sensors enabled by real-time analytics allow utilities to predict peak-hour demand and distribute loads efficiently. Consumers can respond to real-time price signals sent by utilities by varying their levels of energy use during price fluctuations. It helps in grid balancing, outage prevention, and minimization of the requirements for expensive infrastructure upgrades.

- **Dynamic Load Adjustment:** Integration of the IoT further enhances dynamic load management, a core functionality of smart grids, because IoT devices distributed across the grid collect real-time data on the consumption, generation, and transmission of energy. Such information helps utilities dynamically manage energy loads, especially during peak hours, to enhance grid reliability and prevent overloads. For instance, an essential smart grid improved with IoT would show energy demand curves that are much flatter at peak times than conventional grids, indicating better load management [18, 19].
- **Enhanced DR:** Smart grids enabled by transactive IoT offer highly automated and improved DR systems. Equipped with smart meters and IoT devices, energy providers instantly notify consumers of peak hours and current energy pricing. Responding to these signals, responsive devices such as automated industrial equipment and smart thermostats can independently control their operations. This allows them to move nonessential energy use to off-peak hours, improve overall energy management, and enable consumer savings [20].

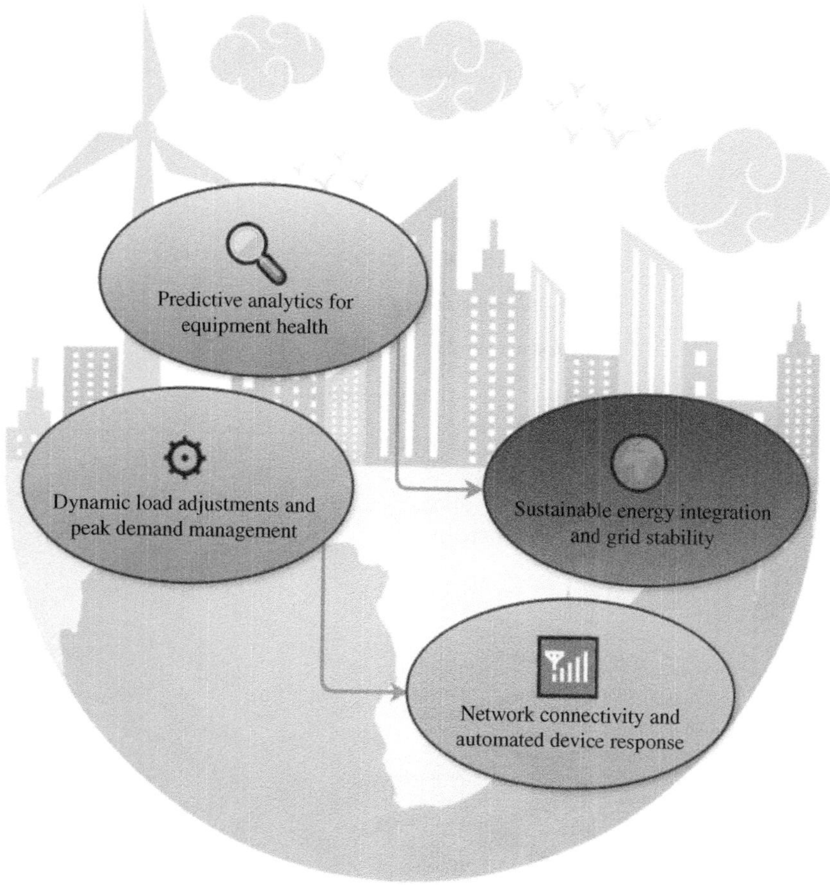

Figure 4.1 Transactive IoT for enhanced grid stability and efficient resource management.

- **Predictive Maintenance:** Another area is predictive maintenance, whereby IoT substantially benefits smart grids. Issues developing inside the microgrid, such as the heating up and degradation of aged transmission lines, can be notified on a timely basis using IoT sensors. This can be useful in situations like electricity disruptions, where pre-emptive actions are critically important [21, 22]. This capability increases grid reliability and service quality while significantly lowering operational costs and extending the infrastructure lifespan.
- **Sustainable Energy Integration:** Smart grids enabled by IoT can handle the efficient integration of renewable energy sources with variability [23]. Real-time energy generation information from IoT devices (renewable energy), wind, and solar – gives the grid the strength to manage its load and store rapid-response solutions that can sustain stability even with high levels of renewable integration. This means that the number of renewable energy sources that increase over time can be integrated into the grid, as exemplified by the rising utilization of renewables in energy production.

4.3.2 Smart Cities

Ideally suited for managing urban energy needs and infrastructure more sustainably and efficiently, the transactive IoT is a powerful enabler. This includes data from residential homes, commercial

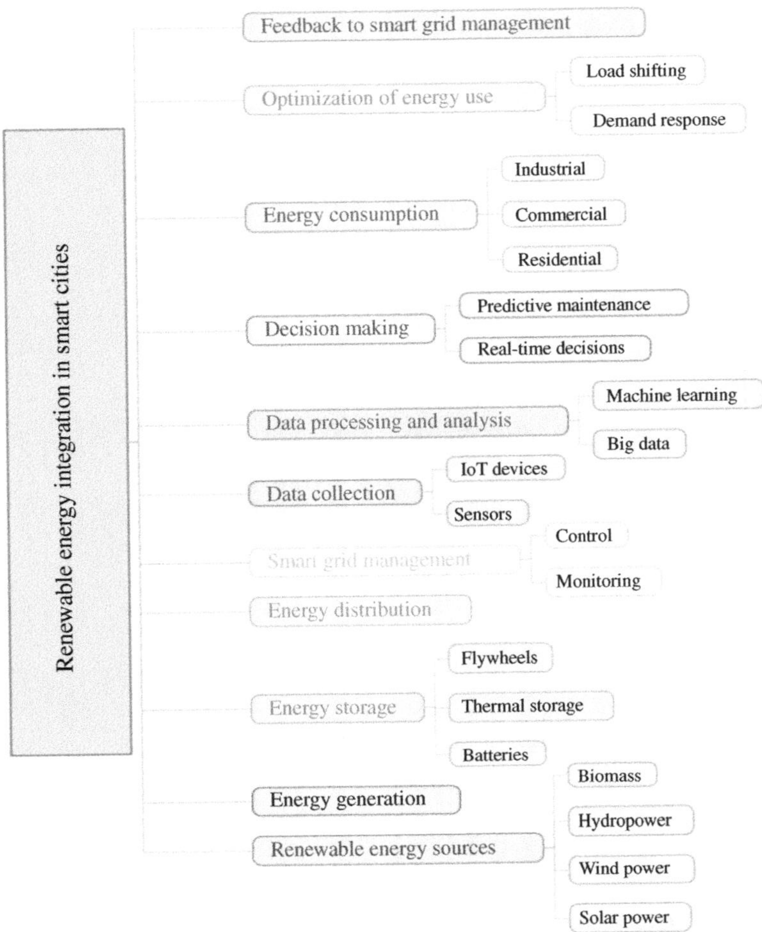

Figure 4.2 Renewable energy integration in smart cities.

spaces, and industrial factories across the city. It can now be analyzed for optimal energy pattern analysis and directly improve energy distribution efficiency, making it possible to collect the data. For example, remotely controlled streetlights and public buildings can be powered down when power demand is low. Transactive IoT also allows for integrating DER, such as rooftop solar on buildings, inside the urban grid, and offers decentralized low-carbon electricity production to support urban sustainability goals. Integrating transactive IoT into smart cities will impact the management of urban energy demands and infrastructure (shown in Figure 4.2). A transactive energy framework enabled by IoT can support cities in optimizing the consumption of energy resources and make infrastructure more resilient and sustainable.

- **Data-Driven Urban Energy Management:** A smart city is a large-scale IoT network that collects large amounts of data regarding energy consumed from buildings, traffic flow, weather conditions, etc. These data are very important and can create a full profile of energy utilized in the city, which is further used for optimizing energy distribution and waste reduction [17, 24]. IoT sensors in buildings detect the levels of occupancy, which are used to adjust the building's heating, ventilation, and air conditioning systems.

- **Integration of Renewable Energy Sources:** Transactive IoT swiftly integrates renewable energy sources into an urban grid. This encourages decentralized power generation through direct real-time energy transactions between producers and consumers, such as solar panels and community wind turbines. As such, it promotes power independence and carbon footprint reduction in cities.
- **Public Infrastructure Optimization:** IoT can make existing smart city infrastructures, such as public transportation and streetlights, efficiently operate energy. Sensors installed inside streetlights enable energy optimization by saving energy in response to the presence of pedestrians or cars on the streets, saving a significant amount of energy from wasting.
- **Interactive Dashboards for Energy Management:** Interactive dashboards could inform citizens and municipal management about current energy generation and consumption. These dashboards may display data from around the city to make adjustments quickly and assist in informed decision-making to save energy.

4.3.3 Case Studies

- **Brooklyn Microgrid:** The value of transactive IoT is best exemplified in Brooklyn, New York, where a blockchain-enabled microgrid enables neighbors to buy and sell solar energy generated within their community, circumventing traditional grid systems. It can reduce energy-related costs and enhance the reliability of the local energy system [25].
- **Power Ledger in Australia:** Using blockchain technology, Power Ledger has created an IoT transactive platform to facilitate household electricity trading. This system allows residents with their solar panels to sell any excess energy they produce back to the grid for other homeowners, increasing the adoption of clean, renewable energy and decreasing reliance on nonrenewable fossil fuels [26].
- **Amsterdam Smart City:** A prototype of transforming Amsterdam into a smart city that illustrates how IoT technologies can be integrated into urban administration to establish more sustainable and citizen-friendly services. Projects like the 3D Print Canal House and the Almere Smart Society demonstrate the city's commitment to innovative, sustainable urban development. Programs such as Vehicle2Grid and City-zen Sustainable District Heating not only enable Amsterdam to meet its sustainability targets but also help the city become a world leader in sustainable living, illustrating how essential technology will become in transforming cities of the future [27].

4.3.4 Future Trends

As the integration of IoT with TESs evolves, numerous future trends and expansions are on the horizon. Such advancements offer sophisticated options that will expand the efficiency, dependability, and sustainability of smart grids, allowing them to be more adaptive to the comprehensive requirements of modern energy management.

4.3.4.1 Advanced Machine Learning (ML) and Artificial Intelligence (AI) Integration

Advanced ML and artificial intelligence (AI) technologies are expected to power transactive IoT systems. These solutions can analyze large amounts of data from IoT devices and predict consumer energy usage, streamline energy distribution across key parts, such as transmission and electricity metering, and identify abnormalities as they occur in real time. For example, DR strategies assisted by AI algorithms can forecast peak demand periods and make automated load changes to meet a balance in other parts of the grid [28]. In addition, ML models are always learning and updateable and can adapt to feedback, thus enabling more optimized and resilient energy systems.

4.3.4.2 Blockchain and Decentralized Energy Trading

As the future of transactive IoT applications unfolds, blockchain technology is set to play a major role [29]. Blockchain technology – a secure framework for decentralized energy trading on the blockchain ensures a high level of transparency and trust, as well as peer-to-peer (P2P) energy trading without an intermediary. This could lower transaction costs and make the energy markets more efficient. Furthermore, the unforgeable recordkeeping ability of blockchain can preserve trust among energy trading participants through an immutable ledger system [30]. As blockchain technology becomes more established, its interconnection with the IoT and TESs is anticipated to grow, fostering the creation of decentralized energy markets.

4.3.4.3 Edge Computing for Real-Time Processing

As mentioned earlier, edge computing is another upcoming trend in most methods. This could reduce latency and bandwidth usage by processing data closer to the source with edge computing, which can be extremely beneficial in real-time decision-making and control. This is significant for time-critical applications, such as automated DR and grid stabilization. Edge-computing devices enable IoT devices to process and act on data locally, thereby improving the speed and reliability of service-oriented architectural systems [31].

4.3.4.4 Enhanced Security Measures

Security is critical when transactive IoT systems become more complex and integrated. Future advancements are expected to fortify cyber security covers against nascent threats. This involves using encryption, secure methods for data communication, and robust ways to authorize the integrity and privacy of data [32]. Furthermore, transactional IoT systems can prevent cyber threats by detecting and responding to malicious activities in real time to improve the security of transaction-enabled services an extra step up.

4.3.4.5 Integration with Renewable Energy Sources

Sustainable energy solutions are leading to integrating renewable energy sources with transactive IoT systems. By using the data from the IoT to manage these highly variable outputs, future applications will focus on optimizing the use of renewables, that is, solar and wind power. This encompasses establishing smarter models to forecast renewable energy and implementing storage options to manage supply and demand [33]. This growth in renewable energy installations will only help to make smart grid systems even more sustainable and resilient.

4.3.4.6 Expansion of Smart City Initiatives

In particular, the rise of smart city initiatives will contribute to developing more transactive IoT applications. This will lead to an increased demand for solutions for managing energy effectively as these cities embrace smart technology. A transactive IoT system will facilitate a full suite of smart city applications, including intelligent street lighting, traffic management, smart buildings, charging infrastructure for electric vehicles (EVs), and much more. Integrating these factors will lead to substantial energy savings, less carbon footprint, and a higher quality of life for citizens living in such cities [34].

4.4 Economic and Market-Based Approaches

4.4.1 Economic Models Used in Transactive IoT Systems

TESs enabled by the IoT have developed advanced economic models that foster efficient and dynamic energy markets. Models of this type are necessary to help solve the complexity of energy supply and demand, enable real-time energy transactions, and implement sustainability. Below, we examine the major economic models for transactive IoT systems and provide deeper insight into their characteristics and use cases.

4.4.1.1 Dynamic Pricing Models

Transactive IoT systems are key enablers for dynamic pricing, which leads to electricity prices fluctuating in real time according to immediate supply and demand conditions. This pricing model drives consumers to change their energy usage toward off-peak hours, reducing the strain on the grid during peak periods. It can be implemented in many ways, including time-of-use (TOU) pricing, critical peak pricing (CPP), dynamic tariffs, and RTP.

- **TOU Pricing:** TOU pricing charges different electricity prices based on demand and supply at the time of day. Consumers are charged higher rates during peak hours and lower rates during off-peak hours [35].
- **Critical Peak Pricing (CPP)**: CPP is combined with TOU and higher pricing during peak hours when electricity demand is higher, encouraging consumers to limit their consumption during critical periods [36].
- **Real-Time Pricing (RTP)**: RTP offers the most variable pricing, with prices changing as often as every hour or even half an hour, depending on real-time market conditions. This model has the highest opportunities for DR and grid efficiency [6].

4.4.1.2 Demand Response Programs

Consumer behavior-based DR programs incentivize limiting electricity usage during peak periods. Participation in these programs can be price-based, with consumers responding to dynamic pricing signals, or incentive-based, where participants receive payments for shedding loads during DR events.

- **Price-Based DR:** Customers are required to adapt their consumption to dynamic pricing signals, such as TOU, CPP, or RTP, to stabilize the grid and diminish system-wide costs [37].
- **Incentive-Based DR:** This program involves direct payments or bill credits to participants to reduce usage during specific events. Utilities or grid operators often run it and may include automated systems that shed the load on behalf of the consumer [5].

4.4.1.3 Peer-to-Peer (P2P) Energy Trading

Enabled by the IoT and blockchain technologies, prosumers can purchase and sell electricity directly with one another. A decentralized energy market is established, in which consumers can actively participate in energy trading and encourage the adoption of renewable sources.

- **Decentralized Marketplaces:** By facilitating secure and transparent transactions with no requirement for intermediaries, blockchain technology reduces transaction costs and makes markets more efficient. A prosumer can carry out P2P trading of any excess energy produced from solar panels or other renewables with neighbors [25].

- **Smart Contracts:** Smart contracts automate the process of transactions, making them trustworthy and eliminating the need for human intervention. Smart contracts are required to manage the complexity of P2P transactions in real time [38].

4.4.1.4 Auction-Based Mechanisms

Transactive IoT systems leverage auction-based mechanisms to allocate energy resources efficiently. These marketplaces conduct open or sealed bid processes, where consumers and producers put forth their price–quantity pairs. A central auctioneer then makes the actual product available to supply and demand following these bids.

- **Day-Ahead and Real-Time Auctions:** Participants in day-ahead auctions submit bids for the energy usage required the following day, enabling market actors to schedule and allocate resources optimally. Real-time auctions are auctioned closer to the delivery time, so they change conditions in real time [39].
- **Market-Clearing Prices:** The market-clearing price is the price set by the auctioneer, at which the total quantity demanded equals the total quality supplied. Such a price guarantees efficient market operation, and no trade takes place other than at the proper value [40].

4.4.1.5 Capacity Markets

Capacity markets are intended to ensure sufficient generating capacity to meet peak demand. These markets pay electricity providers to maintain available capacity if and when demand requires it, which may otherwise be unused, to ensure supply adequacy in the power grid.

- **Capacity Payments:** Providers are paid to be available to provide energy at short notice and guarantee that there will be enough capacity to make up for any shortfall in demand. This mechanism improves grid stability and prevents blackouts [41].
- **Forward Capacity Markets:** These markets start operations years in advance, allowing space for providers to plan and invest in the new capacity needed to satisfy future demand. Forward capacity auctions set the price and quantity of capacity required for future periods [42].

4.4.2 Impact on Consumer Behavior and Energy Market Dynamics

Dynamic pricing models, such as DR-enabled price plans, TOU, and RTP, encourage users to shift their energy demand away from peak times when electricity is more expensive [35]. This shift in behavior not only helps grid stability but also leads to more cost-efficient energy consumption, which reduces the need for additional generation capacity during peak hours. Similarly, there is significant evidence [37] that DR programs that reward customers for reducing their electricity consumption in times of stress, such as peak loads, result in consumers becoming involved proactively in grid management. These price signals and incentives create DRs that allow consumers to be acutely aware of their energy consumption, leading to behavior changes that are not short-lived but permanent, thereby helping energy efficiency and conservation.

Blockchain-enabled P2P energy trading platforms enable consumers by providing them with more capabilities to become prosumers who can produce, consume, and share energy with other individuals directly [25]. Customers can participate financially in renewable energy investments because of the decentralized energy market, promoting both the democratization of energy production and the flow of revenues into local economies. This leads to significant growth in the share of renewable energy supply because consumers are motivated to adopt solar panels and other renewables-becoming prosumers who can trade their excess energy, thus drastically decreasing the time required to recuperate initial investments [38].

Such approaches ensure that the resources are accurately allocated and the grid's dependability is guaranteed through auction-based methods or capacity markets, which helps to determine the dynamics of this energy market landscape. In auction-based systems, bidders convey their willingness to supply energy or pay for it based on what they perceive to be the most efficient solution [40], and market-clearing prices represent a balance between supply and demand. In a competitive environment, energy suppliers feel pressure to innovate and improve efficiency to operate at maximum utilization and meet market demands while making a profit. According to Cramton [41], capacity markets can act as a hedge against the cost of maintaining adequate generation capacity, ensuring that sufficient energy supply is always available to meet peak demands. This stability attracts more investment in energy infrastructure, which makes the grid resilient and reliable.

4.5 Transactive IoT System Architecture

4.5.1 Components of Transactive IoT Systems

Transactive IoT systems require successfully merging physical and software elements crucial for real-time data gathering, communication, analysis, and control (shown in Figure 4.3). This section further breaks down these components, explaining their functions and roles in the operation of TESs.

4.5.1.1 Physical Components

- **Smart Meters and IoT Sensors:** In the conceptualization of transactive IoT systems, the fundamental physical components are the smart meters and IoT sensors. These devices can be deployed on the grid on consumer premises, distribution lines, and substations. It can continuously measure and transmit data and valuable information, including energy consumption, generation, voltage levels, and other important parameters. Real-time data gathering is critical for grid performance monitoring, fault detection, and dynamic pricing architecture [43].
- **Distributed Energy Resources (DERs):** DERs such as solar panels, wind turbines, and battery storage systems are deployed on the grid to enable local energy generation and storage. These resources are attached to IoT sensors and controllers, which observe their performance and cooperate with the management system. DERs contribute to enhanced resilience by utilizing renewable energy sources and decreasing reliance on centralized power plants [44].
- **Advanced Metering Infrastructure (AMI):** This refers to the use of smart meters, data management systems, and communication networks that together assist in a two-way flow of information between utilities and customer AMI, allows utilities to collect more granular usage data, enables DR programs, and engages consumers in energy management [45].
- **Communication Networks:** Reliable communication networks are crucial for ensuring data can be transmitted across the grid. These are built on different infrastructures such as wireless (Wi-Fi, Zigbee), cellular (long-term evolution [LTE], 5G), and wired (fiber optic, power line communication). Data volume, latency requirements, and geographical coverage determine the choice of infrastructure [9].
- **EV Chargers and Smart Appliances:** EV chargers and smart appliances are important for transactive IoT systems, and their use will likely increase over time. These devices are grid-interfaced and IoT-enabled, meaning they can communicate in response to real-time price signals and optimize their operation. EVs chargers, for example, can be set to charge during less-demand peak hours, reducing the price per unit of electricity [33].

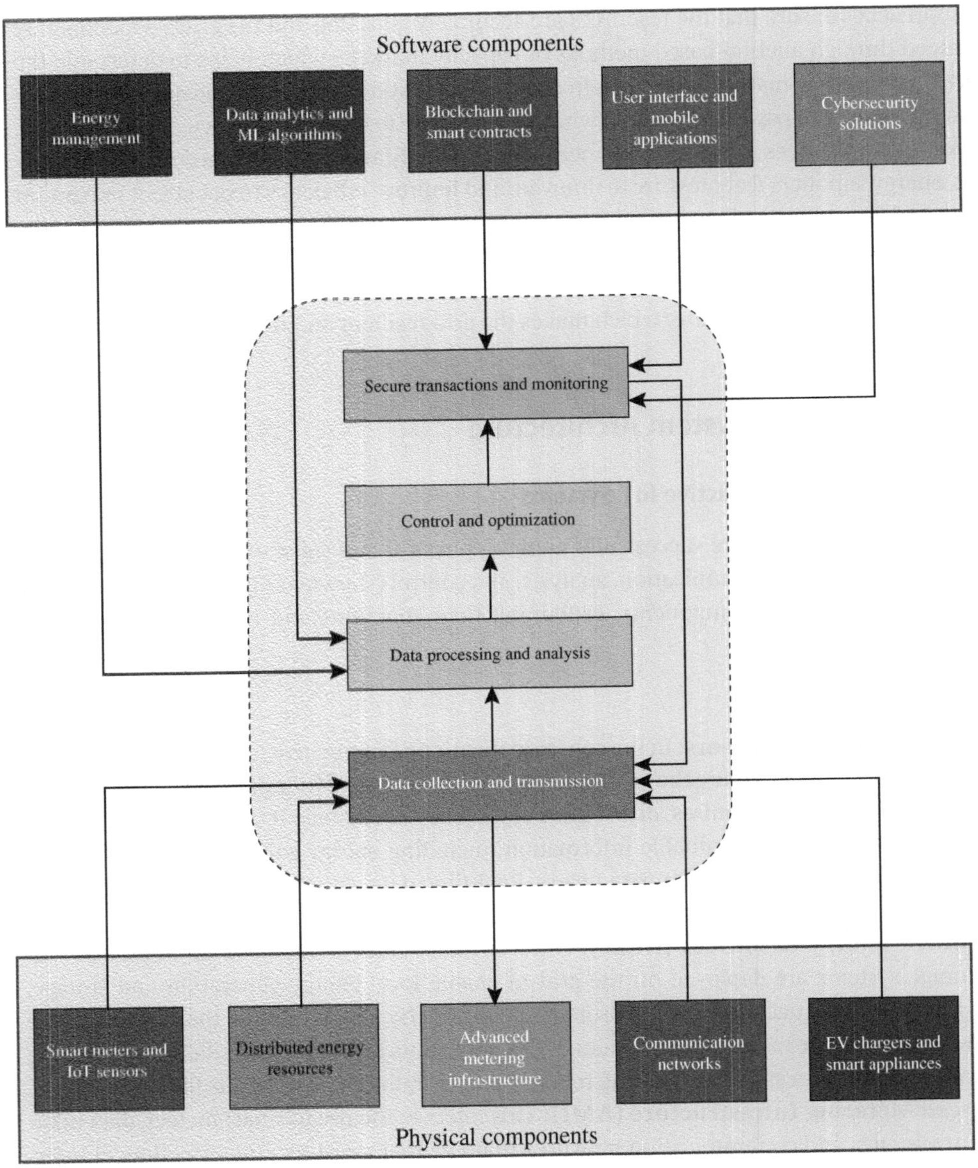

Figure 4.3 Transactive IoT system components.

4.5.1.2 Software Components

- **Energy Management Systems (EMS):** The effective operation of transactive IoT systems demands an EMS software platform. They collect information from multiple sources, analyze it in real time, and produce signals to control the network such that it operates at its best. Functionalities such as load prediction, DR, and energy trade facilitated by the EMS enable efficient and reliable grid operation [14].
- **Data Analytics and ML Algorithms:** Various data analytics and ML algorithms process the enormous amount of data IoT devices generate. Such algorithms anticipate energy demand patterns, identify anomalies, and optimize the grid's operation. ML models constantly learn and improvise using historical data, which can be a powerful tool for making the grid adaptive and resilient [46].

- **Blockchain and Smart Contracts:** Blockchain technology and smart contracts ensure safe, transparent energy transactions over transitive IoT. The blockchain guarantees that transactions are secure and traceable, whereas smart contracts automate the enforcement of an agreement according to pre-programmed triggers. Introducing these technologies accomplishes the decentralization of energy trading and reduces intermediary requirements, increasing market efficiency [16].
- **User Interfaces and Mobile Applications:** These real-time applications and programs give consumers a snapshot of their energy use and costs. These are tools for consumers to follow their load profiles, obtain real-time notifications of dynamic pricing events, and participate in DR programs. These applications encourage energy-saving behaviors by involving consumers and adding to the overall efficiency of the grid [34].
- **Cybersecurity Solutions:** Transactive IoT systems require extensive data exchange and control functions; hence, cybersecurity must be strong enough to prevent potential threats. It aims to protect the privacy of personal and sensitive data using advanced encryption, secure communication protocols, and authentication to keep data confidential. With AI-based security solutions, the network can be monitored, and cyber threats can be identified, detected, and resolved in real time by adding another defensive shield for the grid infrastructure [32].

4.5.1.3 Integration

The physical and software components must be integrated for transactive IoT systems to operate smoothly. Physical (hardware) infrastructure enables the collection and transmission of data, including smart meters, DERs, and communication networks. EMS platforms, data analytics, and blockchain technology process these data and provide real-time control to manage grid operations. For example, a smart meter gathers and transfers household energy consumption data over a secure communication network to an EMS platform. The EMS uses this data for analysis to predict future demand and supply. For this, an EMS can send a message to a smart thermostat to turn down the heating or cooling when the household's consumption is above the normal pattern. Meanwhile, transactions with households and energy providers are logged and performed securely via smart contracts using blockchain technology [16]. Together, these components facilitate the generation of microgrid real-time data acquisition, communication, and control for a dynamic energy management approach. Combining these technologies makes energy grids more robust, resilient, and sustainable, fulfilling consumer demands and changing environments.

4.5.2 Layers of Transactive IoT Systems

A transactive IoT system comprises several interconnected layers, each with specific tasks, which interact with each other to create a complete platform for efficient energy management. These layers are depicted in Figure 4.4.

- **Physical Layer:** Physical components such as energy monitors, thermostats, smart meters, and various sensors are placed under the physical layer. These devices collect data on air quality and electricity usage in real time. This layer includes all DERs that enable localized energy generation and storage, such as solar panels, wind turbines, and battery storage devices. In addition, Wi-Fi-enabled smart home devices and EV chargers are notified in real time by the grid regarding DR events or real-time prices.
- **Communication Layer:** This layer ensures that the data between the devices and systems are transferred accurately. It includes communication modules allowing easy device connections and data transmission, including Wi-Fi, ZigBee, LTE, and fiber optics. As mentioned, the application

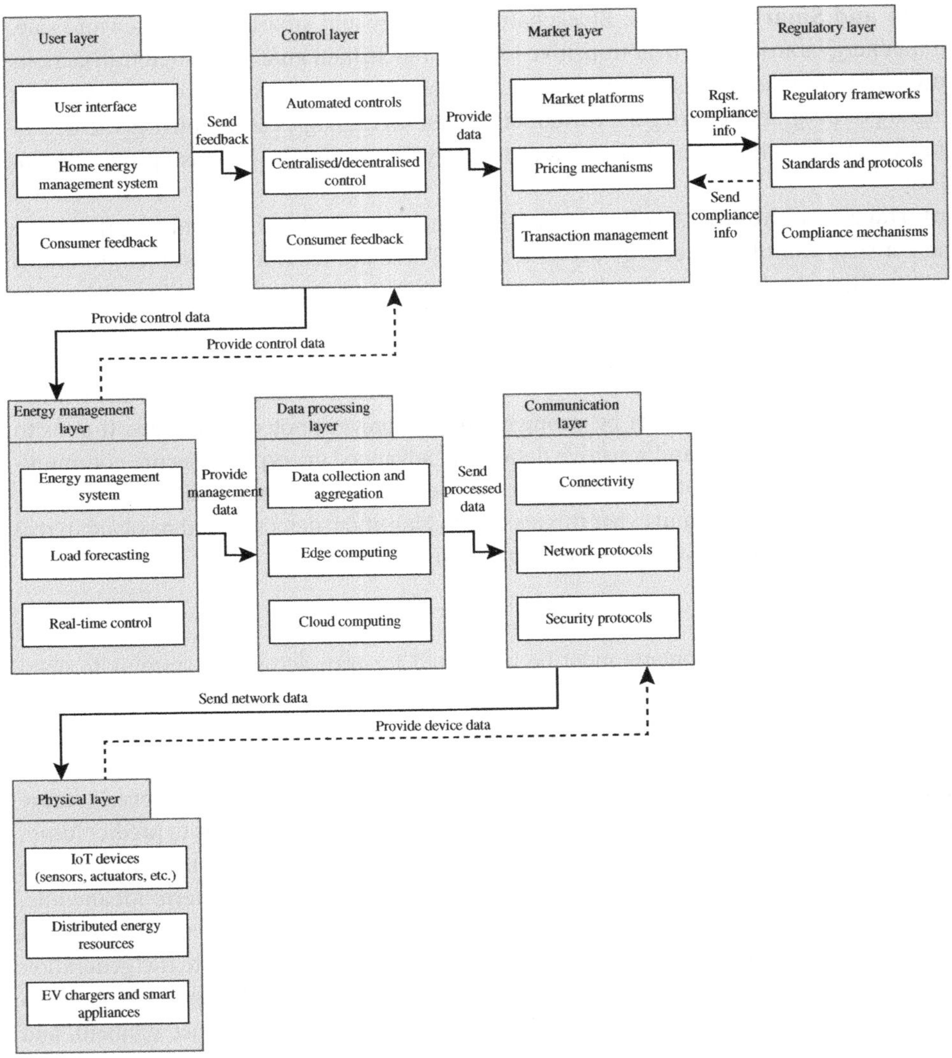

Figure 4.4 Transactive IoT system layers.

layer uses multiple network protocols (transmission control protocol [TCP], Internet protocol [IP], user datagram protocol [UDP], message queuing telemetry transport [MQTT], and constrained application protocol [CoAP]) for secure communication and data transfer. To encrypt and authenticate data, we require security protocols that ensure the integrity of the data and protect it from unauthorized access [43].

- **Data Processing Layer:** This layer is responsible for collecting, aggregating, storing, and analyzing the data. Edge computing allows data processing locally, reducing latency and bandwidth consumption and allowing real-time decisions [31]. Cloud computing platforms provide huge processing and storage capabilities for large-scale data volumes [9]. The data were analyzed using advanced data analytics and ML tools [47] to predict demand, anomaly detection, and grid performance improvement [46].
- **Energy Management Layer:** This layer regulates and optimizes the energy transfer. The EMS integrates data from different sources for the distribution and utilization of energy to enable

decisions that reduce wastage [14]. Optimization algorithms and predictive analytics can allocate resources and maintain the load to make energy consumption more effective. The EMS provides real-time control by changing the energy flow on demand to adapt to market conditions in response to real-time data. Thus, it can be efficiently interfaced with the market, user, and communication levels to provide complete energy management.

- **Regulatory Layer:** The operation of the TESs comes under the regulator layer, which comprises policies, standards, and guidelines of transactive energy concepts. Regulatory frameworks ensure compliance with legal and regulatory responsibilities, such as those pertaining to data protection, information technology (IT) security requirements, or energy market laws. Standards organizations (IEEE, International Electrotechnical Commission [IEC], and National Institute of Standards and Technology [NIST]) have defined guidelines on the properties of hardware devices and communication protocols. Compliance mechanisms are responsible for enforcing adherence to these standards and ensuring the trustworthiness and authenticity of the system [34].
- **Market Layer:** This layer enables economic transactions and market operations. This includes energy trading, auction mechanisms, and P2P transactions, all facilitating a decentralized and efficient energy market. Dynamic pricing models, such as TOU, RTP, and CPP, are implemented to indicate real-time demand and supply conditions [35]. Each transaction is recorded through a secure and transparent recording by transaction management systems, such as blockchain [38]. DR systems automatically modify energy usage and improve the efficiency and stability of the grid by responding to market signals [5].
- **Control Layer:** This layer ensures that appropriate control actions are executed in response to insights derived from data processing. This layer of automated control systems automatically modulates and manages energy flow, such as smart thermostats that adjust heating and cooling based on occupancy levels and temperature. A centralized control area issues commands from a central system to ensure the best energy distribution to avoid local overload. In contrast, decentralized control allows individual devices to make autonomous decisions based on localized data [33]. The system's robustness is also improved when AI and ML algorithms are applied to optimize control actions, thereby enhancing the efficiency and reliability of the entire system [46].
- **User Layer:** The user layer contains interfaces and tools that communicate directly with consumers and prosumers. This layer comprises web and mobile applications that provide real-time information about energy consumption, prices, and market conditions. Home energy management systems (HEMS) are smart systems that enable energy consumers to monitor and manage their energy usage and access feedback, which analyzes the patterns of use by consumers and helps them improve their efficiencies [46]. These tools lead to active energy management and push users toward sustainable consumption.

4.5.3 Security Considerations

Transactive IoT systems must have strong security solutions because of the sensitive data transmission across such networks and critical energy infrastructure. The primary security factors to be considered are authentication and authorization, data privacy and integrity, network trustworthiness, system robustness, and data privacy. Data privacy and integrity are critical for preventing significant disruptions and upholding consumer trust [32]. In addition, secure authentication and authorization mechanisms should be deployed to restrict or prevent unauthorized access to vital infrastructure, allowing only authorized users and devices to access it. IoT devices, which act as potential threat surfaces and entry points for attacks, must utilize secure boots and require regular updates. Secure communication protocols and encryption are required to protect the integrity and confidentiality of the transmitted data at the communication layer, which is vulnerable to various attacks, including denial of service (DoS) or man-in-the-middle (MITM) attacks, so network security has an essential role too.

Countermeasures have been implemented to address these challenges. End-to-end encryption utilizes standards such as advanced encryption standard (AES) and transport layer security (TLS) to ensure that data is kept private and unmodified from the collection point to the destination. Firmware updates patch-known software bugs and security vulnerabilities, and secure boot processes prevent malign firmware from operating. The permissions and capabilities of these services are limited by authorization methods (OAuth) and authentication protocols (e.g., multi-factor authentication [MFA]). Blockchain offers a significant enhancement in security by providing a decentralized and immutable ledger for transactions and communication. This is particularly useful for ensuring transparency and averting manipulation [38]. Smart contracts are used to automate and execute transactions securely based on predefined rules of operation on blockchain networks. Intrusion Detection Systems (IDSs) analyze network data for patterns and anomalies to locate security breaches using ML methods [46]. Network segmentation reduces the attack surface and limits the proliferation of attacks by isolating the essential infrastructure components. Immediate remediation efforts can be executed only if cyber threats are identified before they become a problem; for that purpose, continuous real-time monitoring with analytics is necessary. This level of security has to abide by given security standards and best practices set by organizations such as the IEC and the NIST. Regular security audits and assessments help detect and eliminate vulnerabilities, ensuring the system is safe from constantly mutating threats.

In energy trading platforms, blockchain logs every transaction so that no one can tamper once added to an immutable ledger, ensuring transparency. Smart contracts reduce the risk of manual errors and lessen the requirement for hands-on management, with transactions being automated using real-time data. Blockchain technology decentralizes transaction records in the network and eliminates single points of failure, which greatly enhances the robustness and security of the system [38]. Protecting transactive IoT frameworks requires comprehensive measures, including secure boot mechanisms, encryption, authentication protocols, network segmentation, intrusion detection, and real time. This approach ensures data integrity, protects against cyberattacks, and builds consumer trust for energy systems' secure and reliable functioning.

4.6 Challenges and Solutions

Several challenges exist in deploying transactive IoT systems in energy management, including regulatory, technological, and operational barriers. Interoperability is a problem in the standardization of diverse IoT devices made by different manufacturers, and these vast data volumes place stress on the network, thereby affecting processing capabilities. Such issues require scalable solutions such as edge and cloud computing. Additionally, efficient data management involving lifecycle strategies and sophisticated analytics is critical. Regulations are particularly challenging, dynamic, and inconsistent and present a greater challenge when implementing a system. Interacting with policymakers and participating in standardization processes is key to creating accurate regulatory frameworks. Operationally, it is essential to have redundant systems, backups, and disaster recovery plans to ensure that the system is resilient and reliable. Integrating new IoT systems with legacy energy infrastructure brings challenges, which can be mitigated by upgrading existing systems and adopting stepwise integration approaches. Thus, these barriers must be overcome to capitalize on these complex systems fully.

4.6.1 Challenges Faced When Deploying Transactive IoT

Transactive IoT systems for energy management pose deployment challenges and affect their efficiency, reliability, and sustainability. These can be classified into technical, operational, and regulatory challenges as follows.

4.6.1.1 Technical Challenges

- **Interoperability Issues:** Technical challenges are mostly due to different manufacturers of IoT devices needing to standardize their products. Such devices can operate on different communication protocols and standards; thus, seamless interoperation is a challenge. This lack of interoperability can disrupt data transmission and system integration, thereby decreasing the overall system's efficiency [43].
- **Scalability Concerns:** When the number of connected devices increases, these devices generate more data, which overwhelms network processing capabilities and may lead to performance glitches. Thus, scalability is essential to guarantee the efficiency and responsiveness of transactive IoT systems that have grown and become increasingly complex [31].
- **Data Management:** Another major issue is handling and analyzing the amount of data IoT devices generate. Storing, processing, and analyzing these data becomes easier with effective data management practices. With proper data handling, the advantages of real-time insights and predictive analytics are unlimited [46].
- **Security and Privacy:** Transactive IoT Solutions are susceptible to several cyberattacks, such as Distributed Denial of Service (DDoS) attacks, MITM attacks, and data compromise. Thus, safeguarding the integrity and availability of the system is important. Furthermore, personal data hoovering – say, input on household energy use – raises privacy concerns. Therefore, securing this data and maintaining consumer trust is essential [32].

4.6.1.2 Regulatory Challenges

- **Policy and Compliance:** The regulatory landscape for TESs is still emerging, and inconsistencies or a lack of clear policy can prevent deployment and operation. Complying with regulations can be a difficult and expensive process because the rules regarding energy vary widely between countries, regions, and even localities. These systems must be supported by clear regulatory frameworks [34].

4.6.1.3 Operational Challenges

- **Reliability and Resilience:** Transactive IoT systems are responsible for critical energy grid operations; consequently, maintaining such systems' reliability and resilience is mandatory. Disruptions in such systems can have far-reaching effects on energy distribution and consumer trust. Hence, it is essential to enforce rigorous backup and disaster recovery plans to strengthen system resilience [9].
- **Integration with Existing Infrastructure:** Another challenge is to merge a new IoT system with existing legacy energy infrastructure. Integration can be challenging when there are differences in the type of technology and capabilities [33], which could result in inefficiencies and compatibility problems.

4.6.2 Innovative Solutions and Ongoing Research

Given the complexity of addressing these challenges, several innovative solutions and research initiatives have been investigated. Further research promises to maximize the potential of transactive

IoT systems in energy management and delivery, making energy systems more efficient, reliable, and sustainable.

4.6.2.1 Standardization and Interoperability

Interoperability issues can be addressed by developing unified standards and protocols. Middleware that can translate different protocols and data formats to a common language may assist in improved integration among various devices. This can be performed with the help of industry-wide standards, such as IEEE 2030.5, for smart energy profiles [43].

4.6.2.2 Scalable Computing Solutions

Cloud computing platforms facilitate flexibility to upscale computational power and storage capacity as data throughput increases. Furthermore, edge computing enables local processing, decreasing the load on central servers and improving real-time responsiveness [31].

4.6.2.3 Advanced Data Analytics

The utilization of complex data analytics and ML algorithms will enable large-scale dataset management, along with useful information extraction. Implementing lifecycle management strategies, including data cleaning, validation, and storage, ensures data relevance and integrity [46].

4.6.2.4 Cybersecurity Measures

Cyber threats can be addressed by applying cybersecurity measures, such as end-to-end encryption, secure boot mechanisms, and IDS. Regular security audits and upgrades safeguard the system against ever-evolving threats. This chapter specifically explains the benefits of blockchain technology in facilitating secure transactions and improving transparency [32].

4.6.2.5 Policy Advocacy and Collaboration

Policies and regulations must be developed in such a way as to enable transactive IoT systems while ensuring they are fair, transparent, and efficient. Contribution to industry consortia and developing standards can help craft desirable regulatory models [34].

4.6.2.6 Consumer Education and Engagement

Consumer-centered outreach and engagement programs can raise awareness and advocacy for TESs, ultimately increasing consumer acceptance of smart grid enhancement. Meanwhile, the availability of consumer-friendly user interfaces and tools that facilitate system navigation can also increase consumer involvement, and tools that simplify interactions with the system can also enhance consumer participation [46].

4.6.2.7 Resilient System Design

Robust backup and disaster recovery plans, redundant systems, and fail-over mechanisms can improve the dependability and resilience of transactive IoT systems. In addition, the system's reliability should be guaranteed by regular testing and maintenance [9].

4.6.2.8 Gradual Integration Strategies

Gradual integration is a useful strategy for easing the process of adapting to legacy infrastructure and can help mitigate most obstacles. This can be overcome by interoperability solutions and using compatible technologies to retrofit the existing infrastructure to enable more efficient integration [33].

4.7 Conclusion

As this chapter ends, it is important to consider the insights gained and possible directions that transactive IoT systems within smart grids may take in the future. Integrating IoT with TESs offers an innovative approach to energy management by enhancing sustainability, reliability, and efficiency. This conclusion summarizes the key points discussed, presents a future outlook, and calls for continued innovation to realize the full benefits of such advanced systems.

4.7.1 Summary of Key Points

This chapter explores the integration of TESs with IoT, underscoring its significance in transforming energy management within a grid. This integration fosters efficient real-time data collection, communication, and control mechanisms critical for the dynamic management and distribution of energy.

4.7.1.1 Convergence of Transactive Energy Systems and IoT

TESs use market-based mechanisms to balance the supply and demand of energy, while IoT delivers the necessary real-time data and connectivity. This integration makes energy distribution and management easier, as real-time information and automatic controls can be easily implemented.

4.7.1.2 Significance in Modern Applications

We emphasize the importance of transactive IoT in current applications, such as smart grids and smart cities. Transactive IoT can enhance energy systems' efficiency, reliability, and sustainability. IoT technology in smart grids supports dynamic load management, predictive maintenance, and efficient communication between the grid elements, such as integrating renewable energy sources into the grid. In smart cities, IoT stimulates real-time monitoring and optimization of public infrastructure.

4.7.1.3 Economic and Market-Based Approaches

Economic- and market-based approaches have also been identified as the most relevant elements of IoT data use in energy consumption and market process optimization. IoT technologies that support dynamic pricing and DR programs are critical for encouraging consumers to respond to price signals and effectively participate in energy markets.

4.7.1.4 Transactive IoT System Architecture

Each layer in a transactive IoT architecture has been described, including the physical, communication, and data processing layers. Devices at the physical layer, such as smart meters and sensors, provide real-time IoT data. Communication layers guarantee the secure transmission of data, and the data processing layer is used to analyze these data to facilitate the smooth functioning of transactive IoT systems. These layers work in unison to provide the foundational structures for energy management.

4.7.1.5 Security Considerations

It also referred to security considerations, emphasizing the necessity of strong cybersecurity measures to safeguard critical data and maintain the system's integrity. Cybersecurity measures for ensuring these transactive possibilities are effective and secure from cyberattacks and upholding consumer privacy is vital.

4.7.1.6 Challenges and Innovative Solutions

These challenges include technical, regulatory, and operational issues in deploying transactive IoT systems. While technical challenges involve the interoperability and management of data, regulatory challenges require overcoming evolving policies. The challenges from an operational perspective include ensuring system reliability and integration with the legacy system infrastructure. This study explored new solutions and ongoing research efforts to overcome these barriers, highlighting the need for standardization, scalable computing solutions, advanced data analytics, and strong cybersecurity measures.

4.7.2 Future Outlook

In grid systems, the introduction of transactive IoT has impressive prospects and ample scope for future research. As the systems become sophisticated, they become cost-effective and affordable at the same time. Incorporating IoT devices into such systems promises transparency and wider acceptance of transactive IoT systems. Advanced AI and ML algorithms can enhance energy optimization and advanced data analytics. Reliability and trust will be integral to future technologies; hence, blockchain is essential in introducing security and transparency in decentralized energy systems. Future work is required to improve interoperability standards and security protocols. Developing robust and scalable grid systems is essential to address the increasing diversity of renewable energy resources embedded in such systems. Finally, the research demands consumer-centric solutions, providing access to users and allowing them to interact with their energy usage data in real time.

4.7.3 Call to Action for Continuous Innovation

Before we can completely realize the benefits of transactive IoT systems, there is still a way to go. Such technologies' continuous research, development, and deployment are paramount to achieving this objective. In collaboration with industry, academia must research and invest in IoT-based systems to enhance trust and security and improve efficiency, particularly when dealing with renewable energy optimization. Furthermore, strong cooperation between academia and industry with government agencies is required to develop universal regulatory frameworks. Developing such systems is a technological breakthrough and a clear path toward a sustainable future. To meet the high-tech demands of contemporary society, research must focus on incorporating IoT devices to optimize renewable energy systems and take full advantage of them without a profound impact on the environment. The information provided and insights gained in this chapter clearly understand the positive effects of merging IoT technology with TESs. This integration will transform the renewable energy landscape to create an efficient and much cleaner future energy infrastructure.

References

1 Ashton, K. (2009). That 'internet of things' thing. *RFID Journal* 22: 97–114.
2 Gharavi, H. and Ghafurian, R. (2011). Smart grid: the electric energy system of the future. *Proceedings of the IEEE* 99 (6): 917–921.
3 Depuru, S.S.S.R., Wang, L., Devabhaktuni, V., and Gudi, N. (2011). Smart meters for power grid: challenges, issues, advantages and status. *Renewable and Sustainable Energy Reviews* 15 (6): 2736–2742.

4 Amin, S.M. and Wollenberg, B.F. (2005). Toward a smart grid: power delivery for the 21st century. *IEEE Power and Energy Magazine* 3 (5): 34–41.

5 Siano, P. (2014). Demand response and smart grids—A survey. *Renewable and Sustainable Energy Reviews* 30: 461–478.

6 Albadi, M.H. and El-Saadany, E.F. (2008). A summary of demand response in electricity markets. *Electric Power Systems Research* 78 (11): 1989–1996.

7 Fang, X., Misra, S., Xue, G., and Yang, D. (2011). Smart grid—The new and improved power grid: a survey. *IEEE Communications Surveys & Tutorials* 14 (4): 944–980.

8 Rathnayaka, A.D., Potdar, V.M., Dillon, T., and Kuruppu, S. (2015). Framework to manage multiple goals in community-based energy sharing network in smart grid. *International Journal of Electrical Power & Energy Systems* 73: 615–624.

9 Guo, Y., Wan, Z., and Cheng, X. (2022). When blockchain meets smart grids: a comprehensive survey. *High-Confidence Computing* 2 (2): 100059.

10 Tiwari, A. and Pindoriya, N.M. (2022). Automated demand response in smart distribution grid: a review on metering infrastructure, communication technology and optimization models. *Electric Power Systems Research* 206: 107835.

11 Hossain, E., Khan, I., Un-Noor, F. et al. (2019). Application of big data and machine learning in smart grid, and associated security concerns: a review. *IEEE Access* 7: 13960–13988.

12 Mehmood, M.Y., Oad, A., Abrar, M. et al. (2021). Edge computing for IoT-enabled smart grid. *Security and Communication Networks* 2021 (1): 5524025.

13 Nandury, S.V. and Begum, B.A. (2017). Big data for smart grid operation in smart cities. *2017 International Conference on Wireless Communications, Signal Processing and Networking (WiSPNET)*. pp. 1507–1511. IEEE.

14 Assad, U., Hassan, M.A.S., Farooq, U. et al. (2022). Smart grid, demand response and optimization: a critical review of computational methods. *Energies* 15 (6): 2003.

15 Bhavani, N.G., Kumar, R., Panigrahi, B.S. et al. (2022). Design and implementation of IoT integrated monitoring and control system of renewable energy in smart grid for sustainable computing network. *Sustainable Computing: Informatics and Systems* 35: 100769.

16 Pop, C., Cioara, T., Antal, M. et al. (2018). Blockchain based decentralized management of demand response programs in smart energy grids. *Sensors* 18 (1): 162.

17 Machele, I.L., Onumanyi, A.J., Abu-Mahfouz, A.M., and Kurien, A.M. (2024). Interconnected smart transactive microgrids—A survey on trading, energy management systems, and optimisation approaches. *Journal of Sensor and Actuator Networks* 13 (2): 20.

18 Alaba, F.A., Sani, U., Dada, E.G., and Mohammed, B.H. (2024). AIoT-enabled smart grids: advancing energy efficiency and renewable energy integration. In: *Artificial Intelligence of Things for Achieving Sustainable Development Goals*, 59–79. Cham: Springer Nature Switzerland.

19 Güçyetmez, M. and Farhan, H.S. (2023). Enhancing smart grids with a new IoT and cloud-based smart meter to predict the energy consumption with time series. *Alexandria Engineering Journal* 79: 44–55.

20 Gupta, N., Prusty, B.R., Alrumayh, O. et al. (2022). The role of transactive energy in the future energy industry: a critical review. *Energies* 15 (21): 8047.

21 Bakare, M.S., Abdulkarim, A., Zeeshan, M., and Shuaibu, A.N. (2023). A comprehensive overview on demand side energy management towards smart grids: challenges, solutions, and future direction. *Energy Informatics* 6 (1): 4.

22 Sahoo, S.K., Yanine, F.F., Kulkarni, V., and Kalam, A. (2023). Recent advances in renewable energy automation and energy forecasting. *Frontiers in Energy Research* 11: 1195418.

23 Balam, S.K., Jain, R., Alaric, J.S. et al. (2023). Renewable energy integration of IoT systems for smart grid applications. *2023 4th International Conference on Electronics and Sustainable Communication Systems (ICESC)*. pp. 374–379. IEEE.

24 Marinakis, V. and Doukas, H. (2018). An advanced IoT-based system for intelligent energy management in buildings. *Sensors* 18 (2): 610.

25 Mengelkamp, E., Gärttner, J., Rock, K. et al. (2018). Designing microgrid energy markets: a case study: the Brooklyn microgrid. *Applied Energy* 210: 870–880.

26 PowerLedger. (2017). PowerLedger Whitepaper. Powerledger.io.

27 Madakam, S. and Ramachandran, R. (2016). Amsterdam smart city (ASC): fishing village to sustainable city. *WIT Transactions on Ecology and the Environment* 204: 831–842.

28 Khan, S.U., Khan, N., Ullah, F.U.M. et al. (2023). Towards intelligent building energy management: AI-based framework for power consumption and generation forecasting. *Energy and Buildings* 279: 112705.

29 Al Hwaitat, A.K., Almaiah, M.A., Ali, A. et al. (2023). A new blockchain-based authentication framework for secure IoT networks. *Electronics* 12 (17): 3618.

30 Alam, K.S., Kaif, A.D., and Das, S.K. (2024). A blockchain-based optimal peer-to-peer energy trading framework for decentralized energy management with in a virtual power plant: lab scale studies and large scale proposal. *Applied Energy* 365: 123243.

31 Premsankar, G., Di Francesco, M., and Taleb, T. (2018). Edge computing for the Internet of Things: a case study. *IEEE Internet of Things Journal* 5 (2): 1275–1284; 10 (1): 853–867.

32 Dalipi, F. and Yayilgan, S.Y. (2016). Security and privacy considerations for IoT application on smart grids: survey and research challenges. *2016 IEEE 4th International Conference on Future Internet of Things and Cloud Workshops (FiCloudW)*. pp. 63–68. IEEE.

33 Alotaibi, I., Abido, M.A., Khalid, M., and Savkin, A.V. (2020). A comprehensive review of recent advances in smart grids: a sustainable future with renewable energy resources. *Energies* 13 (23): 6269.

34 Kumar, D. (2020). Urban energy system management for enhanced energy potential for upcoming smart cities. *Energy Exploration & Exploitation* 38 (5): 1968–1982.

35 Faruqui, A. and Sergici, S. (2010). Household response to dynamic pricing of electricity: a survey of the experimental evidence. *Journal of Regulatory Economics* 38 (2): 193–225.

36 Faruqui, A. (2012). The ethics of dynamic pricing. In: *Smart Grid*, 61–83. Academic Press.

37 Palensky, P. and Dietrich, D. (2011). Demand side management: demand response, intelligent energy systems, and smart loads. *IEEE Transactions on Industrial Informatics* 7 (3): 381–388.

38 Andoni, M., Robu, V., Flynn, D. et al. (2019). Blockchain technology in the energy sector: a systematic review of challenges and opportunities. *Renewable and Sustainable Energy Reviews* 100: 143–174.

39 Conejo, A.J., Morales, J.M., and Baringo, L. (2010). Real-time demand response model. *IEEE Transactions on Smart Grid* 1 (3): 236–242.

40 Kirschen, D.S. and Strbac, G. (2018). *Fundamentals of Power System Economics*. Wiley.

41 Cramton, P. (2013). Spectrum auction design. *Review of Industrial Organization* 42: 161–190.

42 Hobbs, B.F., Kasina, S., Ho, J., and Wogrin, S. (2015). Unit commitment approximations in generation and transmission planning: efficiency & accuracy. *Abstract from INFORMS Annual Meeting*, Philadelphia, PA, USA.

43 Gungor, V.C., Sahin, D., Kocak, T. et al. (2011). Smart grid technologies: communication technologies and standards. *IEEE Transactions on Industrial Informatics* 7 (4): 529–539.

44 Lund, H., Østergaard, P.A., Connolly, D., and Mathiesen, B.V. (2017). Smart energy and smart energy systems. *Energy* 137: 556–565.

45 Wang, W., Xu, Y., and Khanna, M. (2011). A survey on the communication architectures in smart grid. *Computer Networks* 55 (15): 3604–3629.

46 Muleta, N. and Badar, A.Q. (2021). Study of energy management system and IOT integration in smart grid. *2021 1st International Conference on Power Electronics and Energy (ICPEE)*. pp. 1–5. IEEE.

47 Lu, X. (2024). Application of artificial intelligence algorithms in precision marketing with flow data analysis models. *2024 5th International Conference on Mobile Computing and Sustainable Informatics (ICMCSI)*. pp. 378–383. IEEE.

5

IoT Devices in Transactive System

G. Jagadish and P. Sriramalakshmi

School of Electrical Engineering, Vellore Institute of Technology, Chennai, Tamil Nadu, India

5.1 Introduction

The growth of industries for the past 100 years has been increasing exponentially, so the consumption of energy has also increased which led to huge power demand, and attention has been given to reduce the power crisis in the future. Smart grid is a network based on two-way digital communication used for supplying electricity to consumers [1]. Smart grid allows us to monitor the amount of energy consumed by the user and analyze the data for better efficiency, reduced energy consumption, and cost. This gives the transparency and reliability of the supply chain. By this, smart grid helps us to overcome the weakness of the conventional electric grid. The transactive energy (TE) system is a network frame to trade the energy [2]. The TE market depends on renewable energy resources (RESs) and distributed energy resources (DERs). Moreover, the market is willing to adopt renewable energy sources in favor of a green and healthy society. DERs are used globally rather than RESs due to their power delivery efficiency; however, the DERs are a variable resource because of its dependability on weather. This leads to an increased penetration in the industry and creates new challenges in the power delivery system. So, this system requires a safe and stable operation which includes the demand response and a transparent secured system.

The energy management system (EMS) is used for power management, intelligent control, and surveillance programs, helping customers with their concerns about the increased requirement for energy in smart urban centers. EMS presents many elements such as data, sensors, security, and connectivity. The data will be derived via Internet of Things (IoT) for the smart EMS. The smart EMS uses less energy which leads to boost the usage, operation, and maintenance. The TE system refers to a framework that enables the buying and selling of energy in decentralized and automated energy sector. It involves sensors, devices, and systems that communicate and make decisions about energy transactions in real time. This allows us to create a more dynamic and responsive energy grid by autonomously allowing the devices to participate in the energy market and also to optimize energy usage, enhance grid stability, and promote efficiency. This allows us to connect and exchange data for a smarter and efficient environment in various domains. The sensor works for data collection and data transmission. In data collection, sensor detects and measures physical parameters like temperature, humidity, light, and motion. In data transmission, the collected data will be transmitted to the IoT device or cloud platform for analysis and decision-making and for automation in some scenarios. The actuators are devices that work based on received data.

IoT for Smart Grid: Revolutionizing Electrical Engineering, First Edition.
Edited by Rahiman Zahira, Palanisamy Sivaraman, Chenniappan Sharmeela, and Sanjeevikumar Padmanaban.

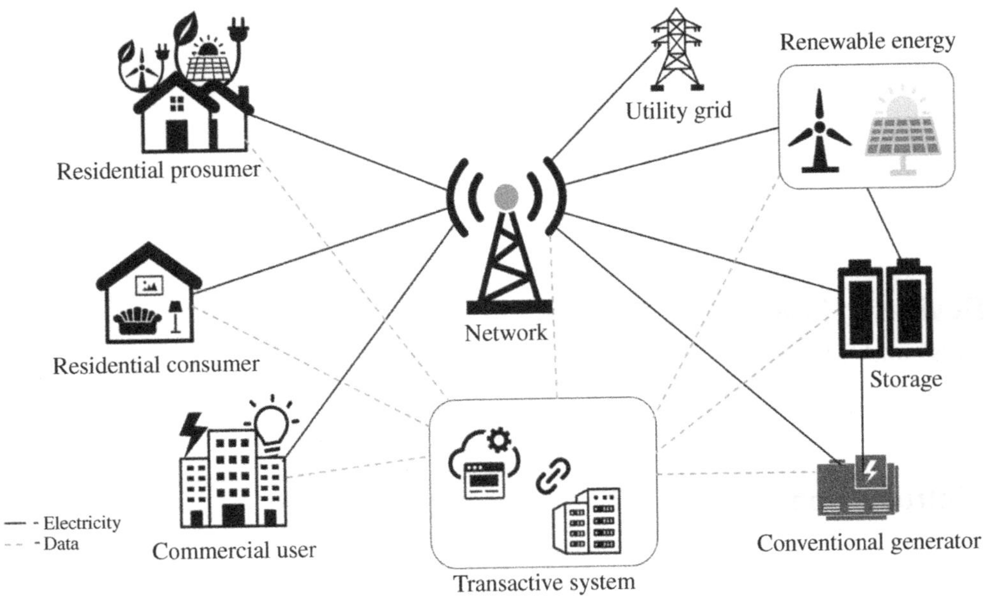

Figure 5.1 Process flow of transactive system [3].

They physically influence the environment by turning on/off, adjusting, or moving the devices. The actuators will interact with the environment by taking actions by adjusting the device in response to the environment.

The sensors and actuators are responsible for the control-based actions in the smart grid. These devices are also used to adjust the energy flows, operation, and configurations of systems for better optimization and efficiency. The transactive system requires data security and connectivity for the data exchange to trade energy. Ensuring the standard and reliable communication infrastructure is crucial for data exchange among the smart grid components. Data security is a concern in the smart grid due to the sensitive nature of the sensor. The sensor data challenges include protecting against cyber threats and ensuring data integration and confidentiality.

Establishing robust encryption, authentication, and continuous monitoring is essential to safeguard the smart grid system from cyber-attacks. Distributed ledgers such as blockchains (BCs) are used to protect the privacy attacks on a transactive system. This chapter explains the integration of IoT devices into TE system, the role of sensors and actuators in transactive models, and various challenges faced in the device connectivity and data security of the IoT devices. By the real-world data and examples, the researchers can obtain insights and strategies of the transactive system to navigate the process of smart grid in IoT for a safe and effortless energy trading to make a sustainable future. The transactive system workflow is explained in Figure 5.1.

5.2 Integration of IoT Devices for Data Collection

IoT device collects the data from sensors and tracks the performance of devices connected to IoT [4]. These data are collected in real time from the sensors that are stored and can be retrieved at any time [4]. The data collection in IoT is differentiated into three types. They are given as follows [4]:

1) Equipment data.
2) Submeter data.
3) Environmental data.

1) **Equipment Data:**
   This data type is used for maintaining the IoT device and the data is collected in real time for predictive maintenance.
2) **Submeter Data:**
   It is a data tool used by the owners to automate the measure of electricity, water, gas, and cable when a building has multiple usages.
3) **Environmental Data:**
   These are the data collected by different IoT sensors that are used to measure and monitor environmental data such as humidity, temperature, movement, and air quality.

5.2.1 Working Layer of Data Collection

The data collection undergoes several layers of process.

a) **Data Layer:**
   Device that is connected to the IoT architecture communicates with each other and forms the primary layer [4]. The IoT devices are categorized into the following: for building unique identifier (UUID), as a transmitting device like Bluetooth, Wi-Fi MAC, etc. [4].
b) **Communication Layer:**
   The communication layer is used for communicating with one another by using the protocols like:
   - **Hypertext Transfer Protocol Secure (HTTPS):** The text-based protocols for lower versions [4].
   - **Message Queuing Telemetry Transport (MQTT):** The protocols are used to handle the embedded system and are optimized to support [4].
   - **Constrained Application Protocol (CoAp):** Protocol based on HTTP semantics and is difficult to connect with firewalls because of poor library browser [4].
c) **IT Edge Layer:**
   Hardware, firmware, and operating systems that are connected to the IoT devices are included in this layer [4].
d) **Event Processing Layer:**
   This layer is responsible for processing and storing data when receiving from the IoT devices. This layer is also responsible for data cleaning, adding meta data, and organizing data insights [4].
e) **Client Communication Layer:**
   Transfers data from IoT devices to the client by acting as a bridge between the backend database and the front-end database [4].

The architecture of the data flow of the transactive system is explained in Figure 5.2.

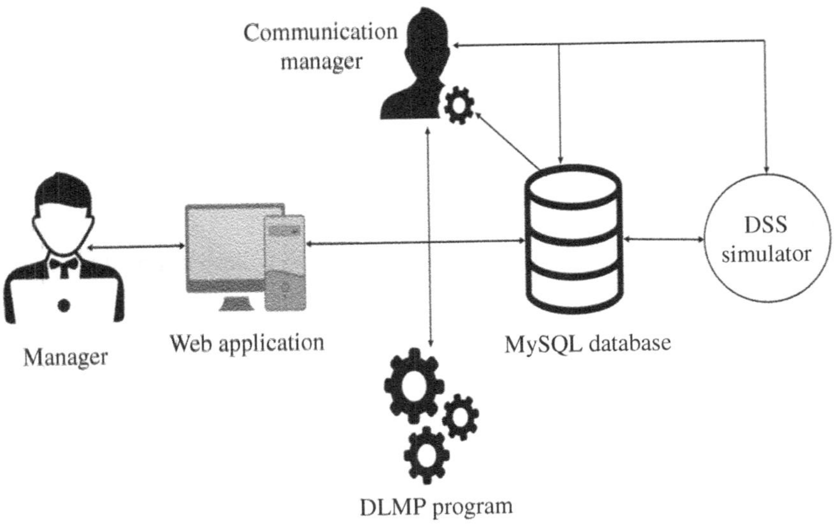

Figure 5.2 Data flow of the architecture of transactive system [5].

5.3 Role of Sensor

The sensor plays a crucial role in enabling the robust and efficient operation of the energy system by providing real-time data and simplifying communication between the components. For the energy control loops, the data is collected from the sensor and delivered to the control algorithm [7]. The control algorithm acts as an actuator in this process.

5.3.1 Local Controls

According to a study of control loop performance assessment, local control loops in HVAC systems are common [8], and the primary control loop in a heating, ventilation, and air conditioning (HVAC) system is proportional control or proportional integral control. In a local control system, the HVAC system usually uses only one sensor which leads to inaccurate sensor reading and rising time, overshoot, and oscillation. When the sensor runs under the fault condition, normal control actions get affected. In the literature, there is no study available to quantify the impacts of sensors on local control systems.

5.3.2 Advanced Control

The advanced control strategies are more powerful so they can outperform the classic feedback control such as proportional integral derivative control and rule-based controls. The model predictive control has been shown in several simulations' studies and in real-world demonstrations to achieve improvements in thermal comfort while reducing the energy consumption of the building to at least 15% [9, 10]. There are also systems that are adaptive, robust, and stochastic for advanced building controls to overcome the problems of weather, activity variation driven by occupancy, and equipment aging. Artificial intelligence (AI) is being used to enhance the controller stability [11, 12]. The thermal flexibility of the building in coordination with other DERs is used to support the grid-level objectives through advanced predictive control techniques [13, 14]. The buildings are used as resources for high-quality data called ancillary. The hardware used in loop algorithms and real building demonstration has the capability for building ventilation systems to offer a good scale of grid services [15, 16]. The responsive building can also lack in defining new tariffs, energy-pricing structures, and contractual energy schemes [17, 18]. As IoT devices use cloud-based servers, the interaction between the buildings and grids becomes effectively easy [19].

5.4 Sensor Types

There are many sensors that are developed with HVAC and control fields and these sensors are classified into many subgroups.

5.4.1 Traditional HVAC Sensors

The HVAC system usually comes in a package unit to build the system where the chillers, boilers, heat pumps, fans, pumps, valves, heat exchangers, filters, dampers, diffusers, ducts, and pipes are commonly present. These transmission line components should be maintained properly for stable and efficient controls [6]. Some HVAC equipment have built-in sensors and controls. This helps monitor the system and it is cost-effective. Also, they avoid the installation of exterior sensors for control optimization. The sensors used in the HVAC system usually monitor temperature, humidity, flow rate, pressure, and gas flow for boilers. These sensors are generally used in control loops that are used to modulate and maintain the control fluctuating at a given set point. The studies show that the consumption of energy in buildings is reduced by 20–40% by using the appropriate HVAC control strategy [20] and the advanced governors are derived from an HVAC system that uses the neural network [21] to minimize the consumption of energy and to maintain the customer's comfort zone. Some sensors are employed to determine the baseline energy consumption and to confirm the enhanced performance using British thermal unit (BTU) and electrical sensors [22].

5.4.2 Occupancy Sensor

The instant system response of the lighting system leads to the development of occupancy-based control. Various building standards and green building rating programs have requested occupancy-based lighting control [25]. The study reports that the potential for energy savings using occupancy information ranges from 20% to 75% [24, 25]. Due to limiting performance of existing occupancy, sensing technology causes faults in on/off switches of the lighting system and there is a report that dissatisfies the occupant of energy wastage [26]. Following the successful implementation of occupancy-based lighting control, efforts have been made to apply occupancy information to HVAC controls. However, due to the slow response of HVAC systems, predictive strategies are often necessary to maintain occupant comfort [27]. Because of this, developing practical and scalable occupancy-based HVAC control has become easier. It is more challenging compared to occupancy-based lighting control. The high-energy-saving potential of occupancy-based building control methods garners significant attention in the control domain of the building. These methods are categorized based on the type of information they provide, such as presence, number of occupants, identity, and location [28, 29]. The information about the presence of occupants makes the system to turn on/off light and HVAC system. Data on the number of occupants can be utilized to adjust ventilation and predict various building loads, which are essential for the optimal control of HVAC control [30]. Recently, vision-based methods have been developed, driven by the rapid advancements in sensing, computing, and computer vision technologies [31, 32] and interactions with furniture, appliances, communication networks, and corresponding plug loads are utilized to estimate occupancy information.

5.4.3 Emerging Sensors

As the technology evolves in the case of IoT, the sensors used in the devices should also be updated to meet the recent technology requirements. Such IoT-based devices should manage sensors to meet the application in demand-controlled ventilators, energy recovery ventilators, and more [33]. To make it more flexible in operations, a dedicated Global Positioning System (GPS) service-based remote is used to control the device and makes the IoT devices futuristic [34]. IoT devices are integrated with advanced control methodologies for building energy management, resulting in energy savings of 30–40% in terms of energy usage costs [35].

5.4.4 Virtual Sensor

Utilizing a dense and complex sensory framework can incur significant expenses in both installation and upkeep. Addressing this challenge involves employing virtual sensing techniques for device detection and value estimation. Making mastery of this approach is increasingly vital in advanced building control design [36]. Model-based virtual sensors are integrated into advanced control systems, such as virtual occupancy and thermal sensors, to ascertain the operation of air function, flow rate, and inlet conditions. The virtual sensor is also used to estimate the value of unknown information whereas the physical sensor requires more data [37].

5.5 Role of Sensors During Data Collection

This section deals with the various roles of sensors during data collection, demand response, and energy efficiency. Also, it helps in the integration of renewable energy, grid stability, and reliability.

5.5.1 Data Collection and Monitoring

Sensor collects real-time data on various parameters such as the following [38]:

- **Energy Consumption:** This monitors the consumption of energy by various appliances, buildings, and industries.
- **Energy Production:** This tracks the output of renewable energy sources like solar panels and wind turbines.
- **Grid Conditions:** This measures the voltage, current, frequency, and other grid parameters to ensure stability and reliability.

5.5.2 Demand Response

Sensors help in power demand response by detecting peak load and load management, where the peak loads identify the period of high-energy demand and signal to reduce power consumption. The load management helps to enable the automated systems to adjust the operation of appliances and equipment based on real-time data to balance supply and demand [38].

5.5.3 Energy Efficiency

By providing detailed energy usage patterns, sensors can identify inefficiencies where it highlights the energy wastage area in a system and helps by suggesting ideas. This is also used to optimize the performance by helping to tune the HVAC system, lighting, and other energy-intensive processes [39].

5.5.4 Integration of Renewable Energy

Sensors are used to monitor the production of renewable energy production by providing data on the performance and output of renewable energy installation, forecasting is used to help predicting the energy production from variable sources like solar and wind and aiding in the betterment of grid [40].

5.5.5 Grid Stability and Reliability

The sensor is used to enhance the grid by identifying the faults and disturbances occurring in the grid to facilitate rapid response and repair. Also, the sensor is used to balance the supply demand ensuring that the energy supply matches the demand closely to maintain stability [41].

5.5.6 User Empowerment

The sensors are used to make informed decisions that allow consumers to understand their energy usage patterns and make more informed decisions about their consumption. This also enables users to participate in energy markets, buying, and selling energy based on real-time data and market conditions [42].

5.6 Role of Actuators

The actuators play a vital role in the management of resources, particularly in energy systems. The transactive system leads to an economic and control mechanism to balance the supply and demand. These actuators are categorized into various types. They are as follows:

- **Intelligent Detection Using Actuator-Assisted Sensor:** This allows the sensor to dynamically adjust and modify both the quality and the quantity of its output. Additionally, integrated sensors derive information such as geometric changes and physical interactions at boundaries [43]. As an illustration, a robot equipped with a gas sensor could identify gas leaks and then cautiously approach the source to shut off the valve [44].
- **Actuator-Assisted Sensor for Intelligent Detection:** To enhance intelligent detection, sensors must pinpoint the desired location and adjust to the surrounding environment, thereby optimizing their capacity for intricate measurements. Integrating actuators with sensors enhances both the precisions and volume of data collected. In soft systems, sensors are supported by actuators in two ways: first, actuators aid in relocating sensors to various positions for data collection, and second, they assist in navigating complex shapes of targeted objects to gather information [43].
- **Sensor-Assisted Actuators for Intelligent Movement:** To navigate its environment effectively, the sensor must possess knowledge of its present location and geometry, enabling it to plan

subsequent movements. Achieving this state necessitates a closed-loop system, where sensor data feeds back to the controller about the current state of actuators, enabling continual adjustment of control inputs. In contrast, open-loop control operates without sensor data, while feedback control offers greater resilience to disturbances, such as alterations in environmental conditions. A range of actuators are available that operate in conjunction with soft strain and force sensors to ensure precise and dependable data acquisition [43].

- **Sensor Actuator for Hybrid of Intelligent Detection and Movement:** In order to accomplish intricate tasks, it is imperative to merge intelligent detection with movement.

5.6.1 The Key Function of Actuators in Transactive System

The following are few main roles of actuators present in the transactive systems.

- **Real-Time Control:**
  The actuators are responsible for the real-time adjustment of the physical system. In the TE system grid, for instance, the output of distributed energy system resources is adjusted like solar panels, batteries, and controllable loads such as HVAC systems and electric vehicles.
- **Market Participation:**
  Transactive system is often sold in markets when the prices fluctuate based on the demand and supply. Actuators respond to these price signals by adjusting their state. For example, a smart thermostat might lower the heating setpoint if energy price rise.
- **Automation and Autonomy:**
  Most of the actuators operate autonomously based on predefined algorithms or AI models that optimize their behavior according to market conditions, user preference, and system constraints.
- **Load Balance and Stability:**
  In the TE system, the actuators help in balancing the load and maintaining the grid stability. The actuators can shift loads, store excess energy, or even curtail generation to match real-time demand with supply.
- **Communication:**
  The actuators communicate with other components of the TE system, including sensors, control centers, and other actuators.

5.6.2 Challenges and Considerations

There are certain challenges and considerations for the use of actuators in TE systems that are as follows:

- **Security and Privacy:**
  To ensure that actuators operate securely and without compromising the user privacy or the system integrated with it.
- **Interoperability Standards:**
  To ensure that actuators communicate with other components from different manufacturers to work together seamlessly [45].
- **Reliability:**
  The actuators should be reliable and capable of operating under various conditions without any interruptions.
- **Scalability:**
  The system should handle a large number of actuators efficiently, as the number of connected devices grows.

5.7 Challenges Faced in Device Connectivity

Transactive system faces several challenges in device connectivity by various issues. These challenges are categorized into technical, operational, and regulatory domains.

- **Interoperability:**
  - The device from different manufacturers uses their own communication method to communicate within them. This makes the seamless integration difficult [45].
  - To overcome this, the consumer should ensure that all devices used should meet the existing standards and can communicate seamlessly [45].
- **Scalability:**
  - If the number of connected devices is high and if there is an increase in a number of devices, network issues may arise, where it affects the performance of the system [46].
  - Data management is the solution for such issues, where it can manage the data created by these devices and helps in processing it [46].
- **Latency and Reliability:**
  - Many transactive system requires real-time communication, which can be a challenge to achieve consistency [46].
  - Continuous and reliable connectivity, especially in remote or challenging environments, is difficult [46].
- **Device Management:**
  - The firmware updates across a large number of devices can be difficult and can cause issues because the updates of the devices are not consistent.
  - The maintenance and troubleshooting of devices in a TE system require a lot of effort and resources.
- **Energy Efficiency:**
  - The balancing of the power consumption during connectivity and performance is difficult, especially for devices using batteries.
- **Laws and Regulations:**
  - Different regions have varying regulations for data privacy, security, and communications, where this can be complicated for device connectivity [47].
  - To meeting these laws and regulations requires a lot of money and administrative efforts [47].
- **Legacy System:**
  - Integrating the new devices with the legacy system may not support modern protocols or standards [48].
  - Updating or fixing the older systems to work with new technologies involves more investment and technical support [48].
- **Complex Architectures:**
  - As the number of devices increases, the system architecture will become more complex for maintenance and troubleshooting.
  - The stakeholders should have coordination, including the device manufacturers, and network providers require a robust framework.

To address these challenges, a significant investment involving technology regulatory support and collaboration between stakeholders is necessary to ensure that the TE system can work at its full potential.

5.8 Challenges in Data Security

The data exchange plays a vital role in the pear-to-pear and other TE trading. For example, peer-to-peer energy trading depends on the participant data, including the consumption and production data. This enables the local market to allocate the energy quantity and the price of energy to be traded. As the peer-to-peer trading carries sensitive data like participants personal data and energy usage patterns, it became essential to protect these data from attackers. There are some challenges faced during data security that are explained below [49].

5.8.1 Challenges Faced During Data Collection

Even though the data collection has a process of different layers, there are difficulties faced during the collection of data and management.

5.8.1.1 Data Security

The data collected in industries like healthcare industries transfers the sensitive information collected by internet of medical things (IoMT) devices that includes protected healthcare information (PHI). The IoT device used in manufacturing industries have access to manufacturing processes as these are sensitive data and securing these data is a challenge for IoT devices [4, 50]. These data can be accessed by using the public internet due to their need of transferring data to cloud platforms that are managed from mobiles and web portals. There are also cases where the data are leaked due to poor passwords and where IoT devices have a default password system to access the data which leads to hacking of device by having a poor password.

5.8.1.2 Data Privacy

The devices used in the TE system should follow various privacy regulation acts like General Data Protection Regulation (GDPR) in EU where it becomes costly to follow these regulations in the transactive system, as the TE system uses a number of devices. Data collected by IoT devices may fall under the purview of privacy regulations such as the EU's GDPR. This encompasses personal details like name, address, phone number, medical information, and other pertinent data [4, 50]. There are also laws for securing these protected data from attack and IoT device users and manufacturers must follow these laws.

- **Consent to Collection:** Obtaining explicit consent from data subjects is necessary for the collection of personal and protected data. However, applying this requirement to IoT devices poses challenges, as these devices often gather data without appropriate permissions.
- **Consent to Processing:** The same law applies, mandating explicit consent from data subjects for data processing. However, monitoring becomes challenging due to the massive volume of data collected and processed by IoT devices [4, 50].
- **Encryption:** This law also states to encrypt the collected data while transferring to protect against unauthorized access. These IoT devices at present lacks at power, and processing the resource makes it difficult for data encryption. This leads to a requirement to protect data in other ways [4, 50].
- **Access Management:** Regulations such as GDPR, Health Insurance Portability and Accountability Act (HIPAA), and similar laws mandate restricting access to data to only those who require it. However, IoT devices processing and distributing data to cloud servers pose challenges in enforcing access limitations [4, 50].

- **Jurisdiction:** The GDPR law restricts to transfer the data to the country that does not have suitable data protection laws. This can be a challenge to the IoT devices where it uses the cloud-based server making it complex to track the flow of data [4, 50].

5.8.1.3 Data Volume

The data collection of IoT devices has become huge over the past five years. In 2019, these IoT devices generated an estimated amount of 18.3 zettabytes of data which is expected to grow to 73.1 zettabytes by 2025 [4, 50]. By having a huge amount of data, it is difficult to store, transfer, and process. The IoT device has a limited amount of internet access making it difficult to transfer huge data and it will become expensive by expanding the cloud to store more data.

5.8.1.4 Data Complexity

IoT devices usually collect data in huge numbers where these devices collect the data as much as possible for gaining more information and send it to the cloud server for the process. This also creates complex datasets by producing huge data. Data generated by IoT devices is disorganized and lacks context, thus necessitating meticulous timestamping, indexing, and correlation with other data sources to render it valuable for decision-making purposes [4, 50]. These huge data make complications in data management and combination. Numerous tools have been developed to address this challenge, but they struggle to keep pace with the sheer volume of data generated by devices. Additionally, solutions capable of handling large data sets often lack the depth of analysis required.

5.8.1.5 Data Protection

Safeguarding consumer data and privacy is paramount, necessitating that any gathered personal information is utilized transparently and with accountability [51]. This also entails incorporating suitable measures to safeguard the privacy rights of individuals, posing a significant challenge to data security [52]. To address this, appropriate measures and technologies should be implemented for privacy and data protection in TE system.

5.8.1.6 Privacy

The privacy is a term which has the ability to control the information that others know about a person and the actions taken by them based on the data they have. In short terms, the privacy is the key element of an individual's personal data [53]. The privacy is an important consideration in data protection as it carries the sensitive information of an individual. Privacy protection in transactive system gives the customer more trust by preventing threats such as unwanted alteration of energy-related data.

5.8.2 Security Threats

Various cybersecurity threats are represented in Figure 5.3.

The system can face various threats, as it has multiple components that interact with each other over different areas if the system is not protected, the following issues may arise.

- **Data Manipulation:** If the attacker gains access to one or more components, then one may manipulate the market/power values which causes disruption in the system. This could spread all over the system as it transfers the data to other components and could cause major damage to the system.
- **Centralized Failure:** The communication in TE system works based on Frame for Network Co-Simulation (FNCS) broker and the system faces a shutdown if the broker fails.

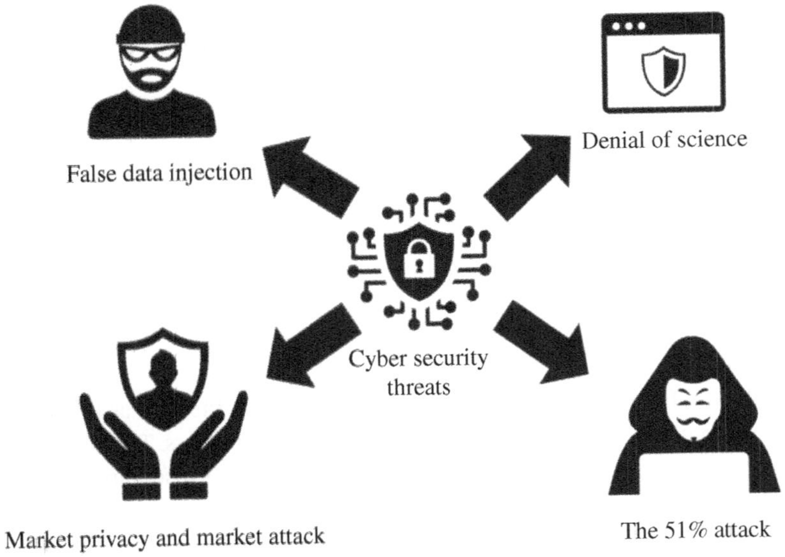

Figure 5.3 Cybersecurity threats [54].

- **Breach of Data:** An attacker may steal the sensitive data of a user if the one gains access to the system and this data can be misused by monitoring the consumers' consumption level.
- **Lack of Availability:** The attacker changes protocols and other details of a TE system which might cause low production of energy, leading to more energy demand and corrupting the entire market.

5.8.3 Decentralized Scheduling

The decentralized scheduling system is used, as the centralized scheduling requires explicit information from the devices used in the system including the data model and behavioral patterns such as production and consumption of energy [55]. This leads to serious privacy and security issues. In addition, all agents present in the system need to communicate with an agent called the central agent; this central agent can cause a breakdown if the central agent becomes failure [56]. The decentralized method is used to address these issues. The decentralized scheduling method does not use explicit data as it uses the local agents which are operated automatically or controlled by the end user. The end user receives the external value signal. Considering this, behavioral patterns are derived and hence the communication among devices becomes flexible with another agent.

5.8.4 False Data Injection

The major cybersecurity issue is the false data injection. The BC handles the data integration and non-denial to the highly sensitive data. False data injection attack usually involves changing data before sending the data to the system or injecting the false data into a system in place of true data [57]. In the TE system, a false data injection will be carried out to the system by generating fake data. This affects the market by causing bad biding or physical damage to the grid [58–60]. The false data injection in the smart power grid also involves physically altering the meter measurements. For this, the attacker requires the knowledge of formatting the power system [61].

5.8.4.1 Denial of Service (DoS)

This is a type of cyberthreat that makes a data traffic and makes it inaccessible. In the TE system, denial of service (DoS) attack can bid randomly and discards throughout the market [58]. This kind of attack is easy to implement, where it requires minimum knowledge about the grid or network configuration. So, this kind of attack is performed by a person who cannot perform complex attacks on the transactive systems. There are some measures that can be used to detect the DoS attack, such as deep learning models [58] which are trained to detect the malicious nodes. The most used DoS attack in this field is the puppet attack, where the malicious node receives a requisition pack, and it returns a route with an empty node at its other end. This damages the node that enters the route discovery which cannot be processed as it has an empty route. This process ends in domino effect by which the node sends route requisition packs to all its friendly nodes. By this, the network is flooded with route requests which causes the DoS [61].

5.8.4.2 The 51% Attack

In this attack, the hacker acquires control over 50% of the system power in a ledger system. This attack enables the hacker to insert false transaction data into the BC, which collapses the market and bidding of a transactive system [62]. As this attack is similar to the false data injection, its execution method relies on threats on the BC network. This makes the difference between the false data injection. These kinds of attacks are unlikely to happen in the global BC such as bitcoin. In this kind of BC, the node comprises many million computational resources. This makes it complicated for the attacker to take over the network. But the transactive system markets are operated in a small-scale manner. With this scale of computing, it is more feasible to implement this attack than in global BC.

5.8.4.3 Market Privacy

The market privacy is a process that is used to protect the consumers data and market data of the TE system. Here the data of consumers must be protected while trading the energy. The transaction history of a prosumer can be used to deal with a lot of information, so their identities should be protected in any cause in the market [63–65]. This system can handle a high level of privacy. For example, the bitcoin provides the total protection of data to its users as it follows the legal regulations, denial, and threat protection. Some system works with less privacy for the integrity. The safety of privacy and trust cannot be limited in any situation. The transactive system should consider all the privacy regulation and should make trade-offs without sacrificing the quality [66]. The market will work based on the value and demand in the global market. The pictorial representation is depicted in Figure 5.4.

5.8.4.4 Market Attacks

The market attacks are performed inside the TE framework. This type of attack is made either by market participants or by outside attackers. This can be differentiated by identifying their location in the TE framework. But these details cannot be found as the technical execution was lacking in the literature. The market participants can attack for their financial growth [56]. Creating a powerful transactive system is required so that the prosumers may try to access the data. These effects may also arise due to errors rather than malicious attacks. Malicious trading is considered a serious threat to TE system that can cause grid instability [54, 67]. The Malicious trading is also used to threaten the market operation and create instability in the market [6, 67]. This attack can include the alteration of price and amount of energy, bids, and offers where this leads to financial instability.

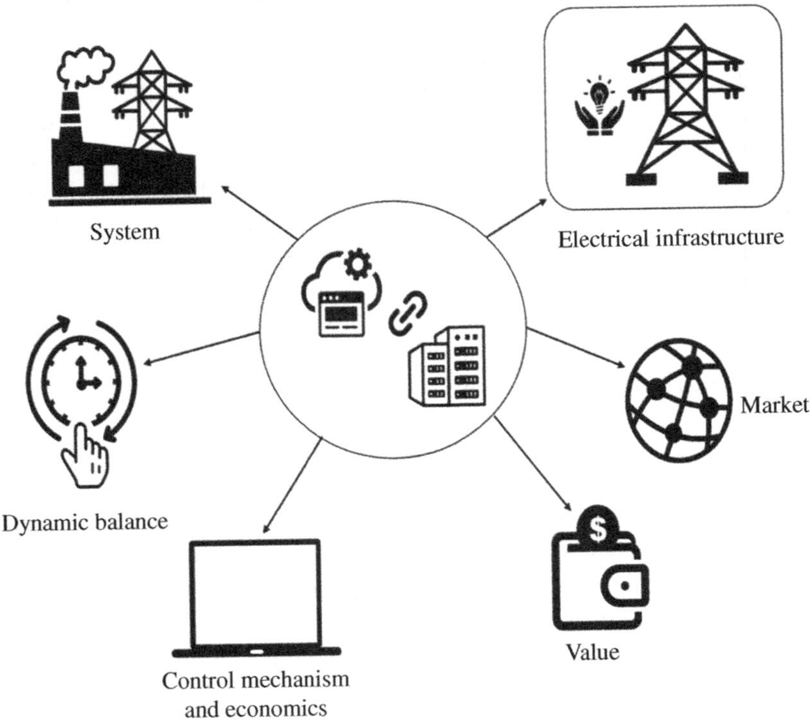

Figure 5.4 Value and demand in the global market [68].

5.8.4.5 Future Scope

The transactive system has become a needed technology in the energy market for its efficiency and robustness. As this technology continues to advance, several key areas likely to shape the evolution of transactive systems:

1) **Enhanced Sensor Technology:** The sensors and actuators used in the transactive system should be accurate and precise. So major development in the sensors and actuators enables finer data collection.
2) **AI and Machine Learning:** As the world is moving toward the AI technology, integrating AI with transactive system enables TE system to work efficiently by analyzing the data collected by sensor.
3) **Interoperability and Standardization:** Developing the standard protocols for devices improves the communication between them. This will help connectivity and integration processes, making it easier to deploy transactive systems across various industries.
4) **Advanced Cybersecurity Measures:** As the data exchange and connectivity increase, robust cybersecurity measures should be advanced in encryption, multifactorial authentication, and anomaly detection systems will be crucial to protect against evolving cyber threats.
5) **Regulatory and Ethical Consideration:** As transactive system becomes more integrated into daily life, regulatory frameworks should be evolved to address data privacy, ownership, and ethical concerns. Policymakers and industries should collaborate to ensure that these systems are used responsibly and equitably.

5.9 Conclusion

This chapter explains the role of sensors and actuators that are playing a major role in collecting and processing data-driven decisions, by being the backbone of the transactive system. These devices enable seamless interactions between physical and digital realms, enhancing efficiency, accuracy, and responsiveness in various applications. This chapter also explains the challenges which are faced during device connectivity and data security. The devices that are connected increase the complexity of maintaining robust and reliable communication networks. As the data flows become more important to the operation, safeguarding this information against breaches and unauthorized access has become more complex. Implementing the security measures such as encryption, authentication, and real-time monitoring will protect sensitive data and maintain system integrity. This chapter provides a comprehensive overview of these dynamics, offering insights into both the technological advancements and the obstacles that need to be navigated in this evolving landscape.

References

1 Techopedia. (2017). Smart grid. https://www.techopedia.com/definition/692/smart-grid (accessed 21 November 2024).

2 IEEE. (2024). Transactive energy management in the smart grid. https://blockchain.ieee.org/verticals/transactive-energy/topics/transactive-energy-management-in-the-smart-grid (accessed 21 November 2024).

3 https://www.researchgate.net/figure/Example-of-microgrid-transactive-energy-system-setup_fig1_336672142.

4 Ubidots. (2024) Here's how IoT data collection works [complete guide]. https://ubidots.com/blog/IoT-data-collection/ (accessed 21 November 2024).

5 https://www.researchgate.net/figure/TPA-System-Architecture-and-High-Level-Data-Flow_fig2_360671526.

6 Laszka, A., Dubey, A., Walker, M., and Schmidt, D. (2017). Providing privacy, safety, and security in IoT-based transactive energy systems using distributed ledgers. *Proceedings of the Seventh International Conference on the Internet of Things*. pp. 1–8.

7 Huang, Q., Amin, W., Umer, K. et al. (2021). A review of transactive energy systems: concept and implementation. *Energy Reports* 7: 7804–7824.

8 Bae, Y., Bhattacharya, S., Cui, B. et al. (2021). Sensor impacts on building and HVAC controls: a critical review for building energy performance. *Advances in Applied Energy* 4: 100068.

9 Li, Y., O'Neill, Z., and Niu, F. (2015). Evaluating control performance on building HVAC controllers. *International Building Performance Simulation Association* 962: 967.

10 Bengea, S.C., Li, P., Sarkar, S. et al. (2015). Fault-tolerant optimal control of a building HVAC system. *Science and Technology for the Built Environment* 21 (6): 734–751.

11 Drgoňa, J., Picard, D., and Helsen, L. (2020). Cloud-based implementation of white-box model predictive control for a GEOTABS office building: a field test demonstration. *Journal of Process Control* 88: 63–77.

12 Drgoňa, J., Picard, D., Kvasnica, M., and Helsen, L. (2018). Approximate model predictive building control via machine learning. *Applied Energy* 218: 199–216.

13 Domahidi, A., Ullmann, F., Morari, M., and Jones, C.N. (2014). Learning decision rules for energy efficient building control. *Journal of Process Control* 24 (6): 763–772.

14 Razmara, M., Bharati, G.R., Hanover, D. et al. (2017). Building-to-grid predictive power flow control for demand response and demand flexibility programs. *Applied Energy* 203: 128–141.

15 Hao, H., Wu, D., Lian, J., and Yang, T. (2017). Optimal coordination of building loads and energy storage for power grid and end user services. *IEEE Transactions on Smart Grid* 9 (5): 4335–4345.

16 Rotger-Griful, S., Chatzivasileiadis, S., Jacobsen, R.H. et al. (2016). Hardware-in-the-loop co-simulation of distribution grid for demand response. *2016 Power Systems Computation Conference (PSCC)*. pp. 1–7. IEEE.

17 Adetola, V., Lin, F., Yuan, S., and Reeve, H. (2018). Building flexibility estimation and control for grid ancillary services.

18 Yoon, A.Y., Kim, Y.J., Zakula, T., and Moon, S.I. (2020). Retail electricity pricing via online-learning of data-driven demand response of HVAC systems. *Applied Energy* 265: 114771.

19 Maasoumy, M. and Sangiovanni-Vincentelli, A.L. (2015). Buildings to grid integration: a dynamic contract approach. *ICCAD*. pp. 473–478.

20 Haack, J., Akyol, B., Allwardt, C. et al. (2016). VOLTTRON™: using distributed control and sensing to integrate buildings and the grid. *2016 IEEE 3rd World Forum on Internet of Things (WF-IoT)*. pp. 228–232. IEEE.

21 Wetter, M. (2004). *Simulation-based Building Energy Optimization*. Berkeley, CA: University of California.

22 Curtiss, P.S., Brandemuehl, M.J., and Kreider, J.F. (1994). Energy management in central HVAC plants using neural networks. *ASHRAE Transactions* 100 (1): 476–493.

23 Narayanan, S., Taylor, R., Yuan, S. et al. (2012). A wireless platform for energy efficient building control retrofits. Estcp ew-0938 final report.

24 Jung, W. and Jazizadeh, F. (2019). Human-in-the-loop HVAC operations: a quantitative review on occupancy, comfort, and energy-efficiency dimensions. *Applied Energy* 239: 1471–1508.

25 Von Neida, B., Maniccia, D., and Tweed, A. (2001). An analysis of the energy and cost savings potential of occupancy sensors for commercial lighting systems. *Journal of the Illuminating Engineering Society* 30 (2): 111–125.

26 Delaney, D.T., O'Hare, G.M., and Ruzzelli, A.G. (2009). Evaluation of energy-efficiency in lighting systems using sensor networks. *Proceedings of the First ACM Workshop on Embedded Sensing Systems for Energy-Efficiency in Buildings*. pp. 61–66.

27 Guo, X., Tiller, D.K., Henze, G.P., and Waters, C.E. (2010). The performance of occupancy-based lighting control systems: a review. *Lighting Research & Technology* 42 (4): 415–431.

28 Mikkilineni, A.K., Dong, J., Kuruganti, T., and Fugate, D. (2019). A novel occupancy detection solution using low-power IR-FPA based wireless occupancy sensor. *Energy and Buildings* 192: 63–74.

29 Labeodan, T., Zeiler, W., Boxem, G., and Zhao, Y. (2015). Occupancy measurement in commercial office buildings for demand-driven control applications—A survey and detection system evaluation. *Energy and Buildings* 93: 303–314.

30 Chen, Z., Jiang, C., and Xie, L. (2018). Building occupancy estimation and detection: a review. *Energy and Buildings* 169: 260–270.

31 Mirakhorli, A. and Dong, B. (2016). Occupancy behavior based model predictive control for building indoor climate—A critical review. *Energy and Buildings* 129: 499–513.

32 Erickson, V.L., Carreira-Perpiñán, M.Á., and Cerpa, A.E. (2011). OBSERVE: occupancy-based system for efficient reduction of HVAC energy. *Proceedings of the 10th ACM/IEEE International Conference on Information Processing in Sensor Networks*. pp. 258–269. IEEE.

33 Brackney, L.J., Florita, A.R., Swindler, A.C. et al. (2012). Design and performance of an image processing occupancy sensor. *Proceedings: The Second International Conference on Building Energy and Environment 2012987 Topic 10. Intelligent Buildings and Advanced Control Techniques*.

34 Minoli, D., Sohraby, K., and Occhiogrosso, B. (2017). IoT considerations, requirements, and architectures for smart buildings—Energy optimization and next-generation building management systems. *IEEE Internet of Things Journal* 4 (1): 269–283.

35 Pan, J., Jain, R., Paul, S. et al. (2015). An internet of things framework for smart energy in buildings: designs, prototype, and experiments. *IEEE Internet of Things Journal* 2 (6): 527–537.

36 Tran, D.H., Sanchez, E., and Nazari, M.H. (2019). Model predictive energy management for building microgrids with IoT-based controllable loads. *2019 North American Power Symposium (NAPS)*. pp. 1–6. IEEE.

37 Borggaard, J., Burns, J.A., Surana, A., and Zietsman, L. (2009). Control, estimation and optimization of energy efficient buildings. *2009 American Control Conference*. pp. 837–841. IEEE.

38 Smarra, F., Jain, A., De Rubeis, T. et al. (2018). Data-driven model predictive control using random forests for building energy optimization and climate control. *Applied Energy* 226: 1252–1272.

39 Awalin, L. and Khairil Rahmat, M. (2020). A recent development of monitoring devices on smart grid. *E3S Web of Conferences*.

40 Al-Kadhim, H.M. and Al-Raweshidy, H.S. (2019). Energy efficient and reliable transport of data in cloud-based IoT. *IEEE Access* 7: 64641–64650.

41 Eltamaly, A.M., Alotaibi, M.A., Alolah, A.I., and Ahmed, M.A. (2021). IoT-based hybrid renewable energy system for smart campus. *Sustainability* 13 (15): 8555.

42 Song, E.Y., FitzPatrick, G.J., and Lee, K.B. (2017). Smart sensors and standard-based interoperability in smart grids. *IEEE Sensors Journal* 17 (23): 7723–7730.

43 Munjin, D. and Morin, J.H. (2011). User empowerment in the internet of things. *arXiv preprint arXiv:1107.3759*.

44 Han, C., Jeong, Y., Ahn, J. et al. (2023). Recent advances in sensor–actuator hybrid soft systems: core advantages, intelligent applications, and future perspectives. *Advanced Science* 10 (35): 2302775.

45 Hawkes, E.W., Blumenschein, L.H., Greer, J.D., and Okamura, A.M. (2017). A soft robot that navigates its environment through growth. *Science Robotics* 2 (8): eaan3028.

46 Widergren, S.E. and Hammerstrom, D.J. (2023). Transactive energy communications interface standards landscape.

47 QualityLogic. (2024). Transactive control and its role in optimizing grid performance. https://www.qualitylogic.com/knowledge-center/what-is-transactive-control-part-1/ (accessed 21 November 2024).

48 Garcia, Y.V., Garzon, O., Delgado, C.J. et al. (2023). Overview on transactive energy—Advantages and challenges for weak power grids. *Energies* 16 (12): 4607.

49 IEEE. (2024). Transactive energy platform: trends, benefits, and challenges of grid modernization. https://blockchain.ieee.org/verticals/transactive-energy/topics/transactive-energy-platform-trends-benefits-and-challenges-of-grid-modernization (accessed 21 November 2024).

50 FirstPoint. (2021). Top 4 challenges in IoT data collection and management. https://www.firstpoint-mg.com/blog/top-4-challenges-in-IoT-data-collection-and-management/ (accessed 21 November 2024).

51 Gerbi Duguma, D., Zhang, J., Aboutalebi, M. et al. (2023). Privacy-preserving transactive energy systems: key topics and open research challenges. *arXiv e-prints*, arXiv-2312.

52 Cavoukian, A., Taylor, S., and Abrams, M.E. (2010). Privacy by design: essential for organizational accountability and strong business practices. *Identity in the Information Society* 3: 405–413.

53 Koops, B.J. and Leenes, R. (2014). Privacy regulation cannot be hardcoded. A critical comment on the 'privacy by design' provision in data-protection law. *International Review of Law, Computers & Technology* 28 (2): 159–171.

54 Sousa-Dias, D., Amyot, D., Rahimi-Kian, A., and Mylopoulos, J. (2023). A review of cybersecurity concerns for transactive energy markets. *Energies* 16 (13): 4838. https://www.mdpi.com/1996-1073/16/13/4838 (accessed 21 November 2024).

55 Peng, L., Feng, W., Yan, Z. et al. (2021). Privacy preservation in permissionless blockchain: a survey. *Digital Communications and Networks* 7 (3): 295–307.

56 Aitzhan, N.Z. and Svetinovic, D. (2016). Security and privacy in decentralized energy trading through multi-signatures, blockchain and anonymous messaging streams. *IEEE Transactions on Dependable and Secure Computing* 15 (5): 840–852.

57 Cai, N. and Mitra, J. (2010). A decentralized control architecture for a microgrid with power electronic interfaces. *North American Power Symposium 2010*. pp. 1–8. IEEE.

58 Saha, S., Ravi, N., Hreinsson, K. et al. (2021). A secure distributed ledger for transactive energy: the electron volt exchange (EVE) blockchain. *Applied Energy* 282: 116208.

59 Barreto, C., Eghtesad, T., Eisele, S. et al. (2020). Cyber-attacks and mitigation in blockchain based transactive energy systems. *2020 IEEE Conference on Industrial Cyberphysical Systems (ICPS)*. Vol. 1, pp. 129–136. IEEE.

60 Mbarek, B., Chren, S., Rossi, B., and Pitner, T. (2020). An enhanced blockchain-based data management scheme for microgrids. *Web, Artificial Intelligence and Network Applications: Proceedings of the Workshops of the 34th International Conference on Advanced Information Networking and Applications (WAINA-2020)*. pp. 766–775. Springer International Publishing.

61 Liu, Y., Ning, P., and Reiter, M.K. (2011). False data injection attacks against state estimation in electric power grids. *ACM Transactions on Information and System Security (TISSEC)* 14 (1): 1–33.

62 Lisovich, M.A., Mulligan, D.K., and Wicker, S.B. (2010). Inferring personal information from demand-response systems. *IEEE Security & Privacy* 8 (1): 11–20.

63 Shuaib, K., Abdella, J.A., Sallabi, F., and Abdel-Hafez, M. (2018). Using blockchains to secure distributed energy exchange. *2018 5th International Conference on Control, Decision and Information Technologies (CoDIT)*. pp. 622–627. IEEE.

64 Chandra, R., Radhakrishnan, K.K., and Panda, S.K. (2023). Privacy protected product differentiation through smart contracts based on bilateral negotiations in peer-to-peer transactive energy markets. *Sustainable Energy, Grids and Networks* 34: 100997.

65 Alqahtani, E. and Mustafa, M.A. (2023). Zone-based privacy-preserving billing for local energy market based on multiparty computation. *2023 IEEE International Conference on Communications, Control, and Computing Technologies for Smart Grids (SmartGridComm)*. pp. 1–7. IEEE.

66 Zhang, Y., Eisele, S., Dubey, A. et al. (2019). Cyber-physical simulation platform for security assessment of transactive energy systems. *2019 7th Workshop on Modeling and Simulation of Cyber-Physical Energy Systems (MSCPES)*. pp. 1–6. IEEE.

67 Li, Z., Bahramirad, S., Paaso, A. et al. (2019). Blockchain for decentralized transactive energy management system in networked microgrids. *The Electricity Journal* 32 (4): 58–72.

68 Gupta, N., Prusty, B.R., Alrumayh, O. et al. (2022). The role of transactive energy in the future energy industry: a critical review. *Energies* 15 (21): 8047. https://www.mdpi.com/1996-1073/15/21/8047 (accessed 10 December 2024).

6

IoT in Power Electronics: Transforming the Future of Energy Management

Dhandapani Lakshmi[1], *Rahiman Zahira*[2], *Vallikanu Pramila*[2], *Gunasekaran Ezhilarasi*[3], *Rajesh K. Padmashini*[1], *Palanisamy Sivaraman*[4], *and Chenniappan Sharmeela*[5]

[1] *Department of EEE, AMET Deemed to be University, Chennai, Tamil Nadu, India*
[2] *Department of Electrical and Electronics Engineering, B.S. Abdur Rahman Crescent Institute of Science and Technology, Chennai, Tamil Nadu, India*
[3] *Department of EEE, Sri Sairam Institute of Technology, Chennai, Tamil Nadu, India*
[4] *Research Scholar, Anna University, Chennai, Tamil Nadu, India*
[5] *Department of EEE, CEG, Anna University, Chennai, Tamil Nadu, India*

6.1 Introduction to IoT in Power Electronics

Power electronics involves the control and conversion of electrical power using semiconductor devices. It is a crucial field for numerous applications, including renewable energy systems, electric vehicles (EVs), industrial automation, and more. The advent of the Internet of Things (IoT) introduces advanced capabilities to these applications, enabling smarter and more efficient energy management solutions [1]. IoT refers to the network of physical devices embedded with sensors, software, and other technologies that connect and exchange data with other devices and systems over the internet. When applied to power electronics, IoT facilitates real-time monitoring, predictive maintenance, automation, and enhanced control of power systems [2].

6.1.1 Applications of IoT in Power Electronics

6.1.1.1 Smart Grid Management

Smart Grids are modern electricity networks that use digital technology to monitor and manage the transport of electricity from all generation sources to meet the varying electricity demands of end users [3]. IoT enhances smart grid management in several ways:

1) **Monitoring and Diagnostics:** IoT devices provide continuous monitoring of grid components, such as transformers, substations, and transmission lines. This real-time data collection allows for quick identification of issues, reducing downtime and maintenance costs.
2) **Load Balancing:** Smart sensors and actuators enable dynamic load balancing, ensuring optimal distribution of electricity across the grid. This helps prevent overloads and reduces energy losses.
3) **Fault Detection and Isolation:** IoT systems can detect faults in the grid, such as short circuits or line breaks, and isolate the affected sections to prevent widespread outages.

IoT for Smart Grid: Revolutionizing Electrical Engineering, First Edition.
Edited by Rahiman Zahira, Palanisamy Sivaraman, Chenniappan Sharmeela, and Sanjeevikumar Padmanaban.

4) **Demand Response:** IoT enables demand response programs, where electricity usage can be adjusted in real time based on supply conditions. This is particularly useful during peak demand periods or when integrating renewable energy sources [4].

6.1.1.2 Energy Management Systems (EMSs)

Energy management systems (EMSs) leverage IoT to optimize energy usage in various settings, from residential to industrial environments:

- **Home Automation:** Smart meters and connected home devices provide detailed insights into energy consumption patterns. Homeowners can monitor usage in real time, receive alerts about unusual consumption, and control devices remotely to save energy.
- **Industrial Automation:** In industrial settings, IoT-enabled EMS can monitor the energy consumption of machinery and equipment. By analyzing this data, companies can identify inefficiencies, optimize processes, and reduce operational costs [5].

6.1.1.3 Renewable Energy Integration

Renewable energy sources, such as solar and wind, are inherently variable and require sophisticated management to ensure reliable power supply [6, 7]. IoT plays a vital role in integrating these sources into the grid:

- **Solar Power Management:** IoT sensors can monitor the performance of solar panels, tracking parameters like temperature, sunlight exposure, and electrical output. This data helps in optimizing panel orientation, detecting faults, and scheduling maintenance.
- **Wind Power Management:** Similar to solar, IoT devices can monitor wind turbines, providing data on wind speed, turbine rotation, and power generation. This enables predictive maintenance and efficient operation.
- **Energy Storage Systems:** IoT can manage energy storage systems, such as batteries, by monitoring charge and discharge cycles, temperature, and overall health. This ensures optimal performance and longevity of the storage devices.

6.1.1.4 Predictive Maintenance

Predictive maintenance is one of the most significant benefits of IoT in power electronics [8]. By continuously monitoring the condition of equipment, IoT systems can predict when maintenance is needed, preventing unexpected failures and reducing downtime:

- **Condition Monitoring:** IoT sensors can track various parameters, such as temperature, vibration, and electrical characteristics of power electronic components. Analyzing this data helps in identifying potential issues before they lead to failure.
- **Operational Efficiency:** Timely maintenance based on predictive analytics extends the lifespan of equipment and improves operational efficiency. This reduces the total cost of ownership and enhances system reliability.

6.1.1.5 Electric Vehicle Infrastructure

The rise of EVs necessitates robust charging infrastructure [9]. IoT enhances the management of this infrastructure, ensuring efficient and user-friendly charging solutions:

- **Charging Stations:** IoT-enabled charging stations can optimize the charging process by monitoring the state of charge, managing power delivery, and providing real-time information to users. They can also integrate with the grid to balance demand and supply.

- **Fleet Management:** For commercial EV fleets, IoT systems can monitor the performance, location, and charging status of each vehicle. This ensures efficient fleet operation, minimizes downtime and reduces operational costs [10].

6.1.2 Benefits of IoT in Power Electronics

The integration of IoT in power electronics offers numerous benefits that enhance efficiency, reliability, and sustainability.

6.1.2.1 Improved Efficiency

Real-time data collection and analysis enable the optimization of power electronic systems, reducing energy consumption and operational costs. By identifying inefficiencies and adjusting operations accordingly, IoT helps achieve higher efficiency levels.

6.1.2.2 Enhanced Reliability

Predictive maintenance and continuous monitoring minimize the risk of unexpected failures, improving the reliability of power systems. This is particularly important in critical applications, such as power grids and industrial processes.

6.1.2.3 Scalability

IoT solutions can be easily scaled to manage large and complex power electronic systems. Whether it is a residential setup, an industrial plant, or a large power grid, IoT provides the flexibility to adapt and expand as needed.

6.1.2.4 Cost Savings

By optimizing energy usage and reducing maintenance costs, IoT in power electronics leads to significant financial savings. This is beneficial for both consumers and providers, improving the overall economics of energy management.

6.1.2.5 Sustainability

Efficient energy management and the integration of renewable energy sources contribute to sustainability goals. IoT helps reduce the carbon footprint of power systems, supporting global efforts to combat climate change [11].

6.1.3 Challenges of IoT in Power Electronics

Despite its numerous benefits, the integration of IoT in power electronics also presents several challenges.

6.1.3.1 Security Concerns

IoT devices can be vulnerable to cyberattacks, which could compromise the operation of power electronic systems and grids. Ensuring robust cybersecurity measures is essential to protecting these systems from potential threats [12].

6.1.3.2 Data Management

The vast amount of data generated by IoT devices requires robust data management and analytics solutions. Handling this data efficiently and extracting valuable insights can be challenging, particularly for large-scale implementations.

6.1.3.3 Interoperability

Ensuring that IoT devices from different manufacturers work seamlessly together can be a challenge. Standardization and interoperability protocols are necessary to enable smooth integration and operation of diverse IoT devices.

6.1.3.4 Initial Investment

The deployment of IoT systems in power electronics can require significant initial investment in hardware, software, and infrastructure. While the long-term benefits often justify these costs, the initial outlay can be a barrier for some organizations.

6.1.4 Future Prospects of IoT in Power Electronics

The future of IoT in power electronics looks promising, with several trends and advancements shaping the landscape [13].

6.1.4.1 Edge Computing

Edge computing involves processing data closer to the source, reducing latency and bandwidth requirements. In power electronics, edge computing can enable faster decision-making and more efficient data handling, particularly for real-time applications.

6.1.4.2 Artificial Intelligence and Machine Learning

Artificial intelligence (AI) and machine learning technologies are increasingly being integrated with IoT systems to enhance their capabilities. In power electronics, AI can improve predictive maintenance, optimize energy management, and enable more sophisticated control algorithms.

6.1.4.3 5G Connectivity

The rollout of 5G networks promises faster and more reliable connectivity for IoT devices. This will enhance the performance of IoT systems in power electronics, enabling real-time communication and more robust data transmission.

6.1.4.4 Enhanced Cybersecurity

As the importance of cybersecurity in IoT systems becomes more evident, advancements in security technologies and protocols are expected. This will help protect power electronic systems from cyber threats and ensure their reliable operation.

6.1.4.5 Integration with Blockchain

Blockchain technology can enhance the security and transparency of IoT systems in power electronics. It can provide a secure and immutable record of data transactions, enhancing trust and reliability in energy management processes. Figure 6.1 shows the ecosystem of the Industrial Internet of Things (IIoT).

6.1.5 Case Studies: IoT in Power Electronics

To illustrate the practical impact of IoT in power electronics, here are a few case studies.

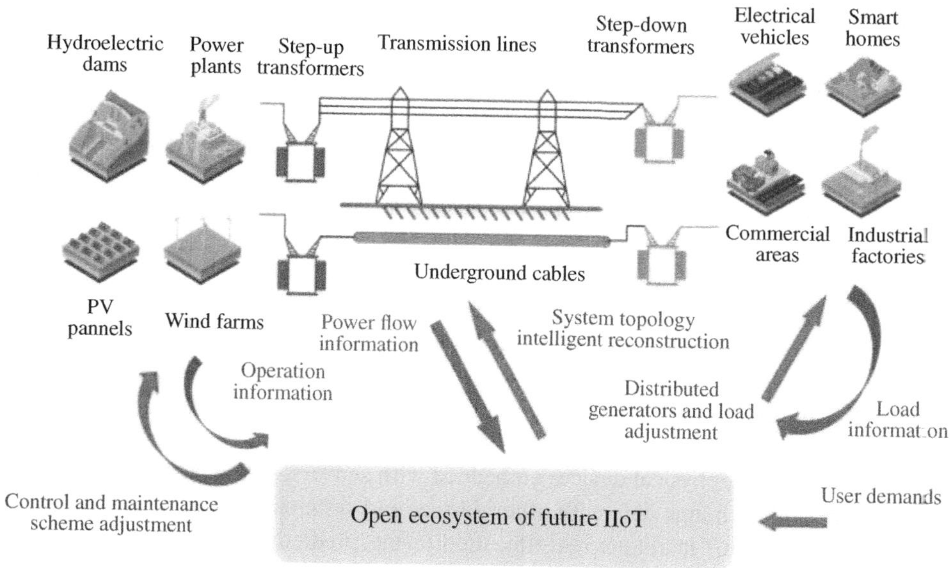

Figure 6.1 Future IIoT.

6.1.5.1 Smart Grid in Denmark

Denmark has implemented one of the most advanced smart grids in the world, leveraging IoT to enhance the management of its electricity network. IoT sensors and devices are used to monitor and control various aspects of the grid, from power generation to distribution. This has resulted in improved efficiency, reduced outages, and better integration of renewable energy sources [14].

6.1.5.2 Solar Power Monitoring in India

In India, IoT is being used to monitor and manage large-scale solar power installations. IoT sensors track the performance of solar panels, providing real-time data on energy production and panel health. This enables operators to optimize energy generation, perform predictive maintenance, and improve the overall efficiency of solar power plants.

6.1.5.3 Predictive Maintenance in Manufacturing

A major manufacturing company in the United States has implemented IoT-based predictive maintenance for its power electronic equipment. By continuously monitoring the condition of machines and analyzing the data with AI algorithms, the company has significantly reduced downtime and maintenance costs. This has led to improved productivity and lower operational expenses.

6.1.5.4 Electric Vehicle Charging Network in Europe

A European consortium has deployed an extensive network of IoT-enabled EV charging stations. These stations provide real-time information on availability, charging status, and pricing to users via a mobile app. The IoT system also optimizes the charging process, balancing the load on the grid and ensuring [15].

6.2 IoT in Power Conversion: Enhancing Efficiency and Reliability

Power conversion plays a critical role in modern electrical systems, enabling the efficient transformation of electrical energy from one form to another. As technology advances, the integration of the IoT into power conversion systems is transforming how these systems are monitored, controlled, and optimized. This comprehensive analysis explores the impact of IoT on power conversion, covering its applications, benefits, challenges, and future prospects [16].

6.2.1 Introduction to Power Conversion and IoT

Power conversion involves changing electrical power from one form to another, such as from alternating current (AC) to direct current (DC), DC to AC, or between different voltage levels. This process is essential in numerous applications, including renewable energy systems, EVs, industrial automation, and consumer electronics [17].

IoT refers to the network of physical devices embedded with sensors, software, and other technologies that connect and exchange data with other devices and systems over the internet. When applied to power conversion, IoT facilitates real-time monitoring, predictive maintenance, automation, and enhanced control of power conversion systems.

6.2.2 Applications of IoT in Power Conversion

6.2.2.1 Renewable Energy Systems

Renewable energy sources, such as solar and wind power, are variable and require efficient power conversion systems to integrate them into the grid or to store the generated energy [18]. IoT enhances the performance and reliability of these systems:

- **Solar Power Conversion:** IoT sensors monitor the performance of photovoltaic (PV) panels and inverters, providing data on parameters such as temperature, sunlight exposure, and electrical output. This data helps optimize panel orientation, detect faults, and schedule maintenance.
- **Wind Power Conversion:** IoT devices track the performance of wind turbines and their associated power conversion systems. By monitoring wind speed, turbine rotation, and power output, operators can optimize turbine operation and perform predictive maintenance.
- **Energy Storage Systems:** IoT enables efficient management of energy storage systems, such as batteries, by monitoring charge and discharge cycles, temperature, and overall health. This ensures optimal performance and longevity of the storage devices.

6.2.2.2 Electric Vehicles (EVs)

The growth of EVs relies on efficient power conversion systems for charging and propulsion [19]. IoT enhances the management and operation of these systems:

- **Battery Management Systems (BMS):** IoT sensors monitor the state of charge, temperature, and health of EV batteries. This data is used to optimize charging and discharging processes, extending battery life and improving safety.
- **Charging Infrastructure:** IoT-enabled charging stations provide real-time information on availability, charging status, and pricing to users. They also manage power delivery, balancing the load on the grid, and ensuring efficient use of available resources.

6.2.2.3 Industrial Automation

In industrial settings, power conversion systems are critical for powering machinery and equipment [20]. IoT improves the efficiency and reliability of these systems:

- **Motor Drives:** IoT sensors monitor the performance of motor drives, tracking parameters such as voltage, current, and temperature. This data helps optimize motor operation, reducing energy consumption and preventing failures.
- **Uninterruptible Power Supplies (UPS):** IoT-enabled UPS systems provide real-time monitoring and diagnostics, ensuring continuous power supply to critical equipment. Predictive maintenance based on IoT data helps prevent unexpected downtime.

6.2.2.4 Consumer Electronics

Power conversion is essential in consumer electronics, from charging devices to powering appliances. IoT enhances the user experience and efficiency of these systems:

- **Smart Chargers:** IoT-enabled chargers for smartphones, laptops, and other devices can optimize charging processes based on the device's battery condition and user preferences. This extends battery life and reduces energy consumption.
- **Home Automation:** IoT devices in smart homes monitor and control power conversion systems for various appliances, optimizing energy usage and providing users with detailed insights into their consumption patterns.

6.2.3 Benefits of IoT in Power Conversion

The integration of IoT in power conversion offers numerous benefits that enhance efficiency, reliability, and sustainability.

6.2.3.1 Improved Efficiency

Real-time data collection and analysis enable the optimization of power conversion systems, reducing energy losses and operational costs. By identifying inefficiencies and adjusting operations accordingly, IoT helps achieve higher efficiency levels.

6.2.3.2 Enhanced Reliability

Predictive maintenance and continuous monitoring minimize the risk of unexpected failures, improving the reliability of power conversion systems. This is particularly important in critical applications, such as renewable energy systems and industrial processes.

6.2.3.3 Scalability

IoT solutions can be easily scaled to manage large and complex power conversion systems. Whether it's a residential setup, an industrial plant, or a large renewable energy farm, IoT provides the flexibility to adapt and expand as needed.

6.2.3.4 Cost Savings

By optimizing energy usage and reducing maintenance costs, IoT in power conversion leads to significant financial savings. This is beneficial for both consumers and providers, improving the overall economics of energy management.

6.2.3.5 Sustainability

Efficient energy management and the integration of renewable energy sources contribute to sustainability goals. IoT helps reduce the carbon footprint of power conversion systems, supporting global efforts to combat climate change.

6.2.4 Challenges of IoT in Power Conversion

Despite its numerous benefits, the integration of IoT in power conversion also presents several challenges [5, 21].

6.2.4.1 Security Concerns

IoT devices can be vulnerable to cyberattacks, which could compromise the operation of power conversion systems. Ensuring robust cybersecurity measures is essential to protect these systems from potential threats.

6.2.4.2 Data Management

The vast amount of data generated by IoT devices requires robust data management and analytics solutions. Handling this data efficiently and extracting valuable insights can be challenging, particularly for large-scale implementations.

6.2.4.3 Interoperability

Ensuring that IoT devices from different manufacturers work seamlessly together can be a challenge. Standardization and interoperability protocols are necessary to enable smooth integration and operation of diverse IoT devices.

6.2.4.4 Initial Investment

The deployment of IoT systems in power conversion can require significant initial investment in hardware, software, and infrastructure. While the long-term benefits often justify these costs, the initial outlay can be a barrier for some organizations.

The IIoT is transforming electrical applications by enabling real-time monitoring, predictive maintenance, automation, and advanced analytics. Key components of IIoT systems, including sensors, communication networks, edge devices, cloud platforms, and advanced analytics, work together to enhance the efficiency, reliability, and sustainability of electrical systems. While challenges such as security and interoperability remain, the benefits of IIoT in electrical applications are substantial, driving continued innovation and improvement in the industry [22]. As technology evolves, the future of IIoT in electrical applications looks promising, with emerging trends and advancements poised to further enhance its capabilities and impact. The components of IIoT are shown in Figure 6.2.

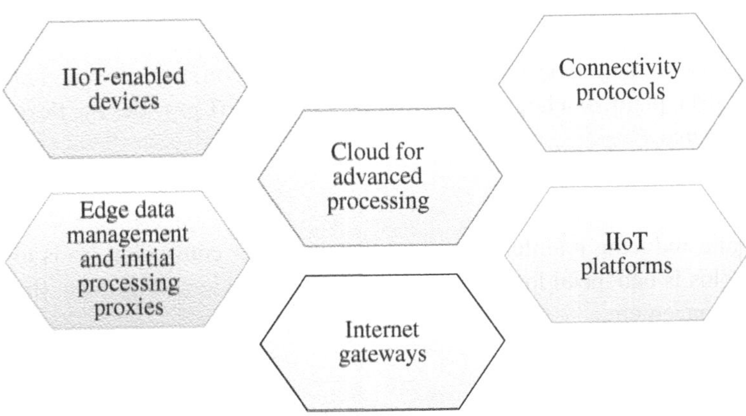

Figure 6.2 Components of IIoT.

6.3 Introduction to IIoT-Driven Automation

IIoT-driven automation refers to the use of interconnected devices, sensors, and systems to automate industrial processes, monitor performance in real time, and enable data-driven decision-making. By leveraging advanced data analytics and connectivity, IIoT enables seamless communication between machines, humans, and enterprise systems, transforming traditional manufacturing and production into smart, connected ecosystems [23].

6.3.1 Components of IIoT-Driven Automation

IIoT-driven automation integrates various components to streamline operations, improve efficiency, and optimize resource utilization across industrial processes.

6.3.1.1 Sensors and Actuators

Sensors and actuators are pivotal components that enable the collection of real-time data from physical assets and the execution of automated actions based on that data.

- **Sensors:** Monitor parameters such as temperature, pressure, humidity, vibration, and flow rates in industrial equipment and environments.
- **Actuators:** Control and manipulate physical processes based on commands received from IIoT systems, optimizing operational parameters for efficiency and productivity.

6.3.1.2 Communication Networks

Communication networks facilitate the seamless exchange of data between devices, sensors, control systems, and cloud platforms, ensuring timely decision-making and response.

- **Wired Networks:** Utilize Ethernet, fieldbus protocols (e.g., Profibus, Modbus), and industrial Ethernet standards (e.g., PROFINET, EtherNet/IP) for reliable and high-speed data transmission within factory floors.
- **Wireless Networks:** Include technologies such as Wi-Fi, Bluetooth, Zigbee, and cellular (4G/5G) for flexible and mobile connectivity across distributed industrial environments.

6.3.1.3 Edge Computing

Edge computing brings data processing and analysis closer to the source of data generation, reducing latency and enabling real-time decision-making at the edge of the network [24].

- **Edge Devices:** Perform local data aggregation, preprocessing, and analysis to filter out noncritical data and reduce bandwidth usage.
- **Edge Gateways:** Facilitate secure connectivity between edge devices and central cloud platforms, enhancing scalability and reliability in IIoT deployments.

6.3.1.4 Cloud Platforms

Cloud platforms serve as centralized repositories for storing, managing, and analyzing large volumes of data generated by IIoT devices and systems [25].

- **Data Storage:** Store historical and real-time data for long-term analysis, compliance reporting, and predictive maintenance.
- **Data Analytics:** Employ machine learning algorithms, AI, and predictive analytics to derive actionable insights, optimize processes, and improve operational efficiency.

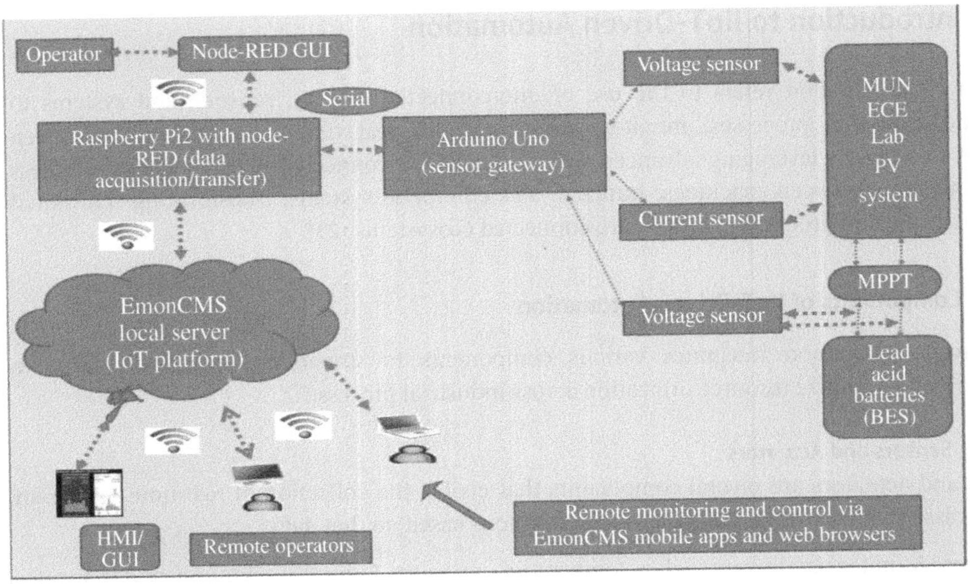

Figure 6.3 ICS and SCADA operation.

6.3.1.5 Industrial Control Systems (ICS) and Supervisory Control and Data Acquisition (SCADA)

ICS and SCADA systems provide centralized control, monitoring, and automation of industrial processes, ensuring seamless integration with IIoT technologies, which is shown in Figure 6.3.

- **SCADA Systems:** Monitor and control industrial processes through graphical user interfaces (GUIs), enabling operators to visualize real-time data and manage operational parameters.
- **Programmable Logic Controllers (PLCs):** Automate control processes and execute predefined logic based on input from sensors and IIoT systems, optimizing workflow efficiency and response times.

6.4 Future Prospects of IoT in Power Conversion

The future of IoT in power conversion looks promising, with several trends and advancements shaping the landscape.

6.4.1 Edge Computing

Edge computing involves processing data closer to the source, reducing latency and bandwidth requirements. In power conversion, edge computing can enable faster decision-making and more efficient data handling, particularly for real-time applications.

6.4.2 Artificial Intelligence and Machine Learning

AI and machine learning technologies are increasingly being integrated with IoT systems to enhance their capabilities. In power conversion, AI can improve predictive maintenance, optimize energy management, and enable more sophisticated control algorithms.

6.4.3 5G Connectivity

The rollout of 5G networks promises faster and more reliable connectivity for IoT devices. This will enhance the performance of IoT systems in power conversion, enabling real-time communication and more robust data transmission.

6.4.4 Enhanced Cybersecurity

As the importance of cybersecurity in IoT systems becomes more evident, advancements in security technologies and protocols are expected. This will help protect power conversion systems from cyber threats and ensure their reliable operation.

6.4.5 Integration with Blockchain

Blockchain technology can enhance the security and transparency of IoT systems in power conversion. It can provide a secure and immutable record of data transactions, enhancing trust and reliability in energy management processes. The blockchain integration is shown in Figure 6.4.

6.4.6 Case Studies: IoT in Power Conversion

To illustrate the practical impact of IoT in power conversion, here are a few case studies.

6.4.6.1 Solar Power Conversion in Australia

In Australia, a large solar power plant has implemented IoT to monitor and manage its power conversion systems. IoT sensors track the performance of PV panels and inverters, providing real-time data on energy production and system health. This has resulted in improved efficiency, reduced maintenance costs, and increased reliability [26].

6.4.6.2 Electric Vehicle Charging Network in California

A network of IoT-enabled EV charging stations has been deployed across California. These stations provide real-time information on availability, charging status, and pricing to users via a mobile app.

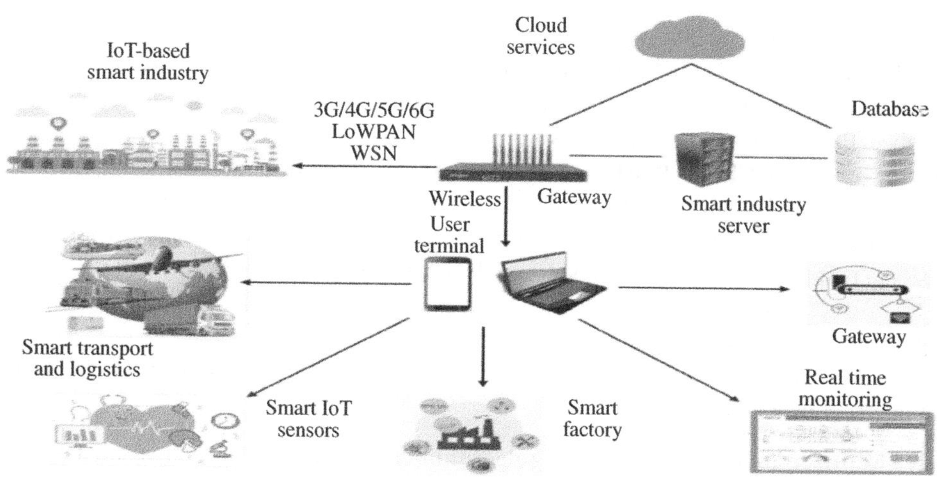

Figure 6.4 Block chain integration.

The IoT system also optimizes the charging process, balancing the load on the grid and ensuring efficient use of available resources.

6.4.6.3 Industrial Motor Drives in Germany

A major manufacturing plant in Germany has integrated IoT with its motor drives to enhance performance and reliability. IoT sensors monitor the condition of the motor drives, providing data on voltage, current, and temperature. This data is used for predictive maintenance, reducing downtime and improving operational efficiency.

6.4.6.4 Smart Home Energy Management in Japan

In Japan, a smart home project has implemented IoT to optimize energy usage in residential buildings. IoT devices monitor and control power conversion systems for various appliances, providing homeowners with detailed insights into their consumption patterns. This has led to significant energy savings and improved user satisfaction.

6.4.7 Technical Aspects of IoT in Power Conversion

To fully understand the integration of IoT in power conversion, it is essential to delve into the technical aspects. These include the components of IoT systems, data communication protocols, and the role of advanced analytics [27].

6.4.7.1 Components of IoT Systems in Power Conversion

IoT systems in power conversion typically comprise several key components:

- **Sensors:** These devices measure various parameters such as voltage, current, temperature, and vibration. They are essential for monitoring the performance and health of power conversion systems.
- **Actuators:** These devices execute control actions based on data from sensors and analytics. In power conversion, actuators can adjust settings on inverters, motor drives, and other equipment to optimize performance.
- **Edge Devices:** These are intermediary devices that aggregate and preprocess data from sensors before sending it to the cloud or central servers. They play a critical role in reducing latency and bandwidth requirements.
- **Communication Networks:** IoT systems rely on robust communication networks to transmit data between sensors, actuators, edge devices, and central servers. Common protocols include Wi-Fi, Zigbee, Bluetooth, and cellular networks.
- **Central Servers and Cloud Platforms:** These systems store and process data, running advanced analytics and machine learning algorithms to extract insights and make decisions.
- **User Interfaces:** These can be dashboards, mobile apps, or other interfaces that provide users with access to real-time data, analytics results, and control functions.

6.4.7.2 Advanced Analytics and Machine Learning

Advanced analytics and machine learning play a crucial role in extracting valuable insights from IoT data in power conversion:

- **Predictive Maintenance:** Machine learning algorithms analyze historical and real-time data to predict when equipment is likely to fail. This enables timely maintenance, reducing downtime and maintenance costs.

- **Energy Optimization:** Advanced analytics can identify patterns and correlations in energy usage data, providing recommendations for optimizing power conversion processes and reducing energy consumption.
- **Anomaly Detection:** Machine learning models can detect anomalies in data that may indicate faults or inefficiencies in power conversion systems. This enables proactive measures to address issues before they escalate.
- **Control Optimization:** AI algorithms can optimize control strategies for power conversion systems, adjusting parameters in real time to maximize efficiency and performance.

6.5 Regulatory and Standardization Considerations

The deployment of IoT in power conversion must consider regulatory and standardization aspects to ensure safety, reliability, and interoperability.

6.5.1 Safety Standards

Power conversion systems must comply with safety standards to protect users and equipment. Regulatory bodies such as the International Electrotechnical Commission (IEC) and the Institute of Electrical and Electronics Engineers (IEEE) provide guidelines for the safe design and operation of power conversion systems [28].

6.5.2 Data Privacy and Security Regulations

IoT systems in power conversion generate and transmit large amounts of data, raising concerns about data privacy and security. Regulations such as the General Data Protection Regulation (GDPR) in Europe and the California Consumer Privacy Act (CCPA) in the United States provide guidelines for protecting personal data.

6.5.3 Interoperability Standards

Ensuring interoperability between IoT devices from different manufacturers is essential for seamless integration and operation. Standards such as the Open Connectivity Foundation (OCF) and the Industrial Internet Consortium (IIC) promote interoperability and provide guidelines for developing IoT solutions.

6.6 IoT in Power Transmission for Long Distance

6.6.1 Introduction to Long-Distance Power Transmission and IoT

Long-distance power transmission involves the transfer of electrical energy from generation sources, often located in remote areas, to consumption centers. High-voltage transmission lines, transformers, and substations are key components of these systems. Efficient and reliable transmission is essential to ensure a stable power supply and minimize energy losses.

IoT refers to the network of physical devices embedded with sensors, software, and other technologies that connect and exchange data with other devices and systems over the internet.

When applied to power transmission, IoT facilitates real-time monitoring, predictive maintenance, automation, and enhanced control of transmission systems [29].

6.6.2 Applications of IoT in Long-Distance Power Transmission

6.6.2.1 Real-Time Monitoring and Diagnostics

Real-time monitoring of transmission lines and equipment is crucial for maintaining the reliability and efficiency of power transmission systems. IoT enhances monitoring and diagnostics in several ways:

- **Transmission Line Monitoring:** IoT sensors can monitor various parameters of transmission lines, such as voltage, current, temperature, and sag. This real-time data helps detect abnormalities and potential faults before they lead to failures.
- **Substation Monitoring:** IoT devices can continuously monitor the performance of substation components, including transformers, circuit breakers, and switchgear. This enables early detection of issues and reduces the risk of equipment failure.
- **Environmental Monitoring:** IoT sensors can track environmental conditions, such as weather, temperature, and humidity, that may impact the performance of transmission lines. This data helps operators anticipate and mitigate adverse effects.

6.6.2.2 Predictive Maintenance

Predictive maintenance is one of the most significant benefits of IoT in power transmission [30]. By continuously monitoring the condition of equipment, IoT systems can predict when maintenance is needed, preventing unexpected failures and reducing downtime:

- **Condition-Based Maintenance:** IoT sensors track various parameters, such as temperature, vibration, and electrical characteristics of transmission equipment. Analyzing this data helps in identifying potential issues before they lead to failure.
- **Maintenance Scheduling:** Timely maintenance based on predictive analytics extends the lifespan of equipment and improves operational efficiency. This reduces the total cost of ownership and enhances system reliability.

6.6.2.3 Fault Detection and Isolation

Faults in transmission systems can lead to widespread outages and significant economic losses. IoT enhances fault detection and isolation, minimizing the impact of faults:

- **Fault Detection:** IoT systems can detect faults in transmission lines and equipment, such as short circuits, ground faults, and line breaks. Real-time data from sensors enables rapid identification of fault locations.
- **Fault Isolation:** Once a fault is detected, IoT devices can automatically isolate the affected section of the transmission system. This prevents the fault from propagating and causing further damage, improving the overall stability of the grid.

6.6.2.4 Load Balancing and Optimization

Efficient load balancing is essential for maintaining the stability and reliability of long-distance power transmission systems. IoT enables dynamic load balancing and optimization:

- **Load Monitoring:** IoT sensors can monitor the load on transmission lines in real-time, providing data on power flow and demand. This helps operators balance the load and prevent overloads.
- **Dynamic Load Management:** IoT systems can dynamically adjust the load on transmission lines by controlling the flow of electricity. This ensures optimal distribution of power and reduces energy losses.
- **Renewable Integration:** IoT facilitates the integration of renewable energy sources, such as wind and solar, into the transmission grid. By monitoring the variable output of renewables, IoT systems can adjust the flow of electricity to maintain grid stability.

6.6.2.5 Asset Management

Effective asset management is crucial for the reliability and longevity of transmission systems. IoT enhances asset management by providing detailed insights into the condition and performance of transmission assets:

- **Asset Tracking:** IoT devices can track the location and status of transmission assets, such as transformers, towers, and conductors. This data helps operators manage assets more effectively and plan maintenance activities.
- **Lifecycle Management:** By monitoring the condition and performance of transmission assets, IoT systems can optimize their lifecycle. This includes scheduling replacements, upgrades, and retirements based on data-driven insights.

6.6.3 Benefits of IoT in Long-Distance Power Transmission

The integration of IoT in long-distance power transmission offers numerous benefits that enhance efficiency, reliability, and sustainability:

6.6.3.1 Improved Efficiency

Real-time data collection and analysis enable the optimization of transmission systems, reducing energy losses and operational costs. By identifying inefficiencies and adjusting operations accordingly, IoT helps achieve higher efficiency levels.

6.6.3.2 Enhanced Reliability

Predictive maintenance and continuous monitoring minimize the risk of unexpected failures, improving the reliability of transmission systems. This is particularly important in critical applications, such as power grids and industrial processes.

6.6.3.3 Scalability

IoT solutions can be easily scaled to manage large and complex transmission systems. Whether it is a regional grid or a national transmission network, IoT provides the flexibility to adapt and expand as needed.

6.6.3.4 Cost Savings

By optimizing energy usage and reducing maintenance costs, IoT in power transmission leads to significant financial savings. This is beneficial for both consumers and providers, improving the overall economics of energy management.

6.6.3.5 Sustainability

Efficient energy management and the integration of renewable energy sources contribute to sustainability goals. IoT helps reduce the carbon footprint of transmission systems, supporting global efforts to combat climate change.

6.6.4 Challenges of IoT in Long-Distance Power Transmission

Despite its numerous benefits, the integration of IoT in power transmission also presents several challenges.

6.6.4.1 Security Concerns

IoT devices can be vulnerable to cyber-attacks, which could compromise the operation of transmission systems. Ensuring robust cybersecurity measures is essential to protecting these systems from potential threats.

6.6.4.2 Data Management

The vast amount of data generated by IoT devices requires robust data management and analytics solutions. Handling this data efficiently and extracting valuable insights can be challenging, particularly for large-scale implementations.

6.6.4.3 Interoperability

Ensuring that IoT devices from different manufacturers work seamlessly together can be a challenge. Standardization and interoperability protocols are necessary to enable smooth integration and operation of diverse IoT devices.

6.6.4.4 Initial Investment

The deployment of IoT systems in power transmission can require significant initial investment in hardware, software, and infrastructure. While the long-term benefits often justify these costs, the initial outlay can be a barrier for some organizations.

6.6.5 Future Prospects of IoT in Long-Distance Power Transmission

The future of IoT in long-distance power transmission looks promising, with several trends and advancements shaping the landscape.

6.6.5.1 Edge Computing

Edge computing involves processing data closer to the source, reducing latency and bandwidth requirements. In power transmission, edge computing can enable faster decision-making and more efficient data handling, particularly for real-time applications.

6.6.5.2 Artificial Intelligence and Machine Learning

AI and machine learning technologies are increasingly being integrated with IoT systems to enhance their capabilities. In power transmission, AI can improve predictive maintenance, optimize energy management, and enable more sophisticated control algorithms.

6.6.5.3 5G Connectivity

The rollout of 5G networks promises faster and more reliable connectivity for IoT devices. This will enhance the performance of IoT systems in power transmission, enabling real-time communication and more robust data transmission.

6.6.5.4 Enhanced Cybersecurity

As the importance of cybersecurity in IoT systems becomes more evident, advancements in security technologies and protocols are expected. This will help protect transmission systems from cyber threats and ensure their reliable operation.

6.6.5.5 Integration with Blockchain

Blockchain technology can enhance the security and transparency of IoT systems in power transmission. It can provide a secure and immutable record of data transactions, enhancing trust and reliability in energy management processes.

6.6.6 Case Studies: IoT in Long-Distance Power Transmission

To illustrate the practical impact of IoT in long-distance power transmission, here are a few case studies.

6.6.6.1 Smart Grid in the United States

A major utility company in the United States has implemented a smart grid project that leverages IoT to enhance the management of its transmission network. IoT sensors and devices are used to monitor transmission lines, substations, and transformers. This has resulted in improved efficiency, reduced outages, and better integration of renewable energy sources.

6.6.6.2 Wind Power Transmission in Europe

In Europe, a large wind power project has integrated IoT to monitor and manage the transmission of electricity from offshore wind farms to the mainland grid. IoT sensors track the performance of transmission lines and transformers, providing real-time data on power flow and system health. This has led to increased reliability and optimized power transmission.

6.6.6.3 High-Voltage Direct Current (HVDC) Transmission in China

China has deployed several HVDC transmission projects that use IoT to monitor and control the long-distance transmission of electricity. IoT devices provide real-time data on voltage, current, and temperature, enabling operators to optimize the performance of HVDC systems. This has resulted in improved efficiency and reduced energy losses.

6.7 Conclusion

The adoption of the IIoT in electrical applications marks a transformative shift toward smarter, more efficient, and reliable systems. By leveraging advanced sensors, robust communication networks, and powerful data analytics, IIoT enables real-time monitoring, predictive maintenance, and enhanced power management. This integration not only optimizes performance but also reduces downtime and ensures compliance with regulatory standards. As IIoT technology continues to evolve, its application in the electrical industry will play a critical role in advancing industrial automation and achieving sustainable energy management. Effective cybersecurity controls also guarantee data integrity and security against attacks. IIoT is going to play an important part in the electrical industry as it develops, contributing to promote sustainable energy practices, increase automation, and encourage innovation – all of which will help the sector become smarter and more robust in the long run.

References

1 Farhangi, H. (2010). The path of the smart grid. *IEEE Power and Energy Magazine* 8 (1): 18–28. https://doi.org/10.1109/MPE.2009.934876.

2 Raza, U., Kulkarni, P., Sooriyabandara, M., and Palade, A. (2017). A review of the role of smart grids in the development of sustainable energy. *Renewable and Sustainable Energy Reviews* 68: 888–898. https://doi.org/10.1016/j.rser.2016.09.105.

3 Ahmad, M.S. and Khan, Z.A. (2017). A review on the role of IoT in energy conservation and management of smart grid. *Renewable and Sustainable Energy Reviews* 72: 734–745. https://doi.org/10.1016/j.rser.2017.01.076.

4 Farhangi, H. (2011). The path of the smart grid: opportunities for Europe. *Energy Policy* 39 (12): 7148–7159. https://doi.org/10.1016/j.enpol.2011.01.018.

5 Gungor, V.C., Sahin, D., Kocak, T. et al. (2011). Smart grid technologies: communication technologies and standards. *IEEE Transactions on Industrial Informatics* 7 (4): 529–539. https://doi.org/10.1109/TII.2011.2167659.

6 Yasin, M.M. and Habib, A. (2015). Smart grid communication infrastructure: a comprehensive survey. *IEEE Communications Surveys & Tutorials* 17 (2): 944–980. https://doi.org/10.1109/COMST.2014.2360902.

7 Li, F. and Hu, J. (2017). A survey on communication architectures in smart grid. *IEEE Access* 5: 9533–9545. https://doi.org/10.1109/ACCESS.2017.2709780.

8 Lu, D., Li, X., Ota, K., and Dong, M. (2017). A survey on the edge computing for the Internet of Things. *IEEE Access* 5: 637–646. https://doi.org/10.1109/ACCESS.2017.2653465.

9 Lopez, D., Parsaei, H., and Zakarian, A. (2017). Smart grid technologies: modeling and control. *IEEE Transactions on Industrial Informatics* 13 (4): 2045–2054. https://doi.org/10.1109/TII.2017.2703470.

10 Aazam, M., Huh, E.N., and Foo, C.Y. (2014). Smart grid communication infrastructure: review of recent advances and future challenges. *IEEE Access* 2: 733–748. https://doi.org/10.1109/ACCESS.2014.2346059.

11 Iqbal, M.M. and Hassan, S.A. (2015). Smart grid communications and networking: technologies and solutions. *IEEE Transactions on Industrial Informatics* 11 (5): 1056–1064. https://doi.org/10.1109/TII.2015.2451565.

12 Wang, P., Zhang, Y., Wang, J., and Xie, C. (2018). A survey on the communication architectures in smart grid. *Journal of Communications and Networks* 20 (6): 598–607. https://doi.org/10.1109/JCN.2018.000119.

13 Hancke, G.P. and Silva, B.J. (2017). The role of the Internet of Things in smart grids. *Journal of Network and Computer Applications* 86: 35–48. https://doi.org/10.1016/j.jnca.2017.03.013.

14 Zhang, Y., Yuan, Y., and Wang, J. (2018). Security and privacy in smart grid: challenges and solutions. *IEEE Access* 6: 7597–7610. https://doi.org/10.1109/ACCESS.2018.2792103.

15 Sun, H., Zhang, J., and Li, J. (2017). Smart grid: vision, opportunities, and challenges. *IEEE Transactions on Industrial Informatics* 13 (4): 1376–1384. https://doi.org/10.1109/TII.2017.2702178.

16 Mahmud, M.A., Zareei, M., Mahmud, M. et al. (2019). A comprehensive review on demand response programs in smart grid: pricing strategies, technological advances, and energy saving assessments. *Sustainable Cities and Society* 45: 607–630. https://doi.org/10.1016/j.scs.2018.11.027.

17 Kim, J.H. and Cho, Y.H. (2019). Demand response framework based on blockchain for smart grid. *Sustainability* 11 (16): 4472. https://doi.org/10.3390/su11164472.

18 Piroozfard, H., Kiah, M.L.M., and Zaidan, A.A. (2019). A review on IoT-based energy-efficient intelligent buildings using machine learning and deep learning paradigms. *Sustainable Cities and Society* 50: 101620. https://doi.org/10.1016/j.scs.2019.101620.

19 Akram, M.U., Javaid, N., Yousaf, S.Z., and Khan, Z.A. (2019). A comprehensive review on the role of IoT in healthcare and smart grid. *Computer Networks* 159: 61–80. https://doi.org/10.1016/j.comnet.2019.04.003.

20 Guerrero, J.M., Vasquez, J.C., Matas, J. et al. (2013). Hierarchical control of droop-controlled AC and DC microgrids—a general approach toward standardization. *IEEE Transactions on Industrial Electronics* 60 (4): 1254–1262. https://doi.org/10.1109/TIE.2012.2199329.

21 Liu, Y., Xiao, L., and Li, L. (2017). Internet of Things in industries: a survey. *IEEE Transactions on Industrial Informatics* 10 (4): 2233–2243. https://doi.org/10.1109/TII.2017.2667462.

22 Gungor, V.C., Sahin, D., Kocak, T. et al. (2013). A survey on smart grid potential applications and communication requirements. *IEEE Transactions on Industrial Informatics* 9 (1): 28–42. https://doi.org/10.1109/TII.2012.2218253.

23 Mahmoud, M.S., Mohamed, A., and El-Saadany, E.F. (2018). A comprehensive review of the applications of blockchain technology in smart grids and buildings. *IEEE Transactions on Industrial Informatics* 15 (5): 2823–2833. https://doi.org/10.1109/TII.2018.2828259.

24 Farris, I., Nicolosi, G., and Rossi, M. (2017). IoT-aided monitoring and energy management for sustainability in smart cities. *IEEE Transactions on Industrial Informatics* 13 (2): 813–820. https://doi.org/10.1109/TII.2016.2648423.

25 Li, K., Zhang, Y., Hui, P. et al. (2016). A survey on the edge computing for the Internet of Things. *IEEE Access* 5: 171–190. https://doi.org/10.1109/ACCESS.2016.2626890.

26 You, C., Zhang, K., Chae, H., and Kim, J. (2017). Energy internet: the architecture and the road ahead. *IEEE Communications Magazine* 55 (1): 62–68. https://doi.org/10.1109/MCOM.2017.1600514.

27 Lin, H., Huang, C., and Chuang, H. (2018). A novel energy management system with hierarchical architecture for smart home energy networks. *IEEE Transactions on Industrial Informatics* 14 (12): 5346–5356. https://doi.org/10.1109/TII.2018.2806215.

28 Liu, M., Zhou, Y., Wang, Z. et al. (2019). Blockchain for secure and efficient data sharing in vehicular edge computing and networks. *IEEE Transactions on Industrial Informatics* 15 (6): 3680–3689. https://doi.org/10.1109/TII.2018.2878342.

29 Talari, S., Datta, S., Yamada, S., and Krishnamurthy, V. (2017). Towards scalable and reliable IOT-based smart grid communications using SDN for smart cities. *IEEE Transactions on Sustainable Computing* 2 (3): 278–290. https://doi.org/10.1109/TSUSC.2017.2694765.

30 Le, T. and Rao, M.V.C. (2019). A review of state-of-the-art energy storage systems for smart grid applications: integration, challenges, and grid impact. *IEEE Transactions on Industrial Informatics* 15 (8): 4749–4766. https://doi.org/10.1109/TII.2018.2878101.

7

Harnessing IoT: Transforming Smart Grid Advancements

Pijush K. Dutta Pramanik[1], Bijoy K. Upadhyaya[2], Ajay Kushwaha[3], and Debashish Bhowmik[4]

[1]*School of Computer Applications and Technology, Galgotias University, Greater Noida, Uttar Pradesh, India*
[2]*Department of Electronics and Communication Engineering, Tripura Institute of Technology, Narsingarh, Tripura, India*
[3]*Electrical and Instrumentation Engineering Department, Thapar Institute of Engineering and Technology, Patiala, Punjab, India*
[4]*Electrical Engineering Department, Mizoram University, Aizawl, Mizoram, India*

7.1 Introduction to Smart Grid and IoT Integration

The integration of the smart grid and the Internet of Things (IoT) has emerged as a transformative force in the energy sector, redefining the way utilities and consumers interact with the power grid. This introductory section lays the groundwork for understanding the core principles and technologies that underpin this convergence. Beginning with the fundamental concepts of IoT and the core tenets of the smart grid, this section delves into the seamless integration of these two paradigms, exploring how sensor networks, data analytics, remote monitoring and control, and smart devices come together to enable unprecedented visibility, responsiveness, and optimization across the power infrastructure. Integrating remote monitoring and control capabilities further empowers utilities to monitor the grid's status and dynamically adjust operations from a centralized control center. Smart devices and appliances, on the other hand, allow for two-way communication and active participation of consumers in demand-side management (DSM), fostering a more collaborative and efficient energy ecosystem.

7.1.1 IoT Fundamentals

The IoT is a concept that allows networking between various types of physical devices or "things" embedded with sensors, actuators, software, electronics, etc., to collect and exchange data over the internet [1]. IoT is a vision where things such as smartphones, wearables, home appliances, and cars become "smart" and are capable of sensing, computing, and communicating using the attached electronic devices with remote servers over the internet [2]. The IoT objects usually encompass microcontrollers, wired or wireless transceivers, sensors, actuators, and protocol stacks, allowing communication over the internet. The target hardware devices are constrained in resources such as memory, processing power, data rate, and energy stores and need to be low cost [3]. IoT devices find applications in various sectors, including energy, healthcare, agriculture, automotive, and smart cities.

The basic block diagram of an IoT system is depicted in Figure 7.1 [4]. The physical thing block encompasses sensors and actuators, commonly called transducers, which are capable of transforming one form of energy into another [5]. A sensor collects information about the environmental parameters being monitored, such as temperature and humidity, and converts it into an

IoT for Smart Grid: Revolutionizing Electrical Engineering, First Edition.
Edited by Rahiman Zahira, Palanisamy Sivaraman, Chenniappan Sharmeela, and Sanjeevikumar Padmanaban.

Figure 7.1 Block diagram of an IoT system.

electrical signal. Such signal, usually of an analog nature, is transformed into digital form using an analog-to-digital converter (ADC) circuit before feeding to a microcontroller. An actuator transforms the electrical signal into another form of energy, for example, mechanical energy, and performs some actions based on instructions received from the microcontroller. IoT gateway provides a bidirectional communication channel for a reliable and timely data flow between physical things and the cloud. It can switch and route data packets and translate between protocols to enable interoperability among heterogeneous protocols and provide network-level security to the data. Cloud computing platforms permit the storage of large volumes of data generated by various sensors and the processing of data from anywhere and anytime [6]. It includes primary infrastructure, servers, and storage units essential for real-time operations and processing in IoT [7]. Various types of data analysis algorithms are applied to the stored data in the cloud as per the demand of the application [8]. Data analytics in IoT uses specialized software and tools such as machine learning (ML) and data mining, to process large volumes of data to extract inside and make appropriate decisions. The user interface allows users to remotely monitor, control, and manage different operations in the IoT. It may include an interactive display unit, keyboard, mouse, button, icon, microphone, speaker, etc.

7.1.2 Basics of Smart Grid

A smart grid is a modern electrical grid that combines an electrical power system, communication system, advanced sensing system, advanced metering system, measurement system, decision support, and human interfaces software and hardware for monitoring, controlling, and managing the generation, delivery, reserve, and utilization of electrical energy. It was proposed by Andres E. Carvallo in 2007 [9] and aims to enhance efficiency, decrease energy utilization and price, and maximize the lucidity and dependability of the system.

The smart grid uses smart meters, DSM, smart incorporation of produced energy, storage and renewable sources management, and systems that constantly supply and apply data from a network [10]. It intelligently puts together the actions of all users associated with it, delivering reliable, cost-effective, and safe electrical power efficiently.

The smart grid power distribution system aims to make the power grid more proficient and consistent, enhancing the security and quality of power supply in accordance with the needs of the digital age. It allows for superior penetration of renewable energy sources, widespread and efficient communication from generators to consumers, modern sensors, efficient speed control, and greater consumer participation. In summary, a smart grid is the future of power grids that utilize modern technologies and improve energy efficiency and reliability [11].

Figure 7.2 shows the simplified block diagram of the smart grid system. It integrates three types of entities: generators, which generate electricity; consumers, which consume electricity; and those that can do both. The blocks on the right are the electricity generators, and the blocks on the left represent the second and third categories. Altogether, these entities form a peer-to-peer (P2P) network to facilitate efficient electricity distribution, maintain reduced losses, and improve the quality of electricity supply.

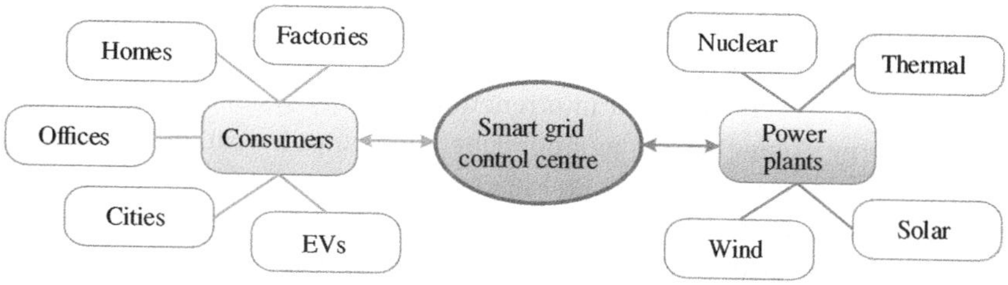

Figure 7.2 Conceptual diagram of a smart grid system.

7.1.3 Integration of Smart Grid and IoT

IoT has revolutionized many domains in which they are adopted. One such promising application area is the power sector, wherein the entire process of power generation, transmission, distribution, consumption, and management can be made more efficient and intelligent with the use of IoT [12, 13]. A smart grid is an advanced technology-enabled electrical grid system that utilizes modern communication, computing, and automation technologies to efficiently manage the generation, distribution, and consumption of electricity [14]. It integrates various components such as sensors, meters, controls, and software applications to optimize the operation of the grid, enhance reliability, reduce energy losses, and accommodate renewable energy sources and electric vehicles (EVs) [15]. Smart grids enable real-time monitoring, analysis, and control of electricity flows, allowing for more effective responses to changes in demand and supply conditions. Low reliability, high outages, low-energy security, high greenhouse gas, and carbon emissions are some of the issues and challenges that the smart grid is expected to address [16]. A smart grid is treated as a communication network on top of the electricity grid to collect and analyze data from various components of a power grid to predict power supply and demand, and it can be used for power management. Advanced metering infrastructure (AMI) is a comprehensive system in a smart grid that enables two-way communication between utility providers and customers [17]. It includes smart meters, communication networks, and data management systems. AMI enhances the efficiency, reliability, and security of the electricity grid.

7.1.3.1 Sensor Networks

Sensor networks play a significant role in integrating smart grid and IoT technologies, permitting real-time monitoring, control, and optimization of electricity generation, transmission, distribution, and consumption. The sensor networks continuously monitor and collect data linked with the grid's performance, energy distribution, and consumption patterns. The following components of sensor networks play a crucial role in integrating IoT with smart grid.

Smart meters: Smart meters are digital devices equipped with sensors and constitute the basis of smart grid networks [18]. They can measure energy consumption in customer households and transfer the usage data to utility firms via wireless or wired networks. Advanced smart meters may also include additional features to monitor power quality, voltage levels, and grid parameters.

Grid monitoring sensors: Sensors deployed across the grid infrastructure can monitor various parameters such as voltage, current, frequency, temperature, and equipment health. These sensors acquire real-time data on grid performance, allowing operators to identify irregularities and prediction failures and improve grid operations.

Distribution automation sensors: Sensors connected to distribution lines and equipment empower automation and optimization of distribution grid operations. These sensors can detect faults, measure power consumption, and identify equipment conditions, enabling fault detection, isolation, and restoration functions and performing load balancing.

Renewable energy sensors: Sensors deployed into renewable energy systems such as wind turbines, solar panels, and energy storage devices monitor generation output, weather conditions, and battery charge status. These sensors empower grid operators to integrate renewable energy resources and optimize their involvement in grid stability and reliability.

Demand response and building automation sensors: Buildings and industrial facilities connected with IoT sensors can detect occupancy, energy consumption patterns, and environmental conditions, which permits the use of a demand response program. This allows consumers to adjust their electricity usage in response to grid signals or pricing incentives.

7.1.3.2 Data Analytics

Sensors installed in smart grids generate large amounts of data. Data analytics tools analyze the data received from different sensors, identify the trends, and present them to the intended subscribers. It helps monitor energy consumption patterns, forecast energy demand accurately, identify anomalies, optimize energy distribution, and maintain grid stability. Data analytics empowers users to predict equipment failures or maintenance needs in advance by analyzing historical data, current conditions, and patterns. It can play a significant role by identifying potential cybersecurity threats and anomalies in network traffic using ML and artificial intelligence (AI), helping to enhance the security of IoT devices and the overall grid infrastructure. With the help of smart meters, data analytics allows us to gain valuable insights into customer energy preferences and energy usage trends. This information can be used to customize services, offer personalized energy-saving tips, and improve customer engagement. In addition, data analytics can optimize the integration of renewable energy sources into the grid by predicting generation output, managing intermittency, and coordinating energy storage systems. This helps maximize the utilization of clean energy resources while maintaining grid stability.

7.1.3.3 Remote Monitoring and Control

Remote monitoring systems, in general, fetch, analyze, transmit, manage, and provide feedback on remote information by utilizing the most advanced science and technology fields of communication technology and other areas. It helps to detect the problems that may lead to downtime of equipment used in the field and provides information about the system's performance. It also combines comprehensive usage of instrumentation, computer software, and electronic technology.

Remote monitoring and control in IoT-enabled smart grids are essential features of modern energy management systems (EMS). IoT devices equipped with sensors are deployed throughout the smart grid infrastructure to measure various parameters such as voltage, current, power factor, temperature, and humidity, generating huge amounts of data. Using data analytics tools, actionable outcomes are triggered through actuators deployed in the field, which can be controlled remotely, automatically, or manually.

7.1.3.4 Smart Devices and Appliances

Smart devices and appliances enable consumers to actively participate in DSM initiatives, support grid stability, and promote energy conservation and sustainability in IoT-based smart grid environments. This necessitates the use of devices such as smart meters – enabling bidirectional energy flow real-time monitoring of energy consumption; smart thermostats – permitting regulation of home heating and cooling appliances as per user preferences and occupancy pattern; smart lighting – fitted with sensors and wireless connectivity features permitting remote on–off, illumination control, as per schedules based on user preferences or environmental conditions. Smart appliances, including household appliances such as washing machines, refrigerators, dishwashers, and ovens, can be equipped with IoT capabilities. These smart appliances can communicate with the smart grid and each other to optimize energy usage, schedule operations during off-peak hours, and even prioritize tasks based on energy availability and cost. EVs are being increasingly considered an integral part of the smart grid ecosystem. EVs can serve as mobile energy storage units, capable of both consuming and supplying electricity to the grid through bidirectional charging capabilities. A smart charging mechanism facilitates EV owners in scheduling charging sessions during off-peak hours or when renewable energy generation is high, balancing the overall load on the grid.

7.2 Architecture of a Smart Grid IoT System

The architecture of an IoT-enabled smart grid system is a sophisticated framework designed to revolutionize traditional energy distribution by leveraging advanced technologies. The architecture involves a hierarchical structure with various layers and subcomponents, each responsible for specific functions and interactions [19].

7.2.1 Device Layer

The device layer in the architecture of the IoT-based smart grid serves as a foundational entity where data is collected, processed, and acted upon. The IoT-based smart grid architecture devices collect diverse datasets pertaining to energy consumption, grid conditions, and environmental factors. This data is the foundation for informed decision-making, enabling utilities to optimize grid operations, enhance reliability, and adapt to dynamic energy demands and environmental conditions. Smart meters, sensors, and actuators constitute the foundational elements of the device layer in an IoT-based smart grid architecture. Together, they enable comprehensive data collection, monitoring, and control capabilities, supporting efficient grid operations, improved reliability, and enhanced grid resilience.

- Smart meters are fundamental components of the device layer, installed at consumer premises to measure electricity consumption with high granularity and accuracy [18]. These meters replace traditional electromechanical meters and enable two-way communication between consumers and utilities. They are used as sensors across the whole distribution grid and collect real-time or interval-based data on energy usage, providing insights into consumption patterns, peak demand periods, and overall energy efficiency. Data received from smart sensors can be utilized for:
  - Faster detection of power outages, quicker response, and timely restoration of power.
  - Sharing information about the power grid's status to customers, cause of outage, estimated restoration time, public safety notice, etc.
  - Enhancing resilience against disruptions, lowering potential outages, and reducing frequency and period of outages by improving the accuracy of the grid asset management and planning.

- Sensors deployed throughout the grid infrastructure gather real-time data on various parameters related to grid conditions, equipment health, and environmental factors [20]. These sensors monitor voltage levels, current flows, power quality, temperature, humidity, and air quality. By continuously monitoring grid conditions and environmental factors, sensors provide valuable insights into grid performance, early detection of faults or anomalies, and assessment of environmental impacts.
  - Power generation infrastructure uses sensors based on temperature, humidity, pressure, vibration, flow, fuel quality, etc. to monitor various aspects of the turbine, boiler, and generator. It enables real-time monitoring and control of critical parameters to optimize the generation process.
  - The transmission network of the smart grid deploys phasor management units, line sensors, transformer monitors, circuit breaker monitors, voltage and current sensors, fault detection and location sensors, etc., to identify and real-time monitor and management of power transmission.
  - The smart grid distribution network uses smart meters, load tap changers on transformers, line fault detectors, distributed energy resource (DER) sensors, power quality sensors, transformer parameter sensors, etc. to deliver a more resilient, efficient, and customer-centric power distribution system.
- Actuators within the device layer enable remote control and automation of grid assets and equipment. These devices respond to commands from control systems by executing specific actions, such as opening or closing switches, adjusting voltage levels, or controlling DERs.
  - The power generation unit primarily uses hydraulic (in hydroelectric power plants) and pneumatic actuators (in steam and gas turbines) as turbine control actuators. Linear and circular actuators help solar plants to optimize sunlight capture. Similarly, in wind energy turbines, pitch and yaw actuators play a crucial role in optimizing the energy capture from varying wind speeds and wind direction.
  - In transmission lines, the electromagnetic actuator enables remote and automated control of circuit breakers and disconnect switches, motorized actuators control switchgear operations, and driven actuators adjust the tap position on transformers to regulate and stabilize voltage levels.
  - The distribution network consists of different actuators, such as the following:
    - Recloser actuators – automatically restore power after temporary faults by opening and closing circuit breakers.
    - Sectionalizer actuators – isolate faulted sections of the distribution network to maintain service in unaffected areas, enhancing grid reliability.
    - Relay actuators – enable remote disconnection and reconnection of the power supply, facilitating demand response and remote management of electricity consumption.
    - Solenoid actuators – switch capacitor banks in and out of the circuit to manage reactive power and improve power factor, enhancing overall grid efficiency.

7.2.2 Communication Layer

Information flow at high speed over a reliable and secure communication network is one of the important requirements of a smart grid to manage complex power systems intelligently and effectively [21, 22]. Smart Grid needs a two-way wide area communications network between different isolated areas, from generation, transmission, and distribution to consumer premises. A reliable communication network is crucial for exchanging data between devices, sensors, and control systems. The communication layer may include wired communication (e.g., Ethernet and power-line) and wireless communication (e.g., wireless fidelity [Wi-Fi], cellular, Zigbee, and low-power wide-area network [LPWAN]) technologies, providing coverage across the entire grid area. The following subsections discuss these two broad communication layers used in IoT-enabled smart grids.

7.2.2.1 Wired Communication Technology

Wired communication technologies provide a reliable and secure communication channel for transmitting data between various components within the grid infrastructure. Some popular wired technologies used in smart grids are discussed as follows:

Fiber optics: Fiber-optic cables offer long-range, high-speed, and high data rate transmission capability, making them most suitable for long-distance communication in smart grid arrangements. Fiber optics provide immunity to electromagnetic interference and offer high levels of security, making them well-suited for critical grid communication tasks such as real-time monitoring, control, and protection. Communications based on fiber-optic cable are primarily used to provide underlying communications supporting various smart grid applications, such as transmission domain communication and substation automation. It offers high bandwidth, that is, 5, 10, 20, or 40 Gbps, and therefore, has the potential to support high-speed data transfer for long distances. Various types of fiber-optic cables, such as underground cables (UGC), optical ground wires (OPGWs), all dielectric self-supporting (ADSS), and submarine cables, are used depending on application and merit. Air-blown fiber (ABF) is a relatively new fiber installation technology that is expected to be implemented in the network system [23].

Power-line communication (PLC): PLC utilizes existing power transmission lines to transmit data signals at higher frequencies alongside the electrical power, enabling communication between smart grid devices without additional communication infrastructure [24]. PLC technology is cost-effective because it exploits existing power distribution infrastructure and can be used for both low-speed and high-speed data transmission, depending on the application requirements. PLC is further divided into two categories, namely narrowband PLC (NB-PLC), operating at 300–500 kHz with a data rate up to 10–500 kbps, and broadband PLC (BB-PLC), which operates between 1.8 and 250 MHz with a data rate up to 300 Mbps.

Ethernet: Ethernet technology is commonly used for local area network (LAN) communication within substations, control centers, and other grid facilities. Ethernet provides high-speed, reliable, and secure communication, making it suitable for connecting devices such as sensors, actuators, controllers, and monitoring equipment within a confined area. Gigabit Ethernet (10 GbE) offers data rates up to 1 Gbps/10 Gbps with a maximum range of 70 km and can be used for high-speed LAN backbones and server connectivity of the smart grid.

7.2.2.2 Wireless Communication Technology

Wireless communication technologies are essential components of smart grid systems, enabling flexible and cost-effective data transmission across various grid assets and infrastructure [25, 26]. Some of the most used wireless communication technologies used in smart grids are presented below [27].

Wi-Fi: Wi-Fi technology provides high-speed wireless communication within a localized area, typically within homes (such as home area networks [HANs]), buildings (such as neighborhood area networks), or substations. Wi-Fi is commonly used for connecting smart meters, sensors, and other grid devices to local networks, enabling real-time data collection, remote monitoring, and control from the grid.

Cellular networks: Cellular networks, including 4G LTE and emerging 5G technology, offer wide-area coverage and high-speed data transmission capabilities, making them suitable for connecting remote and mobile smart grid devices. Cellular networks are used for applications such as AMI, grid monitoring, and asset management. Moreover, cellular networks offer enhanced security features, scalability, and quality of service (QoS) guarantees, which are critical for maintaining the resilience and efficiency of the smart grid. The deployment of 5G networks is expected to provide even greater benefits for smart grids, such as ultrareliable, low-latency communication and massive machine-type communication. These capabilities will support the proliferation of IoT devices within the grid, enabling finer-grained monitoring and control and fostering innovations in grid management and energy efficiency.

LPWAN: LPWAN technologies such as LoRaWAN, Sigfox, and NBIoT are designed for low-power, long-range communication with minimal infrastructure requirements, which is particularly beneficial for rural and remote areas where traditional communication infrastructures may be lacking. LPWANs enable connectivity for battery-powered sensors and devices deployed in remote or hard-to-reach areas, making them ideal for applications such as asset tracking, environmental monitoring, and metering. It allows real-time monitoring and control of DERs, such as solar panels, wind turbines, and energy storage systems, facilitating better integration and management of renewable energy sources within the grid.

Zigbee: Zigbee in smart grids enables real-time communication between smart meters and in-home devices, enabling energy monitoring, demand response, and optimization of grid operations. It is based on the IEEE 802.15.4 standard, provides a low data rate (up to 250 kbps), and offers a very long battery life (up to 10 years). It forms the backbone of HANs, supports demand response programs, and enables centralized control and monitoring of grid sensors and actuators for enhanced grid reliability and efficiency. Zigbee's low-power consumption, mesh networking capabilities, and interoperability make it a key technology for enabling smart grid applications.

Satellite communication: Satellite communication provides ubiquitous coverage and connects smart grid devices in remote or rural areas where terrestrial communication infrastructure is limited or unavailable. Satellite communication is employed for applications such as monitoring off-grid renewable energy systems, remote telemetry, and disaster recovery. It plays an important role in the smart grid by enabling the AMI environment. Satellites can transmit data from smart meters in remote or rural areas back to utility control centers, facilitating accurate billing, outage detection, and analysis of energy usage. This ensures that utilities comprehensively understand energy consumption patterns across their entire service area, which is essential for effective demand response and load management strategies.

Bluetooth: Bluetooth technology, particularly Bluetooth low energy (BLE), is used for short-range wireless communication between devices within close proximity, typically within a few meters. As a key component of HAN within the smart grid, Bluetooth facilitates the seamless integration and interaction of various smart devices, sensors, and appliances. In residential and small commercial settings, Bluetooth enables real-time communication between smart meters, thermostats, lighting systems, and other home automation devices, allowing for detailed monitoring and control of energy consumption. This real-time data exchange helps consumers optimize their energy usage, thereby contributing to energy efficiency and cost savings.

7.2.3 Edge Computing Layer

The role of cloud computing for electrical grid management activities has been well recognized in the past. However, due to the integration of IoT in smart grid management, a huge amount of data is being generated, which demands increased processing costs and delayed service response time. To address the issue, edge computing is introduced in a smart grid architecture to decentralize computing resources to the grid's edge, facilitating real-time data processing and decision-making [28, 29]. Edge devices deployed at substations and distribution points collect and analyze data locally, enabling rapid response to grid events and reducing reliance on a centralized backbone network. This edge computing layer supports distributed control, fault detection, predictive maintenance, and resilience, enhancing grid reliability and efficiency. Integration with central systems ensures coordinated control strategies while maintaining security, scalability, and flexibility within the smart grid ecosystem. Edge computing optimizes bandwidth usage, reduces latency, and enhances overall grid performance by enabling localized processing and control capabilities.

7.2.4 Cloud/Server Layer

The cloud/server layer serves as the backbone of a smart grid, offering scalable computing power, storage capacity, and networking resources [30]. It hosts centralized control systems, data management platforms, and analytical tools that enable grid operators to monitor, analyze, and optimize grid operations in real time. Additionally, the cloud facilitates secure data storage, backup, and disaster recovery, ensuring the integrity and availability of critical grid data. By leveraging cloud services, smart grid deployments can efficiently manage resources, support dynamic scaling, and adapt to changing demands and requirements. Integrating cloud-based applications and services enhances collaboration, innovation, and interoperability across the smart grid ecosystem, driving efficiency, reliability, and sustainability.

7.2.5 Control Center/Management Layer

The control center or management layer in a smart grid architecture serves as the nerve center for monitoring, managing, and controlling grid operations in real time [31]. It integrates data from various sources, applies advanced analytics and control algorithms, and provides tools for decision-making and optimization. Salient features of this layer are as follows:

- **Centralized Oversight:** Serves as the centralized hub for monitoring, managing, and controlling grid operations in real time.
- **Data Integration:** Integrates data from diverse sources such as sensors, meters, and edge devices to provide a comprehensive view of grid performance.
- **Decision Support:** Facilitates decision-making processes by analyzing incoming data and triggering appropriate responses to grid events and anomalies.
- **Functionality:** Manages critical functions, including energy management, demand response, outage management, and grid optimization, to ensure reliability and efficiency.
- **Stakeholder Collaboration:** Supports collaboration among utilities, regulators, and customers by providing visibility into grid operations and fostering transparent communication and decision-making processes.

7.2.6 Cybersecurity Layers

Given the critical nature of smart grid infrastructure, robust cybersecurity measures are essential to protect against cyber threats and ensure data integrity and privacy [32, 33]. This includes encryption, authentication, access control, and regular security audits of IoT devices and communication channels. The following points emphasize the key components and functionalities of cyber security layers in smart grid architecture.

- **Perimeter Defence Mechanisms:** Deploying firewalls, intrusion detection systems (IDS), and intrusion prevention systems (IPS) to protect the perimeter of the smart grid network from unauthorized access and external cyber threats. These mechanisms monitor and filter incoming and outgoing traffic, blocking malicious activity and preventing unauthorized access to critical grid assets.
- **Identity and Access Management (IAM):** Implementing IAM solutions to manage user identities, roles, and permissions within the smart grid ecosystem is crucial in enhancing the security and operational efficiency. IAM systems authenticate users, control access to sensitive grid resources based on their roles and privileges, and enforce security policies to ensure that only authorized personnel can access critical systems and data.
- **Encryption and Secure Communication Protocols:** Utilizing encryption techniques and secure communication protocols to protect data transmitted between grid devices, control systems, and communication networks. Encrypting data-in-motion ensures the confidentiality and integrity of information exchanged within the smart grid, mitigating the risk of data interception and tampering by unauthorized entities.
- **Security Monitoring and Incident Response:** Deploy security monitoring tools such as security information and event management (SIEM) systems to detect and respond to security incidents in real time. These systems continuously monitor network traffic, log data, and system activities to identify suspicious behavior, anomalies, and security breaches. In the event of a security incident, incident response procedures are activated to contain, investigate, and remediate the threat to minimize its impact on grid operations.
- **Regular Patch Management and Vulnerability Assessment:** Implementing patch management processes to ensure that grid devices, software, and firmware are regularly updated with security patches and fixes to address known vulnerabilities. Conducting regular vulnerability assessments and penetration testing helps identify and mitigate security weaknesses in the smart grid infrastructure, reducing the risk of exploitation by cyber attackers.

7.2.7 Integration and Interoperability Layer

To ensure seamless communication, data transmission-reception, and collaboration within the smart grid ecosystem, the role of the integration and interoperability layer is very important [34, 35]. The key aspects of the layer are as follows:

- **Standardization**: Standardization of communication protocols and data formats is essential for seamless integration and interoperability within the smart grid ecosystem. Adopting industry standards ensures compatibility between devices, systems, and applications, enabling efficient data exchange and communication.
- **Application Programming Interface (API) Management**: Effective API management plays a crucial role in facilitating interoperability between different components of the smart grid. APIs serve as the bridge for connecting diverse software applications, services, and data sources, allowing for smooth integration and communication.
- **Data Transformation:** Data transformation and normalization processes are crucial for converting data from varied sources into a consistent format that can be understood and processed by different systems and applications. This ensures data consistency, integrity, and compatibility across the smart grid network.

- **Security Measures:** Implementing robust security measures within the integration and interoperability layer is imperative to safeguard sensitive data, prevent unauthorized access, and protect against cyber threats. Encryption, authentication mechanisms, and access controls play a vital role in ensuring data privacy and integrity.
- **Interoperability Testing:** Regular interoperability testing is essential to validate the compatibility and seamless interaction between different components of the smart grid system. Testing protocols, standards compliance, and integration scenarios help identify and address any interoperability issues proactively, ensuring the smooth operation of the integrated grid infrastructure.

7.2.8 User Interfaces Layer

The user interface layer in a smart grid architecture serves as the primary point of interaction between human users, grid operators, and stakeholders, enabling intuitive visualization, control, and management of grid operations, assets, and services. The key components and their functionalities of the user interface layer are as follows:

- **Web Portals and Applications:** Web portals and applications offer browser-based interfaces for accessing smart grid functionalities from any device with an internet connection. Web portals provide role-based access control, personalized dashboards, and self-service options for users to view their energy consumption, manage preferences, and participate in demand response programs.
- **Mobile Applications:** Mobile applications enable users to access smart grid services and information on smartphones and tablets. Mobile apps provide real-time alerts, notifications, and remote-control capabilities, allowing users to monitor energy usage, receive outage updates, and adjust settings while on the go.
- **Energy Management Applications:** Energy management applications enable users to optimize their energy consumption, reduce costs, and minimize environmental impact. These applications provide energy usage analytics, recommendations for energy-saving measures, and tools for setting energy goals and tracking progress over time.

7.3 Remote Control and Automation in Smart Grids

The automatic smart grid system monitors the total amount of electricity used in the area. Adding monitoring, analysis, control, and communication capabilities to the country's electrical supply infrastructure is the fundamental idea of the smart grid [36].

7.3.1 Smart Grid Components

Smart grids employ bidirectional communication and advanced technologies to boost energy supply security, dependability, and efficiency [37]. These intelligent networks are enabled by a variety of essential components and systems that can be broadly categorized into five main areas:

- **Smart Meters:** Electronic devices that track the amount of electricity used and transmit that data to the distribution center so that it may be monitored and billed to the customer. The smart meter records the energy consumed per hour more precisely and at least once a day [38]. Smart meters facilitate a bidirectional connection between the meter and the central system.

- **AMI:** A key component of the smart grid is the deployment of AMI, which includes smart meters and the supporting communication networks, data management systems, and applications [39]. Smart meters enable two-way communication between the utility and the customer, allowing for real-time monitoring, dynamic pricing, and automated meter reading and billing. This AMI infrastructure is a fundamental enabler of the smart grid's enhanced DSM capabilities.
- **Integrated Communications:** At the core of the smart grid is a robust, two-way communication infrastructure that enables seamless data flow between various grid elements. This includes high-bandwidth communication networks, standardized protocols, and secure data exchange mechanisms. These integrated communications facilitate real-time monitoring, control, and optimization of the grid.
- **Global Positioning Satellite System (GPS):** GPS systems provide accurate time synchronization across the grid, enabling the effective use of phasor measurement units (PMUs) and other time-sensitive technologies [40].
- **Advanced Components:** Smart grid deployment involves the extensive use of intelligent, connected devices and automation systems. This includes smart meters, sensors, actuators, and other state-of-the-art grid components that can continuously monitor, analyze, and respond to grid conditions. These advanced components provide granular visibility into grid operations and enable automated, adaptive control.
- **Enhanced Interfaces and Decision Support:** Smart grids leverage intuitive user interfaces and intelligent decision-support systems to empower grid operators, utility managers, and end-consumers. These enhanced interfaces provide clear, actionable insights derived from the vast amounts of data collected across the grid. Additionally, advanced analytics and AI capabilities help optimize decision-making and automate complex grid operations.
- **Advanced Control Techniques, Sensing, and Measurement:** Smart grids employ a suite of cutting-edge control algorithms, sensing technologies, and measurement systems to maintain the stability, reliability, and efficiency of the electricity network. This includes technologies such as PMUs that provide real-time, synchronized monitoring of the grid's electrical parameters. To obtain a broad picture of the delivery system, the reference phase angle's benefit in relation to a global reference time is helpful [41]. Reducing blackouts and understanding the behavior of the energy system in real time can be accomplished through the appropriate application of this technology. Synchronized measurements in real-time from several distant measurement places in the smart grid can be carried out. Advanced control techniques leverage these precise measurements to enhance grid operations and prevent disruptions.
- **Intelligent Electronic Devices (IEDs):** IEDs are integrated microprocessors that control and monitor various grid components, offering standardized interfaces and communication protocols [42]. IEDs are utilized in conjunction with or as a more contemporary substitute for conventional remote terminal unit (RTU) setup. IEDs provide a standardized set of measurement and control points that are simpler to configure and require less wiring than RTUs because they are integrated with the devices they control. Most IEDs can direct communication with a substation programmable logic controller or the supervisory control and data acquisition (SCADA) system because they feature a communication port and integrated support for common communication protocols. Alternatively, they could be linked to a substation RTU, which serves as a portal to the SCADA server.
- **Electromagnetic Compatibility (EMC):** Ensures grid components can operate effectively without causing or experiencing unacceptable electromagnetic disturbances [43].
- **Power Quality Monitoring (PQM):** Electronic equipment may be susceptible to one or both forms of electromagnetic disturbances, and these disturbances can radiate. PQM refers to the

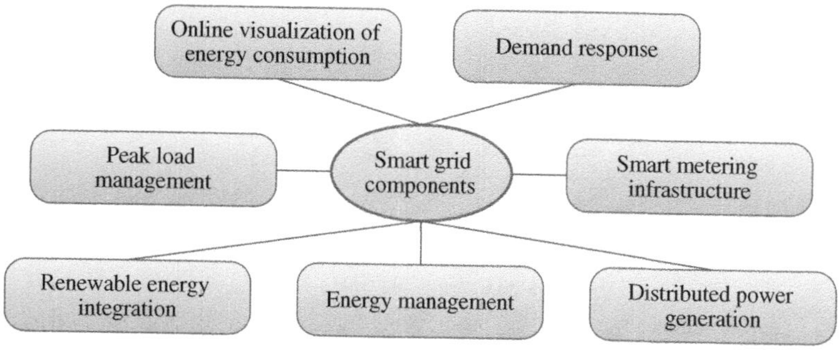

Figure 7.3 Smart grid components.

process that continuously measures and analyzes voltage and current data to provide actionable insights into the health and performance of the power system. Traditionally, the investigation and understanding procedure has been completed physically. However, with smart grids, it is now possible to design and execute an intelligent power system that can automatically examine and understand raw data into meaningful data with the least amount of human intervention.

These smart grid components collaborate to facilitate control strategies, deter energy theft, and assess and manage the condition of the grid equipment. Figure 7.3 shows various smart grid components.

7.3.2 Substation Automation

Substation automation is a crucial component of the IoT-enabled smart grid, enabling the remote supervision, data acquisition, and control of substation operations. Through the integration of advanced communication technologies, such as fiber-optic cables, high-speed communication lines, and modems, substations can be monitored and managed from a centralized location.

At the core of substation automation are microprocessor-based relays, which provide online control and monitoring functionalities [44]. These IEDs eliminate the need for separate SCADA networks, reducing the complexity and cost of the overall system. By transforming the traditional electrical control boards into PC-accessible graphical user interfaces (GUIs), substation automation empowers operators to monitor and manage the power system components with greater efficiency and precision [45]. The IoT-enabled substation automation system leverages the data collected from various IEDs, such as protection relays, meters, and control devices, to provide real-time insights into the performance and health of the power system. This data is then integrated with user-given commands, allowing for seamless remote control and optimization of the substation's operations.

Integrating IoT technologies within substation automation systems enables various advanced capabilities, including predictive maintenance, automated fault detection and isolation, renewable energy integration, and incorporating advanced cybersecurity.

7.3.3 Energy Management System (EMS)

The EMS is an automated toolkit used to monitor, regulate, and enhance the performance of generation and transmission systems in IoT-enabled smart grids. This control system or energy management software is designed to forecast electrical system performance, increase consistency,

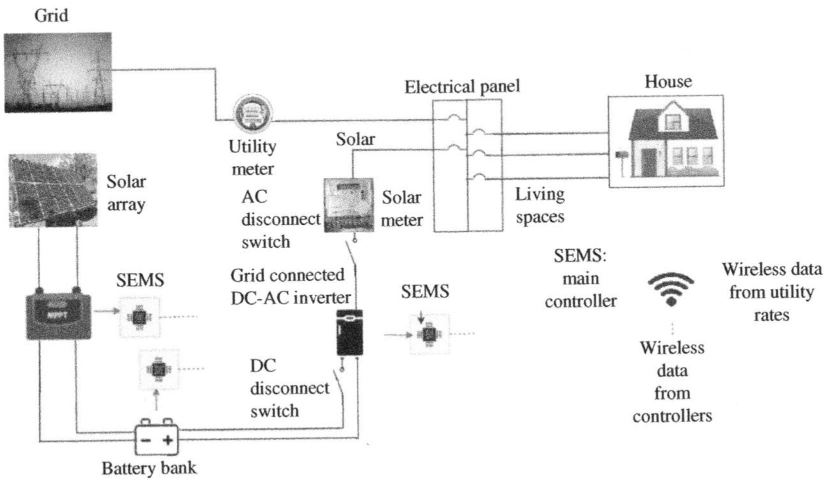

Figure 7.4 EMS architecture design.

and decrease power consumption, ultimately lowering energy expenses. Figure 7.4 shows a basic EMS system.

The core function of the IoT-enabled EMS is to enable real-time communication and control between various components of the smart energy network. By leveraging IoT technologies, the EMS can gather data from sensors and devices across the grid, allowing for precise monitoring and optimization of energy generation, transmission, and consumption [46]. The EMS is also integrated with user profiling and feedback modules, enabling it to tailor energy control and distribution based on consumers' specific needs and preferences. This allows for active optimization of energy usage, going beyond passive consumption tracking to drive meaningful changes in consumer behavior and attitudes toward energy use. Furthermore, the data collected by the IoT-enabled EMS is utilized to train control center operators, conduct engineering studies for future planning and system optimization, and provide detailed analysis of generation and consumption patterns. This comprehensive approach makes the EMS essential to smart grid security, distribution management systems, and substation automation.

7.3.4 Comparing Smart Grid with Traditional Grid

There are many advantages and disadvantages of smart grids over traditional power grids. The smart grid is digitally controlled with the microprocessor's help, making it very quick in response and accurate. Still, the traditional power grid is analog controlled, so it is not that accurate in operation. The smart grid is proactive, whereas the traditional power grid is reactive because it is prone to failures and blackouts. The smart grid offers multiple consumer products compared to the traditional power grid, with a very limited choice of products. In a smart grid, two-way communication is possible, whereas in a traditional power grid, only one-way communication is possible. The use of controllers and sensors in a smart grid is very large compared to that in a traditional power grid in which very few sensors are used. Modern power systems or smart grids are very transparent in operation, that is, nothing is hidden in the operation, maintenance, and billing. However, in the case of the traditional power grid, it is not so. The operation, maintenance, and control of a smart grid are much better than that of a conventional grid [47]. A smart grid ensures the power system's reliable operation.

7.4 Automated Load Shifting Strategies Using IoT

The smart grid initiative has ushered in a new era of automated load shifting strategies enabled by the integration of IoT. This chapter delves into the strategies for automated load shifting, leveraging the real-time data and control capabilities enabled by IoT devices and applications. From understanding electrical load patterns to implementing DSM through load shifting, utilities can now harness the power of IoT to deliver targeted demand response programs and achieve optimal load balancing across the grid. The application of IoT in load shifting empowers utilities to dynamically manage energy consumption and enhance the overall efficiency and sustainability of the power grid.

7.4.1 Electrical Load

Electrical appliances and lights within a home are examples of electrical loads in circuits that require electricity to operate. This contrasts with a power supply source that generates power, such as a generator or battery [48]. Circuit performance with regard to output voltages or currents is impacted by load. An easy illustration is provided by mains power outlets, which produce power at a constant voltage while the load comprises all the electrical equipment connected to the power circuit. A high-power appliance significantly lowers the load impedance when it turns on. The voltages will decrease if the load impedance is not significantly higher than the power supply impedance. Therefore, incandescent lights may decrease substantially when heating equipment is turned on in a home setting.

7.4.2 Load Shifting

Load shifting is an approach to managing power use that moves load demand from peak to off-peak times of the day. Load shifting is not a novel notion; in fact, industrial and commercial establishments have been optimizing energy use and cutting expenses associated with power for many years. An approach for managing electricity load is called load shifting, which involves moving load demand from peak to off-peak hours of the day [49]. Putting it another way, load shifting is just the act of shifting the overall amount of electricity consumed to a different time frame. It would be helpful to define electrical loads and discuss the advantages of moving electricity consumption to a different time frame to thoroughly explain load shifting.

7.4.3 Demand-Side Management Through Load Shifting

DSM [50] is essential for the growth of sustainable cities and communities in the age of smart grids. The practical difficulties encountered when implementing DSM for IoT-enabled household energy management systems (HEMS) through load shifting are discussed in this study. In this context, home appliance characterization, incorporating sporadic renewable energy sources, load classification, diverse limitations, dynamic pricing, and consumer classification have all been covered. Given that DSM is a stochastic optimization problem, a thorough analysis of various optimization strategies for resolving the multi-objective energy management issue has been covered.

7.4.4 Utilizing IoT for Demand Response Programs

Using the IoT to implement demand response programs in smart grids can greatly improve the efficiency and dependability of energy delivery. Utilities may optimize grid operations by integrating

IoT devices such as smart meters and sensors to collect real-time data on energy consumption. These data can subsequently be utilized to execute demand response tactics, such as relocating energy-intensive activities to nonpeak hours or automatically changing thermostats in reaction to grid limitations. Moreover, IoT devices can facilitate more accurate and focused communication with customers, empowering them to make well-informed choices regarding their energy consumption and actively engage in demand response initiatives.

7.4.5 Application of IoT in Load Shifting

Demand-based tariff accounting is used in some nations; more demand corresponds to higher pricing, and vice versa. As a result, the system needs to provide two-way communication between customers and power providers and enable ongoing usage monitoring. IoT is being used more and more to do this [51]. Because it makes it possible to send data from loads and sensors to a cloud, where users can easily access it, the IoT turns out to be the perfect solution for long-distance communications. The IoT integration with energy-related systems, such as electrical loads, EMS, and microgrids, including write renewable energy resources, is called the "Energy Internet."

7.5 IoT Applications for Real-Time Monitoring of Smart Grids

Smart grids' development has revolutionized how electricity is generated, transmitted, and consumed. Smart grids incorporate advanced sensing, communication, and control technologies to distribute energy efficiently, reliably, and sustainably. In this context, IoT applications play a crucial role in the real-time monitoring of smart grids. This chapter discusses the various IoT applications that enable real-time monitoring of smart grids, highlighting their benefits and challenges.

7.5.1 Grid Analytics and Data-Driven Decision-Making

Real-time monitoring of smart grids using IoT produces a substantial volume of data that can be utilized for sophisticated analytics, predictive modeling, and decision-making. This data can include information on energy consumption, grid performance, weather conditions, and other relevant factors that can impact the efficiency and reliability of the grid. With the application of ML algorithms and predictive models, utilities can precisely estimate electricity demand, optimize grid operations, and proactively handle potential problems. Advanced analytics and ML techniques can help utilities find trends, patterns, and outliers in real-time data, leading to data-driven insights. These insights can be used to improve energy distribution, establish efficient demand response programs, and make informed decisions regarding changes to grid infrastructure. Grid analytics can also enhance load forecasting accuracy, identify areas of high energy use, detect power losses, optimize grid topologies, and optimize grid configurations.

Real-time monitoring and analysis of grid data can help utilities make informed decisions, optimize grid operations, prepare for future infrastructure enhancements, identify energy-saving opportunities, and enhance grid reliability and efficiency. Real-time data from IoT devices can be visualized through intuitive dashboards, charts, and maps, enabling utilities to identify trends, patterns, and anomalies. This information can help utilities assess grid operations' efficiency, reliability, and quality, optimize performance, and strive for operational excellence.

Real-time monitoring of smart grids enables utilities to assess grid performance and benchmark it against predefined metrics or industry standards. The real-time data collected through IoT applications helps utilities assess grid operations' efficiency, reliability, and quality. By comparing real-time data against benchmarks, utilities can identify areas for improvement, optimize grid performance, and strive for operational excellence.

7.5.2 Grid Monitoring and Control

The integration of IoT has revolutionized the way utilities monitor and control the power grid, ushering in a new era of real-time visibility, responsiveness, and efficiency. The key aspects of IoT-enabled smart grid monitoring and control are discussed below.

Grid visualization and control: By including IoT in the smart grid, utilities now see and manage their grid infrastructure differently, enabling them with hitherto unheard-of real-time insights and operational effectiveness. Utilities now get a complete, data-driven view of their whole network thanks to IoT device adoption across the grid. Dashboards and intuitive interfaces easily combine and visualize real-time monitoring data from smart meters and IoT sensors, giving utilities a detailed knowledge of the state and performance of their substations, transformers, switches, and other vital grid components. This improved grid visualization helps utilities proactively spot possible bottlenecks, streamline grid operations, and make better judgments on required increases or upgrades. Using the abundance of real-time data, utilities can quickly predict and react to grid circumstances, guaranteeing the dependability and effective distribution of electricity to their consumers. Furthermore, augmented reality (AR) technology has elevated grid visualization. Utilities may engage with their assets in a more understandable and immersive way by overlaying virtual data onto the real-world image of the grid system. AR-enabled grid visualization improves situational awareness, helps with maintenance and repairs, and gives utility workers real-time data that guides their decisions. As a key element of the smart grid, this IoT-driven grid visualization and control capability provides utilities with the tools and insights required to negotiate the changing energy terrain and provide consistent, efficient, and sustainable power to their communities.

Remote grid monitoring and control: IoT applications enable the remote monitoring and control of smart grids in real time. Utilities can obtain real-time data on grid properties, equipment status, and energy flows from any place. The remote monitoring feature allows utilities to promptly address grid events, remotely manage grid operations, and resolve problems without human intervention. Remote grid monitoring and control systems can optimize operating efficiency, decrease maintenance expenses, and boost grid reliability.

Voltage and PQM: Maintaining proper voltage levels and power quality is crucial for ensuring a consistent and reliable energy supply. Real-time monitoring through IoT applications enables utilities to monitor voltage, frequency, and power quality in real time. Grid sensors and devices can accurately detect voltage and frequency levels and promptly relay this information to utility companies. IoT sensors and devices deployed throughout the grid continuously measure voltage variations, harmonic distortions, and other power quality parameters. This information helps utilities identify issues such as voltage sags, swells, or harmonic distortions that can impact grid performance and the quality of power supplied to consumers. Real-time monitoring allows utilities to take corrective actions promptly, ensuring a consistent and reliable power supply.

Distribution monitoring: Distribution monitoring systems that utilize IoT technology offer real-time visibility to the distribution network. These systems employ sensors and communication devices distributed across the grid to monitor factors such as voltage levels, current flows, and power quality. Through the continuous collection and analysis of these data, utilities can identify problems, anticipate possible failures, and enhance the operation of the distribution network. Real-time monitoring enhances grid stability, minimizes downtime, and facilitates proactive maintenance.

7.5.3 Grid Management and Optimization

As the smart grid evolves, intelligent grid management and optimization have become critical for navigating the increasing complexity of DERs, dynamic demand patterns, and two-way power flows. This section briefly discusses the collaborative grid management strategies and integrated technologies that enable utilities to balance supply and demand dynamically, orchestrate demand response and load management programs, and optimize grid operations. From seamless integration with AMI, energy markets, and EVs to innovative load balancing and energy management solutions, the pivotal role of grid optimization in unlocking the full potential of the smart grid is highlighted below.

Collaborative grid management: Smart grids may be monitored in real time using IoT apps. This allows utilities, consumers, and DER providers to work together in grid management. Stakeholders can coordinate their actions to guarantee grid stability, improve energy utilization, and facilitate the integration of DERs by exchanging real-time data on energy generation, consumption, and system conditions. A more decentralized and resilient grid architecture can be achieved through collaborative grid management, which in turn promotes efficiency, transparency, and customer engagement.

Demand response and load management: Leveraging real-time data and connection to maximize energy consumption and lower peak load demands, the smart grid initiative using IoT marks a new era of enhanced demand response management. IoT-based demand response and load control systems have enabled utilities to create seamless communication with smart appliances, thermostats, and other connected devices in consumers' homes. Utilities can dynamically change energy pricing, promote strategic load shifting, and remotely control energy-consuming equipment by tracking real-time data on energy usage and grid conditions. Increased general energy efficiency, cheaper energy prices, and grid stability all depend on this real-time demand response program control of demand. Furthermore, the real-time information gathered by IoT devices for smart grids promotes more precise demand forecasting and focused energy efficiency projects. Using energy consumption pattern analysis, utilities may forecast future demand, maximize energy generation and distribution, and more successfully plan for peak times. This detailed, real-time information also helps utilities spot particular energy savings potential, give consumers individualized comments, and encourage energy-saving practices that help the grid's sustainability be further improved. The fundamental idea of demand response management – where utilities use IoT connectivity to motivate consumers to change their energy consumption patterns in response to grid circumstances or pricing signals – is powering this revolution. A key component of the development of the smart grid is that utilities can use dynamic demand response tactics, interact with consumers, and remotely manage equipment by gathering real-time data on energy consumption, grid load, and pricing to maximize energy use and lower peak demand.

Load balancing and grid optimization: Emphasizing real-time monitoring and data insights to improve the efficiency and stability of the electricity grid, the smart grid has transformed the way utilities approach load balancing and grid optimization. Utilities can constantly track real-time energy supply and demand trends over the grid using IoT applications. Analyzing this granular, data-driven knowledge helps them identify and solve imbalances between energy supply and use, optimizing the efficiency of the electricity distribution. Through this real-time awareness, utilities may guarantee a balanced load distribution over the electrical grid, avoiding the pressure or underutilization of vital grid infrastructure. Dynamic distribution of energy resources depending on demand helps utilities reduce energy waste and raise the general grid's efficiency. Furthermore, the real-time grid condition monitoring made possible by IoT sensors and smart meters gives utilities the information required to maximize grid operations. Clear knowledge of energy consumption patterns at various sites helps utilities to make wise decisions to reduce transmission losses and improve the reliability of the power system. A fundamental part of the smart grid transition, this data-driven approach to load balancing and grid optimization helps utilities run their networks more effectively, consistently, and sustainably – ultimately delivering more value to their consumers and the community.

7.5.4 Smart Grid Planning and Integration

As the smart grid initiative takes shape, utilities must carefully plan to integrate a growing array of DERs, renewable generation, smart buildings, EVs, and advanced technologies. In the following, the critical grid planning, optimization, and system-wide integration strategies, enabling the seamless convergence of these elements into a highly efficient, dynamic, and responsive smart grid, are briefly discussed.

Grid planning and expansion: Real-time monitoring through IoT applications supports grid planning and optimization processes by continuously collecting and analyzing real-time data on energy consumption, demand patterns, and grid performance. This enables utilities to understand future energy needs and plan infrastructure upgrades accordingly. IoT-enabled analytics and optimization tools can help utilities optimize grid configurations, plan for DER integration, and identify opportunities for grid modernization and energy efficiency improvements. Real-time monitoring through IoT applications also aids utilities in grid planning and expansion efforts by providing real-time data on energy consumption patterns, load growth, and grid performance. This information can help utilities identify areas with increased demand or grid constraints, enabling them to plan for grid expansion, upgrade infrastructure, and allocate resources effectively.

Grid optimization for DERs: Real-time monitoring through IoT applications is essential for optimizing grid operations for DERs, including solar panels, wind turbines, energy storage systems, and EVs. IoT devices can provide real-time data on energy generation, consumption, and storage from these distributed resources, enabling utilities to balance supply and demand, manage the integration of DERs, and optimize energy flows within the grid. By continuously monitoring DERs in real time, utilities can maximize their contribution to the grid, improve the efficiency of renewable energy sources, and ensure a reliable and stable power supply. Real-time data on energy generation, consumption, and storage can help utilities manage the intermittency of renewable energy sources, optimize energy flows, and prevent power outages. IoT-based smart grid analytics can also help utilities identify potential issues with DERs, such as equipment failures or performance degradation, allowing for prompt maintenance and repair. This can help reduce downtime, increase energy efficiency, and promote sustainability. In addition, real-time monitoring through IoT applications can help utilities manage the grid integration of DERs, ensuring that these resources are integrated in a way that maximizes their benefits while minimizing their impact on grid stability. This can help utilities improve the reliability and resilience of the grid, reduce greenhouse gas emissions, and promote the adoption of renewable energy sources.

Renewable energy integration: Issues associated with intermittency, unpredictability, and grid stability arise when renewable energy sources are incorporated into the grid at the same time. IoT applications play an important role in real-time monitoring and control of renewable energy sources. Real-time monitoring of smart grids using IoT applications facilitates the integration of renewable energy sources into the grid. Utilities can track weather conditions, monitor generation output from sources such as solar panels and wind turbines, and forecast energy generation patterns by implementing IoT-enabled sensors and monitoring systems at renewable energy sites. Real-time monitoring makes it easier to integrate renewable energy sources efficiently, improves grid management, and maximizes the exploitation of clean energy resources. The real-time data help utilities manage the variability of renewable energy sources, optimize grid operations, and make informed decisions regarding energy storage and grid balancing strategies.

Integration with smart buildings and homes: The utilization of IoT to integrate smart grids with smart buildings and homes presents a multitude of advantages spanning multiple facets of energy management and optimization. Leveraging real-time data provided by IoT from both smart grids and smart buildings/homes allows the optimization of energy usage, reduction of wastage, and the balance of energy demand and supply. This integration also makes it easier to implement demand response capabilities. This means that smart buildings and homes connected to the internet can respond to smart grid signals during peak hours, which helps reduce energy consumption and ensures a stable power supply. An important benefit of this integration is the potential for improved energy efficiency in residential and commercial buildings. IoT devices enhance energy efficiency and reduce waste by automating heating, ventilation, and air conditioning (HVAC), lights, and appliances. This leads to substantial energy conservation and a noteworthy decrease in carbon emissions. Furthermore, IoT enables the seamless integration of renewable energy sources such as solar panels and wind turbines into the smart grid, as well as buildings and homes. Furthermore, with the utilization of sensors, IoT devices can monitor the state of various systems and appliances located within homes and buildings. This allows for proactive maintenance, reducing the likelihood of equipment breakdown and improving overall energy efficiency. Consequently, cost reductions are accomplished, and the management of energy usage is improved. The interoperability provided by the IoT through seamless communication and data exchange between smart grids and buildings/homes enhances energy management capabilities, reduces costs, and increases overall efficiency. Given the scalability of IoT devices, it is possible to include them in smart grids, buildings, and houses of varied sizes and degrees of complexity. This flexibility allows energy infrastructure to adapt to community and commercial needs.

Integration with AMI: One essential IoT use in the context of smart grids is smart metering. It entails the installation of sophisticated meters, which track and log power usage in real time. AMIs consist of smart meters, communication networks, and data management systems. AMI can be seamlessly integrated with smart grid monitoring in real time. IoT applications facilitate the collection of real-time energy consumption data from smart meters, monitoring grid conditions, and communicating with consumers by utilities. By enabling real-time billing, remote meter reading, and demand response functionalities, this integration empowers consumers by providing precise data regarding their energy consumption, thereby fostering the adoption of energy conservation practices.

Integrated energy management: IoT allows the integration of data from several sources, including renewable energy generation, energy storage systems, and energy consumption patterns, to assist comprehensive energy management. By continuously monitoring these components, utilities can effectively manage energy distribution, ensure a balance between energy supply and demand, and fully exploit the potential of renewable energy sources. Integrated energy management allows utilities to make informed decisions regarding energy generation, storage, and distribution. This ultimately results in increased grid efficiency, less dependency on fossil fuels, and improved sustainability.

Integration with energy markets: IoT applications enable the seamless integration of smart grids with energy markets by providing real-time monitoring. IoT devices can monitor real-time energy pricing, market changes, and demand trends. With this data, utilities may better plan for energy trading, optimize energy purchases and sales, and participate in demand response programs. Utilities can control energy supply and demand, optimize revenue, and respond to market changes with the help of real-time integration with energy markets.

Integration with EVs: With the growing adoption of EVs, effectively regulating the demand for charging on the power grid becomes essential. The real-time monitoring of smart grids enables the seamless integration of EVs into the smart grid. IoT allows for tracking the charging status, energy consumption, and location of EVs in real time. Using IoT devices and sensors, EV charging stations can be monitored, real-time data on charging demand can be collected, and EVs can communicate with one another to optimize charging schedules based on grid status. Thanks to this information, utilities can manage EV charging loads, optimize charging schedules, and strike a balance between the energy demand from EVs and the grid capacity. Utilities can also minimize grid congestion and increase the utilization of renewable energy sources for EV charging. In addition, real-time monitoring can enable vehicle-to-grid (V2G) systems, which allow EVs to send power back to the grid during times of high demand or as a backup resource for the grid.

Integration with advanced technologies: Integration with other modern technologies improves IoT-based smart grid monitoring. IoT, AI, and ML can be used to create intelligent EMS that autonomously optimize energy distribution, estimate energy demand, and detect anomalies. For instance, integrating blockchain technology can make energy transactions and grid operations more secure, transparent, and traceable.

7.5.5 Consumer Engagement

The empowered consumer – equipped with real-time knowledge and the capacity to manage their energy use actively – is at the core of the smart grid transition. In this sense, IoT technologies have been revolutionary because they give customers hitherto unheard-of awareness and control over their energy consumption.

Consumers can now instantly track their energy use, expenses, and even environmental impact by means of linked devices and smart home apps. This detailed information helps them to spot areas of efficiency, make wise choices on their energy use, and change their behavior. Smart plugs, EMS, and mobile apps empower users to maximize their energy resources, promoting a sustainable and conserving culture.

Still, the advantages of consumer involvement go beyond personal homes. Utilities can use IoT-enabled platforms to improve service delivery by aggressively, including consumers in the smart grid ecosystem. Utilities may create closer, more responsive relationships with their consumers using real-time outage alerts, tailored energy-saving advice, and smooth remote control of energy consumption. Furthermore, this increased customer involvement can affect the smart grid project in general. Consumers can significantly help the smart grid to be integrated with renewable energy sources, EVs, and changing energy markets as they grow more conscious and invested in the efficiency and sustainability of the system. Their active involvement and comments can help define the grid's future, strengthening its resilience, responsiveness, and fit for the local demands.

The success of the smart grid project depends mostly on its capacity to really empower and involve the consumer as a component of the energy ecosystem. Utilities may open fresh degrees of efficiency, sustainability, and customer happiness by using IoT technologies to support this consumer-centric strategy, enabling a smarter, greener, and more responsive grid.

7.5.6 Security and Regulatory Compliance

As the grid becomes increasingly interconnected and automated, the attack surface for potential cyber threats and regulatory oversight expands significantly. Utilities must prioritize stringent cybersecurity measures and grid protection strategies to safeguard critical infrastructure from malicious actors. Furthermore, the smart grid's integration with DERs, smart devices, and energy markets introduces new regulatory oversight and compliance requirements. Utilities must navigate evolving regulations around data privacy, energy theft detection, and reporting to ensure the integrity and transparency of the grid's operations.

Ultimately, robust security and rigorous regulatory compliance are foundational to building public trust, maintaining grid reliability, and unlocking the full transformative potential of the smart grid initiative. By proactively addressing these challenges, utilities can foster a secure, resilient, and compliant smart grid ecosystem that serves the needs of customers and communities in the IoT era.

The critical cybersecurity measures and grid protection strategies utilities employ to safeguard the system, along with the evolving regulatory compliance and reporting requirements driving the smart grid transformation, are discussed in the following.

Cybersecurity and grid protection: Real-time monitoring of smart grids using IoT applications is crucial for ensuring grid cybersecurity and protection against cyber threats. IoT devices and sensors can monitor network traffic, detect anomalies, and identify potential cybersecurity breaches in real time, enabling utilities to take immediate action to mitigate cyber risks, safeguard critical infrastructure, and maintain the integrity and confidentiality of grid data. IoT-enabled security systems can continuously monitor grid parameters and network traffic, enabling utilities to detect and respond to potential cyber threats in real time. These systems can identify abnormal network behavior, unauthorized access attempts, and anomalous data patterns, ensuring the integrity and security of grid operations. Real-time monitoring through IoT applications also aids in identifying potential vulnerabilities and implementing timely security patches and updates. This can help utilities stay ahead of emerging cyber threats and reduce the risk of cyberattacks on the grid.

Energy theft detection: Energy theft is a significant concern for utilities, leading to revenue losses and increased operational costs. Real-time monitoring through IoT applications helps utilities detect and prevent energy theft. IoT sensors and devices can monitor energy flows, identify abnormal consumption patterns, and flag potential instances of theft or tampering. Smart meters equipped with advanced analytics capabilities can analyze consumption patterns and identify irregularities that may indicate unauthorized or tampered connections. Real-time energy consumption monitoring coupled with anomaly detection algorithms can help utilities promptly identify and take action against energy theft. By mitigating losses and ensuring fair billing practices, real-time energy theft detection enhances revenue protection and promotes a more equitable energy distribution system.

Regulatory compliance and reporting: Utilities can better fulfill their reporting and compliance needs with smart grids when they use IoT applications to monitor them in real time. Utilities can generate precise reports for regulatory agencies by continuously collecting and storing real-time data pertaining to energy consumption, grid performance, and environmental factors. The implementation of real-time monitoring guarantees that utilities possess current and dependable data to adhere to grid reliability, environmental impact, and energy efficiency regulations.

7.5.7 Environmental Monitoring and Sustainability

Real-time monitoring of smart grids using IoT applications facilitates environmental monitoring and sustainability. By providing real-time data on environmental conditions and energy consumption patterns, IoT sensors and devices can help utilities make informed decisions about grid operations, reduce energy waste, and promote sustainability.

One of the key ways that IoT can enhance environmental monitoring and sustainability is through real-time monitoring of energy consumption patterns. IoT sensors and devices can provide real-time data on energy consumption, enabling utilities to identify areas with high energy use and implement energy efficiency measures. This can help utilities reduce energy waste, promote sustainability, and lower greenhouse gas emissions.

IoT sensors measure various environmental parameters, including air quality, temperature, humidity, and noise levels. These collected data can assist utilities in evaluating the environmental impact of energy generation and consumption, identifying pollution sources, and implementing measures to mitigate environmental risks. This also can help utilities make informed decisions about grid operations during extreme weather events or natural disasters. For example, during a heatwave, utilities can use IoT sensors and devices to monitor temperature and humidity levels, adjusting grid operations to reduce the risk of power outages caused by overloaded transformers or other equipment.

In addition, IoT-enabled smart grid systems can help utilities monitor the performance of renewable energy sources, such as solar panels and wind turbines. By providing real-time data on energy generation and environmental conditions, IoT sensors and devices can help utilities optimize the performance of renewable energy sources, reducing greenhouse gas emissions and promoting sustainability.

Furthermore, IoT-enabled smart grid systems can help utilities monitor the environmental impact of their operations. IoT sensors and devices can help utilities identify potential vulnerabilities and implement timely security patches and updates by providing real-time data on energy consumption patterns, grid performance, and environmental conditions. This can help utilities reduce their carbon footprint, promote sustainability, and ensure the safety and reliability of the grid.

Real-time monitoring through IoT applications includes weather monitoring for grid operations. IoT sensors can measure temperature, humidity, wind speed, and solar radiation. These real-time weather data help utilities anticipate weather-related impacts on the grid, such as increased energy demand during heatwaves or potential damage from severe storms. Weather monitoring enables utilities to optimize grid operations, plan maintenance activities, and implement strategies for grid resilience in adverse weather conditions.

7.5.8 Grid Resilience and Disaster Management

As the smart grid integrates more connected devices and renewable energy sources, ensuring grid resilience has become paramount. The integration of IoT in smart grids has the potential to enhance their resilience and disaster management capabilities significantly. The fault detection and diagnosis techniques, as well as the advanced disaster management strategies that enable utilities to proactively mitigate disruptions and rapidly restore power in the face of natural disasters and other emergencies, are briefly discussed in the following.

Fault detection and diagnosis: One of the key ways that IoT can enhance smart grid resilience is through real-time monitoring of grid parameters and network traffic. IoT sensors and devices can continuously monitor grid performance, energy consumption patterns, and environmental conditions, detecting anomalies and potential issues in real time. By utilizing sensors, communication networks, and advanced analytics to identify and diagnose errors as they occur promptly. By continuously monitoring grid parameters and receiving real-time alerts from IoT-enabled sensors, utilities can promptly detect and locate power outages. Moreover, by continuously monitoring voltage variations, current imbalances, and anomalous power consumption patterns, these systems can rapidly detect and pinpoint issues such as short circuits or equipment failures. Facilitated with real-time data, utilities can detect and respond to potential issues before they become major problems and take prompt corrective actions, preventing power outages, reducing downtime, and ensuring the safety and reliability of the grid.

Disaster management: IoT-enabled smart grid systems can help utilities optimize disaster response efforts. IoT sensors and devices can help utilities identify areas with increased demand or grid constraints by providing real-time data on energy consumption patterns, load growth, and grid performance. This information can help utilities allocate resources effectively, prioritizing critical infrastructure and emergency services. In addition, IoT-enabled smart grid systems can help utilities improve their situational awareness during disasters. IoT sensors and devices can provide real-time data on environmental conditions, such as temperature, humidity, and wind speed, enabling utilities to make informed decisions about grid operations during extreme weather events or natural disasters. Furthermore, IoT-enabled smart grid systems can help utilities improve their disaster recovery efforts. In the event of a disaster, real-time monitoring can help utilities quickly identify and isolate issues, reducing the impact of the disaster on the grid and enabling a faster recovery. The power flows can be rerouted dynamically by isolating the faulty grid section and restoring the power in the unaffected areas. The real-time information enables utilities to dispatch repair crews efficiently and effectively. Furthermore, IoT applications provide real-time feedback on power restoration progress, allowing utilities to keep customers informed and estimate the time needed to restore power more accurately. By continuously monitoring grid performance, IoT sensors and devices can help utilities identify potential vulnerabilities and implement timely security patches and updates. This can help utilities reduce the risk of cyberattacks on the grid, ensuring the safety and reliability of the grid during and after a disaster.

7.5.9 Asset Management and Maintenance

The smart grid initiative in the IoT era has revolutionized how utilities manage their physical assets. From real-time asset tracking and microgrid management to predictive maintenance strategies powered by grid-connected energy storage, the IoT-based asset management for a more resilient and efficient power grid is discussed in the following.

Asset tracking and management: Real-time monitoring through IoT enables asset tracking and management in smart grids. IoT sensors and devices can be installed on grid assets, including transformers, switches, and meters, to monitor their location, condition, and real-time performance. Utilities can use these data to plan maintenance tasks, maximize asset utilization, and spot any problems before they result in equipment breakdowns. Asset management and real-time monitoring increase grid reliability, decrease downtime, and prolong the life of grid infrastructure.

Microgrid management: Real-time monitoring through IoT apps is essential to effectively manage microgrids. Microgrids are decentralized energy systems that can function autonomously or in cooperation with the primary power grid. Real-time monitoring of microgrid components, including DERs, energy storage systems, and local demands, is made possible by IoT-based sensors and devices. Operators can utilize this feature to maximize energy generation and consumption efficiency in the microgrid, maintain a balance between supply and demand, and guarantee the grid's stability.

Grid-connected energy storage management: The implementation of real-time monitoring in smart grids allows for the effective management of energy storage devices that are connected to the grid. Energy storage devices can be tracked by IoT, which can gather data on grid conditions, charge levels, and storage capacity in real time. Utility companies can use this data to manage energy storage better, strike a better balance between supply and demand, and offer grid services such as peak shaving and frequency regulation. Monitoring energy storage in real time improves the stability of the power system, facilitates the integration of renewable energy sources, and strengthens the grid's resilience.

Predictive maintenance: Real-time monitoring facilitated by the IoT empowers utilities to execute predictive maintenance tactics for smart grid infrastructure. Utilities can analyze the operation and health of grid equipment, such as transformers, switches, and substations, by continuously collecting data from sensors implanted in these assets. Utilizing advanced analytics and ML algorithms, utilities may proactively predict equipment breakdowns and schedule maintenance by identifying trends, patterns, and anomalies in the collected data. This technique allows utilities to avoid unplanned downtime, decreases costs associated with maintenance, and maximizes the effectiveness of grid assets over their lifetime.

7.6 Challenges in Implementing IoT in Smart Grids

Implementing IoT in smart grids presents several challenges and limitations that must be carefully addressed to ensure successful deployment and operation. Some key challenges and issues and their potential mitigation strategies are discussed below.

Interoperability and standard: Smart grids involve a diverse range of devices, systems, and technologies from multiple vendors, which can lead to interoperability issues. The lack of universally accepted standards can hinder seamless integration and communication between various smart grid components. Developing common standards and protocols is necessary for effective integration. To mitigate this, adopting open standards and protocols for IoT communication, such as message queuing telemetry transport (MQTT), constrained application protocol (CoAP), and OPC unified architecture (OPC UA), is crucial to enable interoperability. Additionally, the development of middleware or gateways that can translate and integrate data from different proprietary systems and collaboration between utilities, technology providers, and standardization bodies to establish common guidelines and frameworks can help address the interoperability and connectivity challenges.

Data security and privacy: The increased connectivity and data exchange in smart grids raises significant security and privacy concerns. IoT devices in smart grids are connected to the internet, making the entire system vulnerable to cyberattacks, especially because IoT devices often have limited security features. Hackers could potentially gain access to sensitive data or disrupt grid operations. Cyberattacks, data breaches, and unauthorized access to sensitive grid information pose serious risks. Protecting data integrity and user privacy and ensuring resilience against cyber threats is critical but challenging in IoT-enabled smart grids. Implementing robust cybersecurity measures, including strong encryption, access controls, and intrusion detection/prevention systems, is essential to mitigate these challenges. Adopting secure communication protocols, such as transport layer security/secure socket layer (TLS/SSL), and end-to-end data encryption, as well as developing comprehensive data governance policies and practices to protect consumer privacy and ensure compliance with regulations, can help address security and privacy concerns.

Scalability: Smart grids involve a vast number of devices (smart meters, sensors, and actuators) spread across a wide geographical area. As the number of IoT devices in smart grids increases, the system must scale up without compromising performance. Ensuring stability as the grid expands is complex and requires robust infrastructure. Adopting a modular and distributed architecture is crucial to mitigate scalability challenges in IoT-based smart grids. This approach leverages edge computing and fog computing principles to process data closer to its source, thereby reducing strain on the core network. Furthermore, utilizing scalable and flexible communication technologies such as 5G, LPWANs, and IPv6 supports the increasing number of IoT devices by providing robust connectivity options. Dynamic resource allocation and load balancing mechanisms ensure efficient management of connected devices and data flows. Scalable data management and storage solutions, such as distributed databases and cloud-based platforms, are essential for handling the expanding volume of data generated within the smart grid ecosystem.

Latency: Smart grids involve integrating many IoT devices, which can strain the network infrastructure and lead to latency issues, especially for real-time applications such as grid control and demand response. To address these challenges, deploying edge and fog computing architectures to process data closer to the source and reduce latency can be beneficial. Additionally, using 5G or other high-bandwidth communication technologies to support the increasing data demands of smart grids and optimizing network topologies and resource allocation to ensure scalability and responsiveness can help mitigate the scalability and latency issues.

Reliability and resilience: IoT devices are susceptible to hardware failures, connectivity issues, and environmental factors. Ensuring continuous operation and resilience of IoT infrastructure under varying conditions (e.g., extreme weather events) is essential. To mitigate this challenge, it is crucial to implement redundant and fault-tolerant communication networks in smart grids that include backup channels and self-healing capabilities. This ensures continuous connectivity and minimizes disruptions. Deploying IoT devices and sensors with durable hardware and resilient software designed to withstand environmental stresses is essential to minimize failures. Developing predictive maintenance and proactive monitoring systems helps detect potential issues early, enabling preemptive maintenance to prevent downtime. Comprehensive disaster recovery and business continuity plans further enhance resilience, ensuring the grid remains operational during extreme weather events or other disruptions.

Energy efficiency and sustainability: IoT devices deployed in smart grids must be energy-efficient to minimize their impact on the grid and reduce overall energy consumption. Many IoT devices have limited power and energy resources. Optimizing the energy consumption of these devices while maintaining their functionality is important for the overall efficiency of the smart grid. Furthermore, while IoT promises efficiency gains, the energy consumed by IoT devices themselves and the environmental impact of manufacturing them must be considered. Balancing the benefits of IoT-enabled smart grids with environmental sustainability goals requires careful planning and lifecycle management of IoT devices. To address this challenge, the design of low-power IoT devices and sensors with energy-efficient hardware and software, the implementation of energy harvesting techniques, such as solar or kinetic energy, to power IoT devices, and the development of intelligent power management algorithms to optimize the energy usage of IoT devices, can be effective strategies.

Regulatory and compliance issues: The deployment of IoT-enabled smart grids may face regulatory and policy hurdles, such as data privacy laws, cybersecurity standards, grid modernization policies, and utility business model changes. IoT deployments in smart grids must comply with these regulatory requirements and standards while addressing the challenges through collaborative efforts between stakeholders, which is essential. However, navigating regulatory landscapes across different regions and ensuring adherence to data protection laws can be complex and time consuming. To mitigate these challenges, active engagement with policymakers, regulators, and industry stakeholders to shape policies and regulations that enable the adoption of IoT in smart grids is crucial. Participation in industry consortia and standards development organizations to influence the creation of guidelines and best practices, as well as compliance with existing regulations and proactive adaptation to evolving policy frameworks, can also help address the regulatory and policy challenges.

Data management and analytics: IoT-enabled smart grids generate large volumes of data that must be collected, processed, and analyzed in real time. Handling the volume, velocity, and variety of data while extracting meaningful insights is a significant challenge. Managing and extracting actionable insights from massive data streams requires robust data management, advanced analytics capabilities, and efficient data storage solutions. Adopting big data technologies, such as distributed processing frameworks (e.g., Apache Spark and Apache Flink) and NoSQL databases, should help handle the volume, velocity, and variety of data that the smart grid generates. Real-time data processing and analytics capabilities, leveraging stream processing engines and ML algorithms, should be implemented to extract insights and enable data-driven decision-making. Furthermore, advanced data visualization and reporting tools should be developed to communicate insights to grid operators and stakeholders effectively. Moreover, data quality, integrity, and security should be ensured throughout the data management pipeline, from data collection to storage and analysis.

Legacy infrastructure integration: Integrating IoT technologies with legacy grid infrastructure and ensuring compatibility can be complex and costly. Retrofitting older grid components with IoT devices and ensuring smooth interoperability poses technical and logistical challenges. Developing solutions that can seamlessly bridge the old and new systems is necessary for a successful smart grid transformation. For seamless integration of IoT and legacy infrastructures in smart grids, it is essential to develop interoperability frameworks and middleware solutions that bridge the gap between IoT systems and legacy grid infrastructure. These solutions facilitate seamless data exchange and coordinated operations. Utilizing open standards and communication protocols such as IEC 61850, open automated demand response (OpenADR), and common information model (CIM) helps integrate diverse systems and avoids vendor lock-in, ensuring compatibility across platforms. Adopting a phased approach to smart grid modernization is beneficial, allowing for the gradual integration of IoT technologies with existing components to minimize disruptions and maximize the value of past investments. Collaborating closely with equipment manufacturers and service providers is crucial for developing retrofit solutions that easily integrate legacy grid components with new IoT-enabled systems, ensuring a smooth transition and enhanced functionality of the smart grid infrastructure.

Skills and workforce gap: Implementing IoT in smart grids requires specialized skills and expertise in data analytics, cybersecurity, and grid automation, which can be a significant challenge for utilities and grid operators. Bridging the skill gap and training the workforce is crucial for successful adoption. Overcoming these challenges will require a comprehensive and collaborative approach involving technology providers, grid operators, policymakers, and end users. However, addressing concerns and resistance among stakeholders (utilities and consumers) regarding IoT systems' privacy, security, and reliability may be an obstacle. Continuous research, innovation, and the development of robust solutions are necessary to unlock the full potential of IoT in smart grid applications. Investing in training and upskilling programs for utility personnel to develop competencies in IoT technologies, data analytics, and system integration is important to address this challenge. Collaboration with educational institutions and industry partners to build a pipeline of skilled professionals in the IoT and smart grid domains and adopt knowledge management strategies to capture and share institutional knowledge within the organization can help mitigate the skills and workforce development challenges.

Cost and return on investment (ROI): Initial setup costs for smart grid infrastructure and ongoing operational expenses can be significant. Demonstrating tangible ROI through improved efficiency, reduced operational costs, and enhanced grid management capabilities is crucial for justifying investments in IoT for smart grids. To mitigate the cost and ROI challenges in smart grids, it is crucial to conduct thorough cost–benefit analyses that quantify the potential benefits of investments. These analyses should include improvements in operational efficiency, reductions in energy losses, enhancements in grid resilience, and improvements in customer engagement. Exploring innovative financing models such as public–private partnerships can help distribute upfront costs and share risks effectively. Implementing energy management and demand response programs that leverage IoT technologies can generate immediate cost savings and demonstrate tangible ROI. Additionally, developing strategies to monetize the data and insights generated by the smart grid, such as offering value-added services to customers or selling aggregated data to third-party providers, can create additional revenue streams and further justify the initial investment in IoT technologies for smart grids.

7.7 Economics of IoT-Enabled Smart Grid

Real-time monitoring of smart grids facilitates real-time energy pricing and tariff management. IoT applications can collect real-time data on energy supply and demand, grid conditions, and market factors. Utilities can leverage these data to implement dynamic pricing models, where energy prices vary based on real-time factors such as peak demand, grid congestion, or renewable energy availability. Real-time energy pricing incentivizes consumers to shift their energy usage to off-peak periods, promote energy conservation, and optimize grid operations.

7.7.1 Pricing Models and Techniques

To maximize grid operations, promote energy saving, and enable the integration of renewable energy sources, the economics of IoT-enabled smart grids is multifaceted, encompassing various costs and pricing models to ensure efficient power distribution and grid stability. Using real-time data, dynamic changes, and customized incentives, the pricing strategies help to match consumer behavior to the demands of the smart grid. Smart grid operators can build a more sustainable and efficient electricity system that benefits consumers and the grid overall by combining strategies, including dynamic pricing, tiered structures, renewable energy incentives, demand response programs, and subscription-based models.

Various pricing models have been used in power grid systems. Some of the prominent pricing models that are suitable for the smart grid are discussed below.

- **Dynamic Pricing:** Dynamic pricing models adjust electricity rates in real time or across predetermined periods to better align consumer behavior with grid optimization and renewable energy integration.
  - **Real-Time Pricing:** This model adjusts electricity prices continuously, often hourly, based on supply and demand. Consumers receive price signals reflecting the current cost of generating and delivering electricity, encouraging them to shift their usage to lower-priced periods. This helps the grid operator balance supply and demand, reduce peak loads, and integrate more renewable energy sources.

- **Time-of-Use (TOU) Pricing:** In this model, electricity prices are predetermined for different periods, such as peak, off-peak, and shoulder hours. Prices are typically higher when demand is highest during peak hours and lower during off-peak hours. This incentivizes consumers to shift their usage to lower-priced periods, reducing strain on the grid during times of high demand.
- **Critical Peak Pricing (CPP):** This is a variation of TOU pricing, where prices are significantly increased during predefined critical peak periods, such as during extreme weather events or when the grid is under severe stress. The high prices during these critical periods are designed to elicit a strong demand reduction response from consumers, helping to maintain grid stability and reliability.

- **Tiered Pricing:** Tiered or block pricing: In this model, consumers are charged a higher rate for electricity consumption above a certain threshold or block of usage. This encourages energy conservation and efficient electricity use, as consumers are incentivized to keep their consumption within the lower-priced tiers.
- **Renewable Energy Pricing:** Pricing models that support renewable energy generation, such as feed-in tariffs and net metering, help incentivize the adoption of distributed renewable energy sources.
  - **Feed-In Tariffs:** This model provides a fixed, guaranteed rate for the electricity consumers generate from their own renewable energy sources, such as rooftop solar. The grid operator or utility company purchases this electricity at a predetermined rate, which helps to incentivize the adoption of distributed renewable generation.
  - **Net Metering:** In this approach, consumers are credited for the excess renewable energy they generate and feed back into the grid. The credit offsets their electricity bills, making it more financially attractive for consumers to invest in on-site renewable generation.
- **Demand Response Pricing:** Demand response pricing approaches, both incentive based and behavior based, aim to reduce peak electricity demand by encouraging consumers to shift or curtail their usage during high grid stress.
  - **Incentive-Based Programs:** These programs offer financial incentives, such as bill credits or direct payments, to consumers who reduce their energy consumption during peak demand periods or when called upon by the grid operator. This helps to shave peak loads and improve grid stability.
  - **Behavioral-Based Programs:** These programs provide consumers with information, feedback, and education to encourage voluntary energy conservation during peak periods. The goal is to influence consumer behavior and shift usage patterns without relying solely on financial incentives.
- **Subscription-Based Pricing:** Subscription-based pricing models, such as flat-rate or fixed-fee structures, provide consumers with more predictable electricity costs while promoting energy efficiency.
  - **Flat-Rate or Fixed-Fee Pricing:** This model involves a predetermined monthly or annual fee for access to a certain amount of electricity, regardless of the consumer's actual usage. This can provide consumers with more predictable and stable electricity costs while still incentivizing energy efficiency and conservation.

The combination and implementation of these pricing models and techniques will depend on various factors, such as the specific regulatory environment, the maturity of the smart grid infrastructure, and the local energy market characteristics. A well-designed pricing strategy that leverages these approaches can help optimize grid operations, promote energy efficiency, and support the integration of renewable energy sources in an IoT-enabled smart grid.

7.7.2 Power Costs

Power pricing includes generator power production cost, transmission cost, and primary and secondary ancillary services cost as main components.

7.7.2.1 Generation Costs

The cost of power generation varies based on the type of power plant. Thermal power plants have a quadratic relationship between cost and power output, while wind and solar power have a more linear cost curve [52]. The "flat" cost curve of renewables makes them economical for trading if the total power output is purchased [53]. At low-power purchase levels, the per-unit cost is higher due to the recovery of capital investment. Flexible plants such as natural gas and diesel can be used for peak load hours as they have steeper cost curves. In recent years, increased renewable installations have driven down bidding prices to be similar to coal-based plants [54, 55]. The relative cost of coal, natural gas, and renewable sources are shown in Figure 7.5. These are approximate graphs; the actual may vary up to 15% [56]. The generation cost is estimated on the levelized cost of energy, which mainly includes capital investment cost, fuel cost, annual operation and maintenance cost, annual interest, and depreciation cost [57].

IoT technologies play a significant role in monitoring and optimizing the performance of different generation sources, including thermal, solar, and wind power plants. IoT sensors and devices provide real-time data on generation efficiency, weather conditions, and equipment health, enabling better forecasting and optimization of supply curves. Fuel prices, plant efficiency, and operational constraints influence the cost of thermal generation. IoT can optimize fuel usage and predict maintenance needs, reducing operational costs. On the other hand, solar and wind power generation sources have variable outputs depending on weather conditions. IoT-enabled predictive analytics improve the accuracy of generation forecasts, optimizing the supply curve and reducing costs associated with over or underproduction.

7.7.2.2 Wheeling Costs

Wheeling costs are the charges associated with transmitting power across the grid. Transmission cost mainly depends on capital investment cost, power to be transferred and distance of

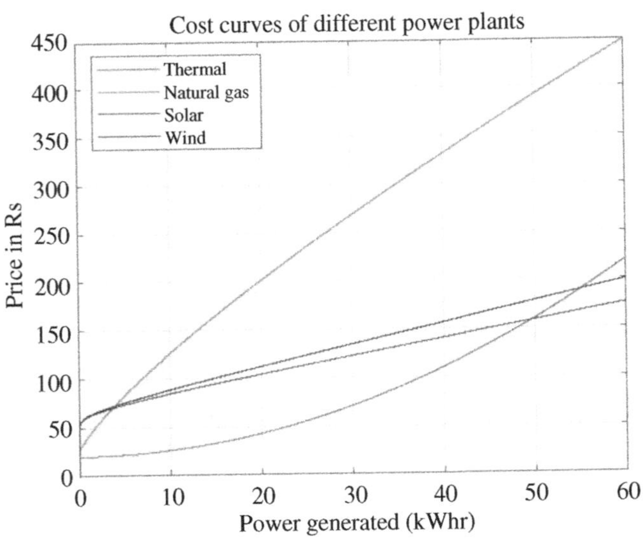

Figure 7.5 Cost curve of different power plants.

transmission, annual operation and maintenance cost, opportunity cost, reinforced cost, etc. [58]. The wheeling or service charges for using a transmission line can be calculated based on the actual loading data of the transmission lines during power transfer. IoT devices in transmission lines and substations provide real-time data on line capacity, congestion, and power flows. This data is crucial for determining the most efficient and cost-effective wheeling charges. These charges often consider the distance the power is being transmitted, either directly or indirectly, through the service charges of all the substations involved.

An IoT-based platform can facilitate coordination between substations for decision-making and monitoring the status of transmission equipment. This can help optimize the transmission system and ensure reliable power delivery.

7.7.2.3 Ancillary Services

In the ancillary services market, the cost per power unit becomes higher as it includes reliability and rapid response requirements. Ancillary services mainly consist of voltage and frequency (primary) support, congestion management, secondary and tertiary support, black start services, etc. The voltage and frequency support is essential for reliable power transmission, and ancillary services are customized as per the state of the transmission network and the dynamic behavior of the customer [59].

IoT-enabled ancillary services also enhance grid stability by enabling real-time monitoring and control of grid parameters. Mainly, ancillary services benefit from IoT for:

- **Active Power:** Active power stabilizes frequency in the grid. IoT sensors monitor frequency deviations and automate responses to stabilize the grid.
- **Reactive Power:** Reactive power provides voltage support. Reactive power support is essential in case of fault or instant load ramp-up. IoT-enabled devices provide real-time voltage support and adjust reactive power supply to maintain voltage levels.

IoT platform may trigger an alarm during transients of frequency and voltage crossing limits and add secondary support at appropriate times. IoT can enhance the responsiveness and reliability of generators by providing real-time monitoring and automated controls as follows:

- **Gas, Diesel, Hydro, and Energy Storage:** These generators offer better ramp control; hence, they are more suitable for secondary and tertiary ancillary services [60]. IoT enhances their performance by providing predictive maintenance and real-time operational data.
- **Thermal Plants:** Thermal plants may also provide reserve and regulation to attain load following. IoT can optimize their reserve and regulation services participation by predicting demand and adjusting generation accordingly.
- **Mixed Generation and Load Management:** A mix of generations, along with load shifting and shedding, can be utilized for ancillary services. IoT facilitates the coordinated operation of different generation sources and DSM through automated load shifting and shedding, improving ancillary service provision.

7.7.2.4 Opportunity Costs

Generators co-optimized for power and ancillary services markets face opportunity costs for not being able to bid their full capacity in the power market. IoT can mitigate these costs by optimizing the dispatch of generation resources and enhancing their participation in multiple markets. Loss of opportunity cost compensates generators for their commitment to ancillary services. IoT provides detailed data on generator availability and performance, ensuring accurate calculation and fair compensation for these costs.

7.7.3 Tariff Calculation

The aggregate of the three costs mentioned in the previous section with a certain rate of return constitutes a tariff for the customer. The tariff calculation is a customized task that depends on generation type, transmission line capacity, distance between generation and receiving node, available transfer capacity and reliability of transmission network, and selection of ancillary services.

Generation cost: Each generation (thermal, solar, wind, etc.) plant has a unique supply curve, which is directly used to calculate the cost of power. The trade in power market is inferred from the supply cost curve.

Wheeling cost: The wheeling cost is calculated using the rolled-in or embedded method, marginal cost method, postage stamp method, and contract path method [58]. The selection of the above methods depends on whether power transactions are either in the bilateral market or spot market with an estimation of the distance between buyer and seller. The roll-in method is applicable in all markets, while marginal cost may be used in spot and over-the-counter (OTC) markets. The postage stamp method is common for local power distribution, and the contract path is used for bilateral trading on dedicated transmission lines [61].

- **Rolled-In Method:** This method distributes the total transmission costs among all users. IoT can optimize this by providing accurate usage data, ensuring fair cost distribution. It is also known as the embedded cost method.
- **Marginal Cost Method:** Calculates the additional cost of transmitting one more unit of electricity. IoT data on real-time congestion and line losses can improve the precision of these calculations, especially in spot and OTC markets.
- **Postage Stamp Method:** Applies a uniform charge regardless of distance. IoT ensures accurate tracking of usage across local areas, supporting fair implementation.
- **Contract Path Method:** Charges are based on a predefined path. IoT devices track the exact path of power flows, ensuring accurate billing in bilateral trading.

Ancillary services: Ancillary services costs are mainly from generators and energy storage batteries. The generators co-optimized for both power and ancillary markets are provided with an additional cost known as loss of opportunity cost [61]. The generator's spinning or reserve power capacity is employed for frequency regulation. In the ancillary services market, the power supply becomes more costly because the system operator decides the quantity of power purchased, and these generators are bound to provide their services.

7.7.4 Pricing Criteria

The pricing model includes all components such as power generation, transmission, and specific rate of return. The pricing must be fair and simple to understand, and it must relate to all real time. IoT could increase the effectiveness of pricing models of smart grid, thereby providing access to real time data of household appliances' power consumption, TOU price pattern throughout the day to the customer, sensor-based measurement module for tracking of power flow, power dispatch of generators, switch gear apparatus status, etc. IoT module may direct EMS and apply on/off control of residential loads, which takes decisions to manage demand-side load and reduce losses. IoT-based systems bring accuracy to data acquisition, and hence, decision-making becomes efficient for the operation of the smart grid.

- Real-time data collection is essential for both suppliers and consumers. A supplier would like to place the bid when the load demand peaks, whereas a consumer would cut its consumption during this period as TOU rates are high. This IoT data-sharing environment makes customers use power intelligently, shifting its burden from peak to off-peak hours. It also brings economic efficiency in trading, better plant load factor of generation, and better utilization of resources.

- The generators are dispatched to follow the demand pattern; hence, the mix of generation plant and their unit commitment will be changed. The price of total generation energy changes according to the different cost curves shown in Figure 7.5. The dynamics of generation units and their dispatch vary depending on the price of the market. The dynamics in pricing and models identify the exact energy dynamics of the grid. Dynamic pricing results in better resource management, and these prices may be communicated through IoT-based modules to all stakeholders except consumers. Dynamic pricing will increase distribution infrastructure in high-demand zones as revenue will fund development.
- Dynamic pricing with a specific rate of return and subsidies on consumer energy consumption by regulation authorities constitutes a tariff. The tariff on energy consumption is communicated by an IoT-based platform to the consumer. The consumer uses the TOU tariff (throughout the day) to restructure its energy consumption pattern. It helps not only the consumer to reduce its bill but also reduce loading on transmission and distribution systems at peak hours. Overall, TOU tariff varies according to real-time consumption of consumers and geographical surplus or deficit zones.
- IoT-based sensor modules communicate the loading data of power lines. This data is used by the load dispatch center (LDC) to manage load flow. In case of full or overloading of lines, the IoT platform sends alarming signals to LDC. The bottleneck that appears in the distribution grid is managed by diverting power to nearby power lines. The increased power loss is accessed by IoT-based communication, which is added to wheeling charges. The congestion in lines reduces transmission efficiency, and this problem can be efficiently eliminated by increasing the transmission capacity of congested lines.
- IoT-based communication of generation, grid data, and consumer behavior is utilized for forecasting the next day or month dynamics. Renewable generation IoT sensor-based data recorded daily, weekly, seasonally, and yearly are combined to forecast upcoming power production. Moreover, weather-based input is mandatory in forecasting models. These inputs can predict the upcoming power trading market process and estimate market risk. This ensures efficient energy management and clearing prices.
- IoT-based communication makes the market "data-driven"; hence, decision-making becomes easy to understand. In such an environment, market abuse becomes inevitable, and market rule violations are tracked. IoT-based platforms provide security to the market and bring transparency.

7.7.5 Consumer and Market-Driven Power Flow

The IoT is critical in enabling consumer and market-driven power flow within the smart grid. Real-time data collection facilitated by IoT-enabled devices allows consumers to closely monitor and manipulate their energy consumption patterns based on available tariffs or TOU pricing. This empowers consumers to optimize their electricity usage and actively reduce their bills.

On the grid management side, the real-time data collected by IoT sensors and devices provides the system operators with granular insights into aggregate demand. This allows the generation to precisely match demand, minimizing transmission losses and wheeling charges. The IoT-based measurement modules enable grid operators to monitor and manage technical constraints such as frequency and voltage issues and congestion to ensure reliable and efficient power delivery.

The IoT-enabled smart grid creates a closed-loop feedback system between the overall inputs and outputs of the power system. The real-time data collected by IoT sensors and devices allows the system to continuously monitor and update the state of generation, power flow, and consumer demand. This feedback loop enables the system to optimize the power flow and generation dynamically to minimize the tariff charged to consumers. The IoT-driven feedback process works through the following steps:

1) IoT devices collect real-time data on consumer energy consumption, market prices, and grid conditions.
2) This data is fed back into the system, providing visibility into the aggregate demand patterns and generation capacity.
3) The system then updates the power generation dispatch to match the dynamic demand while considering factors such as transmission losses and wheeling charges.
4) The updated power flow and generation data are used to calculate the optimal tariff that can be offered to consumers to minimize the overall cost.
5) Consumers then respond to these tariff signals by adjusting their usage patterns, closing the feedback loop.

The IoT-enabled communication infrastructure is also critical for enabling digital power trading practices. It facilitates secure, encrypted data exchange required for single or double-auction-based power trading on spot markets. This includes day-ahead, intraday, and term-ahead markets, where IoT data ensures real-time tracking of market progress and enables efficient market clearing and settlement.

7.7.6 Power Trading Practices

The integration of IoT in smart grids has transformed power trading practices, enabling real-time data exchange, market optimization, and improved grid stability and reliability. IoT-based communication and monitoring are essential for incorporating real-time developments in the bid market to the actual power transfer through the grid, ensuring seamless and efficient power trading. IoT-enabled smart grids have introduced new power trading practices that leverage real-time data and communication to optimize electricity generation, distribution, and consumption. There are three main power trading practices, as discussed below.

Long-term power purchase agreements (PPAs): Long-term PPAs, typically ranging from 20 to 25 years, are established between power generators and utility companies or large consumers. This contract includes annual power purchase, tariff, tariff-increment rate, wheeling charges, power loss, and operational cost allocation. IoT platforms play a crucial role in these agreements by providing real-time data on individual generating units and the loading of transmission lines. This information helps monitor the ramp-up rate, steady-state deviations, and power dispatch to match consumer demand patterns. The real-time data from IoT sensors and devices can be used for pricing, billing, and risk management analysis, enhancing power transactions' overall efficiency and transparency. This practice covers about 88% of India's electricity market share [62].

Power exchange trading: Power exchange trading is based on bid and auction models for defined time frames, including day-ahead, intraday, and term-ahead markets [63]. This covers about 6.3% of India's electricity market share [62]. IoT-based communication and monitoring are essential for tracking the progress of these virtual and physical grid-based markets in real time.

- In the day-ahead power market (DAM), IoT-enabled systems facilitate the rapid acceptance of bids from both customers and suppliers within a 15-minute timeframe. This allows the market

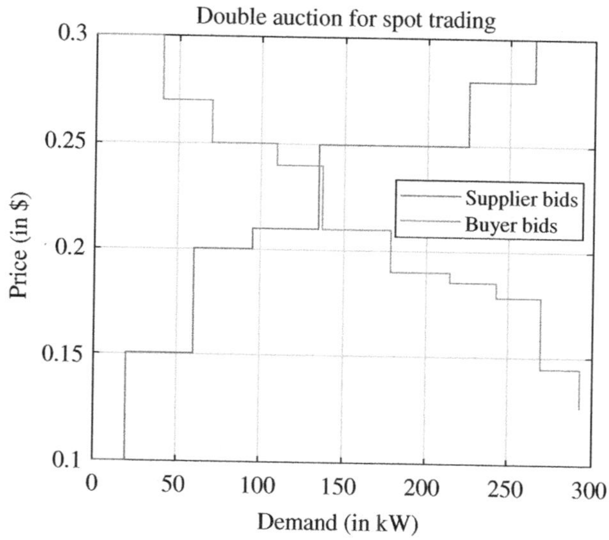

Figure 7.6 Double auction for spot trading.

operator to efficiently run double-auction markets and determine the market clearing volume (MCV) and market clearing price (MCP) for each of the 96 time frames in a day[1] [60]. The typical double-auction process is shown in Figure 7.6. The suppliers and customers whose bids are selected in the auction will then supply or consume power according to their bids. The market operator may be a utility-based entity or an independent organization responsible for overseeing the DAM. Once the DAM is cleared, the system operator will direct the generators and transmission companies to dispatch and transfer the power, respectively, to the end consumers. IoT-based communication and monitoring systems are essential in facilitating the seamless integration of the DAM outcomes with the actual real-time power transfer through the grid. The system operator, whether utility-based or independent, is responsible for managing the power flow and maintaining the stability of the transmission network to ensure the reliable delivery of electricity to consumers based on the DAM outcomes.

- In intraday power markets, trades and actual power transactions are clear within the same day [60]. In this fast-paced market environment, IoT technology plays an essential role in managing short-term market settlement and enabling seamless power transactions in a timely and credible manner. Real-time uncertainties often arise in intraday markets due to the dynamic nature of consumer behavior. As a result, power suppliers may deviate from their contracted dispatch levels to adapt to the changing demand patterns. IoT-based communication and monitoring systems are highly valuable in managing these short-duration imbalances between supply and demand. IoT-enabled systems provide the necessary real-time data and communication channels to stabilize the grid under these uncertain conditions. The power generation is constantly balanced to match the dynamic consumer demand during the actual power transaction process. Any deviations between the contracted power in the bid and the actual power dispatch or consumption are later settled through the imbalance settlement mechanism, as understood from Figure 7.7. The IoT-driven imbalance settlement process is crucial for maintaining grid stability and ensuring the reliability of power supply [64]. By providing real-time data and communication capabilities, IoT technology enables power suppliers and grid operators to

1 https://www.iexindia.com/

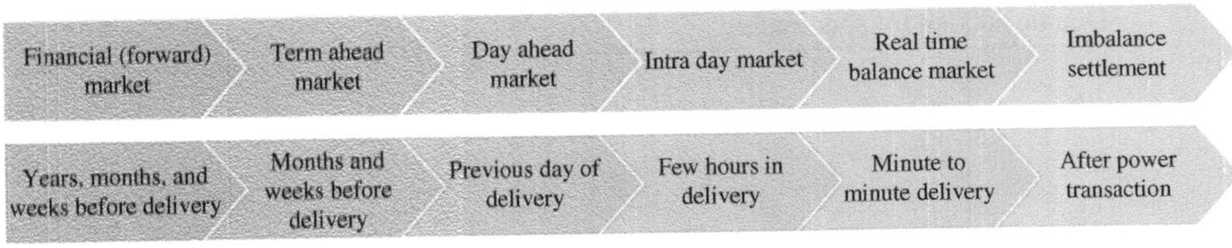

Figure 7.7 Timeline of power trading practices in power exchange.

respond quickly to changes in demand, minimizing the impact of imbalances and ensuring the efficient functioning of the intraday power market.

- In the term-ahead power market (TAM), trading can take place for time horizons up to one week in advance or roughly 10 days ahead of the actual delivery of power [60]. This longer-term market allows suppliers and customers to secure power supply and demand commitments further into the future compared to the DAM. The TAM allows power generators, transmission providers, and large industrial/commercial consumers to strategically plan and hedge their electricity needs. Participants can lock in prices and volumes for the upcoming week or so, providing more certainty in their operations. IoT-enabled systems play a crucial role in the TAM by facilitating data collection, aggregation, and analysis relevant to medium-term power supply and demand forecasts. This includes weather predictions, equipment maintenance schedules, and evolving consumption patterns. The real-time data and advanced analytics capabilities enabled by IoT allow market participants to make more informed bidding decisions in the TAM. This leads to better price discovery and resource optimization in the longer-term power trading horizon.

Bilateral transactions: Bilateral power transactions through traders or utilities involve direct contractual agreements between electricity suppliers and consumers, with or without the involvement of intermediaries. A single party may take on multiple contracts in these bilateral arrangements to meet its specific power requirements. The contract durations in the bilateral power market can range from day-ahead to weekly, monthly, and even multiyear timeframes. This makes it a pool or forward market, compared to an OTC trading model, where the two parties negotiate and agree upon the terms directly [65]. Bilateral power transactions currently account for around 5.6% of India's total electricity market share [62]. IoT technologies play a crucial role in enhancing the transparency and efficiency of these bilateral transactions. IoT-enabled systems provide real-time data and insights on power generation, supply costs, and availability across the entire grid. This information allows the bilateral market participants to make more informed decisions when negotiating contract terms, volumes, and pricing. IoT's enhanced data visibility and analytics capabilities help suppliers better forecast their generation capacity and costs. On the other hand, consumers can better understand their medium-to-long-term power requirements and availability in the grid. This improved transparency and data-driven decision-making in the bilateral power market leads to more efficient matching of supply and demand. It also facilitates trading renewable energy certificates and other flexibility products integral to the evolving power sector.

OTC market also exist for short-term power trade. The differences between OTC market and power exchange or bilateral market are given below:

- The buyer and seller submit their power trading details to an agency or counter. The agency matches requirements from both sides and suggests that both parties meet and finalize the deal. The deal may be finalized through an authorized broker or without any intermediate agency, as shown in Figure 7.8. In power exchange trading, both sides do not know about each other.

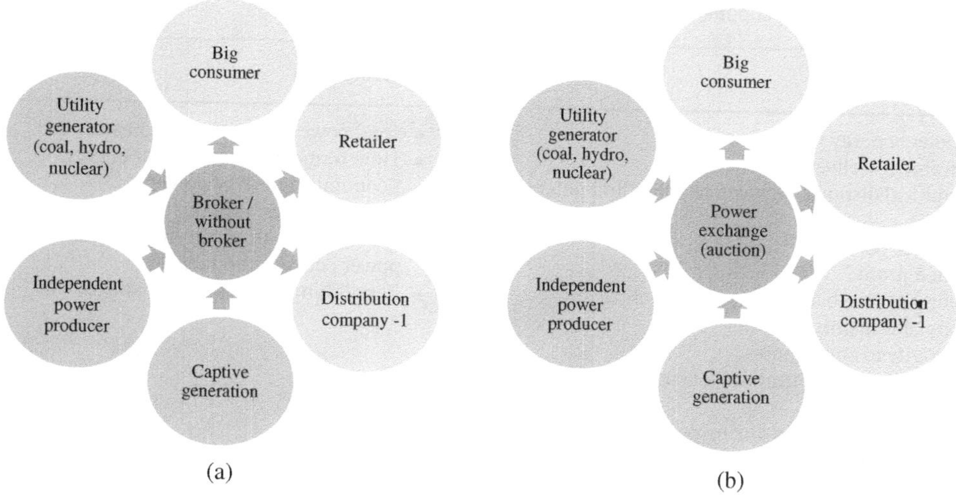

Figure 7.8 (a) Over-the-counter and (b) power exchange trading structure.

- The buyer and seller may make contract for short-term trading because the volume of power is small as compared to the bilateral market. When sellers are not able to sell their power in power exchange (when bid gets rejected), then OTC market is an alternate market for trading.
- When buyer of long-term PPA does not get the full required supply from its conventional supplier due to any reason, then remaining power is bought from power exchange or OTC market.

7.7.7 Real-Time Power Trading

Real-time power trading primarily takes place through power exchanges, which operate as spot markets. These spot markets rely on either single-sided or double-sided auction mechanisms to clear transactions. In a double-auction spot market, both the supply and demand sides actively participate by submitting their bids. Whether conventional or renewable, power generators submit their real-time power output, supply costs, and availability timeframes as part of their bids to the market operator. Similarly, end consumers bid their power consumption requirements, purchase costs, and the desired time frames. The market operator then runs a double-auction settlement process to match the supply and demand bids, determining the MCV and MCP for each trading interval. This double-auction approach in the spot market facilitates competitive price discovery and optimal resource allocation based on real-time supply and demand dynamics.

In contrast, single-sided auction spot markets only accept bids from the supply side, while the demand side is estimated based on historical consumption patterns. The market operator then clears the market by accepting the supplier bids, often resulting in a flat rate charged to all end consumers. This single-sided auction model is less competitive than the double-auction approach, as it does not incorporate the demand-side active participation and price discovery.

IoT-enabled systems play a crucial role in enhancing real-time power trading in single-sided and double-sided spot market models. IoT technologies provide real-time data and analytics capabilities that allow generators, consumers, and the market operator to make more informed bidding and clearing decisions. The real-time visibility of power output, supply costs, and consumption patterns, facilitated by IoT, enables better matching of supply and demand in the spot market. This improves efficiency, reliability, and competitiveness in the real-time power trading ecosystem [66].

Table 7.1 Factors affecting supplier and consumer side trade.

Supplier side	Consumer side
• Total power energy for trading • Time frame to deliver that energy • Geographical distance between supplier and potential consumers • Bidding rate per unit • Estimated profit • Reliability of (bid-based) power dispatch • Capacity of transmission lines • Connectivity to consumers • Real-time energy consumption of consumers (for transparency) • Estimation of power loss during power transmission (by IoT-based metering systems) • Congestion in transmission lines • Provision of ancillary services prior to real power trading by the system operator (SO) • Better performing suppliers must be given incentive	• Total power purchase for consumption • Time frame to consume that energy • Economical tariff of power • Reliable power supply • IoT-based smart meter (for real-time power consumption data) • Easy billing and revenue management

7.7.8 Peer-to-Peer Energy Trading

The integration of real-time monitoring capabilities within the smart grid ecosystem enables the emergence of P2P energy trading. This decentralized approach to energy exchange is facilitated by the advancements in IoT applications, which allow for real-time tracking of energy production and consumption.

The IoT-powered platform provides enhanced control and coordination over energy usage, allowing prosumers (producer or consumer as single entity) to share their excess energy with other consumers or the utility grid. This seamless power exchange on the low-voltage distribution network, involving many decentralized sources and consumers, constitutes a thriving retail energy market within a specific geographical area. The feasibility of power trading between seller and buyer depends upon economic, technical, and physical distance constraints. The factors affecting power transactions in real-time market for both seller and buyer are given in Table 7.1.

The key factors that govern the success of this P2P energy trading ecosystem include:

- **Economic and Transparent Market Settlement:** The market settlement process must be economically viable, with a transparent and easy-to-use billing and payment system. The market should be highly competitive, without intermediaries, allowing new suppliers to join the platform easily.
- **Local Market Operator and Grid Coordination:** The local market system operator will monitor the real-time power transfer and have the authority to select suppliers and re-dispatch power to manage reliable grid operations. The local operator will coordinate with the regional load dispatch center (RLDC) to manage non-P2P power needs through the centralized power market.
- **Reduced Burden on Centralized Infrastructure:** P2P energy trading can reduce the need for investment in high-voltage transmission network expansion, as the power is primarily exchanged within the local distribution network. The reliability of the decentralized system, with diverse suppliers and mutual competition, promotes a local generation-local consumption-based market, which reduces the dependency on centralized generation plants.

- **Improved System Resilience and Sustainability:** In the event of outages or disruptions in the centralized power plants or transmission lines, the local suppliers can step in to uphold the market and manage the load requirements until the centralized systems are restored or alternative suppliers are arranged. This decentralized, low-voltage power market ecosystem adds to the overall sustainability and resilience of the power system.
- **IoT-Based Secure Trading Platform:** The IoT-based secure platform can manage the bidding process, ensure data security, facilitate transparent market settlement, and enable real-time power transfer, thereby building trust among the stakeholders. Integrating smart contracts and blockchain technology with IoT applications can further enhance the security and transparency of the trading process.

7.7.9 Peer-to-Peer Energy Transaction

In the smart grid ecosystem context, P2P energy transactions are a logical progression from the auction-based market settlement processes. This decentralized approach empowers consumers and suppliers to engage in direct energy trading, bypassing the traditional centralized grid infrastructure.

The key premise of P2P energy transactions is that consumers who have bid above the clearing price and suppliers who have bid below the clearing price are allowed to exchange power through the distribution network directly. This dynamic allows for more granular and efficient energy exchanges, leveraging the inherent flexibility of the smart grid.

To enable effective P2P transactions in a decentralized grid, the system requires robust IoT-based communication platforms and real-time data acquisition. These IoT-powered platforms can collect data from various generation nodes, transmission lines, and other grid components, providing a unified view of the grid's power flow status. This real-time visibility is crucial for managing the ongoing power exchanges between multiple suppliers and consumers.

Effective management of P2P transactions in the IoT-enabled smart grid necessitates addressing several key considerations:

- **Power Loss Minimization:** The system should prioritize power transfer routes that minimize energy losses, ensuring the efficient utilization of grid resources.
- **Grid Constraint Compliance:** The P2P transactions must adhere to the technical and operational constraints of the grid, avoiding the violation of grid stability and security parameters.
- **Congestion Management:** In the event of transmission line congestion, the system should be able to dynamically divert power flows to alternative routes, ensuring uninterrupted energy delivery.
- **Proximity-Based Optimization:** Wherever possible, the system should prioritize energy exchanges between suppliers and consumers in close geographic proximity, reducing transmission losses and grid infrastructure requirements.
- **Continuous Grid Monitoring:** Ongoing monitoring and data analysis of the grid's status, including generation, transmission, and consumption, are essential for facilitating seamless P2P transactions and maintaining grid stability.

By addressing these key considerations, integrating P2P energy transactions within the IoT-enabled smart grid ecosystem can unlock significant economic benefits, including improved energy efficiency and reduced operational costs. This decentralized approach empowers both consumers and suppliers, fostering a more dynamic and responsive energy market.

7.7.10 Real-Time Bidding

The integration of IoT technology has enabled novel approaches to real-time bidding and power trading within the smart grid ecosystem. The IoT-based communication infrastructure supports time-frame-based bidding, allowing consumers and suppliers to match the magnitude of power demand and supply accurately. The bidding process can be initiated either manually by the participants or automatically by smart meters. IoT modules can measure the excess available energy, and this information can be directly communicated to the local system operator. The supplier can also share relevant pricing details, streamlining the overall transaction process.

Understanding the need for such real-time bidding mechanisms in the smart grid ecosystem is important. The real-time bidding in smart grids is required for the following factors:

- **Increasing Complexity of the Grid:** As the smart grid evolves, integrating DERs, EVs, and AMI, the need for real-time coordination and optimization becomes increasingly crucial.
- **Accommodating Intermittent Renewable Sources:** The rising share of renewable energy generation, such as solar and wind, necessitates the implementation of real-time bidding mechanisms to manage the fluctuations in power supply and demand.
- **Empowering Consumers and Prosumers:** Real-time bidding empowers consumers and prosumers to participate in the energy market actively, making informed decisions about their energy usage and generation.
- **Supporting DSM:** Real-time bidding enables advanced DSM strategies, allowing consumers to adjust their energy consumption patterns based on real-time pricing signals and grid conditions.
- **Enhancing Grid Resilience:** By facilitating rapid response to changes in the grid, real-time bidding can contribute to the overall resilience of the smart grid, helping to mitigate the impact of disruptions and grid emergencies.

Incorporating real-time bidding in smart grids yields the following benefits:

- **Improved Efficiency:** Real-time bidding allows for the dynamic matching of power supply and demand, leading to more efficient utilization of grid resources and reduced energy wastage.
- **Increased Flexibility:** Participating in real-time bidding empowers consumers and suppliers to respond quickly to changing market conditions, enabling them to optimize their energy transactions.
- **Enhanced Grid Stability:** By facilitating real-time adjustments in power generation and consumption, real-time bidding can help maintain grid stability and mitigate the impact of intermittent renewable energy sources.
- **Reduced Costs:** The competitive nature of real-time bidding can lead to more favorable pricing for both consumers and suppliers, resulting in overall cost savings for the entire energy system.
- **Enabling Renewable Integration:** Real-time bidding can support the seamless integration of renewable energy sources by allowing for the dynamic balancing of supply and demand, addressing the inherent variability of renewable generation.

However, the implementation of IoT-based real-time trading in smart grids faces several challenges, such as the following:

- Ensuring the security of the bidding process and data to prevent unauthorized access or tampering.
- Achieving low latency in the communication system enables quick response times between suppliers and consumers.

- Ensuring the scalability of the communication system to handle the increasing number of participants and transactions.
- Protecting the privacy of supplier and consumer accounts to maintain trust and confidence in the trading platform.
- Maintaining the quality and accuracy of the data used for decision-making, with robust error detection and correction mechanisms.
- Ensuring the reliable operation of the IoT-based system, even during emergency or fault conditions.
- Standardizing the different IoT-based systems and integrating the real-time data from various sources.

To address these challenges, the communication system for P2P trading must be robust and resilient. The IoT-based modules and protocols must be well coordinated, even in a heterogeneous mixture of systems. A coherent flow of data integration across different devices, such as routers, gateways, and protocol converters, is crucial for seamless and transparent decision-making.

The implementation of encryption-based communication, along with reliable IDSs, can help ensure the privacy and security of the trading platform. Accurate data integration is essential for transparent and economically efficient decision-making, where each supplier is provided a fair opportunity to participate in the market.

While the IoT-based communication system is well suited for trading and monitoring applications, its latency and response time limitations may restrict its direct application for critical grid control and protection functions. However, the IoT-based system can contribute to noncritical aspects of power transactions, such as informing about line outages, apparatus malfunctions, and congestion problems. Additionally, IoT-based systems can enhance energy efficiency in the grid through applications such as re-dispatch, countertrade, and price-area management.

7.8 Smart Grid in India

India's power sector is undergoing a transformative journey driven by the country's active embrace of smart grid technologies. With the involvement of diverse stakeholders, India is converting its power infrastructure into a secure, flexible, and sustainable network, delivering digital superiority and reliable energy access for all.

In the past, India faced acute power supply shortages, with the installed generation capacity failing to meet the growing energy demands. This challenge was compounded by high aggregate technical and commercial (ATC) losses, lack of effective measurement systems, power theft, and improper estimation of power losses in the distribution network [67]. As a result, the distribution companies (DISCOMs) were incurring heavy financial losses for decades.

To address these challenges, the Indian government has launched a series of initiatives to reform the country's power system and unleash its untapped potential. The Ministry of Power, Government of India, has spearheaded the National Smart Grid Mission (NSGM [2]) in 2015, the Ujwal Discom Assurance Yojana (UDAY [3]) in 2015, and the Revamped Distribution Sector Scheme (RDSS[4]) in 2021.

---

2 https://www.nsgm.gov.in/
3 https://energyportal.in/power/ujwal-discom-assurance-yojana-uday
4 https://powermin.gov.in/en/content/overview-5

The NSGM is tasked with organizing and monitoring the implementation of smart grid-related policies and initiatives across India [47]. Key objectives of the NSGM include:

- Deploying AMI for real-time monitoring of electrical equipment and consumers.
- Modernizing substations and improving energy efficiency by commissioning smart devices and communication networks.
- Integrating renewable energy sources into the mainstream energy market, optimizing energy trading, and managing peak demand.
- Developing microgrids and exploring demand response capabilities on intelligent communication platforms.

The deployment of AMI has been a significant success, with millions of smart meters installed across the country. This has enabled real-time monitoring, reduced power pilferage, and improved billing and revenue collection. The UDAY and RDSS schemes have further contributed to reducing power losses in the distribution system, narrowing the gap between the average cost of supply (ACS) and the average rate of return (ARR), and enhancing the financial stability of DISCOMs. The ATC losses have been reduced significantly from 25–27% (2014–2015) to 21.5% (2019–2020), and it is 19–20% (2023–2024). ACS–ARR gap also significantly reduced from Rs. 0.77/unit (2014–2015) to Rs. 0.72/unit (2019–2020) [67, 68].

The smart grid initiative has also brought tangible benefits to consumers. Customers can now monitor their power consumption, access TOU tariffs, and view billing details online. The deregulation of the "flat rate" tariff and the introduction of the retail electricity market has empowered consumers to make informed decisions about their power consumption and demand management.

To ensure the success of the smart grid initiative, the government has also focused on training and capacity building for DISCOMs, aggregators, regulatory authorities, and other stakeholders. Multiple training programs have been organized to create awareness and impart skills in AMI, monitoring platforms, sensors, and communication protocols.

The NSGM has initiated pilot projects in 12 centers across India; out of them eleven are already completed [69]. These projects encompass the implementation of AMI, peak load management, outage management, distributed generation, microgrid development, and power quality management. A brief overview of the development of NSGM pilot projects across India is given in Table 7.2.

Beyond the national-level initiatives, India's academic and research institutions also actively contribute to the smart grid transformation. The Indian Institute of Technology (IIT) Kanpur, for instance, is working on several cutting-edge smart grid projects, including the development of a smart city control center, smart homes, advanced information technology (IT) infrastructure, and the integration of renewable energy sources.

Similarly, the Smart Grid Knowledge Center (SGKC) in Manesar is focused on accelerating the adoption of EVs by developing the necessary charging infrastructure. The SGKC is also working on home EMS and enhancing the cybersecurity of the smart grid ecosystem [69]. Additionally, the center is actively engaged in training and capacity-building programs to equip stakeholders with the necessary skills and expertise.

The collaborative efforts between government initiatives, such as the NSGM, and the work being done by academic and research institutions are crucial for the successful implementation and widespread adoption of smart grid technologies across India. This synergetic approach ensures that the smart grid initiative not only modernizes the power infrastructure but also empowers consumers, promotes sustainability, and strengthens the overall resilience of the country's energy system.

Table 7.2 Development of smart grid management infrastructure under NSGM in government sector institutions.

Organization	AMI	Peak load management	Outage management	Distributed generation	Developing microgrid	Power quality management
AVVNL, Ajmer	✓					
APDCL, Assam	✓	✓	✓	✓		
CESC, Mysore	✓	✓	✓	✓	✓	
HPSEB, Himachal Pradesh	✓	✓	✓			✓
PED, Puducherry	✓					
TSECL, Tripura	✓	✓				
TSSPDCL, Telangana	✓	✓	✓			✓
UHBVN, Haryana	✓	✓	✓			
UGVCL, Gujarat	✓	✓	✓			
WBSEDCL, West Bengal	✓	✓	✓			
IIT Kanpur	✓					
SGKC, Manesar	✓		✓	✓	✓	

7.9 Conclusions

The convergence of smart grids and the IoT has ushered in a new era of intelligent and dynamic energy management. Real-time monitoring of various aspects, including smart metering, distribution monitoring, fault detection, demand response, and renewable energy integration, has become a reality. This empowers utilities with a treasure trove of data, enabling them to make data-driven decisions that optimize grid reliability, energy consumption, and operational efficiency.

The transformative power of IoT in smart grids extends far beyond mere monitoring, including load balancing, grid optimization, voltage and PQM, integrated energy management, advanced analytics, predictive modeling, and collaborative grid management. These applications enhance smart grids' efficiency, reliability, sustainability, and resilience, leading to a more intelligent and optimized electricity distribution system.

IoT has also enabled the integration of renewable energy sources, reduced energy waste, and provided real-time monitoring and control capabilities, leading to a more sustainable and efficient energy infrastructure. Real-time environmental monitoring actively supports sustainable energy practices by empowering utilities to make informed decisions to reduce their carbon footprint. Consumer engagement is another key facet of the IoT-enabled smart grid. By providing consumers with real-time data on their energy usage, they are empowered to make informed choices and participate in demand-side response programs, ultimately contributing to a more sustainable energy landscape.

The integration of IoT has revolutionized the power trading landscape within smart grids. Real-time data and communication capabilities optimize power generation, transmission, and consumption, leading to greater efficiency, reliability, and cost-effectiveness for all stakeholders. For long-term PPAs, the IoT platform provides valuable real-time data on individual generating units and transmission line loading. This information can be used to optimize ramp-up rates, monitor deviations, and support pricing, billing, and risk management processes. In the case of bilateral power transactions, IoT-enabled communication allows for direct negotiations between suppliers and consumers, potentially with the involvement of intermediary traders. The real-time data ensures transparency and efficient settlement of these forward market contracts. By embracing the potential of IoT-enabled P2P energy trading, where prosumers can directly sell excess renewable energy to their neighbors, the smart grid ecosystem can unlock significant economic benefits, including improved energy efficiency, reduced grid infrastructure investments, and enhanced system resilience.

However, overcoming challenges such as data security, scalability, and interoperability is crucial for this potential to be fully realized. Robust cybersecurity measures and protocols are essential to safeguard the integrity of the vast data IoT devices collect. Standardization across the industry is paramount to ensure seamless communication and interoperability between diverse devices and systems. Designing scalable architectures that accommodate future growth and evolving needs is critical for long-term success. Additionally, stakeholders can adopt comprehensive data governance, partnerships, and continuous improvement strategies.

As the smart grid continues to evolve, the role of IoT-enabled substation automation and EMS will become increasingly vital, enabling grid operators to manage the complexity of modern power systems, enhance grid resilience, and support the transition toward a more sustainable energy future. Integrating IoT technologies is a key enabler for realizing consumer and market-driven power flow in the smart grid, providing real-time data, secure communication, and enhanced grid management capabilities.

The continuous data flow from IoT-enabled smart grids provides a valuable platform for research and development activities. Researchers can leverage this data to analyze grid performance, develop new algorithms and models, and test innovative grid management strategies. This ongoing data collection, analysis, and improvement cycle is instrumental in advancing smart grid technologies, energy management techniques, and grid resilience practices.

In conclusion, integrating IoT with smart grids presents a compelling vision for the future of energy. By overcoming the existing challenges and harnessing the immense potential of this transformative technology, stakeholders can create a more intelligent, efficient, sustainable, and resilient electricity distribution system. This paves the way for a future where reliable and sustainable energy is accessible to all while minimizing our environmental impact. As India embraces smart grid technologies focusing on IoT, transforming its power sector holds immense promise for grid resilience, renewable energy integration, and consumer empowerment, ultimately leading to a brighter, more sustainable energy future.

References

1 Raj, K. (2017). *Internet of Things: Architecture and Design Principles*. Chennai: McGraw Hill.

2 Pramanik, P.K.D. and Choudhury, P. (2018). IoT data processing: the different archetypes and their security & privacy assessments. In: *Internet of Things (IoT) Security: Fundamentals, Techniques and Applications* (ed. S.K. Shandilya, S.A. Chun, S. Shandilya, and E. Weippl), 37–54. River Publishers.

3 Cirani, S., Ferrari, G., Picone, M., and Veltri, L. (2019). *Internet of Things: Architectures, Protocols and Standards*. Hoboken, NJ: Wiley.

4 Rajiv, B. (2023). What are the major components of Internet of Things. RF Page. https://www.rfpage.com/what-are-the-major-components-of-internet-of-things/ (accessed 9 March 2024).

5 Dunko, G., Misra, J., Robertson, J., and Tom, S. (2017). *A Reference Guide to the Internet of Things*. Raleigh, NC: Bridgera LLC.

6 Pramanik, P.K.D., Pal, S., Brahmachari, A., and Choudhury, P. (2018). Processing IoT data: from cloud to fog. It's time to be down-to-earth. In: *Applications of Security, Mobile, Analytic and Cloud (SMAC) Technologies for Effective Information Processing and Management* (ed. P. Karthikeyan and M. Thangavel), 124–148. IGI Global.

7 Kumar, S., Tiwari, P., and Zymbler, M. (2019). Internet of Things is a revolutionary approach for future technology enhancement: a review. *Journal of Big Data* 6: 111.

8 Pramanik, P.K.D., Pal, S., Mukhopadhyay, M., and Singh, S.P. (2021). Big Data classification: techniques and tools. In: *Applications of Big Data in Healthcare: Theory and Practice* (ed. A. Khanna, D. Gupta, and N. Dey), 1–43. Academic Press.

9 Carvallo, A. and Cooper, J. (2015). *The Advanced Smart Grid: Edge Power Driving Sustainability*. Boston, MA: Artech House.

10 Javaid, N., Hafeez, G., Iqbal, S. et al. (2018). Energy efficient integration of renewable energy sources in the smart grid for demand side management. *IEEE Access* 6: 77077–77096.

11 Gharavi, H. and Ghafurian, R. (2011). Smart grid: the electric energy system of the future. *Proceedings of the IEEE* 99 (6): 917–921.

12 Ben Dhaou, I.S., Kondoro, A., Kakakhel, S.R.U. et al. (2020). Internet of Things technologies for smart grid. In: *Tools and Technologies for the Development of Cyber-Physical Systems* (ed. Sergey Balandin and Ekaterina Balandina), 256–284. IGI Global.

13 Ghasempour, A. (2019). Internet of Things in smart grid: architecture, applications, services, key technologies, and challenges. *Inventions* 4 (22): 1–12.

14 Judge, M.A., Khan, A., Manzoor, A., and Khattak, H.A. (2022). Overview of smart grid implementation: frameworks, impact, performance and challenges. *Journal of Energy Storage* 49: 104056.

15 Dileep, G. (2020). A survey on smart grid technologies and applications. *Renewable Energy* 146: 2589–2625.

16 Kabeyi, M.J.B. and Olanrewaju, O.A. (2023). Smart grid technologies and application in the sustainable energy transition: a review. *International Journal of Sustainable Energy* 42 (1): 685–758.

17 Ghasempour, A. (2019). Internet of Things in smart grid: architecture, applications, services, key technologies, and challenges. *Inventions* 4 (22): 1–12.

18 Avancini, D.B., Rodrigues, J.J., Martins, S.G. et al. (2019). Energy meters evolution in smart grids: a review. *Journal of Cleaner Production* 217: 702–715.

19 Saleem, Y., Crespi, N., Rehmani, M.H., and Copeland, R. (2019). Internet of Things-aided smart grid: technologies, architectures, applications, prototypes, and future research directions. *IEEE Access* 7: 62962–63003.

20 Alonso, M., Amaris, H., Alcala, D., and Florez R, D.M. (2020). Smart sensors for smart grid reliability. *Sensors* 20 (8): 2187.

21 Abrahamsen, F.E., Ai, Y., and Cheffena, M. (2021). Communication technologies for smart grid: a comprehensive survey. *Sensors (Basel)* 21 (8087): 1–24.

22 Ghorbanian, M., Dolatabadi, S.H., Masjedi, M., and Siano, P. (2019). Communication in smart grids: a comprehensive review on the existing and future communication and information infrastructures. *IEEE Systems Journal* 13 (4): 4001–4014.

23 Suhaimy, N., Radzi, N.A.M., Ahmad, W.S.H.M.W. et al. (2022). Current and future communication solutions for smart grids: a review. *IEEE Access* 10: 43639–43668.

24 López, G., Matanza, J., Vega, D.D.L. et al. (2019). The role of power line communications in the smart grid revisited: applications, challenges, and research initiatives. *IEEE Access* 7: 117346–117368.

25 Khan, F., ur Rehman, A., Arif, M. et al. (2016). A survey of communication technologies for smart grid connectivity. *International Conference on Computing, Electronic and Electrical Engineering (ICE Cube)*, Quetta, Pakistan.

26 Pal, R., Chavhan, S., Gupta, D. et al. (2021). A comprehensive review on IoT-based infrastructure for smart grid applications. *IET Renewable Power Generation* 15: 3761–3776.

27 Pramanik, P.K.D., Nayyar, A., and Pareek, G. (2019). WBAN: driving e-healthcare beyond telemedicine to remote health monitoring. architecture and protocols. In: *Telemedicine Technologies: Big Data, Deep Learning, Robotics, Mobile and Remote Applications for Global Healthcare* (ed. D.J. Hemanth and V.E. Balas), 89–119. Elsevier.

28 Minh, Q.N., Nguyen, V.-H., Quy, V.K. et al. (2022). Edge computing for IoT-enabled smart grid: the future of energy. *Energies* 15 (17): 6140.

29 Mehmood, M.Y., Oad, A., Abrar, M. et al. (2021). Edge computing for IoT-enabled smart grid. *Security and Communication Networks* 2021: 5524025.

30 Bagherzadeh, L., Shahinzadeh, H., Shayeghi, H. et al. (2020). Integration of cloud computing and IoT (CloudIoT) in smart grids: benefits, challenges, and solutions. *International Conference on Computational Intelligence for Smart Power System and Sustainable Energy (CISPSSE)*, Odisha, India.

31 Shahinzadeh, H., Moradi, J., Gharehpetian, G.B. et al. (2019). IoT architecture for smart grids *International Conference on Protection & Automation in Power System*, Tehran.

32 Kimani, K., Oduol, V., and Langat, K. (2019). Cyber security challenges for IoT-based smart grid networks. *International Journal of Critical Infrastructure Protection* 25: 36–49.

33 Gunduz, M.Z. and Das, R. (2020). Cyber-security on smart grid: threats and potential solutions. *Computer Networks* 169: 1–17.

34 Alavikia, Z. and Shabro, M. (2022). A comprehensive layered approach for implementing internet of things-enabled smart grid: a survey. *Digital Communications and Networks* 8 (3): 388–410.

35 Tightiz, L. and Yang, H. (2020). A comprehensive review on IoT protocols' features in smart grid communication. *Energies* 13 (11): 2762.

36 Momoh, J.A. (2009). Smart grid design for efficient and flexible power networks operation and control. *Power Systems Conference and Exposition, PSCE '09, IEEE/PES*. pp. 1–8.

37 Gungor, V.C., Sahin, D., Kocak, T. et al. (2011). Smart grid technologies: communication technologies and standards. *IEEE Transactions on Industrial Information* 7 (4): 529–539.

38 Shahid, E.B., Ahmed, Z., Faroqi, A., and Navid-ur-Rehman, R.M. (2012). Implementation of smart system based on smart grid Smart Meter and smart appliances. *Iranian Conference on Smart Grids*, Tehran, Iran.

39 Mohassel, R.R., Fung, A.S., Mohammadi, F., and Raahemifar, K. (2014). A survey on advanced metering infrastructure and its application in smart grids. *27th Canadian Conference on Electrical and Computer Engineering (CCECE)*, Toronto, Canada.

40 Kumar, S. and Moore, K.B. (2002). The evolution of Global Positioning System (GPS) technology. *Journal of Science Education and Technology* 11 (1): 59–80.

41 Farsadi, M., Golahmadi, H., and Shojaei, H. (2009). Phasor Measurement Unit (PMU) allocation in power system with different algorithms. *International Conference on Electrical and Electronics Engineering (ELECO 2009)*, Bursa, Turkey.

42 Hong, J. and Liu, C.-C. (2019). Intelligent electronic devices with collaborative intrusion detection systems. *IEEE Transactions on Smart Grid* 10 (1): 271–281.

43 Steffka, M.A., Paul, C.R., and Scully, R.C. (2022). *Introduction to Electromagnetic Compatibility*, 3e. Wiley.

44 Kabalci, E., Kabalci, Y., and Develi, I. (2012). Modelling and analysis of a power line communication system with QPSK modem for renewable smart grids. *International Journal of Electrical Power & Energy Systems* 34: 19–28.

45 Khanna, A. (2012). Smart grid, smart controllers and home energy automation—creating the infrastructure for future. *Smart Grid and Renewable Energy* 3: 165–174.

46 Lee, D. and Cheng, C.-C. (2016). Energy savings by energy management systems: a review. *Renewable and Sustainable Energy Reviews* 56: 760–774.

47 Alward, Y., Joshi, S.N., and Singh, P. (2018). A review on control and automation based smart grid system and its impact on conventional grid. *International Journal of Engineering Research and Technology* 7 (4): 404–407.

48 Kuster, C., Rezgui, Y., and Mourshed, M. (2017). Electrical load forecasting models: a critical systematic review. *Sustainable Cities and Society* 35: 257–270.

49 Wang, Y., Saad, W., Mandayam, N.B., and Poor, H.V. (2016). Load shifting in the smart grid: to participate or not? *IEEE Transactions on Smart Grid* 7 (6): 2604–2614.

50 Gellings, C.W. and Parmenter, K.E. (2016). Demand-side management. In: *Energy Management and Conservation Handbook*, 2e (ed. F. Kreith and D.Y. Goswami). Boca Raton, FL: CRC Press.

51 Sharda, S., Singh, M., and Sharma, K. (2021). Demand side management through load shifting in IoT based HEMS: overview, challenges and opportunities. *Sustainable Cities and Society* 65: 102517.

52 Wood, A.J., Wollenberg, B.F., and Shebl, G.B. (2014). *Power Generation Operation and Control*. Hoboken, NJ: Wiley.

53 Denholm, P. and Margolis, R. (2008). Supply curves for rooftop solar PV-generated electricity for the United States. https://www.nrel.gov/docs/fy09osti/44073.pdf (accessed 26 May 2024).

54 Solar beats coal cost: implications energy and natural resources. (2017). KPMG. https://assets.kpmg.com/content/dam/kpmg/in/pdf/2017/09/Solar-beats-coal-cost.pdf (accessed 26 May 2024).

55 Solar Energy Corporation of India Limited A Government of India Enterprise Schedule-A Central Public Sector Undertaking. (2022). Manoj Mathur. https://cer.iitk.ac.in/assets/downloads/FoR_CBP13/presentations/Competitive_Bidding_manoj_mathur.pdf (accessed 28 May 2024).

56 The Economic Times. (2024). Torrent Power, Lanco emerge as lowest bidders in gas-fired power auction. https://economictimes.indiatimes.com/industry/energy/power/torrent-power-lanco-emerge-as-lowest-bidders-in-gas-fired-power-auction/articleshow/107883033.cms?from=mdr (accessed 19 June 2024).

57 CFI Team. (2024). Levelized cost of energy (LCOE). Corporate Finance Institute. https://corporatefinanceinstitute.com/resources/valuation/levelized-cost-of-energy-lcoe/ (accessed 12 May 2024).

58 Bhattacharya, K., Bollenand, M., and Daalder, J. (2001). *Operation of Restructured Power Systems*. Norwell, MA: Kluwer.

59 T. P. Database. (2023). Power trading in India- Way Forward. https://www.primedatabase.com/article/2020/Article-C.S.Verma.pdf (accessed 25 April 2024).

60 IEX. (2024). Indian Energy Exchange. https://www.iexindia.com/ (accessed 15 April 2024).

61 Kirschen, D. and Strbac, G. (2004). *Fundamentals of Power System Economics*. Hoboken, NJ: Wiley.

62 Economics Division. (2023). Report on short-term power market in India: 2022-23. Central Electricity Regulatory Commission.

63 Bajaj, R. (2019). Ancillary services and intra-day power trading markets. https://eal.iitk.ac.in/assets/docs/Rohit_Bajaj_Ancillary.pdf (accessed 22 April 2024).

64 Ruska, M. and Similä, L. (2011). Electricity markets in Europe Business environment for smart grids. https://publications.vtt.fi/pdf/tiedotteet/2011/T2590.pdf (Accessed 15 April 2024).

65 Energy Traders Europe. (2022). EFET insight into forward trading in wholesale electricity markets. https://efet.org/files/documents/20220216%20EFET_Insight_01_forward_trading.pdf (accessed 10 April 2024).

66 Stoft, S. (2002). *Power System Economics: Designing Markets for Electricity*. New York: Wiley.

67 Regy, P., Sarwal, R., Stranger, C. et al. (2021). *Turning Around the Power Distribution Sector: Learnings and Best Practices from Reforms*. NITI Aayog, RMI, and RMI India.

68 Kumar, A. (2021). Impact of Ujwal Discom Assurance Yojana (UDAY). PRS Legislative Research https://prsindia.org/policy/discussion-papers/impact-of-ujwal-discom-assurance-yojana-uday (accessed 19 June 2024).

69 Ministry of Power, Government of India. (2023). SG Pilot Projects: Smart Grid Pilot Projects under IPDS. https://www.nsgm.gov.in/en/sg-pilot (accessed 2 May 2024).

8

Cybersecurity Challenges in Smart Grid IoT

Zain Buksh[1], Neeraj A. Sharma[1], Rishal Chand[1], Jashnil Kumar[2], and A. B. M. Shawkat Ali[3]

[1]*Department of Computer Science and Mathematics, The University of Fiji, Lautoka, Fiji*
[2]*Department of Computing Sciences and Information Systems, Fiji National University, Natabua, Lautoka, Fiji*
[3]*Bangladesh University of Business and Technology (BUBT), Dhaka, Bangladesh*

8.1 Introduction

Energy infrastructure expanded to smart grids enabled the progression in the power generation and distribution/consumption transformation [1]. Combining sophisticated digital technology with fully computerized systems, smart grids are one way to make our energy supply networks far more sustainable, effective, and reliable [2]. At the same time, another aspect of the Internet of Things (IoT) developed a more complex and highly connected energy ecosystem (the "Internet of Everything") that has revolutionized smart grids by embedding sensors, communication devices, and big data analytics in grid infrastructure [3].

In this regard, the combination of smart grids and IoT is being implemented to solve many of the issues currently affecting the energy sector [4]. Some examples include efficient management of energy, monitoring, and a quick response time in case of system breakdown or outage. smart grid could use IoT-enabled devices to collect and analyze huge amounts of data from various components of the grid. Additionally, it will open even more grid visibility, and predictive maintenance and force infiltration into better demand forecasting.

This increased connectedness and reliance on digital technology have raised serious cybersecurity concerns. The advancement and networking of smart grids increase their vulnerability to being tampered with by hackers. The IoT systems in a smart grid can influence national security, put a stop to the flow of power, and result in colossal financial losses [5]. In a smart grid IoT context, cybersecurity could be described as a need for continuous and reliable operation of critical infrastructure and goes beyond simple data protection. This chapter will contain a comprehensive analysis of all the cybersecurity challenges that the systems of smart grid IoT will need to face. We start by presenting the research background, which covers the basics of cybersecurity, the architecture of the smart grid, and specific security issues arising from their integration with the IoT. The subsequent section in this chapter will address further the various issues of cybersecurity, including risk assessment and management, data integrity and privacy, as well as authentication and access control.

This chapter also focuses on the technical treatments and techniques applied, such as intrusion detection systems, strong authentication mechanisms, and ways of encryption. We shall also look at the significance of compliance and how independent audits lead to grid security.

IoT for Smart Grid: Revolutionizing Electrical Engineering, First Edition.
Edited by Rahiman Zahira, Palanisamy Sivaraman, Chenniappan Sharmeela, and Sanjeevikumar Padmanaban.

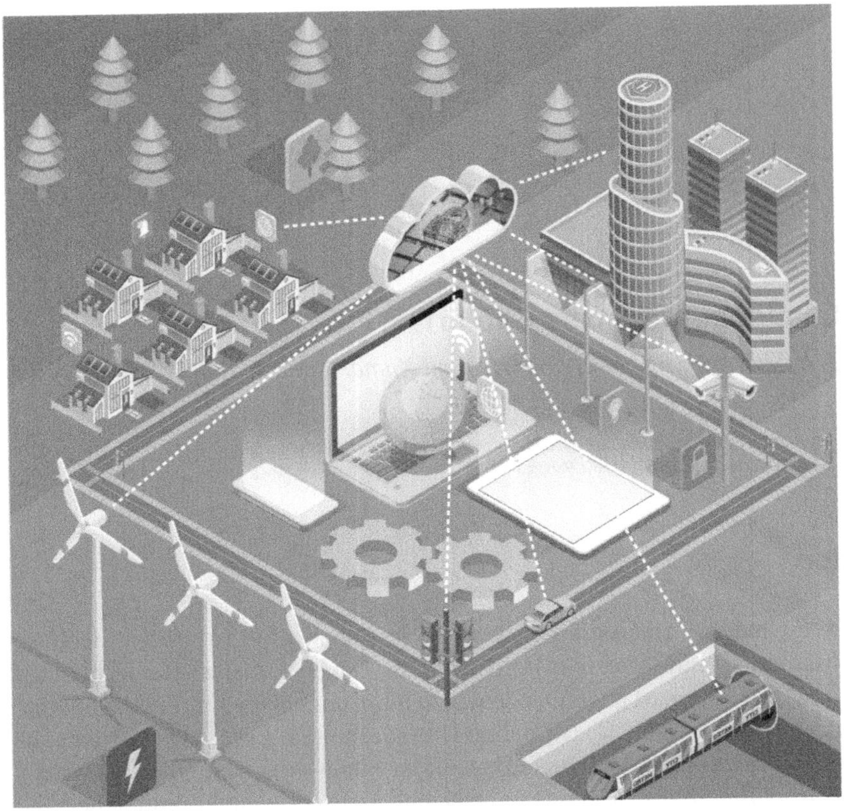

Figure 8.1 Model cybersecurity smart grid.

This chapter looks deeply into human factors, including threats from the inside, to demonstrate the need for all-rounded cybersecurity. Figure 8.1 depicts an illustration of smart city infrastructure, showing connected systems such as residential areas, wind turbines, a cloud computing network, and various forms of urban and transportation technology, all integrated through digital connectivity.

The utility of several mitigation techniques and the practical ramifications of these cybersecurity issues will be portrayed through case studies and real-world examples. We review cyber events that have occurred in smart grid IoT implementations to derive lessons for current and future procedures. In conclusion, we present future trends and issues, forecasting changes in cyber threats and threats of IoT growth in smart grid applications over time. This chapter will address the need for a more secure smart grid IoT system through collaboration and partnerships, and the one following it will deal with research proposals. This chapter will provide researchers, industry players, and policymakers with the information and resources they need in dealing with cybersecurity issues in smart grid IoT and ensuring dependability and resilience in modern energy infrastructure through a detailed analysis of those factors.

8.1.1 Overview

The combination of smart grids and the IoT is a transformative leap forward into the energy sector with the mission of boosting efficiency, reliability, and sustainability in the consumption and distribution of power [6]. Digital technology alone is a way of monitoring and managing the production and distribution of electricity more effectively [7]. IoT involves a network of devices and sensors communicating with each other through the collection, transmission, and analysis of real-time information across grid infrastructure [8]. This allows a better response, prescribed maintenance, and dynamic management of the grid, which means a more resilient and adaptive system. There are equally many cybersecurity challenges with increased interconnectivity and dependency on digital technologies that need to be put in place with security measures to track the campaign from potential casualties [9]. This chapter addresses the critical cybersecurity challenges that smart grid IoT systems bring; highlights the main risks and potential solutions, and details a perspective of the grid security landscape into the future.

8.1.2 Scope

Here is a related works table that summarizes and compares the main themes, findings, and methodologies of our paper alongside other significant studies in the field of IoT and smart grid cybersecurity (Table 8.1).

This chapter looks at the inherent cybersecurity challenges of the smart grid IoT systems, delving so deep that the current threat landscape is well understood, and the strategies for mitigation can be well outlined. This chapter begins with a general overview of smart grids and their integration with the IoT, their synergistic benefits, and the critical role of cybersecurity in safeguarding such

Table 8.1 Related works.

Category	Our Paper (2024)	Paper 1	Paper 2	Paper 3
Paper reference	—	Kimani et al. [5]	Gunduz and Das [10]	Tufail et al. [11]
Key focus	IoT and cybersecurity in smart grids	Security challenges for IoT in smart grids	IoT's role in energy efficiency and management	Privacy and security in smart grid systems
Main findings	Emerging threats and future trends in cybersecurity	Vulnerability of IoT devices in smart grids	Impact of IoT on energy management efficiency	High vulnerability to data breaches in smart grids
Methodologies	Case analysis of IoT security incidents	Systematic literature review and expert interviews	Quantitative analysis of energy data from IoT devices	Survey on privacy concerns and data security practices
Identified challenges	Large attack surfaces, complex network dependencies	Insufficient security protocols for device protection	Integration challenges with existing energy systems	Regulatory gaps affecting data privacy
Proposed solutions	Adoption of advanced cryptographic measures, IoT security standards	Implementation of unified security frameworks	Use of AI for predictive energy management	Strengthening privacy laws, implementing secure communication protocols
Future research directions	Development of adaptive security architectures and policy	Standardization of security measures across devices	Optimization of IoT integration with renewable energy sources	Development of AI-driven privacy protection techniques

highly advanced systems. The section then outlines the research background from the very basic concept of security and smart grid IoT, contrasting and comparing challenges and requirements posed by the intersection of both. The central part of this chapter dismantles several dimensions of cybersecurity issues, among them the architecture issues related to smart grid IoT; data flow vulnerabilities; problems associated with data integrity and confidentiality; authentication and issues of access control; comprehensive risk assessment; and risk mitigation strategies. We move on to the technological cures proposed: Encryption techniques, strong authentication mechanisms, the necessity of threat and incident response, human factors, and insider threats. This chapter provides a wide range of case studies and examples from real-world practice that reflect on past security strategies as well as cybersecurity incidents. The final section covers emerging trends and technologies that are going to shape the future of smart grid IoT security, forecasts the evolution of cyber threats, and gives future research recommendations. Therefore, this chapter will provide the reader with a basic and practical understanding of the smart grid IoT cyber challenges and solutions to build a platform for further research in this critical field.

8.2 Research Background

Understanding this intersection between cybersecurity and smart grid IoT requires a deep examination of each independently and how they uniquely interact with one another. The state of cybersecurity today is such that practices and technologies used toward the protection of systems, networks, and data from several classes of cyber threats are traditionally important but today more important than ever before. The levels of insecurity observed vary from data breaches and malware to highly advanced exploits against critical infrastructure. In reality, the increase in numbers and sophistication of cyberattacks fully justifies the need to implement strong cyber defense to avoid unauthorized access to sensitive information and ensure unbroken digital service [12].

Smart grids represent the modernization of the existing power grids through enhanced information and communication technologies to create an intelligent energy network. This entails features of real-time monitoring of power distribution, efficient energy transmission, and a high level of integration with renewable energy sources [13]. Bringing IoT into smart grids contributes to this feature by dint of sensors, actuators, and communication devices across the grid. These devices result in massive volumes of data, analyzed for grid optimization, prediction of maintenance requirements, and responsive action to fluctuations in energy demand [14].

However, the integration of IoT and smart grid leads to enormous cybersecurity issues. These systems are highly interconnected and will provide various entry points, exposing the whole system to potential cyberattacks. For instance, compromised IoT devices serve as gateways for cyberattackers to penetrate the bigger grid network to cause a disrupted power supply that leads to severe economic consequences on a global scale [15]. In addition, these heterogeneities in IoT devices and the lack of standardized security protocols further compound these vulnerabilities [16]. The present chapter brings out the most important dimensions of smart grid IoT cybersecurity and elaborates on the associated risks and their mitigation strategies. It begins with the basics of cybersecurity and smart grid IoT, thereby outlining the salient features and requirements concerning each domain. The narrative digresses into the particular cybersecurity challenges that come with their integration: Data integrity, confidentiality issues, authentication and access control issues, and the need for a full assessment of risk.

It also surveys the technological solutions and best practices, giving insight into how encryption techniques can help fortify the security within smart grid IoT, along with robust authentication mechanisms and intrusion detection systems. Other factors include regulatory compliance and third-party audits for the realization that security practices must conform to the industry standard and be enhanced throughout. Human factors involved, including those of insider threats, lead to a more comprehensive approach toward cybersecurity.

This chapter provides practical implications, evident from case studies and real-world examples of cyber challenges and the effectiveness of diversified strategies adopted for mitigation. The analysis of incidents and security implementations from them, which turned out to be successful, can assist readers in learning lessons and gaining insights into future activities. It forms an analysis of emerging technologies and upcoming trends, understanding how these might evolve in a way that can actually further spiral out cyber threats and thereby advise the present research and cooperative work in the respective domain. At length, it is in this chapter that the reader will be explained the issues related to cybersecurity challenges in a smart grid IoT and the possible strategies to work with to ensure the resilience and reliability of modern energy infrastructure.

8.2.1 Cybersecurity

In an increasingly digital world, cybersecurity has become a paramount concern as the rise of connected devices and systems creates remarkable opportunities while also rising as a major risk. Cybersecurity refers to the background, technologies, and practices that are aimed at protecting computers, networks, programs, and data from damage and unauthorized access and attacks. The higher the advance in sophistication and pervasiveness of cyber threats, the more critical stringent cybersecurity measures are needed in the present world. The degree of significance of cybersecurity is now more than ever because malicious practices and hacking are continuously growing in complexity. Along with the rise of cyber threats, there are discrete types of malicious activities including hacking, phishing, malware distribution, and ransomware attacks that stand out from their diversity in their posed threats to individuals, businesses, and governments [17].

Cybersecurity is an ever-changing field that evolves continuously to emerging threats. It is greatly driven by the emerging technologies that enhance the complexity of cyberattacks. For instance, attackers now have access to sophisticated tools to carry out zero-day attacks, social engineering, and state-sponsored hacking, thus breaching security defenses. This often requires that cybersecurity professionals think and evolve constantly to deploy multilayered defense strategies and use tools of the trade, such as firewalls, intrusion detection system / intrusion prevention system (IDS/IPS), and encryption technologies, to protect digital assets, as argued [18].

One of the crucial concepts of cybersecurity is knowing the motivation and tactics of your cyber adversaries because cybercriminals range from individual hackers and organized crime groups to nation-state actors, each with specific objectives and capabilities. For example, an individual hacker could be motivated by private gain or to create a reputation, while nation-state actors might have a political objective that is very strategic. This diversity in threat actors dictates a comprehensive and adaptable approach to cybersecurity [19].

In addition, the fast growth of the IoT is quickly enlarging the attack surface. The IoT spans from smart home appliances to industrial sensors, with weak processing power and memory, making it hard to implement ordinary security measures. Hence, cybercriminals use such devices to gain entry into networks, creating massive breaches and disturbances. Security for IoT devices can only be achieved using creativity as space and operability are a challenge [20].

The human factor is another important component of cybersecurity. Basically, lack of strong password usage, falling to phishing, or lack of training in dealing with security issues account for most breaches. Thus, training and a culture of security awareness and education should be high in the order of priorities for any organization. Having regular training and clear policies on security will help reduce such threats [21]. Cybersecurity is fundamentally dynamic, multifaceted, and essential to the integrity of digital data and systems. Threats change, and with them, the tactics and tools used to counter them also change.

8.2.2 Smart Grid IoT

Smart grids involve going a step further in the management and distribution of electric power through the use of new and advanced technologies that will lead to a more efficient, reliable power energy infrastructure that can service rural and remote areas, where electrification has not been cost-effective, or where electrification has not reached. The core of such transformation is the integration of the IoT into the grid by endowing the installation of sensors and actuators, among other communication devices throughout the grid. By so doing, it will enable real-time monitoring, data collection, and dynamic management of the electric grid to realize improved operational efficiency and responsiveness to fluctuations in demand [22].

Smart grids, whose strength comes from the improved IoT, have the following advantages over conventional power grids: support bidirectional communication between the energy provider and consumers, which enhances better control over energy flow and usage. This aspect is special in integrating clean energy, primarily from solar and wind, which are variable in nature and require more adaptive management strategies. Smart grids, through IoT-embedded devices, will vary with such fluctuations to avert adverse effects on the stability and reliability of the energy supply [14].

Data from IoT devices across the grid are key in predictive maintenance and operational optimization. The grid, therefore, monitors its health and performance for its components so that utility providers have prior information on potential challenges that they can rectify before they escalate into failures and outages. Such proactive strategies boost the reliability of the grid, reduce the cost of maintenance, and increase the infrastructure lifespan [23].

The integration of IoT in smart grids also brings forward new challenges, mostly in light of cybersecurity. The interconnectivity introduced in the IoT devices brings in multiple entry points for potential cyber threats, hence making the grid more resilient to other hostile activities. These attacks may range from data breaches and unauthorized access to more severe disruptions, including, but not limited to, denial-of-service (DoS) and grid manipulation. The critical importance of electricity to societal functions means that any disruption has far-reaching impacts and, therefore, a critical need for cyber protection [24].

The heterogeneity of IoT devices opens up a very serious security challenge. The devices commonly come from different manufacturers, each with diverse security protocols and capabilities. Ensuring both interoperability and a high level of security is not an impossible task, but it most certainly does call for a harmonized effort by the industry. Standardization of security protocols and the enforcement of broad security frameworks become imperative in addressing these risks [16].

Besides the technological challenges presented, smart grid and IoT integration also pose regulatory and policy-related issues. Furthermore, governments and regulators gain in importance in laying standards for security and ensuring compliance. Organizations that will develop and enforce regulations to cater specifically to the security needs of IoT systems will be critical to safeguarding smart grids from cybersecurity threats. There needs to be collaboration from different stakeholders, such as utility providers, technology companies, and regulatory agencies, to enable an environment for energy infrastructures that can be efficiently, securely, and resiliently operated [25].

IoT integration into smart grids is to assure several benefits in energy management and efficiency. However, it introduces complex problems concerning cybersecurity. These problems are only to be addressed by a multifaceted approach, which can include technological innovation, regulatory oversight, and industrial collaboration.

8.2.3 Cybersecurity Versus Smart Grid IoT

IoT technology in smart grids marks the beginning of a new era in energy management, promising positive improvements in efficiency, reliability, and sustainability. At the same time, it creates complexity in cybersecurity that requires the exploration of innovative solutions and puts them under constant vigilance. The cybersecurity industry for smart grid IoT is considerably different from the regular security industry, due in part to the critical nature of the infrastructure and the serious costs of security breaches.

The introduction of IoT devices deepens this complexity by integrating thousands of sensors and actuators into the grid, each of which represents a potential vulnerability. Unlike most information technology (IT) environments, where confidentiality and data integrity are the primary concerns, smart grid IoT systems give priority to the availability and reliability of the services provided by the infrastructure. A breach in a smart grid IoT system could give room to power outages, and in worst-case scenarios, catastrophic failures affecting large geographic areas, highlighting the need for very strong security protocols that are tailor-made to the operational necessities of the energy space [26].

The challenges in securing smart grid IoT systems are multifaceted. First, the problem of scale is present; the number of devices connected to a grid introduces a significant multiplier into the attack surface. Each one of the thousands of in-home displays or smart meters has to be secured, a humongous process complicated by the range of hardware and software platforms on the field. Second, many of these devices are low-power and low-capacity devices. Third, traditional cybersecurity measures, such as fancy encryption algorithms or heavy antivirus software, are not practical on such devices, given their limited computational capabilities and storage capacities [27].

Since cybersecurity solutions are applied in smart grid operations, they must be low-latency and highly reactive. In other words, any protection that adds great time to data processing is impractical, that is, it will be just impossible because grid operations cannot afford to be disrupted. This is the reason there is a requirement for the development of lightweight but effective security solutions, capable of operating over the IoT devices embedded in the smart grid in an efficacious manner [28].

These requirements are the lead behind the most advanced research efforts focused on the development of new cryptographic techniques, secure communication protocols, and robust identity and access management systems designed from the very beginning toward smart grid IoT environments. For example, recent work has proposed blockchain to offer enhanced security and trust in IoT-based smart grids, where the distributed nature of blockchain may potentially add a trustable layer by organizing its transactions and data exchanges without creating a single point of failure for cyberattacks and thus providing resilience against those attacks [29].

Furthermore, more and more work is being completed in applying artificial intelligence (AI) and machine learning (ML) approaches to predict the appearance of cybersecurity threats and to react to them in real time. It includes the application of AI-based anomaly detection systems that are based on monitoring network traffic for possibly unusual packet and transaction patterns pointing to a cybersecurity incident. It will help to detect and respond to threats faster and more accurately, which in turn is very important to ensure the integrity and availability of smart grid operations [30].

Moreover, IoT-enabled smart grid system cybersecurity is a fast-evolving and crucially important area that addresses all the peculiar problems connected with the integration of IoT technology

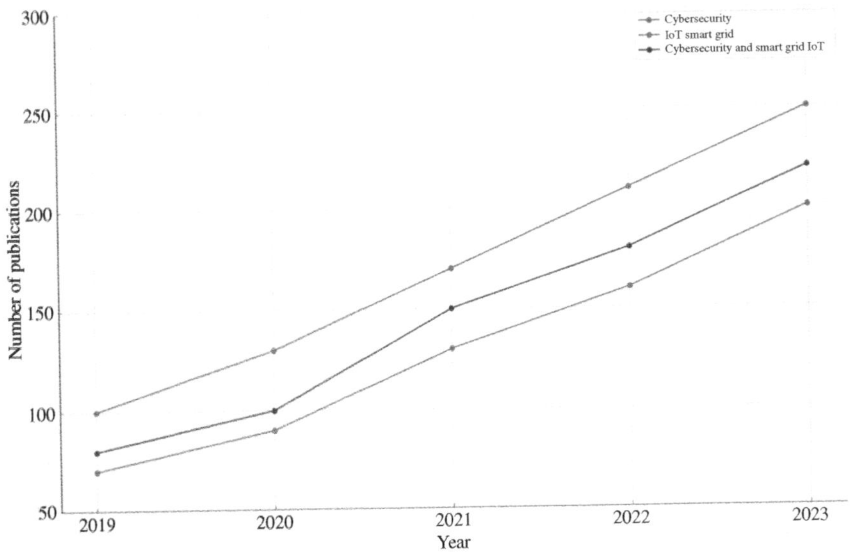

Figure 8.2 Research trends (2019–2023).

into critical energy infrastructure. It is solutions of such that need to be able to balance between high-tech innovations and properly instituted policies so that security measures are not too prohibitive to the functioning of the smart grid.

The trend of research in the field of cybersecurity, IoT smart grid, and the overlap of cybersecurity and smart grid IoT should be as per Figure 8.2 from 2019 to 2023. Overall, the data seems representative of dramatically increased work in the areas by increasing interest and progress in the field. The green line represents the research activity of cybersecurity in terms of the trend represented and shows an exceptional trend upward in a situation denoting escalated importance consonant with the rise in complexity and frequency of cyber threats. The red line of IoT smart grid research also shows a noticeable uptrend and, in this case, too, indicates a growing role of IoT in making grid infrastructure and energy management effective. The blue line represents trends related to the conducted research on the intersection of these two – that of the critical necessity of cybersecurity for smart grid IoT Systems. Figure 8.3 shows representative results in the high rate of growth in cybersecurity research and depicts the increased need and awareness to set the priorities of the security of digital infrastructures against potential threats. This figure points to the dynamic nature of research in these fields and the absolute necessity of continuing and increasing these efforts to enhance resilience, efficiency, and security in these energy systems of the future.

The distribution of research publications in 2023 is depicted in Figure 8.4: cybersecurity, IoT smart grid, and the intersection of cybersecurity with smart grid IoT. In the pie chart, it is clearly shown that about 38.5% of the research publications are in the field of cybersecurity. This leads to a very high level of importance in the cybersecurity domain in this digital era. Research in IoT smart grid sums to 30.8%; this can further indicate efforts in the ongoing developments of the integration of IoT technologies into an energy management system. Lastly, cybersecurity's meeting point with the smart grid IoT amounts to approximately 30.7% of the research published, which shows exactly how serious these issues are by requiring more attention to ensure that the interconnected grid infrastructures are secured. This further verifies the result is balanced across research communities checking into 2023.

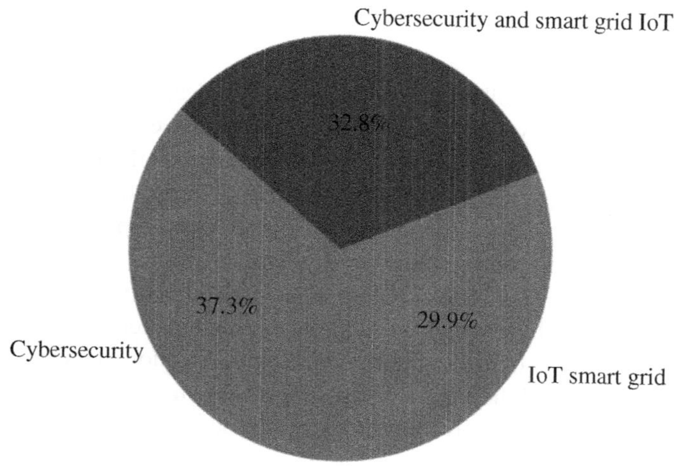

Figure 8.3 Publication distribution (2023).

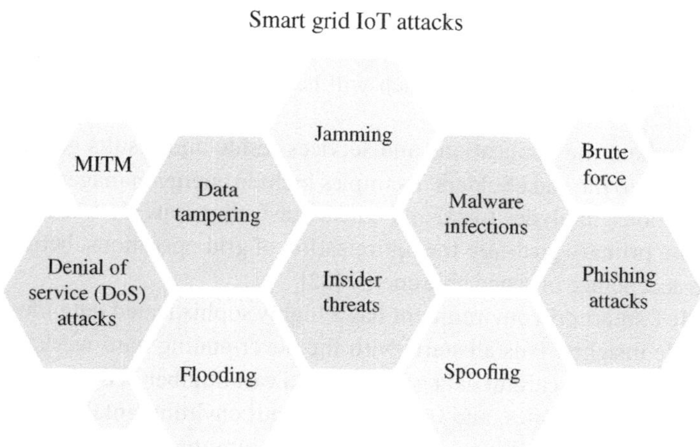

Figure 8.4 Smart grid IoT attacks.

8.3 Cybersecurity Challenges in Smart Grid IoT

8.3.1 Fundamentals of Smart Grid IoT Security

Smart grid IoT security faces several challenges due to the interconnected nature of the grid and the potential vulnerabilities of IoT devices. The smart grid IoT architecture involves the integration of various components such as smart meters, sensors, communication networks, and control systems. This interconnectedness creates a larger attack surface and increases the potential for cyberattacks [5, 31].

8.3.1.1 Smart Grid IoT Architecture

The smart grid is an advancement from the traditional electric grid with improved advanced integration of information and communications technology (ICT) features in realizing efficiency, reliability, and sustainability in providing electricity services. The IoT brings to realization the paramount importance of enabling seamless connectivity and interaction among the various components of the grid. The smart grid IoT architecture has four layers: physical, communication, data, and application [31, 32].

Physical Layer: It refers to the layer that harbors all devices sensing data physically, and taking actions. Today, the definition of the physical layer goes much beyond the first layer to include, besides the actual deployment of the sensors and actuators, the device all over the grid for large-scale monitoring and control of parameters at respective points of the grid, such as voltage, current, and power quality. Intelligent meters facilitate real-time consumption data and, therefore, enable dynamic pricing and demand response strategies [32].

Communications Layer: This allows for communications between devices and control centers while also, at the same time, being serviced by some of the communication technologies that include the wired wide networks, such as fiber optics and power line communication, the wireless networks, including Wi-Fi and Zigbee, and the hybrid systems. The layer ensures that data transmitted at the physical layer is gathered and reliably transferred to data management systems in a secure manner [32].

Data Layer: This layer generally consists of storing, processing, and analyzing the voluminous data that the grid generates. The grid uses databases, cloud computing facilities, and big data analytic tools. Thus, data management becomes crucial and critical for any actionable insight and goes a long way in making informed decisions. For example, predictive analytics of the dataset shall assist in predicting future electricity demand trends, which will help manage the grid proactively [31, 32].

Application Layer: This layer is where the applications and services reside and results generated from processed data are served to the stakeholders. Examples include energy management systems, grid monitoring applications, analytics tools, and consumer-facing apps for tracking energy usage. In other words, the primary goals are the optimization of grid operations, better customer engagement, and the acceptance of renewable energy [32].

Data Flow: Data in the complex IoT smart grid environment has a highly sophisticated data lifecycle from its origin to actionable insights. This all starts with the never-ending hard work of smart meters and sensors to carry out such careful data gathering of a vast number of data: data on trends of energy use, grid health diagnostics, and information about environmental conditions such as temperature and humidity. With this raw data, complex communication networks have been developed to convey this data into centralized or distributed data centers, where the data is processed [31, 32].

Identification of Attack: While IoT-based smart grids offer significant value, their interconnected nature makes them vulnerable to an increasing number of cyber threats. This effectively summarizes the broad spectrums of possible attack vectors that can target smart grid IoT environments [31, 33]. Table 8.2 gives an overview.

Some common attack vectors that pose significant risks to smart grid IoT security include:

DoS Attacks: Attackers may launch DoS attacks aimed at disrupting the normal operation of smart grids by flooding IoT devices, communication networks, or control systems with excessive traffic, rendering them unresponsive or unavailable [33].

Data Tampering: Malicious actors may intercept and manipulate data transmitted within the smart grid infrastructure, leading to incorrect readings, unauthorized access, or malicious commands being executed, potentially causing disruptions or safety hazards.

Insider Threats: Employees or insiders with access to smart grid systems may misuse their privileges to compromise security, leak sensitive information, or sabotage critical infrastructure components, posing significant risks to grid operation and integrity [33].

Table 8.2 Analysis of attacks.

Attack vector	Objectives/Purpose	Targeting layers	Impacts	Security requirements
Denial-of-service	Disrupt normal smart grid operation by flooding with excessive traffic	IoT devices, communication networks	Unresponsiveness, unavailability	Robust traffic filtering, bandwidth management, redundancy in critical systems
Data tampering	Manipulate transmitted data to cause incorrect readings or unauthorized access	Smart grid infrastructure	Incorrect readings, unauthorized access	Data encryption, integrity verification, access control mechanisms
Insider threats	Misuse of privileges by insiders to compromise security or sabotage infrastructure	Employees with access to systems	Compromised security, data leakage	Strict access controls, employee monitoring, regular security training
Malware infections	Spread malware to compromise devices, steal data, or facilitate unauthorized access	IoT devices, network infrastructure	Compromised functionality, data theft	Regular security updates, malware detection systems, network segmentation
Phishing attacks	Deceive personnel to gain unauthorized access to systems or sensitive information	Personnel responsible for management	Data breaches, system compromise	Employee awareness training, email filtering systems, multifactor authentication

Malware Infections: IoT devices within smart grids are susceptible to malware infections, which can propagate across the network, compromise device functionality, steal sensitive data, or facilitate unauthorized access to control systems, undermining the overall security posture of the grid.

Phishing Attacks: Attackers may employ phishing techniques to deceive personnel responsible for smart grid management and gain unauthorized access to sensitive systems, credentials, or information, potentially leading to data breaches or system compromise [31].

Smart grid IoT security faces a myriad of threats. Jamming disrupts data transmission, compromising availability, while spoofing involves impersonating nodes to compromise various layers of security [33]. Injection introduces false data packets, violating integrity, while flooding overwhelms networks, causing availability loss. Man-in-the-middle (MITM) attacks intercept and modify data, compromising confidentiality and integrity. Social engineering manipulates users, stealing data, while eavesdropping compromises confidentiality by monitoring network traffic [31]. Intrusion illegally accesses networks, compromising integrity and confidentiality. Brute force attacks attempt to crack passwords, further compromising integrity and confidentiality. Finally, time synchronization attacks disrupt node synchronization, undermining both integrity and availability. Smart grid IoT attacks are depicted in Figure 8.5.

History of Attacks: The history of smart grid attacks underscores the diverse array of deliberate and unintentional incidents targeting critical infrastructure. These attacks exploit vulnerabilities [31, 33] in both technological and human aspects of smart grid systems:

- **Intentional Internal Attacks:** These involve physical or privileged access to devices, leading to switch attacks, integrity violations, and privacy breaches. For instance, in 2006, engineers disrupted traffic lights in Los Angeles.

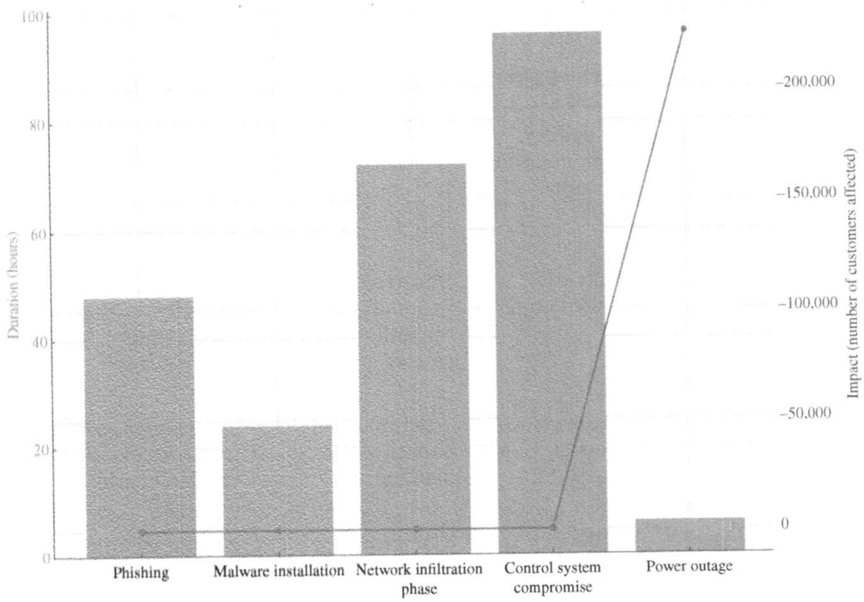

Figure 8.5 Analysis of Ukraine cyberattack phases and impact.

- **Accidental Incidents:** Despite automation, manual interventions by employees can disrupt smart grid operations. Common examples include misconfigurations and errors during maintenance, such as the 2018 blackout caused by an employee in training at Michigan Utility Consumers Energy.
- **External Attacks:** Attackers target availability, integrity, and confidentiality through various means, including scanning networks, exploiting vulnerabilities in wireless technologies, and conducting man-in-the-middle attacks [34].
- **Monitoring System and Supervisory Control and Data Acquisition (SCADA) Attacks:** SCADA systems, crucial for real-time monitoring and control, are vulnerable. Notable incidents include the Slammer Worm attack at an Ohio Nuclear plant and the Stuxnet attack on an Iranian nuclear facility [34].
- **Distributed Denial of Service (DDoS) Attacks:** These aim to disrupt system availability. A notable incident occurred in 2019 when a DDoS attack on a western US utility interfered with grid operations in multiple states [5].
- **Malware Attacks:** Malicious software such as viruses, trojans, and ransomware are employed to disrupt operations. In 2015, hackers seized control of Ukraine's power grid using Black Energy Malware, causing a widespread blackout.
- **Social Engineering Attacks:** Attackers exploit human vulnerabilities, such as phishing, to gain access to confidential information and systems. In the Ukrainian attack, malicious documents were delivered via email, leading to the installation of Black Energy 3 Malware.

Cyberattacks on smart grid infrastructures present formidable challenges to both data integrity and confidentiality. Spoofing, injection, and intrusion attacks directly jeopardize data integrity by tampering with or corrupting critical information, potentially resulting in erroneous system operations or unauthorized access to sensitive data. These breaches not only undermine the reliability and trustworthiness of data but also compromise the overall functionality and safety of smart grid networks [31]. For example, spoofing, injection, and intrusion cyberattacks compromise information directly by changing or polluting critical information, leading to malfunctioning or illegally accessing data within the system. Such types of intrusions compromise not only the reliability and trust of the data but also the overall functionality and safety of intelligent grid networks. These, on the same note, compromise the confidentiality of data through eavesdropping and man-in-the-middle attacks, which secretly intercept sensitive information that is being exchanged between legitimate nodes. This kind of unauthorized access is a breach of the privacy requirement and through such, susceptible information is divulged to rogue elements that would raise the chances of data breaches and privacy infringement. Some of those listed include spoofing attacks, injection attacks, intrusion attacks, eavesdropping attacks, and man-in-the-middle attacks [33].

Moreover, there are other potential passive threats, such as a brute force attack, which may compromise integrity and confidentiality to a very high level. A brute-force approach to crack passwords or paraphrases allows access to confidential data and violates confidentiality and integrity. These attacks such as those from an inside job, industrial espionage, or cyber terrorism do not just threaten data security but also break down any confidence and trust in smart grid operation. As such, the protection of intelligent grid infrastructures from cyberattacks requires security that is active and passive; further ensuring integrity and confidentiality in data criticality during exchanges. Given that the attacks include brute-force attacks [35].

Securing the smart grid IoT ecosystem is imperative for ensuring the reliability, efficiency, and resilience of modern energy management. It can be said that this feature of cybersecurity to support the IoT ecosystem of the smart grid can enable a reliable, efficient, and resilient modern energy management system. That is, based on an understanding of the inherently complex data flow, potential attack vectors, and the attack history, important stakeholders can proactively develop security strategies for critical infrastructures [36].

Integration of IoT technologies in the smart grid brings excellent opportunities for the optimization and automation of system processes. However, it will also be followed closely by new challenges in cybersecurity that need to be addressed by measures such as solid encryption protocols, intrusion detection systems, and employee training programs [37, 38].

8.3.2 Cybersecurity Challenges

8.3.2.1 Data Integrity and Confidentiality Concerns, Authentication, and Access Control Issues

Cybersecurity stands in the middle of all technological landscapes and is responsible for taking care of digital assets and ensuring that the systems or networks are trustworthy. This landscape, however, is replete with its share of challenges, out of which two major concerns are very much visible: one concerning the integrity and confidentiality issue of the data and the other concerning the authentication and access control issues.

Data Integrity and Confidentiality Concerns: The dynamic environment of smart grids and IoT ecosystems is essential. Each interconnection of devices and systems would, in turn, be exposed to a wide array of cybersecurity threats, such as spoofing, injection, and intrusion attacks. In this type of attack, a person impersonates an authentic node, thereby compromising the authenticity of data exchange, possibly causing corrupt data and unauthorized access. Similar types of attacks are identical in threats to injection attacks, in which information is interfered with to turn off operations or malicious nodes are imposed in the network. Both kinds of manipulations over the critical data flow in intelligent grid infrastructures are considered data integrity attacks. Eavesdropping and MITM attacks compromise the data confidentiality requirement by intercepting and accessing sensitive information against the will of a party, hence, creating a privacy threat in networks. Brute force further raises the concern of password cracking and gaining valuable data, which can lead to an information confidentiality and integrity breach. These kinds of challenges that are posed by the scenarios demand solid encryption mechanisms, secure storage protocols, and effective data validation techniques to ensure that the data exists in a safe environment with its confidentiality and integrity maintained during its lifespan [35].

Authentication and Access Control Issues: The significant cybersecurity risk lowers guidelines in smart grids and IoT environments, including authentication and access control mechanisms. Weak authentication mechanisms and a lack of use of multifactors for authentication expose the system to unauthorized access attempts. Things like poor permissions management and basic misconfigurations only worsen an already risky landscape, allowing unwarranted access to critical resources. Strong authentication in identity access controls would require biometrics and token-based authentication. Coupled with robust authorization frameworks involving role-based access controls and the concept of least privilege, it would come in handy. These user profile reviews have to be completed in periodic sessions of deep log monitoring systems, hence, detecting and responding in time to unauthorized access attempts to safeguard sensitive data integrity and confidentiality within intelligent grid infrastructures.

Power Constraints: IoT devices in intelligent grid networks are mostly battery powered, which raises a very challenging question of long-term durability. Conventional batteries usually have a short life span and, indeed need frequent replacement; the cost of maintenance rises, and service disruption can occur. This, therefore, calls for research in the area of high-power compact batteries and wireless power solutions. This way, more efficient and durable battery technologies might be developed to improve a lifetime and functional operation for at least a lifetime period with the assurance of good long-term operation without the requirement for frequent maintenance [5].

Consumer Illiteracy: Intelligent grid networks are weak in various areas, such as the weak consumer awareness of understanding what IoT technologies are and the associated security measures. Many users are ignorant of the dangers that their IoT devices might present, while on the other hand, many are unaware of actions to take to secure these devices from the cyber threat. It is, therefore, sufficient to institute education programs to enable the user to gain knowledge and skills in the process of putting security measures in place and thus reducing the risks of danger. This will further enhance knowledge and literacy in clear and accessible ways of IoT security solutions, training, and resources.

Weak Regulations: Inadequate legislation and regulatory frameworks undermine the security of IoT devices and smart grid networks. The challenges lie in inadequacies in legislation and regulatory frameworks, which sometimes make IoT devices completely devoid of necessary security measures, or they might be victims of any kind of malpractice. Enhancements should be provided, mainly covering adequate laws and regulations regarding security in IoT devices, mechanisms for monitoring compliance, enforcement mechanisms, and reinforcement of safety and resilience in intelligent grid networks [33, 39].

Fear of Reputational Damage: Under-reporting of incidents in intelligent grid networks is often associated with concerns about brand reputation: companies would hardly ever report on critical breaches or any vulnerability that may have a negative implication for their reputation in the market. The opaque approach, coupled with a lack of cooperation, hardly helps in any way to efficiently address the impending cybersecurity threats that could be developed with a perspective of evading risks to reputation. It fosters information sharing among stakeholders and assures a culture of responsibility and accountability, which enhances the incident response within the intelligent grid networks [40].

Inhibitory Price: High device installation costs and pricing of technologies related to IoT are the barriers to the uptake of technologies within intelligent grid networks. This is already being made evident by the prohibitive upfront investment required in deploying IoT devices and infrastructure by some of the larger rail operators. Cost reduction through economies of scale is necessary to make the investments in IoT technologies perpetuated through grants or subsidies where needed to make devices accessible and affordable to consumers and businesses. Alternative areas of research could include business models of leasing or subscription, which will provide for reduced upfront fees and expedite the roll-out of IoT-based intelligent grid solutions [40].

High-Speed Networks: Seamless connectivity and low latency are essential for the proliferation of IoT networks in smart grids. Relationships between devices in a massive IoT deployment in smart grids require low latency, seamless connectivity, and support of the deployment of IoT devices and their applications on a large scale. However, the requirements needed to support the deployment of IoT devices and their applications on a large scale are hardly possible with the network infrastructures currently in use. Next-generation and new networking technologies, such as 5G, will boost network speed, capacity, and reliability to ensure the serving of IoT-based grids. This would translate to upgrading and expanding the infrastructural network, which is necessary to make sure that the connectivity within the innovative grid network is solid and reliable [41, 42].

Privacy Issues: There are concerns for privacy rights and data protection arising out of the IoT technologies deployed in smart grid networks. This is so because large-scale data collection and analysis generated by IoT devices may give away information on sensitive behaviors, preferences, or activities of the user. However, mechanisms should be defined for the building of trust and confidence in the IoT-enabled solutions of smart grids together with the user so that the choices of the user in terms of privacy are respected. Examples of leadership using IoT-enabled intelligent grid solutions can help instill a sense of trust and confidence. Another thing can be added: compliance with data protection laws and standards, together with transparency and control concerning the collection and utilization of data, would be the approaches to mitigating privacy risks and ensuring responsible data handling practices in smart networks [33].

Spectrum Scarcity: Wireless spectrum limitations pose challenges for the growth and expansion of IoT systems in smart grids. The limitation in the wireless spectrum can be a challenge to optimize the growth and proliferation of IoT systems in smart grids. As the number of devices attached to it increases, spectrum demand could quickly outstrip availability, resulting in interference and congestion problems. Another set of solutions may lie in technology such as cognitive radio, which can access and optimize dynamic spectrum. Above all, it must also address concerns related to interference and be compatible with the current spectrum allocations so that it can be largely accepted and adopted in smart grid network service.

Talent Gap: A shortage of skilled professionals in IoT device management and cybersecurity hampers efforts to secure smart grid networks effectively. One significant limitation of this intelligent grid network security setup is the large gap in human resources that are highly skilled in the management and cybersecurity issues of IoT devices. This means that as IoT deployments become more complex and large-scale, the demand for specialists in device management, network security, and data analytics will increase. Such findings call for an investment in education and training by these workforce development programs and initiatives to bridge the talent gap and develop a pipeline of these professionals in possession of the necessary skills to drive innovation and assure robust measures of cybersecurity in intelligent grid networks [35].

It firmly remains a challenge to cybersecurity given data integrity, confidentiality, data authentication, and access control, which the remedy for lies in the combination of technical, policy, and awareness-based solutions. Therefore, the vital part of security is the exhibition of the best strategies in security, a driving force for control culture, and proactive risk management in continuous inspection and audit. In essence, then, robust security and the organization can effectively elude threats on cybersecurity and therefore guarantee data within a smart grid and IoT to be confidential and integral against cyberattacks.

8.3.3 Risk Assessment and Management

8.3.3.1 Risk Assessment and Strategies for Risk Mitigation

The risk assessment process is comprehensive; it identifies, analyzes, and evaluates all possible risks that may impede the availability, integrity, and confidentiality of the systems and data in smart grid and IoT systems in a systemic way. This process usually includes:

Comprehensive Risk Assessment: A comprehensive risk assessment involves systematically identifying, analyzing, and evaluating potential risks that may impact the confidentiality, integrity, and availability of systems and data within the smart grid and IoT environments [39, 41, 42]. This process typically includes the following steps:

1) **Identification of Assets:** Identify and list all assets in an intelligent grid or IoT list, including hardware devices, software applications, data repositories, and networks.
2) **Threat Identification:** Identify potential threats and vulnerabilities that can be exploited against a system, which might take shape in a malware attack, an insider threat, a physical access attack, or a supply chain threat, for instance.
3) **Vulnerability Assessment:** It must conduct vulnerability assessments and identify all known vulnerabilities in the hardware, software, and network infrastructure components that may be preyed upon by potential threat actors.
4) **Impact Analysis:** Determine the possible impact of threats and vulnerabilities identified on the critical assets and functions of the smart grid or IoT environment, which may include data loss, downtime, financial loss, regulation fines, and impact on reputation.
5) **Likelihood Assessment:** Estimate the probability of the threat that has been identified to exploit the weakness and impact different consequences that are under consideration while taking into account available historical information, threat intelligence, and best practices within the industry [41].
6) **Risk Prioritization:** Prioritize the risk listed based on the likelihood and potential impact and give special consideration to areas of high risk that are most likely to be a barrier to attaining the objectives of the organization and its mission-critical processes.

Strategies for Risk Mitigation: After proper identification and prioritization of the risks, organizations are safely able to take up the matter of risk mitigation with the help of devising and executing ways by which the likelihood and consequences of the risks may be reduced [41, 42]. Effective risk mitigation strategies may include:

1) **Implementing Security Controls:** Deploying technical controls, such as firewalls, intrusion detection systems, antivirus software, encryption protocols, and access control mechanisms, to mitigate known vulnerabilities and prevent unauthorized access to critical assets. Put in place technical controls, including a firewall, intrusion detection system, antivirus software, encryption protocols, and access mechanisms that the control is trying to police against known threats and prevent unauthorized access to critical assets [41].
2) **Security Awareness Training:** Provides involved cybersecurity training and awareness programs for all employees and stakeholders to sensitize them about common cyber threats, best practices in data protection, and the procedures for reporting security incidences.
3) **Regular Vulnerability Patching:** Proactive implementation of the patch management process to keep software, firmware, operating systems, and other software applications updated regularly to avoid exploitation by threat actors from known vulnerabilities.
4) **Incident Response Planning:** Developing and implementing an incident response plan to effectively detect, respond to, and recover from security incidents and data breaches, minimizing the impact on business operations and mitigating potential financial and reputational losses.

5) **Continuous Monitoring and Evaluation:** Develop and implement an incident response plan that can rapidly detect, respond to, and recover from an incident and data breach with as little impact on business operations, financials, and reputation loss as possible.
6) **Third-Party Risk Management:** Continually, monitoring and evaluation procedures to keep up with the changing threat landscape and to assess the effectiveness of security controls in dynamically adjusting the mitigation measures. Take third-party risk management seriously. There should be solid processes for third-party risk management, assuring related assessment and mitigation of third-party risks from vendors, suppliers, or service providers in respect to the sensitivity of data being availed, or by extension, and their criticality in service provision within the smart grid or IoT ecosystem [41].

By conducting comprehensive risk assessments and implementing effective risk mitigation strategies, organizations can proactively identify and address potential cybersecurity risks, safeguarding the integrity, confidentiality, and availability of their systems and data within smart grid and IoT environments.

8.3.4 Technological Solutions and Best Practices

8.3.4.1 Encryption Techniques

It plays a crucial role when it comes to secure smart grid IoT data with integrity and confidentiality. Advanced encryption ensures the security of sensitive data at rest and in transit, making it quite unlikely that an unauthorized party will access and alter sensitive data. Advanced Encryption Standard (AES), extensively popular in the field of encryption, is followed because of its extreme security in all forms of advanced applications. Due to the symmetric nature of AES encryption, it is efficient and very suitable for large bulk data. It is feasible for use in smart grid environments that are supposed to continuously generate and transmit data securely [43].

Public key infrastructure (PKI), therefore, is the other significant aspect in securing communication in smart grid IoT. Public key infrastructures help enable secure communication, and device authentication, and allow key exchange through the use of asymmetric encryption. This technology is especially effective for IoT devices that require secure communication channels when transferring crucial information throughout the grid [44].

This concept is further reinforced with the application of end-to-end encryption (E2EE) to ensure that the data from the sending point is encrypted to the receiving point, leaving no room for intermediaries to intercept or access it using methods applied to ensure the data confidentially flows along the different nodes within the smart grid [45].

8.3.4.2 Robust Authentication Mechanisms

It should be ensured that smart grid IoT systems are accessed and controlled only by valid entities through robust authentication mechanisms. Multifactor authentication (MFA) can be considered one of the best practices in this regard. With MFA, the user is expected to offer at least two forms of proof "something they know (password)," something they have (token or smartphone), and something they are (biometric verification) before the system permits access. Access by unauthorized personnel due to compromised credentials is very well addressed by this layered approach.

It is biometric verification that adequately offers the uniqueness of personal identification through one's biological traits, such as fingerprints, facial recognition, or iris scanning. This is because of the almost impossibility of biometric data replication, making this method pretty strong, especially in protecting very critical control systems and sensitive data in a smart grid environment [46].

Digital certificates are one of the key digital documents used in the framework of PKI for device and user authentication. These authenticate all entities on the network; therefore, communication and data transmission are only allowed to identify devices. The system uses digital certificates to prevent man-in-the-middle attacks and many other forms of unauthorized access [47].

Additionally, role-based access control (RBAC) would ensure that people have only what they need to do their job in terms of information and the systems that hold that information. Job function access is reduced, resulting in an overall drop in activities associated with insider threats and lessening potential damage from a compromised account [48].

8.3.5 Threat Detection and Incident Response

A host of cybersecurity challenges crop up when IoT devices are integrated into the power grid, shaping the smart grid. To deal with these issues, robust strategies for threat detection and incident response (TDIR) must be followed because, through early and precise detection, utilities can secure their critical infrastructure and reduce the impacts of cyberattacks.

8.3.5.1 Importance of Threat Detection

Effective threat detection serves as the foundation of smart grid cybersecurity. Early identification of cyberattacks enables a swift response, minimizing potential damage and safeguarding critical infrastructure [49]. Prompt detection also expedites recovery by facilitating a targeted restoration effort focused on the affected areas. Continuously monitoring systems provide invaluable insights into attacker behavior and emerging threats, empowering utilities to proactively strengthen their defenses [50]. Furthermore, robust threat detection underpins a strong TDIR strategy, ensuring compliance with regulations focused on grid security and reliability. These detection systems also contribute to streamlined forensic investigations by capturing valuable data such as logs and network traffic recordings that aid in identifying the root cause of the attack and informing future defensive improvements. By prioritizing threats based on severity and fostering industry-wide collaboration through threat intelligence sharing, utilities can optimize resource allocation and collectively develop stronger defense against a constantly evolving cyber threat landscape [51].

8.3.5.2 Incident Response Strategies

Effective incident response strategies are essential in the smart grid landscape due to the ever-present threat of cyberattacks. These strategies, outlined in incident response plans, guide utilities through the complex process of detecting, containing, eradicating, and recovering from cyber threats, aligning with principles from industry standards such as those outlined.

The establishment of professional incident response teams made up of people with cybersecurity, IT operations, and engineering backgrounds is fundamental to these tactics. These teams are crucial in governing all essentials of a cyberattack, such as finding, containment, elimination, and retrieval. Strong communication systems, between the team and external stakeholders, such as regulatory agencies, are vital for ensuring a coordinated and efficient response [49].

Implementing well-defined incident response plans is a critical component of incident response techniques. These plans define precise activities to be completed in the event of a cyberattack, such as isolating compromised systems, gathering forensic evidence, reducing the effect, and restoring impacted systems. The SysAdmin, Audit, Network, and Security (SANS) Institute and the National Institute of Standards and Technology (NIST) both emphasize the importance of regular testing and upgrading these strategies to maintain their efficacy against emerging cyber threats [52].

Good incident response strategies involve follow-up measures, beginning with activities in the aftermath of the incident: that is, postmortem analysis and the root cause of the attack, identification of the holes exposed, and distribution of mitigation measures toward the incident, and then improvement of future incident mitigation, as recommended by leading industry experts. Sharing of threat intelligence gleaned from incidents builds a collaborative ecosystem in the smart grid and raises the bar on collective defense [53].

8.3.6 Regulatory Compliance and Standards

The dynamic landscape of smart grid technologies necessitates a robust regulatory framework to ensure grid security and reliability. Regulatory standards, such as those set by the North American Electric Reliability Corporation (NERC) and detailed in their critical infrastructure protection (CIP) standards, mandate cybersecurity controls for utilities. Compliance not only meets regulatory requirements but also enhances overall security [54]. NERC's CIP standards cover diverse areas such as physical security, system security, incident reporting, vulnerability assessments, prioritizing cybersecurity measures, and risk management.

The NIST provides a voluntary Cybersecurity Framework (CSF) [55], offering a risk-based approach tailored to smart grid utilities' needs. While not mandatory, adopting the CSF aligns utilities with best practices.

The International Electrotechnical Commission (IEC) offers internationally developed standards, with the IEC 62443 series focused on security in industrial automation and control systems, as a basis for smart grid security. Smart grid cybersecurity is further strengthened by third-party audits. Independent firms perform audits to evaluate the level of compliance with security standards and identify vulnerabilities [56]. Nowadays, third-party certifications, such as IEC 62443 and ISO/IEC 27001, take place, meaning that the existence of both recognized standards and, through the help of the certifications, increases confidence for stakeholders and competitiveness in the market [57]. Third-party audits and certifications complement each other in terms of a comprehensive approach to smart grid cybersecurity. While audits provide feedback for ongoing improvement activities, certifications formalize commitments made in this area, enhance the protective posture, and therefore contribute to a more secure smart grid ecosystem.

8.3.7 Human Factors and Insider Threats

The human element and insider threats play critical roles in both strengthening and potentially weakening the cybersecurity posture of a smart grid. Understanding these factors is essential for developing effective security strategies.

8.3.7.1 Role of Human Factors

Cybercriminals and hackers exploit human weaknesses through social engineering tactics, concentrating on utility personnel who have access to critical systems. The smart grid sector increases vulnerabilities related to social engineering attacks through phishing emails and other social engineering attempts, which increase with the increasingly remote nature of work. Cognitive biases such as availability and confirmation biases may lead to security failure by causing a person to ignore a threat or not act on suspicious activity. These types of human risks can be mitigated by raising awareness among the employees. Such programs should cover the existing cyber threats, social engineering practices, and how to behave safely in a smart grid environment. Cultivating a culture of security awareness encourages employees to report suspicious behavior, which further strengthens the human firewall against cyberattacks. Securing a smart grid will require the interaction of security experts and, essentially, all staff in a utility. Organizations can design strategies to mitigate the risks to have the maximum leverage of human intelligence and minimize the potential error by taking into consideration human factors, which contribute to security vulnerabilities [58].

8.3.7.2 Insider Threats and Mitigation

The reasons leading to insider threats in the smart grid industry could vary from unauthorized access to data, data exfiltration, misuse, and eventually sabotage in such a way that it could lead to wide power blackouts. The reasons behind insider threats could be wide: from financial gain and revenge to ideological beliefs or simple carelessness. Serious consideration of this principle will, therefore, mitigate the potential risks: users should access only those levels that are enough for their work and do not cause much damage in case the credentials are compromised. It makes it, therefore, very paramount to have the right kind of background checks during the hiring of employees and contractors. Equally important is adequate staff training on security awareness to be educated about the threat of malicious insiders. User behavior needs to be established by searching for any suspicion in behavior; this can be completed by analyzing access logs, data movement, and system changes to detect a potential insider threat. Data loss prevention (DLP) solutions supplement this protection through the oversight role in data movements. A security culture needs to be established within the organization. Organizations can considerably improve their protection from insider threats by fostering a situation where employees can observe and report any suspicious incidents [59].

8.4 Case Studies and Real-World Examples

The investigation of the practical world in relation to cybersecurity incidents in smart grid IoT is very important. This provides information on the practical implications and an understanding of the importance of use cases for better security measures in the future. This section highlights specific breaches, raising tactics, and subsequent effects on grid operations. Viewing such incidents helps us discover vulnerabilities and extract valuable lessons useful in building such robust security strategies. The description provided through the detailed case studies will give insight into common attack vectors and the effectiveness of different defensive mechanisms, thus helping to bridge the gap between theoretical understanding and practical implementation. It is important to develop a system more secure and resilient within smart grid IoT capabilities in such a way that it can adapt to a moving threat scenario.

8.4.1 Analysis of Past Cybersecurity Incidents

The introduction of IoT technology in smart grids has drastically redefined energy management to be efficient, reliable, and sustainable. Consequently, such technological advancement is also coupled with correspondingly high-ranking cybersecurity risks. The outcomes of past cybersecurity incidents on smart grid IoT systems are particularly informative on vulnerabilities and defense mechanisms that are important to protect smart grid infrastructures. Here, we go through some of the more significant breaches, deriving key lessons that can help inculcate better security strategies in the future.

8.4.1.1 Notable Breaches in Smart Grid IoT Deployments

Case 1: One of the most impactful incidents happened in the year 2015 on Ukrainian soil: a cyberattack against the power grid that resulted in massive outages, with almost 225,000 people going dark from the power network. Hackers first planted a malware trap in the network using spear-phishing emails and then used that as an entry point to move into the control systems. The breach showed how disastrous a cyberattack could be against critical infrastructure, revealing the very limited cybersecurity defense that grids had [60].

Another advanced attack is the cyber campaign Dragonfly, which operated against the energy industries in Europe and North America from 2011 to 2014. The attackers used sophisticated techniques, such as the Havex malware through the infiltration of industrial control systems. This revealed that emerging cyber threats is getting advanced and so is the stronger call toward better defenses [61].

For example, the 2018 Triton malware attack on a petrochemical plant in Saudi Arabia demonstrated succinct safety-system risks. Designed to manipulate industrial safety controllers, Triton could have caused catastrophic physical damage had it not been detected in good time. This incident showed that a cyberattack, when causing harm, can go beyond data theft and disruptions to impose very strong criteria in critical security for smart grid IoT systems [62].

8.4.1.2 Analysis

Figure 8.6 shows a blow-by-blow account of the Ukraine cyberattack – depicting a sequence of steps and the impacts one would expect said steps to have on the power grid. The bar and line graphs

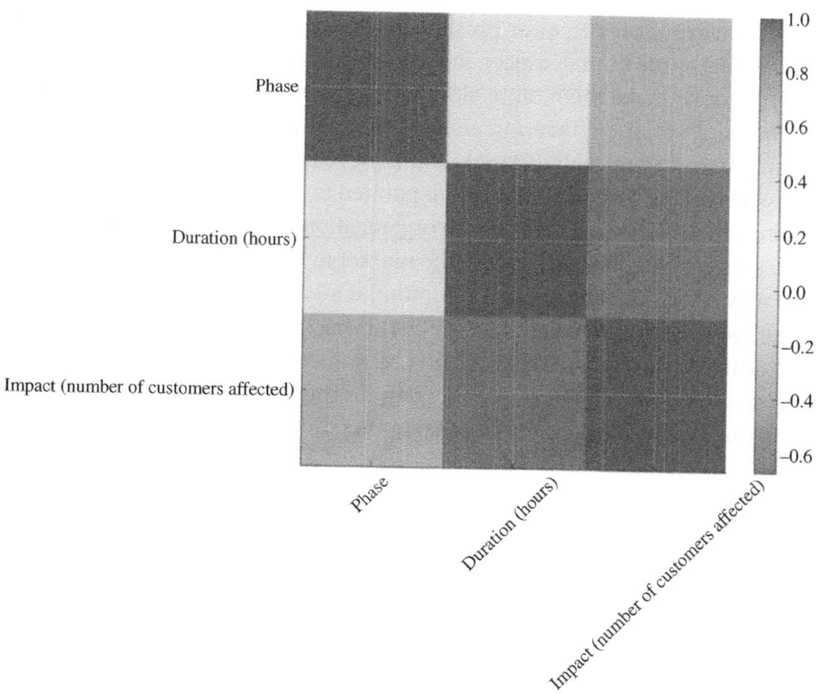

Figure 8.6 Correlation matrix of attack phases.

show the time length of the different steps and the impact in terms of customers affected. Phishing was followed by malware installations, network infiltration, control system compromise, and finally, a major power outage at the station serving 225,000 customers. These initial steps resulted in no apparent effects, but that fact was of the utmost importance because it laid the preconditions for the final, most damaging step. This then implies that much emphasis must be put on the early phases of the attacks to avert severe consequences.

The heatmap in the above Figure 8.6 outlines, by phase, and correlations among those in an attack set into sharp relief how initial breaches could develop into severe outcomes. Strongly correlated attack phases, for example, network infiltration and control system compromise or any other may signal that one attack phase will lead to another, hence, early detection and response is the key. Such then understanding shows the multistage nature of critical infrastructure attacks and the need for all-around security measures, encompassing all phases. By understanding the relationships, organizations can then be in a position to come up with better-baked strategies of system protection and mitigation strategies for subsequent attacks on the system.

8.4.1.3 Lessons Learned from Previous Incidents

By analyzing these breaches, we can identify several important lessons for enhancing cybersecurity in smart grid IoT deployments:

- **Multilayered Security:** The Ukraine power grid attack demonstrated the need for multiple layers of security. It would be incorrect to rely on one layer, such as firewalls or antivirus software. Strong security strategies should incorporate network segmentation, intrusion detection systems, continuous monitoring, and the many other methods designed to pinpoint and resolve any threats before they are blown out of proportion.
- **Employee Training and Awareness:** The spear-phishing technique used in the Ukraine attack showed vulnerabilities relating to the human aspect. Regular training and awareness sessions for employees have to be initiated to make them more alert in identifying and reacting to phishing techniques and other forms of social engineering attacks. In most cases, the greatest defense against cybercrimes is employees; thus, their alertness could prevent the first stage of the attack.
- **Advanced Threat Detection:** The Dragonfly campaign pointed to the fact that these are times that require advanced threat detection mechanisms. An organization should engage in the use of advanced analytics, ML plus ensign tools available in the market to be in a position to trace any abnormal patterns that may reveal themselves as a symptom of a long-drawn cyberattack. This would serve to minimize the level of havoc to be caused by this hack through early detection.
- **Recurring Security Audits/Penetration Testing:** It is of immense importance for security audits and penetration tests to be carried out time and again for the continuous spotting of vulnerabilities in advance. Proactively take measures to ensure that different security protocols are effective against constantly changing threats.
- **Access Control:** Robust access controls should be put in place to prevent unauthorized access to important systems. The Dragonfly event revealed, by way of example, how critical such stringent access management and policy that supports role-based access controls and MFA is.
- **Safety and Security System Integration:** With the malware attack on Triton, the integration of safety and security systems reaches unparalleled levels. In this way, when security extends to the safety system, attackers will be unable to cause physical harm. What is important for protection in the case of a safety breach are fault-tolerant safety protocols and continuous monitoring of the safety systems in place.
- **Incident Response Planning:** Adequate planning to respond to incidents is indispensable. This consists of the development and periodical updating of incident response plans and the holding

of drills to ensure awareness among all concerned personnel of their respective roles in the event of a cyber incident. Coordinated responses will produce a significant decrease in the risk of such an attack.

Accordingly, this leads to achievement in cybersecurity postures of organizations that would later bring about the securing of IoT systems developed for smart grids. Such lessons underscore the fact that strong and inclusive cybersecurity, covering both technological and human factor areas, is really necessary. In a smart grid, resilience is achieved through continuous watch by leveraging advanced technology and a security awareness and preparedness culture.

8.4.2 Successful Cases of Cybersecurity Strategies

This will help explain the successful implementation of cybersecurity strategies toward having strong defense measured against potential threats to the IoT smart grid system. This section will provide some case studies on some practical implementation of security measures and some reviews of organizations recognized for their strong cybersecurity posture. We shall analyze these cases to draw best practices and strategies that can be employed to secure smart grid IoT infrastructures.

8.4.2.1 Case Study of Effective Security Measures

Case Study 1: The North American Electric Reliability Corporation (NERC)

NERC is the pioneer in applying advanced and very tough cybersecurity measures meant for the protection of bulk power systems throughout North America. NERC's CIP standards spell out actions as quite needed that are to be taken to defend critical assets and associated systems. These standards include the identification of assets and security management controls, personnel and training, incident reporting, and recovery plans.

The most notable application of such measures was the 2019 GridEx V exercise, a biennial event by NERC for testing and improving resilience in the electricity sector regarding both cyber and physical threats. More than 6,500 participants from utilities, government agencies, and law enforcement took part in the exercise. The test reflects the application of NERC's approach to working with the CIP standards, and accordingly, NERC is very committed to securing the grid to be better prepared with improved response capabilities [63].

Case Study 2: Pacific Gas and Electric Company (PG&E)

In addition, some of the effective cybersecurity strategies that have been operational in its operations have been indicated by Pacific Gas and Electric (PG&E) Company. The cybersecurity program at PG&E is designed for the detection and response of anomalies by continuous monitoring with threat intelligence and incident response. Some of the strategies PG&E has used include infusing advanced analytics and ML for real-time detection and response to anomalies.

In 2018, for example, PG&E proactively stopped a cyberattack that aimed to breach its SCADA system. Its strong incident response plan, with near-real-time monitoring and advanced alerting threats, had therefore been able to counteract successfully before great damage had occurred. This is a case in point for being proactive with technology and cybersecurity [64].

8.4.2.2 Case Studies of Robust Cybersecurity Measures

Case Study 3: Southern California Edison (SCE)

Southern California Edison (SCE) is noted for a comprehensive CSF – having a risk management approach, employee training, alignment with regulatory compliance – and using multiple layered efforts in securing its systems: applying firewalls, intrusion detection systems, and encryption to harden state-of-the-art infrastructure. It also emphasizes employee awareness and training to mitigate some of the risk factors due to human error.

SCE has also enhanced its cybersecurity posture by working within industry standards and regulatory frameworks. The firm usually goes through third-party audits and collaborates with the industry to learn new emerging threats and practices. Such commitment to improvement and compliance has made SCE an exemplary model for cybersecurity across the utility sector [65].

Case Study 4: Duke Energy

The other very strong applied CSF is Duke Energy. Three key principles underpin the approach to cybersecurity at Duke Energy: prevention, detection, and response. Motivated by investments in sophisticated advanced technology, the corporation focuses on state-of-the-art AI and ML to enhance threat detection. This cybersecurity-protected entity again has arrangements with its respective operations center for the monitoring and response of threats every minute, every hour, and every day.

One of the other main focuses for Duke is partnering with various governmental agencies and industry groups to make recommendations on intelligence gathered on threats and to collaborate on continued efforts to further develop cybersecurity initiatives. This proactive approach has put Duke Energy consistently ahead of the curve on emerging threats and continues to develop as increased security measures come into place. The collateral emphasis on collaboration and technology integration is a best practice for other organizations in this sector [66].

Through the above case studies and examples, it emerges those effective cybersecurity strategies for smart grid IoT systems require a mix of stringent standards, cutting-edge technology, real-time monitoring, and prompt incident response. NERC, PG&E, SCE, and Duke Energy have already proven the importance of a multilayered strategy with a heavy emphasis on technology and human elements in ensuring the protection of critical infrastructures. The above examples, with the assistance of key takeaways, derive best practices useful for the improvement of the cybersecurity posture for smart grid IoT systems across the globe.

8.4.3 Evaluating Existing Solutions

The landscape of cybersecurity for the smart grid IoT system is dynamic and keeps changing at all times. This makes it so important that the security solutions at hand are robust and adaptive. Therefore, let us consider some current solutions, describe them, and take a look at their positive and negative sides in relation to the further improvement strategy within the general smart grid IoT system.

8.4.3.1 Network Segmentation and Firewalls

Strengths: Network segmentation is the practice of breaking down a network and creating small, isolated sections – secured using a firewall. This controls lateral movement in the network and limits the level of widescale damage that would be possible because firewalls would be acting like check filters, limiting both incoming and outgoing traffic according to preset security rules, hence barring unauthorized access [67].

Limitations: However, mechanisms concerning issues around network segmentation and firewalls are pretty good in terms of access control. More sophisticated attackers expose the weaknesses of segmented networks when they tunnel through or encrypt traffic in order to bypass the firewalls. Firewalls and network segmentation are operational challenges, considering the amount of work that goes into maintaining and configuring them [67].

8.4.3.2 Intrusion Detection and Prevention Systems (IDPS)

Strengths: Intrusion detection and prevention systems (IDPS) is intended to detect and prevent such malicious activity within the network. It is designed to monitor, in turn taking further action on potentially threatening blocks or suspicious patterns within network traffic. More advanced IDPS systems use ML and behavioral analysis to boost detection capabilities and cut down on false positives [68].

Limitations: The issues of concern, therefore, with the IDPS are the occurrence of false positives and false negatives. False positives will, therefore, cause unnecessary disruption while false negatives will let the threat through without detection. Besides, IDPS requires instances of continuous updating to be in a position to identify new threats, meaning it demands a large number of resources and expertise [68].

8.4.3.3 Advanced Encryption Techniques

Strengths: Encryption is a fundamental security feature that protects data at rest and during transmission by converting data to an unreadable format that can only be accessed by an individual with a decryption key. AES and Public Key Infrastructure are examples of more sophisticated encryption techniques [69].

Limitations: However, this encryption process limitation can introduce latency, degrading system performance, especially in real-time environments such as the smart grid. Moreover, encryption requires very secure key management, which is a challenging task and requires strict protocols to guard against unauthorized access [69].

8.4.3.4 Multifactor Authentication (MFA)

Strengths: MFA inherently improves security by the imposition of numerous levels of verification barring access. In the process, the threat of unauthorized access is brought down to the minimum, if the credentials were to be compromised. Some of the most popular ways of going about MFA include password-and-token combinations, biometric verification, and one-time password (OTP) [70].

Limitations: MFA implementation can be challenging, more so in legacy systems never designed with such in mind. MFA can also be quite inconvenient for the users, mostly causing pushback and therefore a reduction in productivity. For that reason, the reliability of the system through which MFA is implemented has to be assured, and the MFA system must be user-friendly to be effective [70].

8.4.3.5 Regular Security Audits and Penetration Testing

Strengths: Regular security audits and penetration testing are part of such proactive practices to find and mitigate vulnerabilities before being exploited. Such exercises are conducive to full-scale security postural assessments, simulated attacks to identify weaknesses, and actionable recommendations toward improvement [71].

Limitations: These require specialized skills and resources in terms of the capability to carry out effective security audits and penetration testing, which can prove quite costly. In addition, such assessments give a picture of the security posture at any given instance; hence, for effectiveness, the assessment requires continuous tests [71].

8.4.3.6 Incident Response Planning and Drills

Strengths: By defining roles and responsibilities and putting procedures in place for responding to an attack, effective incident response planning is achieved. This makes such a team, which is in charge of the operations for incident response, well prepared for real-life scenarios, hence, reducing the impact of breaches [72].

Limitations: Building up and sustaining a good incident response plan could, thus, be a resource-draining process. That is to say, ensuring that all those involved are rightly trained and that the plan is put up to date in line with emerging threats and changes to the infrastructure are the mainstay premises of the plan [72].

Existing solutions for the security of smart grid IoT systems are evaluated to reveal a mix of strengths and limitations. Although it is a good mix of protection mechanisms from network segmentation, firewalls, IDPS, encryption, MFA, security audits, and incident response planning, this hard challenge goes on to pose. It is imperative to develop certain solutions to these challenges that the combination of technological advances is brought in order, continuously monitor periodic updating, and all-inclusive training programs. These evaluations may be used to devise better means of making smart grid IoT systems more resilient to cyberattacks.

8.5 Future Trends and Considerations

8.5.1 Emerging Technologies Impacting Smart Grid IoT Security

In addition to that, the emergence of AI and ML in cybersecurity is a breakthrough in this smart grid IoT security. These technologies enable real-time detailed analysis of bountiful data for detecting patterns and anomalies that indicate potential security threats, thus ameliorating threat detection. In automation with response to known threats and prediction, future vulnerability will happen through adaptively learning from previous incidents while adapting and improving over time. This capability to adjust and learn over time makes AI and ML a required security-enabling tool of smart grid IoT under dynamic and complex environmental conditions [73].

Another critical technological development is the application of blockchain technology in smart grid security. It is both decentralized and cryptographically secure, which makes it appropriate for ensuring integrity and transparency in transactions within the grid. It will provide a safe, tamper-proof record of energy generation, distribution, and utilization to minimize any cases of fraud and hence foster trust among stakeholders. On top of fostering trust and managing fraud, the technologically enacted blockchain fosters secure peer-to-peer energy trading to help create energy networks that are resilient and decentralized [74].

8.5.2 Anticipated Cybersecurity Challenges

Cybersecurity challenges that smart grid IoT systems will face will also change as they mature. An expected challenge, in this respect, is the evolution of cyber threats. Attackers have continually developed new techniques and tools for the exploitation of vulnerabilities, and their evolution into new threats necessitates equally ratcheted up defensive measures. Phishing, ransomware, and advanced persistent threat (APT) will likely become more pervasive and harder to detect, so they will represent significant risks for smart grid IoT systems. The problem of risk that has arisen with the proliferation of the IoT and the interconnectivity of its devices is troubling precisely because the more the number of devices integrated into the grid, the bigger an attack surface available for potential cyber threats and the number of their entry points. The securing of these diverse devices often resource constrained is, in fact, a difficult problem. First, in the case of compromise within any devices or segment, its effect may cascade to disrupt the whole grid. This problem, in fact, with all the diverse devices often resource restricted has now put on the smart grid IoT systems the burden of ensuring the integrity of channels and ensuring data integrity [75].

8.5.3 Recommendations for Future Research

Research in smart grid IoT security should be centered on AI and ML algorithm development, both advanced and tailor made, for inclusion in cybersecurity activities for smart grids in responding to these emerging challenges. This research will highlight the fine-tuning of the developed algorithms for better accuracy and efficiency in threat detection and mitigation. Work being completed to integrate AI and ML with other technologies, such as blockchain, would be another source of innovation [76].

This recommendation will bring the strengths of the approaches together. Collaborative efforts and partnerships are other important domains in smart grid IoT development that will push forward the cybersecurity of smart grid IoT with the kind of blockchain. Collaboration efforts among academia, industry, and governments will facilitate the sharing of knowledge, and the development of standards, and best practices. Public–private partnership in the sharing of threat intelligence and resources improves the collective ability to respond to cyber threats. International cooperation is also important because cyber threats do not respect borders, and the energy infrastructures are interconnected.

Moreover, it is through the development and evolution of cyber threats by integrating emerging technologies that the future of smart grid IoT security is deeply linked. This work relocates research efforts within advanced AI and ML applications, explores the potential of blockchain, and fosters collaborative partnerships such that developed security strategies are robust and ensure the resilience and reliability of smart grid IoT systems in the face of evolving challenges.

8.6 Conclusions

The infusion of IoT technologies within smart grids is a transformative paradigm in the context of contemporary energy management, professing superior efficiency, reliability, and sustainability. However, the integration poses serious challenges in terms of cybersecurity and mandates a multidimensional approach to assure the safety and resilience of this most critical infrastructure. Throughout this chapter, we have tried to cover the various aspects of security in smart grid IoT systems, from threat landscape analysis and review of significant security incidents to the determination of effective defense strategies and forecasting future trends.

Critical and in-depth analysis of past cybersecurity incidents, such as the Ukraine power grid attack and the Dragonfly campaign, exposes the intensity and seriousness of the threats looming over smart grid IoT systems. These are very strong examples of the need for comprehensive security approaches from the side of technology and human elements. Multilevel security approaches with network segmentation, advanced intrusion detection systems, powerful encryption techniques, and MFA have time and again turned out to be critical in risk mitigation and the protection of critical assets. Moreover, critical for sustaining a proactive security posture is regular security auditing, penetration testing, and incident response planning.

Thus, these provide promising ways of adaptation for the reinforcement of smart grid IoT security with emergent technologies such as AI, ML, and blockchain. AI and ML offer advanced capabilities in threat detection, automated response, and predictive analytics, allowing dynamic and adaptive defense against the rapidly changing cyber threat landscape. Blockchain is a decentralized, cryptographically secured structure that gives strong answers to the problems of data integrity, transparency, and trust in energy transactions. Integration of such technologies can develop tremendously the security framework for smart grid IoT systems.

However, there might be a brighter future for smart grid IoT security. All smart grid security strategy is a continuous course of action toward the innovation and adaptation of the strategies against the rapidly evolving cyber threats, with the attack surface expanding more and more due to the integration of IoT devices. Future research is anticipated to develop further advanced AI and ML algorithms that would fit cybersecurity needs, investigate synergistic integrations of emerging technologies, and foster enhanced collaboration across academia, industry, and the government sector to develop more resilient energy infrastructures that can withstand and rapidly recover from cyber incidents.

To conclude, smart grid IoT security is an ongoing and multifaceted comprehensive process from advanced technology and monitoring to training and observance with governmental and regional regulations. The track, based on previous incidents and effective implementation throughout the implementation process, will provide a path on which the development of an effective security strategy can be based. All of this will have to be finished via a combined partnership approach and emerging technologies that can implement smart grid IoT resilient and reliable against changes in the cyber threat landscape. It will have to be a proactive, integrated way to fully support the continued growth and maintenance of modern energy infrastructure.

References

1 Baležentis, T. and Štreimikienė, D. (2019). Sustainability in the electricity sector through advanced technologies: energy mix transition and smart grid technology in China. *Energies* 12 (6): 1142. https://doi.org/10.3390/en12061142.

2 Ahmad, T. and Zhang, D. (2021). Using the Internet of Things in smart energy systems and networks. *Sustainable Cities and Society* 68: 102783. https://doi.org/10.1016/j.scs.2021.102783.

3 Hussain, F. (2017). Internet of everything. In: *Internet of Things*, 1–11. Springer https://doi.org/10.1007/978-3-319-55405-1_1.

4 Abir, S.M., Anwar, A., Choi, J., and Kayes, A.S. (2021). IoT-enabled smart energy grid: applications and challenges. *IEEE Access* 9: 50961–50981. https://doi.org/10.1109/ACCESS.2021.3067331.

5 Kimani, K., Oduol, V., and Langat, K. (2019). Cyber security challenges for IoT-based smart grid networks. *International Journal of Critical Infrastructure Protection* 25: 36–49. https://doi.org/10.1016/j.ijcip.2019.01.001.

6 Kuguoglu, B.K., van der Voort, H., and Janssen, M. (2021). The giant leap for smart cities: scaling up smart city artificial intelligence of things (AIoT) initiatives. *Sustainability* 13 (21): 12295. https://doi.org/10.3390/su132112295.

7 Markovic, D.S., Zivkovic, D., Branovic, I. et al. (2013). Smart power grid and cloud computing. *Renewable and Sustainable Energy Reviews* 24: 566–577. https://doi.org/10.1016/j.rser.2013.03.068.

8 Berte, D.-R. (2018). Defining the IoT. *Proceedings of the International Conference on Business Excellence* 12: 118–128. https://doi.org/10.2478/picbe-2018-0013.

9 Culot, G., Fattori, F., Podrecca, M., and Sartor, M. (2019). Addressing industry 4.0 cybersecurity challenges. *IEEE Engineering Management Review* 47 (3): 79–86. https://doi.org/10.1109/EMR.2019.2927559.

10 Gunduz, M.Z. and Das, R. (2020). Cyber-security on smart grid: threats and potential solutions. *Computer Networks* 169: 107094. https://doi.org/10.1016/j.comnet.2019.107094.

11 Tufail, S., Parvez, I., Batool, S., and Sarwat, A. (2021). A survey on cybersecurity challenges, detection, and mitigation techniques for the smart grid. *MDPI: Energies* 14 (18): 5894 https://doi.org/10.3390/en14185894.

12 Anderson, R. and Moore, T. (2006). The economics of information security. *Science* 314 (5799): 610–613.

13 Fang, X., Misra, S., Xue, G., and Yang, D. (2024). Smart grid — the new and improved power grid: a survey. *IEEE Communications Surveys & Tutorials* 14 (4): 944–980. https://doi.org/10.1109/SURV.2011.101911.00087.

14 Gungor, V.C., Sahin, D., Kocak, T. et al. (2011). Smart grid technologies: communication technologies and standards. *IEEE Transactions on Industrial Informatics* 7 (4): 529–539. https://doi.org/10.1109/TII.2011.2166794.

15 He, H. and Yan, J. (2016). Cyber-physical attacks and defences in the smart grid: a survey. *IET Cyber-Physical Systems: Theory & Applications* 1: 13–27. https://doi.org/10.1049/iet-cps.2016.0019.

16 Umar Farooq, M., Muhammad, W., Anjum, K., and Sadia, M. (2015). A critical analysis on the security concerns of Internet of Things (IoT). *International Journal of Computer Applications* 111: 1. https://doi.org/10.5120/19547-1280.

17 Singer, P. and Friedman, A. (2014). *Cybersecurity and Cyberwar: What Everyone Needs to Know*. Oxford University Press ISBN: 9780199918119.

18 Whitman, M.E. and Herbert, J.M. (2021). *Principles of Information Security*, 7e. Cengage.

19 Geers, K. (2011). *Strategic Cyber Security*. NATO Cooperative Cyber Defence Centre of Excellence.

20 Roman, R., Zhou, J., and Lopez, J. (2013). On the features and challenges of security and privacy in distributed internet of things. *Computer Networks* 57 (10): 2266–2279. https://doi.org/10.1016/j.comnet.2012.12.018.

21 Sasse, M.A., Brostoff, S., and Weirich, D. (2001). Transforming the 'weakest link' — a human/computer interaction approach to usable and effective security. *BT Technology Journal* 19: 122–131. https://doi.org/10.1023/A:1011902718709.

22 Amin, S.M. and Wollenberg, B. (2005). Toward a smart grid: power delivery for the 21st century. *IEEE Power and Energy Magazine* 3 (5): 34–41. https://doi.org/10.1109/MPAE.2005.1507024.

23 Vincenzo, G., Flavia, G., Gianluca, F. et al. (2015). *Smart Grid Projects in Europe - Lessons Learned and Current Developments*, EUR 24856 EN, OP LD-NA-24856-EN-N. JRC Publications Repository https://doi.org/10.2790/32946.

24 Yan, Y., Qian, Y., Sharif, H., and Tipper, D. (2013). A survey on smart grid communication infrastructures: motivations, requirements and challenges. *IEEE Communications Surveys & Tutorials* 15 (1): 5–20. https://doi.org/10.1109/SURV.2012.021312.00034.

25 Weerakkody, S. and Sinopoli, B. (2019). Challenges and opportunities: cyber-physical security in the smart grid. In: *Smart Grid Control* (ed. J. Stoustrup, A. Annaswamy, A. Chakrabortty, Z. Qu). Cham: Springer. https://doi.org/10.1007/978-3-319-98310-3_16.

26 Rawat, D.B. and Bajracharya, C. (2015). Cyber security for smart grid systems: Status, challenges and perspectives. *SoutheastCon 2015*. Fort Lauderdale, FL: IEEE. https://doi.org/10.1109/SECON.2015.7132891.

27 Jadoon, A.K., Wang, L., Li, T., and Zia, M.A. (2018). Lightweight cryptographic techniques for automotive cybersecurity. *Hindawi: Wireless Communications and Mobile Computing* 2018: 1640167. https://doi.org/10.1155/2018/1640167.

28 Bekkali, A.E., Essaaidi, M., and Boulmalf, M. (2023). A blockchain-based architecture and framework for cybersecure smart cities. *IEEE Access* 11: 76359–76370. https://doi.org/10.1109/ACCESS.2023.3296482.

29 Al Barazanchi, I., Murthy, A., Rababah, A.A. et al. (2022). Blockchain technology - based solutions for IOT security. *Iraqi Journal For Computer Science and Mathematics* 3 (1): 53–63. https://doi.org/10.52866/ijcsm.2022.01.01.006.

30 Chen, M., Mao, S., and Liu, Y. (2014). Big data: a survey. *Mobile Networks and Applications* 19: 171–209. https://doi.org/10.1007/s11036-013-0489-0.

31 Shahinzadeh, H., Moradi, J., Gharehpetian, G.B. et al. (2019). IoT architecture for smart grids. *International Conference on Protection and Automation of Power System (IPAPS)*. pp. 22–30. Iran: IEEE.

32 Ghasempour, A. (2019). Internet of Things in smart grid: architecture, applications, services, key technologies, and challenges. *Inventions* 4: 22.

33 Turki Alsuwian, A.S. (2022). Smart grid cyber security enhancement: challenges and solutions—a review. *Sustainability* 14 (21): 14226.

34 Radziwill, Y. (2015). *Cyber-Attacks and the Exploitable Imperfections of International Law*. Brill, ISBN: 978-90-04-29830-9.

35 Khoei, T.T., Slimane, H.O., and Kaabouch, N. (2022). A comprehensive survey on the cyber-security of smart grids: cyber-attacks, detection, countermeasure techniques, and future directions. *Arxiv: Cryptography and Security* 1–20. arXiv:2207.07738.

36 Saadat, S., Bahizad, S., Ahmed, T. and Maingot, S. (2020). Smart grid and cybersecurity challenges. *2020 5th IEEE Workshop on the Electronic Grid (eGRID)*. Aachen, Germany: IEEE. https://doi.org/10.1109/eGRID48559.2020.9330660.

37 Abbas, A.K. (2024). Cybersecurity challenges in smart grids: a focus on information technology. *Journal of Electrical Systems* 20 (6s): 2024. https://doi.org/10.52783/jes.2922.

38 Mrabet, Z.E., Ghazi, H.E., Kaabouch, N., and Ghazi, H.E. (2018). Cyber-security in smart grid: survey and challenges. *Arxiv: Cryptography and Security*, arXiv:1809.02609.

39 Verma, R. (2024). Cybersecurity challenges in the era of digital transformation. In: *Transdisciplinary Threads Crafting the Future Through Multidisciplinary Research Volume - 1* (ed. R.P. Mahurkar, C. Hargunani, S. Chaudhary, et al.), 187 https://doi.org/10.25215/9392917848.20. Infinity Publication Pvt. Ltd.

40 AlSalem, T.S., Almaiah, M.A., and Lutfi, A. (2023). Cybersecurity risk analysis in the IoT: a systematic review. *MDPI: Electronics* 12 (18): 3958. https://doi.org/10.3390/electronics12183958.

41 Erkuden Rios, A.R. (2020). Continuous quantitative risk management in smart grids using attack defense trees. *Sensors* 20 (16): 4404.

42 Vikas Lamba, N.Š. (2019). Recommendations for smart grid security risk management. *Cyber-Physical Systems* 5: 1–27.

43 Daemen, J. and Rijmen, V. (2020). *The Design of Rijndael: The Advanced Encryption Standard (AES)*. Springer.

44 Adams, C. and Lloyd, S. (2003). *Understanding PKI: Concepts, Standards, and Deployment Considerations*. Addison-Wesley Professional.

45 Schneier, B. (2015). *Applied Cryptography: Protocols, Algorithms and Source Code in C, 20th Anniversarye* ISBN: 978-1-119-09672-6. Wiley.

46 Jain, A.K., Ross, A.A., and Nandakumar, K. (2011). *Introduction to Biometrics*. Springer.

47 Housley, R. (2002). Internet X.509 public key infrastructure certificate and certificate revocation list (CRL) profile. Retrieved from Data Tracker. https://datatracker.ietf.org/doc/html/rfc3280

48 Sandhu, R., Coyne, E., Feinstein, H., and Youman, C. (1996). Role-based access control models. *IEEE Computer* 29 (2): 38–47. https://doi.org/10.1109/2.485845.

49 NERC. (2006). Cyber security (permanent). NERC. https://www.nerc.com/pa/Stand/Pages/Cyber-Security-Permanent.aspx

50 Wanda, P. and Jie, H.J. (2019). A survey of intrusion detection system. *International Journal of Informatics and Computation (IJICOM)* ISSN: 2685-8711, https://doi.org/10.35842/ijiccm.v1i1.7.

51 William, S. (2017). *Cryptography and Network Security - Principles and Practice*, 7e. Pearson.

52 NIST. (2018). Framework for improving critical infrastructure cybersecurity. National Institute of Standards and Technology. https://nvlpubs.nist.gov/nistpubs/cswp/nist.cswp.04162018.pdf

53 NIST. (2012). Computer security incident handling guide. National Institute of Standards and Technology (.gov). https://doi.org/10.6028/NIST.SP.800-61r2.

54 NERC. (2024). Critical infrastructure protection (CIP) standards. North American Electric Reliability Corporation (NERC). https://www.nerc.com/pa/Stand/Pages/default.aspx

55 NIST. (2024). Cybersecurity framework. NIST.GOV. https://www.nist.gov/cyberframework

56 NIST. (2022). Guide to intrusion detection and prevention systems (IDPS). Computer Security Resource Center. https://doi.org/10.6028/NIST.SP.800-94.

57 ISO. (2022). ISO/IEC 27001:2022. International Organization for Standardization. https://www.iso.org/standard/27001

58 ENISA. (2019). ENISA report: cybersecurity culture guidelines: behavioral aspects of cybersecurity. European Union Agency for Cybersecurity (ENISA). https://www.enisa.europa.eu/publications/cybersecurity-culture-guidelines-behavioural-aspects-of-cybersecurity

59 Theis, M.C., Trzeciak, R.F., Costa, D.L. et al. (2019). *Common Sense Guide to Mitigating Insider Threats*, Technical Report CMU/SEI-2018-TR-010, 6e. Carnegie Mellon University https://doi.org/10.1184/R1/12363665.v1.

60 Park, D. and Walstrom, M. (2017). Cyberattack on critical infrastructure: Russia and the Ukrainian power grid attacks. University of Washington. https://jsis.washington.edu/news/cyberattack-critical-infrastructure-russia-ukrainian-power-grid-attacks/

61 Symantec. (2017). Dragonfly: western energy sector targeted by sophisticated attack group. Symantec.com: https://symantec-enterprise-blogs.security.com/threat-intelligence/dragonfly-energy-sector-cyber-attacks.

62 MIT. (2019). Triton is the world's most murderous malware, and it's spreading. MIT Technology Review. https://www.technologyreview.com/2019/03/05/103328/cybersecurity-critical-infrastructure-triton-malware/

63 NERC. (2020). GridEx V: grid security exercise lessons learned report. North American Electric Reliability Corporation. https://www.nerc.com/pa/CI/ESISAC/GridEx/GridEx%20V%20Public%20Report.pdf.

64 PG&E. (2023). Vulnerability disclosure policy. Pacific Gas & Electric. https://www.pge.com/en/about/company-information/vulnerability-disclosure-policy.html.

65 SCE. (2020). Cybersecurity: protecting the grid. SCE Report. Southern California Edison (SCE). https://www.sce.com/sites/default/files/inline-files/PolicyonInfoSecurityCybersecurityPrivacy.pdf.

66 FPSC. (2023). Cyber and physical security protections. The Florida Public Service Commission. https://www.floridapsc.com/pscfiles/website-files/PDF/Publications/Reports/General/Electricgas/2023_FPL_Cybersecurity.pdf

67 Mhaskar, N., Alabbad, M., and Khedri, R. (2021). A formal approach to network segmentation. *Computers & Security* 103: 102162. https://doi.org/10.1016/j.cose.2020.102162.

68 NIST. (2022). Special Publication 800-161: Supply Chain Risk Management Practices for Federal Information Systems and Organizations. National Institute of Standards and Technology (NIST). https://doi.org/10.6028/NIST.SP.800-161r1.

69 Meng, Y., Liangliang, Z., Yao, R. et al. (2021). Research on fast encryption method for smart energy management system in smart gird. *International Conference on Communications, Information System and Computer Engineering (CISCE)*. Beijing, China: IEEE. https://doi.org/10.1109/CISCE52179.2021.9445871.

70 Hassan, M., Metwally, K., and Elshafey, M.A. (2024). ZF-DDOS: an enhanced statistical-based DDoS detection approach using integrated Z-score and fast-entropy measures. *6th International Conference on Computing and Informatics (ICCI)*. New Cairo - Cairo, Egypt: IEEE. https://doi.org/10.1109/ICCI61671.2024.10485097.

71 Bryushinin, A.O., Dushkin, A.V., and Melshiyan, M.A. (2022). Automation of the information collection process by osint methods for penetration testing during information security audit. *2022 Conference of Russian Young Researchers in Electrical and Electronic Engineering (ElConRus)*. Saint Petersburg, Russian Federation: IEEE, https://doi.org/10.1109/ElConRus54750.2022.9755812.

72 Hou, Z.S., Zhang, Q.Y., Geng, B. et al. (2020). An emergency drill command system facing the power grid emergency operation site. *4th Annual International Conference on Data Science and Business Analytics (ICDSBA)*. Changsha, China: IEEE. https://doi.org/10.1109/ICDSBA51020.2020.00072.

73 Zheng, Y., Pal, A., Abuadbba, S. et al. (2020). Towards IoT security automation and orchestration. *Second IEEE International Conference on Trust, Privacy and Security in Intelligent Systems and Applications (TPS-ISA)*. Atlanta, GA: IEEE. https://doi.org/10.1109/TPS-ISA50397.2020.00018.

74 Hasan, M.K., Alkhalifah, A., Islam, S. et al. (2022). Blockchain technology on smart grid, energy trading, and big data: security issues, challenges, and recommendations. *Wireless Communications and Mobile Computing* 2022: 9065768. https://doi.org/10.1155/2022/9065768.

75 Goudarzi, A., Ghayoor, F., Waseem, M. et al. (2022). A survey on IoT-enabled smart grids: emerging, applications, challenges, and outlook. *Energies* 15 (19): 6984. https://doi.org/10.3390/en15196984.

76 Omitaomu, O.A. and Niu, H. (2021). Artificial intelligence techniques in smart grid: a survey. *Smart Cities* 4 (2): 548–568. https://doi.org/10.3390/smartcities4020029.

9

IoT-Based Monitoring for Substations

Rajesh K. Padmashini[1], Dhandapani Lakshmi[1], Rajasekharan Rajasree[1], Janarthanan N. Rajesh Kumar[2], Rahiman Zahira[3], Palanisamy Sivaraman[4], and Chenniappan Sharmeela[5]

[1]*Department of EEE, AMET Deemed to be University, Chennai, Tamil Nadu, India*
[2]*Department of Computer Science and Engineering, Sree Sastha Institute of Engineering and Technology, Chembarambakkam, Tamil Nadu, India*
[3]*Department of Electrical and Electronics Engineering, B.S. Abdur Rahman Crescent Institute of Science and Technology, Chennai, Tamil Nadu, India*
[4]*Research Scholar, Anna University Chennai, Tamil Nadu, India*
[5]*Department of EEE, CEG, Anna university, Chennai, Tamil Nadu, India*

9.1 Introduction to IoT-Based Monitoring for Substations

The substation plays a crucial role in the transportation of power. Power transformers are used to accomplish switching, protection, and voltage level conversion from high to low and vice versa. Substations, which are essential parts of the power system, are commonly composed of transformers, circuit breakers (CBs), relays, lightning arresters (LAs), current transformers (CTs), potential transformers (PTs), isolators, capacitors, and other miscellaneous equipment [1, 2].

To put it another way, a substation is a collection of equipment that is used to adjust several characteristics of an electric supply, including the frequency, power factor, voltage level, and conversion from alternating current (AC) to direct current (DC) [3]. Substations are usually monitored and controlled by costly programmable logic controllers (PLCs) and supervisory control and data acquisition (SCADA) systems; however, these approaches are more labor-intensive and require more maintenance [4]. Under traditional substation protection, relay operation typically necessitates the use of pilot wires, especially in differential protection. This raises the possibility of a sudden cessation of relay operation and raises the relay's capital cost. IoT-based substation monitoring and controlling provides a viable way to counter the disadvantages because it has a completely automated system that guarantees a greater level of reliability and, as a result, an increase in system performance through the efficient use of the equipment. The phrases "Internet" and "Things," where "Things" is any device that is connected to the internet, are the source of the abbreviation "Internet of Things," or IoT [5].

The IoT is a system of networked computing devices, mechanical and digital machinery, products, animals, or people. Each device has a unique identification and the ability to transfer data across a network without requiring human-to-human or human-to-computer interaction. The IoT has grown to become a part of the larger class of cyber-physical systems with the addition of sensors and actuators in recent years [6, 7]. This class of systems also includes technologies such as

IoT for Smart Grid: Revolutionizing Electrical Engineering, First Edition.
Edited by Rahiman Zahira, Palanisamy Sivaraman, Chenniappan Sharmeela, and Sanjeevikumar Padmanaban.

virtual power plants, smart grids, smart homes, intelligent transportation, and smart cities. The IoT is thought to be the key technology for establishing a smart substation because of its affordable, networkable microcontroller modules.

The availability of low-cost, networkable microcontroller modules has made the IoT the technology of choice for developing a smart substation. Although IoT standards such as constrained application protocol (CoAP) [8], message queuing telemetry transport (MQTT) [9], and extensible messaging presence protocol (XMPP) [10] have been proposed, the technology itself is still in its early stages of development.

The qualities, as well as the advantages and disadvantages, of these procedures vary. Smart technologies are gradually replacing antiquated ones as a result of the global technological revolution [11–13]. In the power sector, IoT technology is becoming more and more attractive. By 2030, it is estimated that 30–60 billion things will be available online worldwide [14, 15]. Another concept utilizing IoT technology is a remotely controlled smart grid system [16–19]. Substation automation (SA) [20] and IoT-based remote health monitoring of electrical equipment are other popular topics [21–25].

9.2 Components of Substation Automation and Monitoring

The SA consists of the following components which are shown in Figure 9.1.

9.2.1 Data Communication

The key component of any SA system is data communication, which serves as the system's structural glue. Electrical protection and local control will continue to operate in the absence of connectivity, and the local device may store some data, but the SA system cannot function as a whole. Communications will take shape according to the architecture that is employed, and architecture may then influence the type of communication that is selected.

9.2.1.1 Electrical Protection

To safeguard persons and equipment from harm and minimize damage in the case of an electrical malfunction, electrical protection remains an essential component of each electrical switchgear

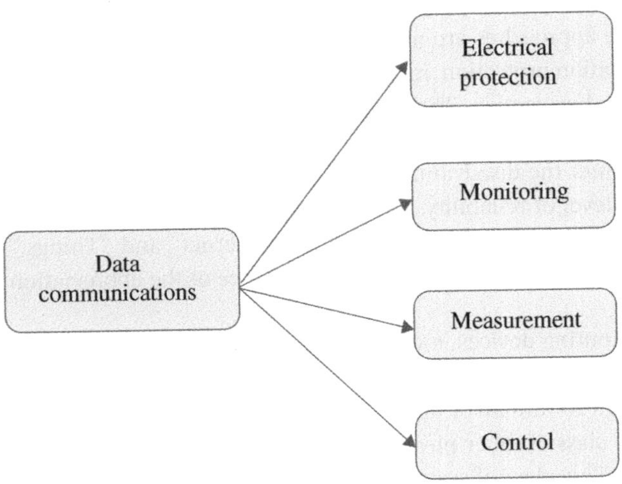

Figure 9.1 Components of substation automation and monitoring.

panel. Although it is an essential component of SA under normal circumstances, electrical protection is a local function and should be able to operate independently of the system if needed. In any SA system, the functions of electrical protection should never be compromised or limited.

9.2.1.2 Monitoring

Identifying what happened where, when, and in what order might help with fault analyses. This is a useful tool for enhancing protection and power system efficiency. The information gathered from condition monitoring can be used to inform preventative maintenance operations.

9.2.1.3 Measurement

A central control room or central database is usually used to display and store the vast amount of real-time data that is collected from the substation or switchgear panel. Electrical, analog, and disturbance records for fault analyses are all included in the measurement. This reduces the need for staff to visit a substation in order to get information, increasing worker security and reducing staff burdens. The vast amount of real-time data collected will be of great use in conducting network studies, such as load flow analyses, planning, and avoiding significant disruptions in the power network that may result in substantial losses in production.

9.2.1.4 Control

Control can be performed either locally or remotely. The control device may reasonably perform the operations on its own, such as bay interlocking, sequence switching, and synchronizing checks, which are referred to as local control. There is very little chance of human error and very little chance of human interference. It should be possible for local control to operate independently of the rest of the SA system.

Direct commands, such as opening or closing a CB, can be supplied to the remote-controlled devices. The SCADA station(s) can be contacted to request specific information, and the system allows for the modification of relay settings. This greatly improves switching operations and saves time by removing the need for employees to travel to the substation, which is especially useful in emergency scenarios. Significant production losses may be avoided, and workers are given a safer working environment. Furthermore, having a comprehensive view of the power network across the entire plant or factory, the operator or engineer at the SCADA terminal may make better decisions.

9.3 Architecture of Substation Automation

The architecture of SA is shown in Figure 9.2.

Figure 9.2 Architecture of substation automation.

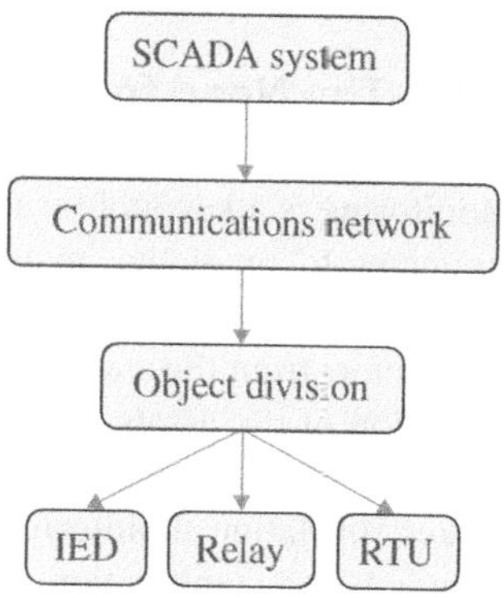

9.3.1 SCADA System

The virtual brain of the SA system is the SCADA master station or master stations. The SCADA master gathers data and information from the field, makes decisions about how to use it, stores it (either raw or through processing), and sends commands and/or requests to the remote devices. Consequently, the entire SA system is under the effective control of the SCADA master. All that is needed to operate a SCADA master these days is a SCADA software package and an advanced, dependable PC or workstation, along with any necessary peripheral and support hardware.

In a power transmission network, a SCADA master station can be installed in each substation (station level), with all of the substation SCADA stations connected to a local area network (LAN) or wide-area network (WAN) (network level); alternatively, a single SCADA master station can be directly in charge of multiple substations, doing away with the need for a station level.

9.3.2 Communications Network

SA mainly depends on the communications network for the neural network. Among various field instruments intelligent electronic devices (IEDs), and SCADA systems, will make the network sure that commands, processed data, and raw data are transmitted quickly, effectively, and error free. Although some copper wiring will remain between the various equipment inside a substation, fiber–optic cables are typically used as the physical medium in modern networks.

In order to fulfill the requirements of a communication network, it has to operate as a separate, smart subsystem and not just a collection of copper and fiber optic cables.

The SCADA station level, which may include a remote SCADA master station in a central control room or a SCADA master station located within the substation, is interfaced with the bay level over the communication network.

9.3.3 Object Division

IEDs, third-generation microprocessor-based relays, and/or remote terminal units (RTUs) form the object division of the SA system. In addition to the digital inputs from the SCADA Master other field devices IED's and auxiliary contacts also receive analog inputs from the transducers, voltage transformers (VT's), and CT's in the various switchgear panels. They can carry out complex mathematical and logical calculations and output data to the switchgear, SCADA Master, IEDs, and other field instruments to execute an operation. The bay level (local intelligence in the form of IEDs, RTUs, etc.) and process level (field information from CTs, VTs, etc.) form the component division.

9.4 The Need for IoT in Substation Monitoring

Monitoring is a key tool for utilities in order to shift from corrective or time-based maintenance strategies to predictive maintenance [26]. It is the first step toward the implementation of condition-based monitoring (CBM) strategy. Monitoring basically consists of acquiring significant parameters from the assets of interest. The collected data allow for analyses and diagnosing of the condition of the assets, which is of great use as a support to the decision-making maintenance schedule and then, reducing failures and breakdowns. The huge amount of features to consider makes substation monitoring complex. It can be both online (continuous) and offline (discontinuous), or a combination of both, depending on the asset and the diagnosis to perform. Proper sensors, data acquisition, and software for processing are also needed. Moreover, substations are built with assets from different manufacturers, each with its own communication protocols. A monitoring system can be structured in three levels, – Level 1: Data acquisition from each asset through appropriate sensors. Level 2: Data storage and processing at substation level. Level 3: Integration of the data from different IoT-based substation monitoring addresses several critical needs and offers numerous benefits that traditional monitoring systems cannot provide. Here are the primary needs and advantages of implementing IoT-based substation monitoring.

9.4.1 Components of IoT-Based Substation Monitoring System

1) **Sensors**
   - Measure various parameters such as voltage, current, temperature, humidity, and vibration.
   - Installed on critical substation components such as transformers, CBs, and bus bars.
2) **Microcontroller/Embedded Systems**
   - Devices such as Arduino, Raspberry Pi, or industrial-grade PLCs to collect and process sensor data.
3) **Communication Modules**
   - Modules for data transmission such as Wi-Fi, global system for mobile communication (GSM), LoRa, Zigbee, or Ethernet.
4) **Cloud Platform**
   - Cloud services such as Amazon Web Services (AWS) IoT, Microsoft Azure IoT, or Google Cloud IoT for data storage, processing, and analytics.
5) **Data Analytics and Machine Learning**
   - Tools and algorithms for analyzing data, detecting anomalies, and predicting failures.
6) **Visualization and Dashboard**
   - Software for real-time data visualization, trend analysis, and alert management.
7) **Alert and Notification System**
   - Automated systems for sending alerts and notifications via email, short message service (SMS), or push notifications when predefined thresholds are exceeded.

9.5 Automation and Control in Substation Environment

Automation and control in a substation environment significantly enhance the efficiency, reliability, and safety of electrical power distribution. Integrating advanced technologies such as IoT and SCADA. A comprehensive overview is shown below.

9.5.1 Key Components of Substation Automation and Control

1) **IEDs**
   - Devices such as protective relays, CBs, and transformers are equipped with communication capabilities.
   - Monitor and control electrical parameters, providing protection and control functions.
2) **SCADA System**
   - Centralized system that collects data from field devices, processes it and displays it to operators.
   - Allows for remote control and automation of substation operations.

3) **PLCs**
   - Industrial computers monitor inputs and make decisions based on programmed logic to control outputs.
   - Used for automating processes and controlling equipment.
4) **Communication Networks**
   - Fiber optics, Ethernet, and wireless communication systems connect various substation components.
   - Ensures reliable and secure data transmission.
5) **Sensors and Actuators**
   - Sensors measure parameters such as current, voltage, temperature, and humidity.
   - Actuators control physical devices such as CBs and transformers based on sensor inputs.
6) **Human–Machine Interface (HMI)**
   - Interface for operators to monitor and control substation equipment.
   - Displays real-time data and allows for manual intervention if needed.

9.5.2 Functions of Substation Automation and Control

SA is the cutting edge technology in electrical engineering [27]. It means having an intelligent, interactive power distribution network including:

1) Increased performance and reliability of electrical protection.
2) Advanced disturbance and event recording capabilities, aiding in detailed electrical fault analysis.
3) Display of real-time substation information in a control center.
4) Remote switching and advanced supervisory control.
5) Increased integrity and safety of the electrical power network including advanced interlocking functions.
6) Advanced automation functions such as intelligent load shedding.

Thus, primary control in the substation is of two categories:

1) Normal routine operation by operator's command with the aid of analog and digital control system.
2) Automatic operation by action of protective relays, control systems, and personal computer (PC).

The automated substation functioning can be treated as the integration of two subsystems, as discussed below.

9.5.2.1 Control System

The task of control system in a substation includes data collection, scanning, event reporting, and recording; voltage control, power control, frequency control, other automatic and semiautomatic controls, etc. [28].

The various switching actions such as auto reclosing of line CBs, operation of sectionalizing switches, onload tap changers are performed by remote command from the control room. The other sequential operations such as load transfer from one bus to another and load shedding are also taken care of by the control center.

The various switching actions such as auto reclosing of line CBs, operation of sectionalizing switches, onload tap changers are performed by remote command from control room. The other sequential operations such as load transfer from one bus to another and load shedding are also taken care of by the control center.

9.5.2.2 Protective System

The task of the protective system includes sensing abnormal condition, annunciation of abnormal condition, alarm, automatic tripping, backup protection, and protective signaling.

The above two systems work in close cooperation with each other. Most of the above functions—such as automatic switching sequences, sequential event recording, compiling of energy, and other reports—are integrated with software in the substation computer. This software is of modular design, which facilitates the addition of new functions.

The communication between CBs, autoreclosers, and sectionalizing switches in the primary and secondary distribution circuits located in the field and the PC in the distribution substation control room is through radio telecontrol or fiber optic channel or power line carrier channel as is feasible.

9.5.3 Benefits of Substation Automation and Control

Continuous monitoring of electrical parameters and equipment status in real time to ensure smooth operation and performance analysis through data collection from sensors and IEDs. Remote control capabilities for CBs, transformers, and other devices, along with automatic control mechanisms that are triggered by predefined logic and conditions to enhance efficiency. Automatic detection and isolation of faults for equipment protection and safety measures, coupled with rapid responses to abnormal conditions to prevent equipment damage and potential outages. Data logging of historical data for analysis and reporting purposes, including the generation of performance evaluation reports to ensure regulatory compliance. Management of alarms and notifications for abnormal conditions with a focus on prioritization and categorization to guarantee a quick and efficient response to issues. Load management strategies for optimizing load distribution to prevent overload and ensure balanced system operation, along with dynamic load adjustments based on real-time information. Energy management techniques for monitoring and optimizing energy consumption and efficiency, such as integration with renewable energy sources and storage systems for sustainability and cost-effectiveness.

9.6 Substation Automation and Monitoring

An SA system is a collection of hardware and software components that are used to monitor and control an electrical system, both locally and remotely [29]. A SA system also automates some repetitive, tedious, and error-prone activities to increase the overall efficiency and productivity of the system.

9.6.1 Traditional Substations

High availability and constant operation of an electrical substation have always been the focus of an electrical company. More faults mean more interruption of service to clients and it translates to less revenue that is not desirable to any company. From the early age of electrical systems, engineers and operators have always been interested in collecting useful information on different devices in a substation so they can better evaluate the health of their system, predict potential problems, and – in case of a fault – analyze and troubleshoot the problem as soon as possible to protect their high value assets and to improve their continuous service to their clients.

Early substations consisted of mechanical relays and meters that barely supported recording and had no means of communication. Fault recorders were capturing information mainly in the form of paper charts, so reading and analyzing the information was not a straightforward process.

Lack of communication caused any maintenance or troubleshooting to be costly and lengthy because personnel had to be sent to substations that were often far away and hard to reach.

9.6.2 Modern Substations

With the introduction of microprocessor technology, digital protection and control devices became more intelligent. IEDs being implemented in substations today contain valuable information, both operational and nonoperational, needed by many user groups within the utility [30]. An IED is any device that incorporates one or more processors with the capability to receive or send data/control from or to an external source (e.g., electronic multifunction meters, digital relays, and controllers). IED technology can help utilities improve reliability, gain operational efficiencies, and enable asset management programs including predictive maintenance, life extensions, and improved planning.

IEDs are a key component of substation integration and automation technology. Substation integration involves integrating protection, control, and data acquisition functions into a minimal number of platforms to reduce capital and operating costs, reduce panel and control room space, and eliminate redundant equipment and databases. Automation involves the deployment of substation and feeder operating functions and applications, ranging from SCADA and alarm processing to integrated volt/var control, in order to optimize the management of capital assets and enhance operation and maintenance (O&M) efficiencies with minimal human intervention.

IEDs facilitate the exchange of both operational and nonoperational data. Operational data also called SCADA data, are instantaneous values of power system analog and status points such as volts, amps, megawatt (MW), mega volt ampere reactive (MVAR), circuit breaker (CB) status, and switch position. This data is time-critical and is used to monitor and control the power system (e.g., opening CBs, changing tap settings, and equipment failure indication). Nonoperational data consists of files and waveforms, such as event summaries, oscillographic event reports, or sequential events records, in addition to SCADA-like points (e.g., status and analog points) that have a logical state or a numerical value. This data is not needed by the SCADA dispatchers to monitor and control the power system.

9.6.3 Apparatus and Components, Basic Functions, and Classification (ABC) of Substation Automation

SA is the act of automatically controlling the substation via instrumentation and control devices. SA refers to using data from IED's, control and automation capabilities within the substation, and control commands from remote users using SCADA to control power-system (switch yard) devices [31]. SA system is commonly used to control, protect, and monitor a substation. However, over the years advances in electronics, information, and communications technology have brought about sweeping changes in the way substations are operated. The advent of software-based SA systems connected by serial links rather than rigid parallel copper wiring gradually became the norm rather than the exception. Though successful and widely accepted, these systems were based on either the manufacturers' own proprietary communication solutions or the defined use of communication standards from other application domains, such as DNP3 or IEC 60870-5-104.

SA is a system to enable an electric utility to remotely monitor, control, and coordinate the distribution components installed in the substation. Substation automation system (SAS) is based on a decentralized architecture and a concept of bay-oriented and distributed intelligence, for safety and availability reasons.

9.7 Examples

9.7.1 Components of Substation

The power grid is an essential component of the systems used in the generation, transmission, and distribution of electricity. Electrical substations must be used for any power system functioning. These are essential parts that the substations need in order to generate power. Substations can modify the voltage and frequency levels to change the amount of electricity required to power clients. Electrical substations come in many different varieties; some of the most common ones are generating, pole-mounted, indoor, outdoor, converter, distribution, transmission, and switching substations. In some instances, such as thermal plants, numerous hydroelectric power plants, and wind farm electricity generation systems, the collector substation is visible. It helps transfer power from multiple turbines into a single transmission unit.

Electrical power can be transferred from the producing units to the distribution units using a range of electrical substation components that are coupled together in the substation, such as an isolator, a bus bar, and a power transformer. The electrical substation components are needed for the substation installation. The equipment in the substation plays the following major functions. An electrical substation's design requires careful engineering planning. Important steps in the substation design process include switching system design, equipment placement and planning, component ordering and selection, engineering support, structural design, electrical layout design, relay protection, and key apparatus ratings.

9.7.2 IoT-Based Monitoring and Control of a Power Transformer

Transformers are critical components in any electrical power system, playing a vital role in the transmission and distribution of electricity. To ensure the reliable and efficient operation of these transformers, advanced monitoring and control systems have been developed that leverage a variety of sensor technologies. Temperature sensors, for instance, are used to closely monitor the temperature of the transformer oil and windings, providing early warning signs of potential overheating that could lead to catastrophic failures. Vibration sensors, on the other hand, are employed to detect any abnormal vibrations that may indicate developing mechanical issues within the transformer. Similarly, gas sensors are utilized to measure the levels of dissolved gases in the transformer oil, as changes in these gas concentrations can signify potential faults developing inside the unit. Current and voltage sensors further monitor the load conditions on the transformer, allowing operators to identify any abnormalities that could compromise performance or lifespan. Humidity sensors round out the sensor suite, tracking the moisture content within the transformer oil to ensure optimal dielectric properties are maintained. All of this sensor data is then collected and transmitted to a centralized data acquisition system, which ensures the integrity and accuracy of the information before relaying it to the central monitoring system. This monitoring system, in turn, leverages advanced analytics powered by machine learning and artificial intelligence algorithms to analyze historical trends, predict potential failures, and provide recommended maintenance actions to operators. The system also incorporates secure communication networks, whether wired or wireless, to transmit the sensor data while safeguarding against cyber threats. Finally, the control system integrates with the monitoring platform, enabling remote adjustments to transformer operations, such as tap changer settings, as well as the ability to initiate emergency shutdowns in response to critical faults detected by the system. Through this comprehensive, technology-driven approach to transformer monitoring and control, power utilities can maximize the lifespan and reliability of these essential grid assets.

9.7.3 IoT-Based Monitoring and Control of Voltage Transformer

VT, also known as a PT, is a type of transformer used in electrical power systems to step down high voltage levels to lower, safer levels for measurement and monitoring purposes. This allows for the accurate measurement of high voltages using standard low-voltage instruments. VTs are essential components in power system protection and control schemes. Voltage sensors are utilized to monitor voltage levels on the primary and secondary sides of the PT, while temperature sensors are used to monitor the temperature of the PT and its surroundings. Humidity sensors are employed to monitor environmental conditions that could impact the PT, and vibration sensors are used to detect mechanical issues or wear and tear. Smart meters are also used for accurate voltage and energy measurement, and gas sensors are utilized to detect any gas leaks, particularly for oil-filled transformer.

IoT gateways are employed to gather data from various sensors and transmit it to the cloud, using communication protocols such as MQTT, CoAP, or hypertext transfer protocol (HTTP) for data transmission. Network connectivity options include Ethernet, Wi-Fi, Cellular, or low-power wide-area network (LPWAN) (e.g., long range wide area network [LoRaWAN] and narrowband Internet of Things [NB-IoT]).

Cloud platforms such as AWS IoT, Azure IoT, and Google Cloud IoT are used for data storage, processing, and analytics. Edge computing is also utilized for real-time data processing and decision-making at the edge. SCADA integration is used for supervisory control and data acquisition systems for remote monitoring and control, while automated control systems are employed for real-time adjustments and emergency responses. Sensors collect real-time data on voltage, temperature, humidity, and vibrations, which are then sent to IoT gateways using wired or wireless communication. IoT gateways process the data and transmit it to the cloud or edge computing devices, using secure communication protocols to ensure data integrity and security. In the cloud, data is stored and processed, with machine learning algorithms analyzing the data to predict faults, optimize performance, and provide insights. Edge computing devices handle real-time processing for critical functions, reducing latency. A central dashboard provides a real-time overview of the PT's status, accessible via various devices.

9.7.4 IoT-Based Monitoring and Control of Current Transformer

A CT is a type of transformer used primarily to measure AC. It produces a reduced current accurately proportional to the current in the circuit, which can be conveniently connected to measuring and recording instruments. CTs are crucial for the monitoring and protection of electrical power systems. Current sensors are utilized to monitor the current on the primary and secondary sides of the CT, while temperature sensors are employed to monitor the temperature of the CT and its surroundings. Humidity sensors are used to monitor environmental conditions, and vibration sensors are implemented to detect mechanical issues. Smart meters are utilized for accurate current and energy measurement. IoT gateways are employed to collect data from various sensors and transmit it to the cloud. Communication protocols such as MQTT, CoAP, or HTTP are used for data transmission, and network connectivity options include Ethernet, Wi-Fi, Cellular, or LPWAN (e.g., LoRaWAN and NB-IoT). Cloud platforms such as AWS IoT, Azure IoT, and Google Cloud IoT are utilized for data storage, processing, and analytics. Edge computing is employed for real-time data processing and decision-making at the edge. SCADA integration is used for supervisory control and data acquisition systems for remote monitoring and control, while automated control systems are implemented for real-time adjustments and emergency responses.

9.7.5 IoT-Based Monitoring and Control of Circuit Breaker

A CB is an essential electrical device designed to protect an electrical circuit from damage caused by overload or short circuit. Its basic function is to interrupt current flow after a fault is detected. Integrating IoT technology into CB monitoring revolutionizes traditional CBs with features such as real-time data, remote monitoring, and predictive maintenance. The process involves various components such as sensors for measuring different parameters, a microcontroller for data collection and processing, a communication module for transmitting data to the cloud platform, a cloud platform for storing and analyzing data, and software or a dashboard for visualizing data and managing alerts. The architecture includes data collection by sensors, data processing by a microcontroller, data transmission to the cloud platform through various communication modules, cloud storage and analysis, and visualization with real-time alerts.

9.7.6 IoT-Based Monitoring and Control of Lightning Arrester

LA is a device used in electrical power systems and telecommunications systems to protect the insulation and conductors of the system from the damaging effects of lightning. LAs are critical for safeguarding electrical equipment from transient over-voltage events caused by lightning strikes or switching surges. Sensors are utilized to monitor various aspects such as voltage levels, discharge currents, temperature, humidity, and operational status of the arrester. Communication technologies such as IoT gateways are employed to gather data from sensors and transmit it to the cloud using protocols, such as MQTT, CoAP, or HTTP, with network connectivity options such as Ethernet, Wi-Fi, Cellular, or LPWAN. Data processing and storage involve cloud platforms, such as AWS IoT, Azure IoT, and Google Cloud IoT for storage, processing, and analytics, along with edge computing for real-time decision-making. Control systems include SCADA integration for remote monitoring and control, as well as automated control systems for real-time adjustments and emergency responses.

9.8 Others

9.8.1 Substation Integration of Renewable Energy

Sensors and IEDs are utilized to monitor and control the integration of solar or wind energy at substations. Communication is established between renewable energy sources and the substation control center to track energy production, automatically adjust load and generation, forecast energy production, and enable remote control of renewable energy equipment. The benefits include optimized energy usage, improved grid stability, and accurate reporting of renewable energy contributions for regulatory purposes.

9.8.2 Smart Substation with Advanced Metering Infrastructure (AMI)

Deploy smart meters at customer endpoints and key substation nodes, utilizing a mix of wired and wireless communication for data transmission. Integrate smart meters with the SCADA system to collect real-time consumption data, implement demand response programs, and detect faults in the distribution network. Provide accurate billing, use data analytics for load forecasting, and enhance grid planning. This setup enables monitoring of electricity consumption, aggregation of data at substations, reduction of peak load through demand response, and identification of faults in the distribution network using smart meter data. The benefits include improved customer service, operational efficiency, and data-driven decision-making for better planning and operational strategies.

9.8.3 Substation Automation for Industrial Plants

Install sensors for monitoring electrical parameters and environmental conditions in an industrial substation, utilizing PLCs for local control and IEDs for protection and monitoring, while establishing a robust communication network with fiber optics or industrial Ethernet. Ensure continuous real-time monitoring of power quality, load, and environmental conditions, implement automated control for load shedding, generator control, and power factor correction, and utilize data analytics for predictive maintenance and scheduling. Provide plant operators with real-time visualization and control interfaces through HMI and dashboards, resulting in reduced downtime, enhanced power quality, increased operational efficiency, improved safety, and decreased manual interventions and maintenance costs in industrial operations.

9.9 Conclusion

The continuous power will be able to provide because of "Substation Monitoring and Control using IOT." Additionally, real-time parameter monitoring is carried out to guarantee the safety of the substation's equipment. The system's design makes remote substation control simple. It makes two-way exchanges possible. The substation can communicate with the service provider to indicate the type of fault that has been connected to it. The operation of smart substations is more responsive to changing operating conditions and, as a result, is more suited for the complicated operating conditions and growing use of renewable energy found in modern smart grids.

References

1 McDonald, J.D. (2012). *Electric Power Substations Engineering*, 3e, 22-1–22-30. Boca Raton, FL: CRC Press.

2 Begovic, M.M. (ed.) (2013). *Electrical Transmission Systems and Smart Grids: Selected Entries from the Encyclopedia of Sustainability Science and Technology*. Springer, ISBN: 978-1-4614-5830-2.

3 Kezunovic, M., Guan, Y., Guo, C., and Ghavami, M. (2010). The 21st century substation design: vision of the future. *iREP Symposium on Bulk Power System Dynamics and Control (iREP) - VIII (iREP)*. pp. 1–8.

4 Yi, Y.-H., Wang, L-T., and Tao, Y.-J. (2011). Research of network transmission of process bus based upon IEC 61850. *International Conference on Advanced Power System Automation and Protection (APAP)*. pp. 1578–1582.

5 Leonardi, A., Mathioudakis, K., Wiesmaier, A., and Zeiger, F. (2014). Towards the smart grid: substation automation architecture and technologies. *Advances in Electrical Engineering* 2014: 1–13.

6 Li, F., Qiao, W., Sun, H. et al. (2010). Smart transmission grid: vision and framework. *IEEE Transactions on Smart Grid* 1 (2): 168–177.

7 Moore, R., Midence, R., and Goraj, M. (2010). Practical experience with IEEE 1588 high precision time synchronization in electrical substation based on IEC 61850 process bus. *Proc. IEEE PES General Meeting*, Minneapolis. pp. 1–4.

8 State Grid Cooperation of China (SGCC) (2014). 2014-2020 rolling planning of smart grid in SGCC (in Chinese).

9 Huang, Q., Jing, S., and Zhen, W. (2015). *Innovative Testing and Measurement Solutions for Smart Grid*. Wiley-IEEE Press.

10 Q/GDW Z 410-2010 (2010). *Technical Guide for Smart Electric Equipment*. State Grid Cooperation of China (SGCC) (in Chinese).

11 Fan, C. (2012). The data acquisition in smart substation of China. In: *Data Acquisition Applications* (ed. Z. Karakehayov), 123–164. InTech, ISBN 978-953-51-0713-2.

12 Weifeng, L., Shuangle, Z., Weiping, M., and Ming, A. (2011). Application limitation of electronic current and voltage transformers in digital substations. *IEEE Power Engineering and Automation Conference (PEAM)*. pp. 496–499.

13 Li, H. and Wang, L. (2011). Research on technologies in smart substation. *International Conference on Smart Grid and Clean Energy Technologies (ICSGCE)*. pp. 113–119.

14 Li, H. (2012). Research on technologies of intelligent equipment in smart substation. *IEEE PES ISGT Asia*. pp. 1–5.

15 Zheng, Y., Wang, D., Zhou, Z., and Cao, T. (2014). Hierarchical protection control system of smart substations. *Journal of Modern Power Systems and Clean Energy* 2 (3): 282–288.

16 Zhou, F., Cheng, Y., Xiao, J., and Du, J. (2014). Research of digital metering system and calibration technology of smart substation. *IEEE Workshop on Advanced Research and Technology in Industry Applications (WARTIA)*. pp. 1240–1242.

17 Tang, C., Guo, H., Zhang, J., and He, M. (2011). A new centralized high voltage power metering terminal in smart substations. *Automation of Electric Power Systems* 35 (5): 87–90. (in Chinese).

18 Li, L., Ota, K., and Dong, M. (2017). When weather matters: IoT-based electrical load forecasting for smart grid. *IEEE Communications Magazine* 55 (10): 46–51.

19 Singh, M.J., Agarwal, P., and Padmanabh, K. (2016). Load forecasting at distribution transformer using IoT based smart meter data from 6000 Irish homes. *Proceedings of the International Conference on Contemporary Computing and Informatics (IC3I)*, Noida. pp. 758–763.

20 Sinnapolu, G. and Alawneh, S. (2018). Integrating wearables with cloud-based communication for health monitoring and emergency assistance. *Internet of Things* 1–2 (1): 40–45.

21 Chandra, A.A., Jannif, N.I., Prakash, S., and Padiachy, V. (2017). Cloud based real-time monitoring and control of diesel generator using the IoT technology. *Proceedings of the International Conference on Electrical Machines and Systems (ICEMS)*, Sydney. pp. 1–5.

22 Hai, Q., Zheng, W., Yan, Z. (2013). Status monitoring and early warning system for power distribution network based on IoT technology. *Proceedings of the International Conference on Computer Science and Network Technology (ICCSNT)*, Dalian. pp. 641–645.

23 Li, R., Liu, J., and Li, X. (2012). A networking scheme for transmission line on-line monitoring system based on IoT. *Proceedings of the International Conference on Computing Technology and Information Management (NCM and ICNIT)*, Seoul. pp. 180–184.

24 Shen, X., Cao, M., Lu, Y., and Zhang, L. (2016). Life cycle management system of power transmission and transformation equipment based on internet of things. *Proceedings of the International Conference on Electricity Distribution (CICED)*, Xi'an. pp. 1–5.

25 Tom, R.J. and Sankaranarayanan, S. (2017). IoT based SCADA integrated with fog for power distribution automation. *Proceedings of the Iberian Conference on Information Systems and Technologies (CISTI)*, Lisbon. pp. 1–4.

26 Farrero, J., Villafafila-Robles, R., Velasquez, J.L. et al. (2009). Full substation monitoring. *CIRED 2009 - 20th International Conference and Exhibition on Electricity Distribution - Part 1*, Prague, Czech Republic. pp. 1–3. https://doi.org/10.1049/cp.2009.0699

27 Saravanan. A., Farook. S., Kathir. I., Pushpa. S., Padmashini. R.K., Logeswaran. T., Ravishankar. S., and Rajaram. A. (2024). Adaptive solar power generation forecasting using enhanced hybrid function networks with weather modulation. *International Journal of Renewable Energy Research (IJRER)* 14 (2).

28 Karthick Manoj, R and Sasilatha, T. (2024). Automated plant disease detection using efficient deep ensemble learning model for smart agriculture. *Using Traditional Design Methods to Enhance AI-Driven Decision Making*. IGI Global. pp. 18–336.

29 Priya. S., Vinoth Kumar. P., Sridevi Batumalay V., M., and Gunapriya D. (2024). Convolutional neural network for battery system monitoring and SOC estimation for Ev applications to achieve sustainability. *Journal of Applied Data Sciences* 5 (4): 1802–1813.

30 Priya, S., Sagayaraj, R., Sujith, S., and Malathi, S. (2023). An efficient monitoring scheme for standalone solar PV system using IoT. *2023 - 8th International Conference on Communication and Electronics Systems (ICCES)*, Coimbatore, India. pp. 391–395. https://ieeexplore.ieee.org/document/10192806.

31 Ghurde, C., Bharit, R., Patil, A., Bhosale, S., Yadav, K., Nachane, P., and Lohote, G. (2020). Substation automation operation and protection. *International Journal for Research in Applied Science & Engineering Technology (IJRASET)* 8 (V), ISSN: 2321–9653.

10

IoT Application in Condition Monitoring and Fault Diagnosis in Electrical Systems

Ravichandran Karthick Manoj[1], *Dhandapani Lakshmi*[1], *Rajasekharan Rajasree*[1], *Sukumaran Aasha Nandhini*[2], *Palanisamy Sivaraman*[3], and *Rahiman Zahira*[4]

[1]*Department of EEE, AMET Deemed to be University, Chennai, Tamil Nadu, India*
[2]*Department of Electronics and Communication Engineering, Sri Sivasubramaniya Nadar College of Engineering, Chennai, Tamil Nadu, India*
[3]*Research Scholar, Anna University, Chennai, Tamil Nadu, India*
[4]*Department of Electrical and Electronics Engineering, B.S. Abdur Rahman Crescent Institute of Science and Technology, Chennai, Tamil Nadu, India*

10.1 Introduction

The integration of Internet of Things (IoT) technologies with condition monitoring (CM) and fault diagnosis systems represents a significant advancement in the management and maintenance of electrical systems. IoT applications in this domain offer unprecedented capabilities for real-time monitoring, predictive maintenance (PdM), data-driven decision-making, optimizing maintenance strategies, and minimizing downtime. IoT technologies encompass a network of interconnected devices [1], sensors, and systems that collect and exchange data over the internet. When applied to CM and fault diagnosis in electrical systems, IoT enables continuous monitoring of equipment parameters, early detection of abnormalities, and proactive maintenance interventions. This introduction explores the key aspects of IoT application in CM and fault diagnosis in electrical systems highlighting its benefits and potential impact on industrial operations. Electrical machine CM and problem diagnosis are becoming more and more common due to the fact that electrical machines are becoming more and more important in both industry and daily life. Reactive and preventive maintenance are the two basic categories of conventional maintenance procedures. Reactive maintenance is necessary after a failure has already happened, whereas preventive maintenance focuses mostly on a system's regular overhaul and whether or not maintenance is necessary [1].

Figure 10.1 represents the overview of CM in electrical systems. Reactive maintenance involves fixing an equipment that has already broken which impedes progress. PdM is a superior option because it allows for continuous machine health monitoring and the selection of only malfunctioning units for maintenance. The machine can be fixed before a disastrous scenario arises because the issue can be found early on. However, depending on the machine type, the drive control mechanism and the load behavior predictive approaches can be fairly complex. PdM of electrical machinery involves a wide range of research topics because of this. Signal processing, statistical data analysis, mathematical modeling, artificial intelligence (AI), design, optimization of sensors, and processing boards are a few examples of these domains. The latest developments and difficulties in the electrical machine status monitoring in this chapter provide a brief overview

IoT for Smart Grid: Revolutionizing Electrical Engineering, First Edition.
Edited by Rahiman Zahira, Palanisamy Sivaraman, Chenniappan Sharmeela, and Sanjeevikumar Padmanaban.

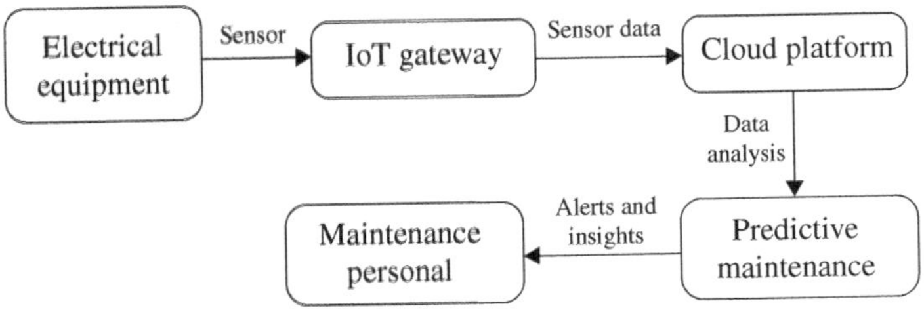

Figure 10.1 Overview of condition monitoring in electrical systems.

of the field. Numerous citations are included in a summary of a broad range of diagnostic fields and their accompanying features [2].

10.2 Importance of Condition Monitoring (CM) in Electrical Systems

CM holds paramount importance in ensuring the reliability, safety, and efficiency of electrical systems across various industries. By continuously monitoring the performance of electrical equipment, CM enables proactive maintenance strategies that prevent unexpected failures and optimize operational efficiency. Figure 10.2 represents the IoT CM in the electrical system.

In industries such as power generation, manufacturing, and transportation where uninterrupted operation is critical, CM plays a vital role in minimizing downtime and maximizing productivity. For instance, in power generation plants, CM helps identify potential faults in turbines, generators, and transformers, allowing timely maintenance to prevent costly outages. Similarly, in manufacturing facilities, CM ensures the smooth operation of machinery by detecting early signs of malfunction, thereby preventing production disruptions [3].

CM contributes to safety enhancement by identifying potential hazards such as overheating, insulation breakdowns, and electrical arcing. CM helps to mitigate the risk of accidents and ensures compliance with safety standards and regulations. Additionally, CM aids in optimizing maintenance activities and resource allocation by focusing efforts on equipment that requires immediate attention based on its condition thereby reducing unnecessary maintenance costs.

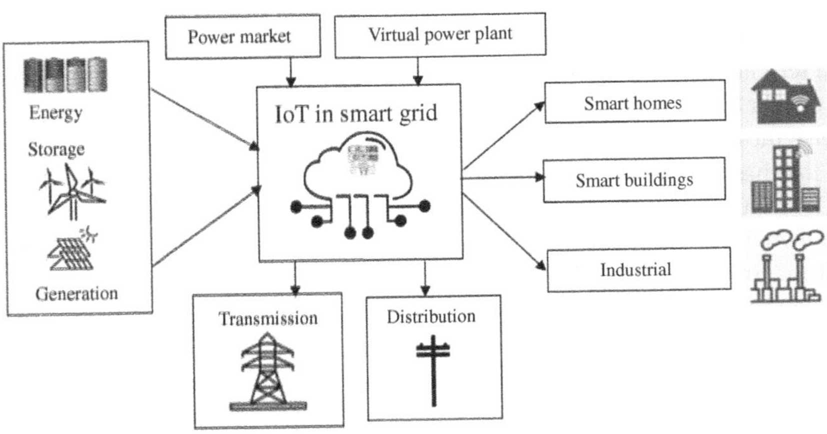

Figure 10.2 IoT condition monitoring in electrical systems.

10.3 Enhancing Reliability and Performance of Condition Monitoring

Enhancing the reliability and performance of CM in electrical systems involves the strategic implementation of advanced technologies and proactive methodologies. High-precision sensors play a crucial role, as they can detect minute changes in operational parameters such as temperature, vibration, and electrical characteristics, providing accurate data essential for early fault detection. Integrating wireless sensor networks (WSNs) further improves flexibility and reduces installation costs, enabling comprehensive monitoring of remote or hard-to-reach equipment. Centralized data management systems that support real-time data acquisition and processing are vital for immediate anomaly detection and response, reducing the risk of unexpected failures.

Advanced data analysis techniques, including machine learning (ML) and AI, enhance the predictive capabilities of CM systems. These technologies analyze historical and real-time data to forecast equipment failures accurately, enabling PdM models that optimize maintenance schedules and reduce costs. Utilizing big data analytics and advanced data visualization tools helps maintenance teams quickly understand complex data, facilitating informed decision-making [4].

The integration of edge computing and the IoT further enhances the efficiency and reliability of CM. Edge computing processes data close to the source, reducing latency and enabling faster decision-making. IoT devices improve connectivity and data sharing among different monitoring systems, ensuring seamless communication and data exchange.

10.4 Proactive Maintenance Strategies Enabled by Condition Monitoring

Adopting condition-based maintenance (CBM) and risk-based maintenance (RBM) strategies ensures that maintenance activities are performed based on the actual condition and criticality of the equipment, thus reducing unnecessary maintenance and focusing resources on critical assets. Regular training programs and skill development for maintenance personnel are essential to keep them updated on the latest technologies and best practices, ensuring effective monitoring and maintenance [5].

Finally, adherence to industry standards and regulatory compliance is crucial. Ensuring that CM practices follow standards such as ISO 17359 improves reliability and ensures the implementation of best practices. Standardized procedures for monitoring and maintenance activities ensure consistency and reliability in the monitoring process, ultimately leading to enhanced performance and longevity of electrical systems.

Implementing CM enables organizations to adopt proactive maintenance strategies, leading to a myriad of benefits including reduced downtime, minimized repair costs, and prevention of catastrophic failures. By continuously monitoring the health and performance of electrical systems, CM provides early detection of potential faults or abnormalities, allowing maintenance teams to intervene before issues escalate into costly failures.

One of the key advantages of CM is its ability to facilitate proactive maintenance rather than reactive approaches. Instead of waiting for equipment to fail before taking action, CM allows maintenance teams to anticipate and address issues before they cause downtime. By identifying early warning signs such as abnormal vibrations, temperature fluctuations, or electrical anomalies, maintenance can be scheduled during planned downtimes or during periods of low production, minimizing disruptions to operations [6].

Moreover, the proactive nature of CM helps in minimizing repair costs. By addressing issues at an early stage, before they result in significant damage or component failure, organizations can avoid costly repairs or replacements. For example, detecting a minor bearing wear in a motor through vibration analysis allows for timely lubrication or replacement, preventing further damage to the motor and avoiding the need for a costly motor replacement.

Furthermore, CM plays a crucial role in preventing catastrophic failures that can have severe consequences on both operations and safety. By continuously monitoring critical equipment such as transformers, generators, or high-voltage components, CM can detect potential faults or degradation in insulation, which if left unaddressed, could lead to equipment failures, fires, or even explosions. By taking preventive actions based on early warning signs provided by CM, organizations can mitigate the risk of such catastrophic events, ensuring the safety of personnel and protecting valuable assets.

10.5 Methods of Condition Monitoring

CM encompasses various methods aimed at continuously assessing the health and performance of electrical systems. These methods enable proactive maintenance strategies by detecting early signs of deterioration or potential faults, thereby preventing unexpected failures and optimizing equipment performance. This comprehensive exploration will delve into the seven key methods of CM in electrical systems, highlighting their principles, applications, and benefits.

i) **Vibration Analysis:** The vibration levels of rotating machinery are measured through vibration analysis. Vibration patterns that change may indicate a sign of imbalances, damaged components, or misaligned equipment. Vibrations at different locations are measured by sensors which are mounted on or close to rotating machinery. The vibration sensors identify abnormal patterns by analyzing waveform, frequency, and amplitude [7]. These patterns can evaluate equipment deterioration or identify possible defects with the help of IoT techniques [7].
ii) **Oil Analysis:** Lubricating oil samples are evaluated using oil analysis to check for corrosion, wear, and contamination. This nondestructive method may provide information on the current condition of components like gears, bearings, and hydraulic systems. For performing an oil analysis, samples are taken directly from running machinery and equipment are analyzed in the laboratory setting. In general, an analysis is composed of multiple evaluation levels such as particle counting, chromatography, and spectroscopy. Viscosity and other indicators of malfunctioning or maintenance-needed equipment are measured by these techniques in addition to identifying pollutants, particles, metals, and other elements [8].
iii) **Thermography:** Equipment that produces abnormal temperature changes has been subjected to thermography. Temperature variations may indicate mechanical faults, lubrication problems, or overheating and these variations can be detected by thermal cameras. Thermal cameras are used in thermography to find infrared radiation if a product is generating. Surface temperatures are represented visually from this radiation and observers can easily identify abnormal temperature patterns with infrared thermography which indicate problems with the device or helps to detect future equipment failure.
iv) **Acoustic Analysis:** To record the sound waves produced by any devices or machinery, acoustic analysis is used. Changes in sound patterns can be detected by microphones or acoustic sensors and these changes can indicate abnormal equipment function. Sensors use acoustic

analysis to record the sound waves produced by machinery when it is operating. Advanced signal processing is included in acoustic analysis to differentiate between normal and abnormal working disturbances. Differences in the sound's frequency, loudness, or pattern may indicate faulty equipment or an upcoming breakdown.

v) **Electrical Monitoring:** Continuous and real-time analysis of important electrical characteristics such as voltage, current, and induction is provided through electrical monitoring. Changes in these characteristics indicate electrical overload, deteriorating equipment, or other major safety and fire risks that can be identified using electrical monitoring. In electrical panels and other devices, sensors and meters are connected and these sensors provide information for the voltage and current associated with the machinery, as well as for other characteristics like induction, capacitance, and insulation quality. After that, these data can be analyzed for any variations from typical operating circumstances [7, 8].

10.6 Implementation of Vibration Analysis

Vibration analysis works by detecting, measuring, and analyzing mechanical vibrations produced by rotating machinery. These vibrations are indicative of the health and condition of the equipment and can reveal potential faults or abnormalities [9]. The following are the vibration analysis working details.

10.6.1 Sensor Placement

Vibration sensors, typically accelerometers, are strategically placed on the equipment being monitored. These sensors are attached to key components such as motors, generators, turbines, or bearings. The placement of sensors is critical to ensure accurate measurement of vibrations relevant to the operation of the equipment.

10.6.2 Measurement of Vibrations

The vibration sensors continuously measure the mechanical vibrations produced by the rotating machinery. As the equipment operates, it generates vibrations due to factors such as unbalance, misalignment, bearing wear, structural defects, or resonance. These vibrations manifest as periodic oscillations in the equipment's components.

10.6.3 Signal Conditioning

The electrical signals generated by the vibration sensors are conditioned using signal conditioning circuits. These circuits filter, amplify, and process the signals to ensure they are suitable for analysis. Signal conditioning helps improve the quality of the vibration signals and remove any unwanted noise or interference.

10.6.4 Data Acquisition

The conditioned vibration signals are then fed into a data acquisition system (DAS). The DAS digitizes the analog signals and records them as digital data. The digitized vibration data typically includes information such as amplitude, frequency, phase, and time-domain characteristics.

10.6.5 Analysis and Interpretation

The digitized vibration data is analyzed using specialized software designed for vibration analysis. This software performs various types of analysis to interpret the vibration signals and identify patterns or anomalies [10]. The analysis techniques may include:

- **Time-domain Analysis:** Examining the vibration signals in the time domain to identify changes in amplitude, frequency, or waveform shape over time.
- **Frequency-domain Analysis:** Decomposing the vibration signals into their frequency components using techniques such as Fourier analysis. This helps identify dominant frequencies associated with specific fault conditions.
- **Waveform Analysis:** Analyzing the shape and characteristics of the vibration waveform to identify irregularities or abnormalities.
- **Trend Analysis:** Monitoring changes in vibration levels over time to detect trends or patterns indicative of deteriorating equipment health.

10.6.6 Diagnostic Tools and Reporting

The analysis software may include diagnostic tools such as spectrum analysis, waterfall plots, or trend analysis to further interpret the vibration data. These tools help identify specific fault conditions or anomalies and provide insights into the health of the equipment. The results of the analysis are typically presented in the form of reports or visualizations that can be used by maintenance personnel to make informed decisions about maintenance actions.

10.6.7 Maintenance Actions

Based on the findings of the vibration analysis, maintenance personnel can take appropriate actions to address identified issues. These actions may include corrective maintenance to repair or replace faulty components, preventive maintenance to mitigate potential failures, or PdM to schedule maintenance activities based on the predicted health of the equipment.

10.7 Vibration

Vibration refers to the oscillation or movement of an object or system about a reference point. In the context of machinery and equipment, vibration is often caused by dynamic forces acting within the system [11]. These forces can arise from various sources such as rotating or reciprocating components, unbalanced masses, misalignment, mechanical defects, or external disturbances.

10.7.1 Types of Vibration

10.7.1.1 Free Vibration

Free vibration occurs when a mechanical system is disturbed from its equilibrium position and allowed to vibrate freely without any external forces acting on it. The vibration continues until the energy dissipates due to damping effects. Examples include a guitar string vibrating after being plucked or a pendulum swinging back and forth.

10.7.1.2 Forced Vibration

Forced vibration occurs when an external force or excitation is applied to a mechanical system, causing it to vibrate at a specific frequency. The frequency of the forced vibration is determined by the frequency of the excitation force. Examples include machinery vibrations induced by rotating components, such as motors or turbines.

10.7.1.3 Resonant Vibration

Resonant vibration occurs when a mechanical system is subjected to an external force at or near its natural frequency. This causes the system to resonate, resulting in large amplitude vibrations. Resonance can lead to excessive stress and premature failure of components. It is important to avoid operating machinery at or near its resonant frequencies.

10.7.1.4 Random Vibration

Random vibration refers to vibrations that do not have a predictable pattern or frequency. These vibrations arise from random or stochastic forces such as mechanical noise, turbulence, or environmental vibrations. Random vibrations can be challenging to predict and analyze but are common in real-world applications.

10.7.1.5 Torsional Vibration

Torsional vibration occurs when a twisting or torsional force is applied to a mechanical system, causing it to oscillate about its axis of rotation. This type of vibration is common in rotating machinery such as engines, crankshafts, or shaft-driven systems. Excessive torsional vibration can lead to fatigue failure of components.

10.7.1.6 Longitudinal and Transverse Vibration

Longitudinal vibration occurs when the displacement of the vibrating object is parallel to the direction of the applied force, while transverse vibration occurs when the displacement is perpendicular to the direction of the force. These types of vibrations are often observed in beams, shafts, or other structural components.

10.7.2 Methods of Vibration Measurement: Tools and Techniques

Vibration measurement involves the quantification of mechanical oscillations or movements of an object or system. Accurate measurement of vibration is essential for monitoring the health and performance of machinery and equipment [10]. Here are the common methods used to measure vibration:

10.7.2.1 Accelerometers

Accelerometers are sensors that measure acceleration, including the acceleration caused by vibration. They typically consist of a piezoelectric or micro-electro-mechanical system (MEMS) sensor element that generates an electrical signal proportional to the acceleration experienced by the sensor. Accelerometers are widely used for vibration measurement due to their small size, high sensitivity, and ability to measure vibrations across a wide frequency range.

10.7.2.2 Velocity Sensors

Velocity sensors measure the rate of change of displacement of a vibrating object over time. These sensors are commonly used in vibration measurement because velocity is directly proportional to vibration amplitude and is often preferred for low-frequency measurements. Velocity sensors can be electromagnetic, piezoelectric, or capacitance based.

10.7.2.3 Displacement Sensors

Displacement sensors measure the absolute displacement of a vibrating object from its equilibrium position. These sensors provide accurate measurements of vibration amplitudes but are less commonly used compared to accelerometers and velocity sensors. Displacement sensors include linear variable differential transformers (LVDTs), capacitance sensors, and laser displacement sensors.

10.7.2.4 Proximity Probes

Proximity probes are used to measure the radial vibration of rotating machinery such as turbines, compressors, and pumps. These sensors detect changes in the gap between the probe tip and a target surface (such as a rotating shaft) caused by shaft vibration. Proximity probes are particularly useful for monitoring critical equipment in industrial environments.

10.7.2.5 Seismic Sensors

Seismic sensors, also known as geophones, are used for measuring ground vibrations, especially in applications such as earthquake monitoring, structural health monitoring, and seismology. These sensors detect vibrations in the Earth's surface caused by seismic waves and ground motion.

10.7.2.6 Laser Doppler Vibrometers (LDVs)

Laser Doppler vibrometers (LDVs) are noncontact instruments that use laser beams to measure the velocity of a vibrating surface. These devices are highly accurate and suitable for measuring vibrations on delicate or inaccessible surfaces. LDVs are often used in research and development, as well as in industries such as aerospace and automotive.

10.7.2.7 Strain Gauges

Strain gauges measure the deformation or strain of a vibrating object caused by mechanical stress. While primarily used for structural health monitoring and material testing, strain gauges can also indirectly measure vibration by detecting changes in strain induced by dynamic loads.

10.7.2.8 Microphones (for Sound Vibration)

In some cases, microphones can be used to measure sound vibrations, which are essentially air pressure fluctuations caused by sound waves. While not as common in industrial vibration monitoring, microphones are widely used in acoustics, noise measurement, and audio engineering.

10.7.3 Characteristics of Vibration

The characteristics of vibration refer to the properties and attributes that describe the motion or oscillation of an object or system. These characteristics provide valuable insights into the nature and behavior of the vibration, allowing for its analysis, interpretation, and eventual diagnosis of potential issues [7, 10]. Here are the key characteristics of vibration:

10.7.3.1 Amplitude

Amplitude refers to the maximum displacement or magnitude of the vibration waveform from its equilibrium position. It represents the intensity or strength of the vibration and is typically measured in units such as millimeters (mm) or meters (m). Amplitude provides information about the energy content of the vibration and its potential to cause damage to machinery or structures.

10.7.3.2 Frequency

Frequency refers to the rate at which the vibration oscillates or cycles per unit of time and is measured in Hertz (Hz). It represents the number of complete vibrations occurring in one second. Frequency is inversely related to the time period of the vibration (T), where frequency (f) = $1/T$. The frequency spectrum of a vibration signal provides information about the dominant frequencies present and can help identify potential sources of vibration [12].

10.7.3.3 Phase

Phase refers to the relative timing or alignment of two or more vibrating objects or waveforms. It describes the relationship between the positions of objects or waveforms at a specific point in time and is typically measured in degrees or radians. Phase difference can provide information about the synchronization or interaction between vibrating components in a system.

10.7.3.4 Direction

Direction refers to the orientation or axis along which the vibration occurs. Vibration can occur in one, two, or three dimensions, depending on the number of axes involved. Understanding the direction of vibration is essential for diagnosing issues such as misalignment, imbalance, or structural resonances in machinery and structures.

10.7.3.5 Damping

Damping refers to the dissipation of energy in a vibrating system, resulting in a gradual decrease in vibration amplitude over time. It represents the system's ability to resist or absorb mechanical energy and is influenced by factors such as material properties, friction, and damping mechanisms. Damping affects the decay rate and stability of vibrations and can be categorized as underdamped, critically damped, or overdamped.

10.7.3.6 Harmonics

Harmonics are integer multiples of the fundamental frequency present in a vibration signal. They result from nonlinearities or resonances in the vibrating system and can contribute to the complexity of the vibration waveform. Harmonics can provide valuable information about the condition of machinery and the presence of defects such as gear tooth wear or bearing faults.

10.7.3.7 Crest Factor

Crest factor is the ratio of peak amplitude to root mean square (RMS) amplitude of a vibration signal. It provides information about the shape and peakiness of the vibration waveform and is used to assess the severity of vibration. A high crest factor indicates a highly peaked waveform with sharp peaks, while a low crest factor indicates a more sinusoidal waveform.

10.8 What Can Vibration Analysis Detect?

Vibration analysis is a powerful technique used to detect various mechanical and structural issues in machinery and equipment. Here are some common types of faults and abnormalities that vibration analysis can detect:

10.8.1 Unbalance

Unbalance occurs when the mass distribution of a rotating component, such as a rotor or fan, is not uniform. This leads to vibration due to centrifugal forces acting on the unbalanced mass. Vibration analysis can detect unbalance by identifying characteristic frequency components in the vibration spectrum [2].

10.8.2 Misalignment

Misalignment occurs when the rotational axes of coupled components, such as shafts and bearings, are not properly aligned. This results in uneven loading, increased friction, and vibration. Vibration analysis can detect misalignment by identifying frequency components related to shaft rotation and bearing faults.

10.8.3 Bearing Faults

Bearing faults, such as fatigue, wear, or lubrication issues, can cause vibration due to irregular motion of the bearing elements. Vibration analysis can detect bearing faults by identifying characteristic frequency components associated with rolling element defects, cage faults, or lubrication problems.

10.8.4 Mechanical Looseness

Mechanical looseness refers to excessive clearance or play in mechanical components, leading to erratic motion and vibration. Vibration analysis can detect mechanical looseness by identifying changes in vibration patterns, such as increased broadband vibration across multiple frequency bands.

10.8.5 Resonance

Resonance occurs when a mechanical system is excited at its natural frequency, resulting in amplified vibration amplitudes. Vibration analysis can detect resonance by identifying frequency components corresponding to the natural frequencies of the system and assessing their amplitudes and phase relationships.

10.8.6 Gear Problems

Gear problems, such as tooth wear, pitting, or misalignment, can cause vibration and noise in gear-driven machinery. Vibration analysis can detect gear problems by analyzing frequency components related to gear meshing frequencies, sidebands, and modulation effects.

10.8.7 Electrical Faults

Electrical faults, such as rotor bar defects, stator winding faults, or insulation degradation, can cause vibration in electric motors and generators. Vibration analysis can detect electrical faults by identifying frequency components associated with electromagnetic forces and air-gap eccentricity.

10.8.8 Structural Resonance

Structural resonance occurs when mechanical vibrations excite natural frequencies of a structure, leading to excessive vibration amplitudes and potential structural damage. Vibration analysis can detect structural resonance by analyzing frequency components corresponding to the natural frequencies of the structure.

10.8.9 Lubrication Issues

Lubrication issues, such as insufficient lubrication, contamination, or degradation of lubricants, can cause friction, heat, and vibration in bearings and gears. Vibration analysis can detect lubrication issues by identifying changes in vibration patterns and frequency components associated with bearing and gear faults.

10.9 Block Diagram of Vibration Monitoring System

A vibration monitoring system is a comprehensive solution designed to continuously monitor the mechanical condition of machinery and equipment by analyzing vibration signals [3]. It helps in detecting various faults, abnormalities, and performance issues to ensure the reliability, safety, and efficiency of industrial assets [13].

Figure 10.3 represents the various components of vibration monitoring system. The system typically consists of several key components:

10.9.1 Vibration Sensors

Vibration sensors, such as accelerometers or velocity sensors, are installed at critical points on the machinery or equipment to measure mechanical vibrations. These sensors convert mechanical motion into electrical signals, which are then transmitted to the monitoring system for analysis.

10.9.2 Signal Conditioning

The electrical signals from the vibration sensors are conditioned using signal conditioning circuits. These circuits filter, amplify, and preprocess the signals to ensure they are suitable for further analysis. Signal conditioning helps improve the quality and integrity of the vibration signals.

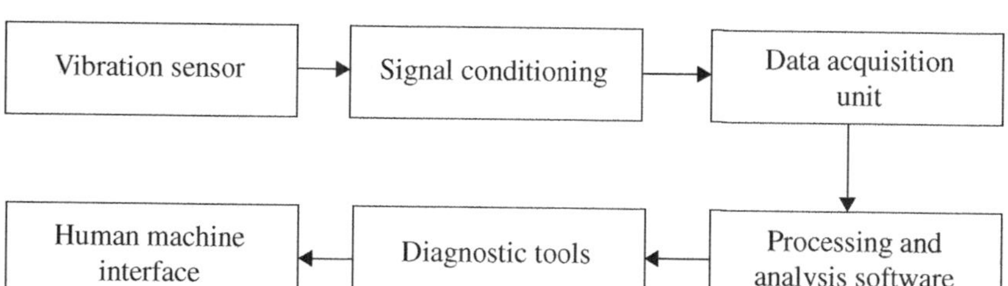

Figure 10.3 Block diagram of vibration monitoring.

10.9.3 Data Acquisition Unit (DAQ)

The conditioned vibration signals are then fed into a data acquisition unit (DAQ), which digitizes and records the signals. The DAQ unit may include multiple channels for simultaneous monitoring of multiple sensors. It also performs tasks such as signal averaging, sampling, and synchronization.

10.9.4 Processing and Analysis Software

The digitized vibration data is analyzed using specialized processing and analysis software. This software applies various algorithms and techniques to interpret the vibration signals, identify patterns, and detect anomalies indicative of potential faults or abnormalities. Analysis methods include time-domain analysis, frequency-domain analysis, and waveform analysis.

10.9.5 Diagnostic Tools

The analysis software may include diagnostic tools such as spectrum analysis, waveform analysis, and trend analysis. These tools provide insights into the health and condition of the machinery, helping in the diagnosis of specific faults or issues. They also enable users to set alarm thresholds and prioritize maintenance activities based on the severity of detected abnormalities.

10.9.6 Human–Machine Interface (HMI)

The monitoring system typically features a human–machine interface (HMI) for user interaction and visualization of the collected data. The HMI may consist of a graphical user interface (GUI) or a dashboard displaying real-time vibration data, trend plots, alarms, and diagnostic information. It allows operators and maintenance personnel to monitor the condition of machinery and take appropriate actions.

10.10 Industrial Applications of Vibration Analysis

Vibration analysis is widely used across various industries to monitor the health and performance of machinery and equipment. Here are some key industrial applications of vibration analysis:

10.10.1 Manufacturing

In manufacturing plants, vibration analysis is used to monitor the condition of rotating equipment such as motors, pumps, compressors, and gearboxes. By detecting faults such as unbalance, misalignment, bearing defects, and mechanical looseness, vibration analysis helps prevent unexpected downtime, optimize maintenance schedules, and improve overall equipment reliability and efficiency.

10.10.2 Power Generation

In power generation facilities, vibration analysis is essential for monitoring the condition of turbines, generators, and auxiliary equipment. By identifying abnormalities such as rotor unbalance, shaft misalignment, and bearing faults, vibration analysis helps ensure the reliability and performance of power generation assets, minimizing the risk of costly failures and unplanned outages.

10.10.3 Oil and Gas

In the oil and gas industry, vibration analysis is used to monitor critical machinery such as pumps, compressors, and drilling rigs. By detecting issues like cavitation, gear wear, bearing defects, and structural resonance, vibration analysis helps optimize equipment performance, enhance safety, and minimize maintenance costs in oil refineries, petrochemical plants, and offshore platforms.

10.10.4 Aerospace

In the aerospace sector, vibration analysis plays a vital role in monitoring the health of aircraft engines, turbines, and structural components. By detecting faults such as blade damage, bearing defects, and engine imbalance, vibration analysis helps ensure the safety and reliability of aerospace systems, reducing the risk of in-flight failures and maintenance-related incidents.

10.10.5 Automotive

In the automotive industry, vibration analysis is used to monitor the condition of vehicle components such as engines, transmissions, and suspension systems. By detecting faults like bearing wear, gear damage, and drivetrain issues, vibration analysis helps automotive manufacturers ensure the quality, performance, and reliability of vehicles, improving customer satisfaction and brand reputation.

10.10.6 Mining and Minerals

In mining operations, vibration analysis is employed to monitor heavy machinery such as crushers, conveyors, and vibrating screens. By identifying faults such as gear wear, bearing defects, and structural fatigue, vibration analysis helps optimize equipment reliability, minimize downtime, and improve productivity in mining and mineral processing facilities.

10.10.7 Rail Transportation

In the rail industry, vibration analysis is used to monitor the condition of locomotives, railcars, and track infrastructure. By detecting faults such as wheel defects, axle wear, and track irregularities, vibration analysis helps ensure the safety, efficiency, and reliability of rail transportation services, reducing the risk of accidents and service disruptions.

10.10.8 Marine and Shipping

In maritime operations, vibration analysis is employed to monitor marine propulsion systems, ship engines, and onboard equipment. By detecting faults such as shaft misalignment, propeller damage, and bearing defects, vibration analysis helps ensure the safety, performance, and operational efficiency of ships and marine vessels, reducing fuel consumption and maintenance costs.

10.11 Advantages of Vibration Analysis for Condition Monitoring in Electrical Systems

1) **Early Fault Detection:** Vibration analysis can detect mechanical faults in electrical equipment such as motors, generators, and transformers at an early stage, allowing for timely intervention before the faults escalate into major failures.
2) **Comprehensive Assessment:** Vibration analysis provides a comprehensive assessment of the health and condition of electrical equipment by monitoring mechanical components such as bearings, rotors, and shafts, which are critical for the reliable operation of electrical systems.
3) **Predictive Maintenance:** By identifying abnormal vibration patterns indicative of potential faults or degradation, vibration analysis enables PdM strategies, allowing maintenance to be scheduled based on equipment condition rather than fixed time intervals, reducing downtime and maintenance costs.
4) **Improved Reliability:** Implementing vibration analysis for CM enhances the reliability and availability of electrical systems by identifying and addressing mechanical issues that can lead to unexpected failures, unplanned downtime, and costly repairs.
5) **Increased Safety:** Detecting and addressing mechanical faults through vibration analysis enhances the safety of electrical systems by mitigating the risk of equipment failures that could result in accidents, injuries, or damage to personnel and property.
6) **Optimized Performance:** By monitoring vibration levels and trends, vibration analysis helps optimize the performance of electrical equipment by identifying opportunities for efficiency improvements, reducing energy consumption, and extending equipment lifespan.

10.12 Disadvantages of Vibration Analysis for Condition Monitoring in Electrical Systems

1) **Complexity:** Vibration analysis requires specialized knowledge, expertise, and equipment for data acquisition, signal processing, and analysis, which can be complex and challenging to implement, particularly for organizations with limited resources or experience in vibration monitoring.
2) **Interpretation Challenges:** Interpreting vibration data and identifying the root causes of abnormalities may require advanced analytical skills and experience, making it difficult for inexperienced personnel to effectively diagnose and address mechanical issues detected through vibration analysis.
3) **Cost of Implementation:** Implementing a vibration analysis program for CM involves upfront costs for equipment, software, training, and personnel, which can be significant and may pose a barrier to adoption for some organizations, particularly small or medium-sized enterprises with limited budgets.
4) **False Alarms:** Vibration analysis may generate false alarms or false-positive results if not properly calibrated, leading to unnecessary maintenance interventions, increased downtime, and resource wastage, undermining the credibility and effectiveness of the CM program.
5) **Limited Effectiveness for Certain Components:** Vibration analysis may be less effective for monitoring certain electrical components, such as solid-state devices or components with

minimal mechanical motion, where other CM techniques such as thermography or electrical testing may be more suitable.

6) **Environmental Factors:** Environmental factors such as temperature, humidity, and operating conditions can influence vibration measurements and may require careful consideration and calibration to ensure the accuracy and reliability of vibration analysis results.

10.13 Importance of Fault Diagnosis in Electrical System

Fault diagnosis in electrical systems is crucial for maintaining operational reliability, safety, and efficiency [14]. Here is why fault diagnosis holds significant importance in electrical systems:

10.13.1 Ensuring Safety

Fault diagnosis helps identify potential electrical hazards such as short circuits, overloads, ground faults, and insulation failures. By promptly detecting and addressing these faults, fault diagnosis reduces the risk of electrical fires, shocks, and accidents, ensuring the safety of personnel, equipment, and facilities.

10.13.2 Preventing Equipment Damage

Electrical faults, if left undetected, can lead to severe damage to electrical components, including motors, transformers, switches, and control systems. Fault diagnosis enables early detection of issues such as overheating, insulation breakdown, and voltage irregularities, preventing equipment damage and extending lifespan.

10.13.3 Minimizing Downtime

Electrical faults can cause unexpected downtime, production interruptions, and operational disruptions, leading to productivity losses and revenue impacts. Fault diagnosis facilitates proactive maintenance by identifying potential faults before they escalate into major failures, minimizing unplanned downtime and optimizing equipment uptime.

10.13.4 Optimizing Maintenance

Fault diagnosis allows for the implementation of CBM strategies, where maintenance activities are scheduled based on the actual condition of electrical equipment rather than fixed time intervals. This approach optimizes maintenance resources, reduces unnecessary maintenance tasks, and extends the life of electrical assets.

10.13.5 Improving Reliability

Reliable electrical systems are essential for the continuous operation of critical infrastructure, industrial processes, and commercial facilities. Fault diagnosis helps maintain the reliability of electrical systems by identifying and addressing faults that could lead to equipment failures, power outages, or system malfunctions.

10.13.6 Enhancing Energy Efficiency

Electrical faults such as voltage fluctuations, imbalances, and power factor issues can reduce energy efficiency and increase operating costs. Fault diagnosis helps identify energy-related faults and inefficiencies, allowing for corrective actions to optimize energy usage, improve power quality, and reduce energy consumption.

10.13.7 Compliance with Regulations

Compliance with electrical safety standards, codes, and regulations is essential for ensuring legal and regulatory compliance, as well as maintaining certification and accreditation. Fault diagnosis helps organizations identify and rectify noncompliant conditions, ensuring adherence to industry standards and regulatory requirements.

10.13.8 Preserving Assets and Investments

Electrical systems represent significant investments for organizations, and proper maintenance is essential to protect these assets and maximize their value. Fault diagnosis helps preserve the integrity and performance of electrical assets, safeguarding investments and ensuring a positive return on investment (ROI) over the long term.

10.14 Integration with IoT of Conditional Monitoring Electrical System

10.14.1 Sensor Deployment

The integration process begins with the deployment of IoT-enabled sensors throughout the electrical infrastructure. These sensors are strategically placed to monitor key parameters such as voltage, current, temperature, humidity, vibration, and insulation resistance. The sensors are connected to the IoT gateway via wired or wireless communication protocols, such as Wi-Fi, Bluetooth, Zigbee, or LoRaWAN [15].

10.14.2 Data Acquisition

The sensors continuously collect data on equipment condition, performance, and environmental factors in real time. The collected data is transmitted to the IoT gateway, where it is aggregated and processed before being sent to the cloud-based platform for further analysis. Data acquisition may also involve the use of data loggers or programmable logic controllers (PLCs) to capture information from legacy equipment or analog signals.

10.14.3 Data Transmission

The IoT gateway acts as a bridge between the sensors and the cloud-based platform, facilitating secure data transmission over the internet or local network. The gateway preprocesses the data, applies data encryption and compression techniques to ensure data security and efficiency, and then transmits the data to the cloud server using standard communication protocols such as message queing telemetry transport (MQTT), hyper text transfer protocol (HTTP), or constrained application protocol (CoAP).

10.14.4 Cloud-Based Platform

In the cloud-based platform, the received data is stored, processed, and analyzed using advanced analytics algorithms and ML models [2]. The platform provides a scalable and flexible environment for data management, storage, and computation, enabling real-time monitoring, PdM, and data-driven insights. Cloud-based platforms may leverage scalable cloud services such as Amazon Web Services (AWS), Microsoft Azure, or Google Cloud Platform (GCP).

10.14.5 Data Analysis and Visualization

The cloud-based platform performs data analysis to identify patterns, trends, and anomalies in the collected data. Advanced analytics techniques, including statistical analysis, pattern recognition, and anomaly detection, are applied to extract actionable insights and PdM indicators. The analyzed data is visualized through interactive dashboards, charts, graphs, and reports, providing maintenance personnel with actionable information for decision-making.

10.15 Real-Time Monitoring and Predictive Maintenance

10.15.1 Real-Time Monitoring

IoT-enabled sensors continuously monitor equipment parameters such as temperature, voltage, current, and vibration levels in real time [5]. The collected data is transmitted to the cloud-based platform for analysis, allowing maintenance personnel to monitor equipment health and performance remotely from anywhere with internet connectivity. Real-time monitoring enables early detection of abnormalities, faults, or anomalies, facilitating timely intervention and preventive maintenance actions.

10.15.2 Predictive Maintenance

The cloud-based platform applies PdM algorithms to analyze historical data, performance trends, and equipment health metrics. ML models are trained to predict potential equipment failures before they occur based on patterns indicative of degradation or impending faults. PdM indicators are generated, and automated alerts are sent to maintenance teams when abnormal conditions are detected, enabling proactive measures to prevent downtime and minimize disruptions.

10.15.3 Root Cause Analysis

In addition to predicting equipment failures, the cloud-based platform performs root cause analysis to identify underlying factors contributing to equipment degradation or malfunctions. By correlating data from multiple sensors and historical records, maintenance personnel can pinpoint the root causes of faults and develop targeted mitigation strategies. Root cause analysis enables organizations to address systemic issues and improve overall equipment reliability and performance.

10.15.4 Condition-Based Alerts

Condition-based alerts are generated based on predefined thresholds or abnormal conditions detected in the equipment. These alerts are sent to maintenance personnel via email, SMS, or mobile notifications, prompting them to investigate and address potential issues promptly. Condition-based alerts ensure timely intervention and preventive maintenance actions, minimizing downtime and optimizing equipment uptime.

10.15.5 Continuous Improvement

The integration of IoT-enabled CM systems enables continuous improvement through feedback loops and iterative optimization. Maintenance personnel analyze the effectiveness of maintenance actions, review performance metrics, and refine PdM models based on observed outcomes. Continuous improvement processes drive operational excellence, enhance reliability, and maximize the value of CM investments.

10.16 Energy Management and Asset Performance Optimization

10.16.1 Energy Monitoring

IoT-enabled sensors monitor energy consumption, power quality, and demand patterns in real time, providing organizations with insights into energy usage and efficiency. The collected energy data is transmitted to the cloud-based platform for analysis, where energy management algorithms identify opportunities for optimization [5]. Energy monitoring enables organizations to implement energy-saving measures, optimize power distribution, and reduce operating costs.

10.16.2 Asset Performance Optimization

The cloud-based platform performs comprehensive asset performance optimization by analyzing equipment health, reliability, and lifecycle data. Maintenance personnel can prioritize maintenance activities based on equipment criticality, risk factors, and performance metrics. Asset performance optimization enables organizations to maximize asset uptime, minimize downtime, and optimize asset performance across the entire lifecycle, resulting in improved operational efficiency and cost-effectiveness.

10.16.3 Integration with Enterprise Systems

IoT-enabled CM systems can be seamlessly integrated with enterprise asset management (EAM) systems, supervisory control and data acquisition (SCADA) systems, and other business applications. This integration enables centralized data management, workflow automation, and collaboration across departments. Maintenance data collected from IoT sensors is linked to asset records, work orders, and maintenance schedules within the EAM system, streamlining maintenance processes and improving asset visibility.

10.17 Safety, Compliance, and Future Trends

10.17.1 Safety and Compliance

IoT integration enhances safety and compliance in electrical systems by providing early warning alerts for potential hazards or faults. Real-time monitoring and PdM capabilities enable organizations to identify and address safety risks promptly, ensuring compliance with regulatory requirements and industry standards. By proactively managing equipment health and performance, organizations can minimize the risk of accidents, injuries, and regulatory violations.

10.17.2 Regulatory Compliance

IoT-enabled CM systems help organizations comply with regulatory requirements, industry standards, and quality management systems.

10.18 Future Trends in IoT Application in Condition Monitoring and Fault Diagnosis in Electrical Systems

Future trends in IoT application for CM and fault diagnosis [4] in electrical systems are expected to bring about significant advancements in PdM, data analytics, and automation [16]. Here are some anticipated trends:

1) **Edge Computing:** Edge computing involves processing data closer to its source, reducing latency and bandwidth requirements. In the context of IoT for CM, edge computing enables real-time data analysis and decision-making at the edge devices or gateways. This trend will lead to faster response times and improved efficiency in identifying and addressing equipment faults.
2) **AI and ML:** AI and ML algorithms are increasingly being integrated into IoT systems for PdM and fault diagnosis. These algorithms can analyze large volumes of sensor data to identify patterns, anomalies, and correlations that may indicate potential equipment failures. As AI capabilities continue to evolve, they will become more sophisticated in predicting and preventing faults in electrical systems.
3) **Digital Twins:** Digital twins are virtual replicas of physical assets or systems that mimic their behavior in real time. In the context of electrical systems, digital twins can be used to simulate the performance of equipment and predict potential faults or failures. By combining IoT data with digital twin models, organizations can gain deeper insights into equipment health and optimize maintenance strategies.
4) **Blockchain Technology:** Blockchain technology offers secure and decentralized data storage and transaction capabilities. In the context of IoT for CM, blockchain can be used to ensure the integrity and immutability of sensor data, maintenance records, and diagnostic information. This trend will enhance data security, transparency, and trustworthiness in CM systems.
5) **Predictive Analytics as a Service (PAaaS):** PAaaS is a cloud-based service that provides PdM capabilities on-demand. Organizations can leverage PAaaS platforms to access advanced analytics algorithms and ML models for predicting equipment failures and optimizing maintenance schedules. This trend will democratize access to PdM technologies, particularly for small and medium-sized enterprises (SMEs).

6) **Integration with Industry 4.0 Technologies:** IoT for CM will continue to integrate with other Industry 4.0 technologies such as the Industrial Internet of Things (IIoT), Cyber-Physical Systems (CPS), and Big Data analytics. This integration will enable seamless communication and interoperability between different systems, leading to more holistic approaches to CM and fault diagnosis in electrical systems.
7) **Human–Machine Collaboration:** As IoT systems become more advanced, there will be a greater emphasis on human–machine collaboration in CM and fault diagnosis. Maintenance personnel will work alongside AI-driven algorithms and analytics tools to interpret data, make decisions, and implement corrective actions. This collaborative approach will leverage the strengths of both humans and machines, resulting in more effective maintenance strategies.
8) **Augmented Reality (AR) and Virtual Reality (VR):** AR and VR technologies will play an increasingly important role in training, visualization, and remote assistance for CM and fault diagnosis. Maintenance personnel can use AR/VR devices to visualize equipment data, simulate maintenance procedures, and receive guidance from experts in real time. This trend will improve the efficiency and accuracy of maintenance tasks, particularly in complex or hazardous environments.

References

1 Huang, M., Liu, Z., and Tao, Y. (2020). Mechanical fault diagnosis and prediction in IoT based on multi-source sensing data fusion. *Simulation Modelling Practice and Theory* 102: 101981.

2 Liu, D., Cui, L., and Wang, H. (2023). Rotating machinery fault diagnosis under time-varying speeds: a review. *IEEE Sensors Journal* 23: 29969.

3 Swain, A., Abdellatif, E., Mousa, A., and Pong, P.W.T. (2022). Sensor technologies for transmission and distribution systems: a review of the latest developments. *Energies* 15 (19): 7339.

4 Costa, F.F., de Almeida, L.A.L., Naidu, S.R., and Braga-Filho, E.R. (2004). Improving the signal data acquisition in condition monitoring of electrical machines. *IEEE Transactions on Instrumentation and Measurement* 53 (4): 1015–1019.

5 Rastegari, A. (2015). Strategic maintenance development focusing on use of condition based maintenance in manufacturing industry. Diss. Mälardalen University.

6 Peng, Y., Dong, M., and Zuo, M.J. (2010). Current status of machine prognostics in condition-based maintenance: a review. *The International Journal of Advanced Manufacturing Technology* 50: 297–313.

7 Vishwakarma, M., Purohit, R., Harshlata, V., and Rajput, P. (2017). Vibration analysis & condition monitoring for rotating machines: a review. *Materials Today Proceedings* 4 (2): 2659–2664.

8 Furse, C.M., Kafal, M., Razzaghi, R., and Shin, Y.-J. (2020). Fault diagnosis for electrical systems and power networks: a review. *IEEE Sensors Journal* 21 (2): 888–906.

9 Yuvaraj, T., Lakshmi, D., Sivarajeswari, S. et al. (2024). Application of Salp Swarm Algorithm for optimal placement of DSTATCOM in the radial distribution networks. *AIP Conference Proceedings* 3044 (1): 050002.

10 Khan, N., Rafiq, F., Abedin, F., and Khan, F.U. (2019). IoT based health monitoring system for electrical motors. *2019 15th International Conference on Emerging Technologies (ICET)*. IEEE.

11 Lakshmi, D., Ezhilarasi, G., Rekha, K., and Jegadeeswari, G. (2021). Automatic detection and power shutdown for gas leakage and its monitoring system. *International Journal of Aquatic Science* 12 (3): 333–340.

12 Lakshmi, D., Zahira, R., Ravi, C.N. et al. (2023). Application of optimization technique in modern hybrid power systems. In: *IoT, Machine Learning and Blockchain Technologies for Renewable Energy and Modern Hybrid Power Systems* (ed. C. Sharmeela, P. Sanjeevikumar, P. Sivaraman, and M. Joseph), 149–171. River Publishers.

13 Zahira, R., Lakshmi, D., Ezhilarasi, G. et al. (2022). Stand-alone microgrid concept for rural electrification: a review. In: *Residential Microgrids and Rural Electrifications* (ed. S. Padmanaban, C. Sharmeela, P. Sivaraman, and J.B. Holm-Nielsen), 109–130. Academic Press.

14 Raja, H.A., Vaimann, T., Rassõlkin, A. et al. (2021). IoT based tools for data acquisition in electrical machines and robotics. *2021 IEEE 19th International Power Electronics and Motion Control Conference (PEMC)*. IEEE.

15 Raja, H.A., Vaimann, T., Rassõlkin, A., and Kallaste, A. (2022). Condition monitoring and fault detection for electrical machines using IoT. *Proceedings of the Future Technologies Conference*. Cham: Springer International Publishing.

16 Gundewar, S.K. and Kane, P.V. (2021). Condition monitoring and fault diagnosis of induction motor. *Journal of Vibration Engineering & Technologies* 9: 643–674.

11

IoT-Powered Robust Anomaly Detection and CNN-Enabled Predictive Maintenance to Enhance Solar PV System Performance

Kumaresa P. Punitha

Department of Electrical and Electronics Engineering, P S R Engineering College, Anna University, Chennai, Tamil Nadu, India

11.1 Introduction

In recent years, the integration of Internet of Things (IoT) technology into various sectors has revolutionized traditional systems, and the realm of renewable energy is no exception. Solar power, in particular, stands as a pillar of sustainable energy sources, yet its efficiency and maintenance remain pivotal concerns.

The integration of solar photovoltaic (PV) systems into the energy grid has seen substantial growth due to its renewable and sustainable nature. However, like any complex system, solar PV arrays are susceptible to faults and performance degradation over time. Timely detection and accurate diagnosis of these faults are critical for ensuring the reliability and efficiency of solar power generation [1]. Traditional methods of fault detection and classification in PV systems often rely on manual inspection or rule-based algorithms, which may lack the ability to handle the complexity and variability of real-world scenarios. In recent years, machine learning techniques, particularly convolutional neural networks (CNNs), have shown promise in automating the fault detection and classification process by leveraging the power of data-driven approaches. The research [2] describes a method for detecting, classifying, and locating string to string (SS), string to ground (SG) and open circuit (OC) faults utilizing multi-output deep learning (DL) techniques, including CNNs, long short-term memory (LSTM), and bi-directional long short-term memory (Bi-LSTM) networks. This work [3] would have used a normal bibliographic scan to diagnose a solar PV issue. This study [4] introduces a unique fault detection algorithm based on machine learning and applies it to the identification of defects in heating, ventilation, and air conditioning (HVAC) systems. The study [5] investigates the sensitive factors for the seven most common chiller problems using global sensitivity analysis (GSA) with a Random Forest (RF) meta-model. Fault identification and diagnosis for grid-connected PV systems using the C4.5 decision tree method are reported in [6]. To improve the solar system's robustness, the papers [7, 8] present a trained CNN-based fault detection approach based on PV module photos. The fault detection approach for a large-scale PV plant based on frequency response analysis is discussed in [9]. A cascading neural network methodology for defect identification and diagnosis in solar thermal plants is described in [10]. References [11] and [12] provide an overview of fault diagnosis in microgrids with solar PV system integration. Reference [13] discusses fault detection and diagnostics in large solar thermal systems. The Improved Real Coded Genetic Algorithm, a mathematical optimizer, is used here [14] to forecast the likely fault pattern. This article [15] discusses fault diagnostics for a solar-assisted heat pump (SAHP) system

IoT for Smart Grid: Revolutionizing Electrical Engineering, First Edition.
Edited by Rahiman Zahira, Palanisamy Sivaraman, Chenniappan Sharmeela, and Sanjeevikumar Padmanaban.

using limited data and expert knowledge. Researchers [16, 17] suggest an intelligent method for failure detection and classification (FDC) in solar-powered systems. The majority of the publications reviewed above did not include data-gathering equipment setups. References [18–20] propose a low-cost IoT solution for real-time problem diagnostics of PV modules. However, correction, solving, or enhancing efficiency is missing from all publications.

This chapter aims to tackle the challenges of solar panel system efficiency and maintenance by leveraging the capabilities of IoT devices and advanced data analytic techniques to promptly identify anomalies. The integration of hardware components, including the ESP32 microcontroller, DHT11 sensor for environmental data monitoring, and voltage and current sensors, establishes the foundation for real-time monitoring of a two-panel, 20 W solar array. The utilization of a CNN algorithm enables continuous analysis of critical performance parameters. This proactive approach not only facilitates swift detection of potential issues, such as degradation, shading, or electrical faults but also enables immediate intervention, minimizing system downtime and maximizing power output. By ensuring the efficient operation of solar panel systems, this endeavor strives to make a significant contribution to sustainable energy generation and promote the widespread adoption of renewable energy sources.

11.2 IoT Application in Condition Monitoring

Condition monitoring traditionally relied on periodic manual inspections or the installation of dedicated monitoring systems, which were often costly and limited in their capabilities. IoT applications in condition monitoring have disrupted this paradigm by offering cost-effective, scalable, and intelligent solutions. By leveraging wireless sensor networks, cloud computing, and advanced data analytics, IoT-enabled condition monitoring systems can continuously gather and process data from various sources, providing comprehensive insights into the health and performance of monitored assets.

The integration of IoT in condition monitoring has enabled the collection of diverse data streams, including vibration, temperature, pressure, acoustics, and other relevant parameters. This wealth of data, coupled with powerful data analytic techniques, such as machine learning and artificial intelligence, enables the early detection of anomalies, identification of potential failures, and optimization of maintenance schedules. Predictive algorithms can analyze historical and real-time data to forecast equipment degradation, enabling proactive maintenance interventions before catastrophic failures occur.

IoT applications in condition monitoring have found widespread adoption across various industries, including manufacturing, energy, transportation, and healthcare. In the manufacturing sector, IoT-powered condition monitoring systems monitor critical machinery, such as motors, pumps, and compressors, ensuring optimal performance and minimizing unplanned downtime. In the energy sector, these systems are employed to monitor the health of wind turbines, solar panels, and power generation equipment, improving reliability and maximizing energy yield.

Furthermore, IoT condition monitoring solutions offer remote accessibility, allowing real-time monitoring and control from centralized locations, reducing the need for on-site personnel and enabling faster response times. The integration of IoT with cloud computing and data visualization tools provides stakeholders with intuitive dashboards and real-time alerts, facilitating informed decision-making and streamlining maintenance operations.

11.3 IoT Application in Fault Prediction

The reliability and efficient operation of electrical systems are critical for ensuring uninterrupted power supply and minimizing the risk of costly downtime. Electrical faults, such as insulation failures, short circuits, and overloads, can lead to severe consequences, including equipment damage, safety hazards, and financial losses. Traditional fault detection and maintenance approaches often rely on periodic inspections or reactive measures after a fault has already occurred, resulting in unplanned outages and increased maintenance costs. The advent of the IoT has revolutionized the way electrical systems are monitored and maintained. IoT applications in fault prediction of electrical systems offer a proactive approach to identifying and mitigating potential faults before they occur, enabling predictive maintenance strategies and enhancing system reliability. IoT-enabled fault prediction systems leverage a network of intelligent sensors and devices strategically placed throughout the electrical infrastructure. These sensors continuously monitor various parameters, such as voltage, current, temperature, vibration, and insulation resistance, generating a wealth of real-time data. This data is then transmitted through secure communication channels to cloud-based platforms or local gateways for further processing and analysis. The integration of IoT with advanced data analytic techniques, including machine learning and artificial intelligence, plays a pivotal role in fault prediction. By analyzing historical data patterns and correlating them with known fault signatures, predictive models can be trained to recognize early warning signs of potential failures. These models can then be applied to real-time sensor data, enabling the early detection of anomalies and the prediction of impending faults. IoT-based fault prediction systems offer several advantages over traditional approaches. Real-time monitoring and data analysis allow for continuous system assessment, enabling timely interventions and minimizing the risk of catastrophic failures. Predictive maintenance strategies facilitated by fault prediction can significantly reduce maintenance costs by avoiding unplanned downtime and optimizing maintenance schedules. Furthermore, IoT-enabled fault prediction systems can be scaled and adapted to various electrical systems, ranging from residential installations to large-scale industrial facilities and utility grids. The ability to remotely monitor and diagnose faults also enhances safety by reducing the need for personnel to physically inspect potential fault locations, minimizing exposure to hazardous environments.

11.4 Overview of Solar PV System Faults

Solar PV systems, despite their robust design, are susceptible to various faults that can impact their performance and energy yield. These faults can arise from different components of the system, including the PV modules, inverters, wiring, and electrical connections. Understanding the types of faults and their root causes is crucial for effective monitoring, maintenance, and fault mitigation strategies.

PV Module Faults (F1)

- **Hotspots:** Localized areas of overheating within a PV module caused by cell mismatches, shading, or defects.
- **Delamination:** Separation of the PV module's layers due to environmental stress or manufacturing defects.

- **Potential Induced Degradation (PID):** Voltage-induced degradation of the PV module's performance due to leakage currents.
- **Soiling:** Accumulation of dust, dirt, or other contaminants on the module surface, reducing light transmission.
- **Mechanical Damage:** Cracks, scratches, or other physical damage to the PV module caused by environmental factors or mishandling.

Inverter Faults (F2)

- **Overheating:** Excessive temperature buildup within the inverter due to poor ventilation or component failure.
- **Electrical Faults:** Short circuits, open circuits, or ground faults within the inverter or its connections.
- **Software/Firmware Issues:** Bugs or compatibility issues with the inverter's firmware or control software.

Wiring and Connection Faults (F3)

- **Loose Connections:** Poor electrical connections lead to increased resistance and power losses.
- **Cable Damage:** Physical damage to cables or wiring due to environmental factors or improper installation.
- **Corrosion:** Oxidation or degradation of electrical connections due to exposure to moisture or harsh environmental conditions.

Shading Faults (F4)

- **Partial Shading:** Obstructions casting shadows on portions of the PV array, reducing energy generation.
- **Complete Shading:** Full obstruction of solar irradiance on the PV array due to nearby structures or objects.

Tracking System Faults (for Tracking Systems) (F5)

- **Mechanical Failures:** Issues with the tracking mechanism, such as motor failures or misalignment.
- **Control System Faults:** Errors in the tracking system's control algorithms or sensor inputs.

Ground Faults (F6)

- Ground faults occur when an unintended connection is made between the electrical conductors of the solar PV system and the ground.
- Ground faults can result in electrical shorts, fire hazards, and damage to system components.

Communication Errors (F7)

- Communication errors between sensors, data acquisition systems, and monitoring devices can lead to data loss, inaccurate measurements, and impaired system performance.
- Reliable communication is crucial for effective fault detection and monitoring.

Environmental Factors (F8)

- Environmental factors such as temperature variations, humidity, dust, ultraviolet (UV) radiation, and debris accumulation can impact the performance and reliability of solar PV systems.
- Monitoring and mitigating the effects of these factors are essential for maintaining optimal system operation.

Early detection and prediction of such anomalies are essential for timely maintenance and maximizing energy yield.

11.5 Need for IoT and CNN Algorithm for Anomaly Detection of Solar PV System

The integration of IoT and CNNs for anomaly detection in solar PV systems offers several advantages and addresses critical needs in monitoring and maintaining these systems. Here are some key reasons highlighting the need for IoT and CNN in solar PV anomaly detection:

- **Real-Time Monitoring and Data Collection**
  IoT technologies enable the deployment of a network of sensors and devices that can continuously monitor various parameters of the solar PV system, such as voltage, current, temperature, and irradiance levels. This real-time data collection is crucial for detecting anomalies and performance deviations as they occur, enabling timely intervention and minimizing energy losses.
- **Remote Monitoring and Accessibility**
  IoT systems allow remote monitoring and control of solar PV installations, reducing the need for on-site personnel and enabling centralized management of distributed systems. This remote accessibility enables efficient monitoring of large-scale solar farms or geographically dispersed installations, facilitating proactive maintenance and rapid fault diagnosis.
- **Scalability and Cost-Effectiveness**
  IoT-based monitoring solutions are highly scalable and can be deployed across a wide range of solar PV system sizes, from residential installations to utility-scale power plants. Additionally, IoT technologies often leverage low-cost sensors and wireless communication protocols, making them a cost-effective solution for anomaly detection compared to traditional monitoring systems.
- **Data Processing and Pattern Recognition**
  CNNs excel at extracting complex patterns and features from high-dimensional data, such as images and sensor data streams. By integrating CNNs with IoT-collected data, anomaly detection models can learn the intrinsic patterns and correlations among various sensor measurements, enabling accurate identification of anomalous behaviors or degradation trends.
- **Automated Fault Diagnosis**
  CNNs can be trained on historical data and known fault signatures to automatically classify and diagnose different types of faults or anomalies in solar PV systems. This automated fault diagnosis capability can significantly reduce the time and effort required for manual inspections and troubleshooting, leading to faster response times and minimizing downtime.
- **Predictive Maintenance**
  By leveraging the predictive capabilities of CNNs, IoT-based monitoring systems can anticipate potential faults or performance degradation before they occur. This enables proactive maintenance strategies, reducing unplanned downtime and extending the lifespan of solar PV systems.
- **Visual Data Analysis**
  CNNs are particularly well suited for analyzing visual data, such as images or video footage of solar PV installations. IoT-enabled cameras or drones can capture visual data, which can be processed by CNNs to detect anomalies such as soiling, hotspots, or physical damage to the modules, providing comprehensive monitoring beyond just electrical parameters.

The combination of IoT and CNN technologies offers a powerful solution for comprehensive monitoring, anomaly detection, and predictive maintenance of solar PV systems, ultimately improving their reliability, efficiency, and overall performance. As these technologies continue to evolve, their integration will become increasingly crucial for optimizing the operation and maintenance of renewable energy systems.

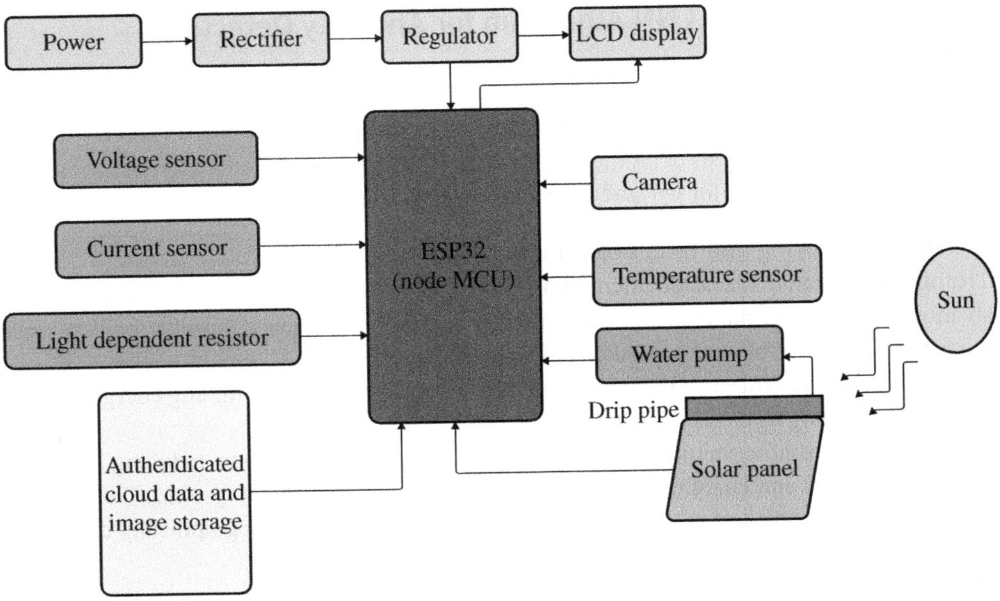

Figure 11.1 Block diagram representation of the solar PV system.

11.6 System Description

The proposed system for data collection setup comprises ESP32 node IoT controllers, a current sensor, a voltage sensor, a temperature controller, light-dependent resistor, and a camera. The system current, voltage, temperature, and light intensity are collected by ESP32 node and stored in cloud-based Google Sheet which is allowed to view authenticated personnel. If the temperature falls above set threshold values, a water pump activates, drip pipe drops the water on the panel for cooling and cleaning the dust on the panels to boost efficiency. If voltage, current, temperature, or light intensity falls above/below set threshold levels, the camera captures panel images, relayed to Google Sheets for further analysis (Figure 11.1).

11.7 Proposed Algorithm

11.7.1 Deep Learning

DL is a broad sub field of machine learning that deals with artificial neural networks that have multiple layers (hence the term "deep"). It is a technique for implementing machine learning models that are inspired by the structure and function of the human brain. DL models can learn hierarchical representations of data and perform tasks such as image recognition, natural language processing, and speech recognition.

11.7.2 Convolution Neural Networks (CNNs)

CNNs are a specific type of DL model that is primarily used for processing data that has a grid-like topology, such as images, videos, or audio spectrograms. CNNs are inspired by the visual cortex of the brain and are particularly well suited for tasks such as image classification, object detection, and face recognition. The key difference between CNNs and other DL models lies in their architecture. CNNs have a unique structure that includes:

Convolutional Layers: These layers apply a set of learnable filters to the input data, allowing the network to detect features or patterns within the data.
Pooling Layers: These layers downsample the feature maps obtained from the convolutional layers, reducing the spatial dimensions while retaining the most important information.
Fully Connected Layers: These are the same as in other neural networks, where all the neurons are connected to the neurons in the previous and next layers.
The combination of convolutional and pooling layers allows CNNs to learn spatial and temporal hierarchies of features, making them highly effective for tasks involving images, videos, and other grid-like data.

The proposed CNN-based framework leverages sensor data collected from a solar PV system, including voltage, current, temperature, and irradiance measurements. These multivariate time-series data are preprocessed and transformed into image-like representations, allowing the CNNs to exploit their superior feature extraction capabilities for pattern recognition and anomaly detection. The CNN architecture consists of multiple convolutional and pooling layers, followed by fully connected layers. The model is trained on a comprehensive dataset containing both normal and anomalous operating conditions, enabling it to learn the intrinsic patterns and correlations among the various sensor data streams. During the inference phase, the trained CNN model analyzes real-time sensor data and classifies the operating condition as normal or anomalous. Furthermore, the model's predictive capabilities are leveraged to forecast potential anomalies and performance degradation, enabling proactive maintenance and minimizing downtime.

The proposed CNN-based approach is evaluated using real-world data collected from a solar PV system, and its performance is compared against traditional machine learning techniques. The results demonstrate the superiority of the CNN model in accurately detecting and predicting anomalies, achieving high accuracy, precision, and recall. This study contributes to the development of intelligent monitoring and predictive maintenance systems for solar PV installations, ultimately enhancing their reliability, efficiency, and overall performance. The proposed CNN-based framework can be easily adapted to other renewable energy systems, making it a valuable asset in the pursuit of sustainable and reliable energy solutions.

11.8 Results and Discussion

11.8.1 IoT-Powered Data Collection

Figure 11.2 shows the components and connections of the hardware setup. The two solar panels of 20 W convert sunlight into direct current (DC) electricity. An ESP32 controller monitors the current, voltage, light intensity, and temperature of the solar panel, including the load using sensors, and displays the data on a cloud service using Wi-Fi. It also controls the water pump and the camera using digital output pins. A current sensor that measures the current flowing through the solar panel, or the load. A voltage sensor that measures the voltage across the solar panel, or the load. A temperature sensor that measures the temperature of the solar panel. An light dependent resistor (LDR) that measures the intensity of the sunlight on the solar panel. A camera that captures images of the solar panel when the monitored parameters are abnormal, and sends them to the cloud service via Wi-Fi. A cloud service that stores and displays the data and images from the ESP32 controller on a web page or a mobile app is shown in Figure 11.3. Likewise, daily 65 numbers of

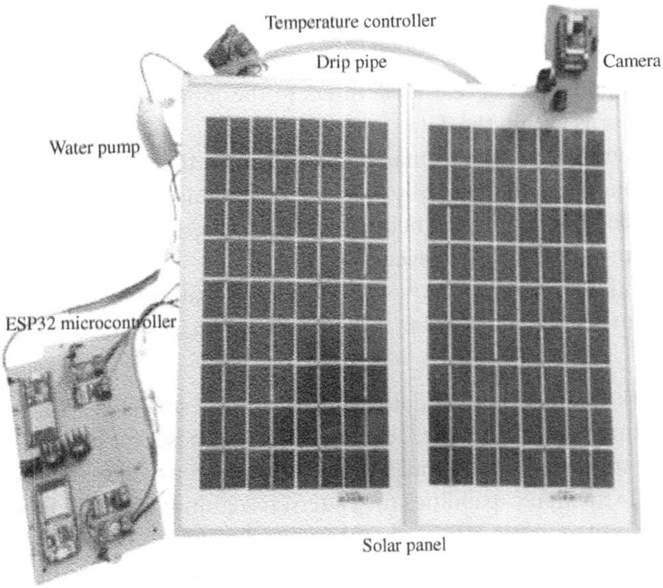

Figure 11.2 Proto-type setup for smart solar PV system.

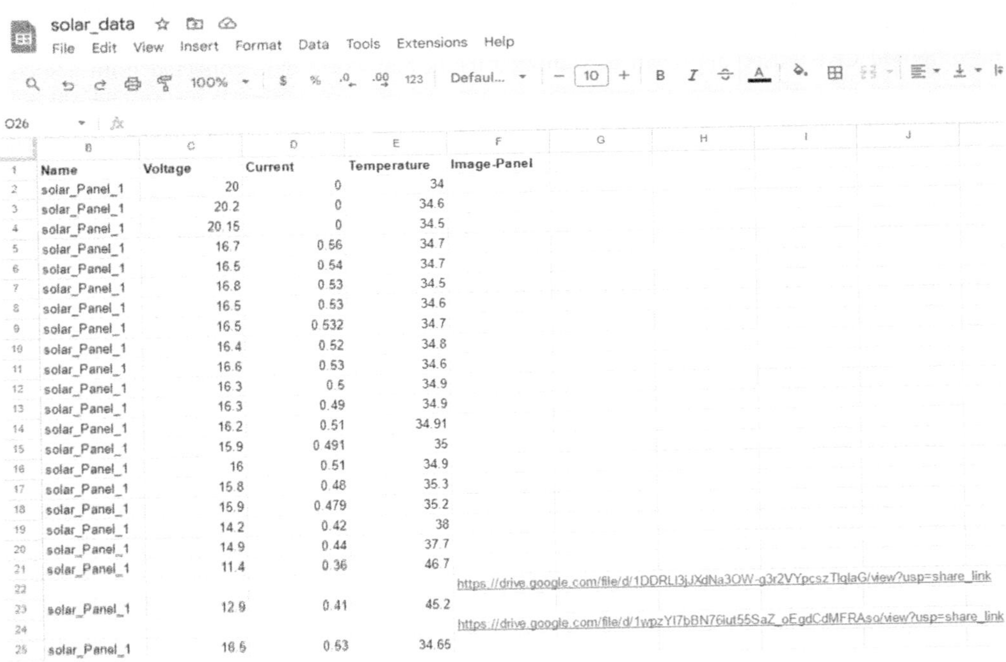

	B	C	D	E	F
1	Name	Voltage	Current	Temperature	Image-Panel
2	solar_Panel_1	20	0	34	
3	solar_Panel_1	20.2	0	34.6	
4	solar_Panel_1	20.15	0	34.5	
5	solar_Panel_1	16.7	0.56	34.7	
6	solar_Panel_1	16.5	0.54	34.7	
7	solar_Panel_1	16.8	0.53	34.5	
8	solar_Panel_1	16.5	0.53	34.6	
9	solar_Panel_1	16.5	0.532	34.7	
10	solar_Panel_1	16.4	0.52	34.8	
11	solar_Panel_1	16.6	0.53	34.6	
12	solar_Panel_1	16.3	0.5	34.9	
13	solar_Panel_1	16.3	0.49	34.9	
14	solar_Panel_1	16.2	0.51	34.91	
15	solar_Panel_1	15.9	0.491	35	
16	solar_Panel_1	16	0.51	34.9	
17	solar_Panel_1	15.8	0.48	35.3	
18	solar_Panel_1	15.9	0.479	35.2	
19	solar_Panel_1	14.2	0.42	38	
20	solar_Panel_1	14.9	0.44	37.7	
21	solar_Panel_1	11.4	0.36	46.7	
22					https://drive.google.com/file/d/1DDRLI3jJXdNa3OW-g3r2VYpcszTlqlaG/view?usp=share_link
23	solar_Panel_1	12.9	0.41	45.2	
24					https://drive.google.com/file/d/1wpzYI7bBN76iut55SaZ_oEgdCdMFRAso/view?usp=share_link
25	solar_Panel_1	16.5	0.53	34.65	

Figure 11.3 Output data collection time, panel, voltage, current, and temperature as well as capture image in cloud server.

data and for a month 2000 data are collected for the network. A water pump that sprays water on the solar panel in the predetermined time period or when the temperature is high, tc clean and improve the efficiency of the system.

11.8.2 Utilization of CNN for Classification and Prediction

The flowchart for the CNN processing steps for solar PV anomaly detection and classification is depicted in Figure 11.4. The pie chart in Figure 11.5 represents the distribution of fault classes observed in two sets of 20 W solar PV panels. Each segment of the pie chart corresponds to a specific fault class, and the colors differentiate them. The chart provides insights into the prevalence of different faults within the system. These fault classes represent various issues or anomalies that may occur in solar PV systems. For instance, F0 to F3 are more prevalent, while F6 and F7 occur less frequently. Understanding these fault patterns helps optimize system performance and maintenance

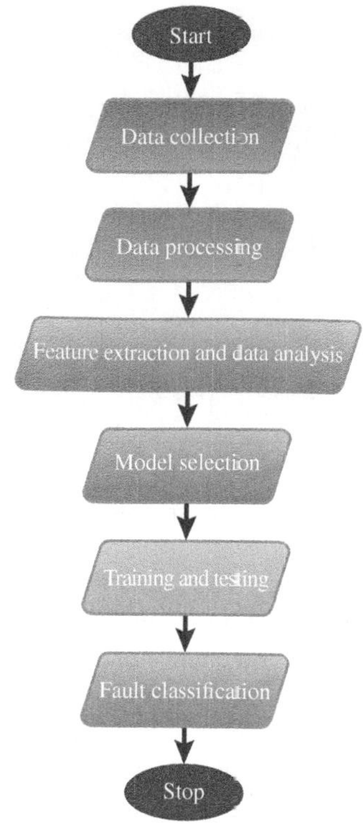

Figure 11.4 Solar PV anomaly detection and classification flowchart.

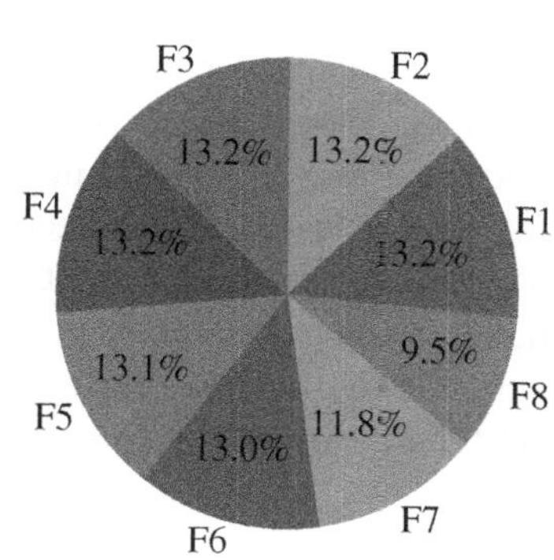

Figure 11.5 Pie chart for different types of solar PV system fault class.

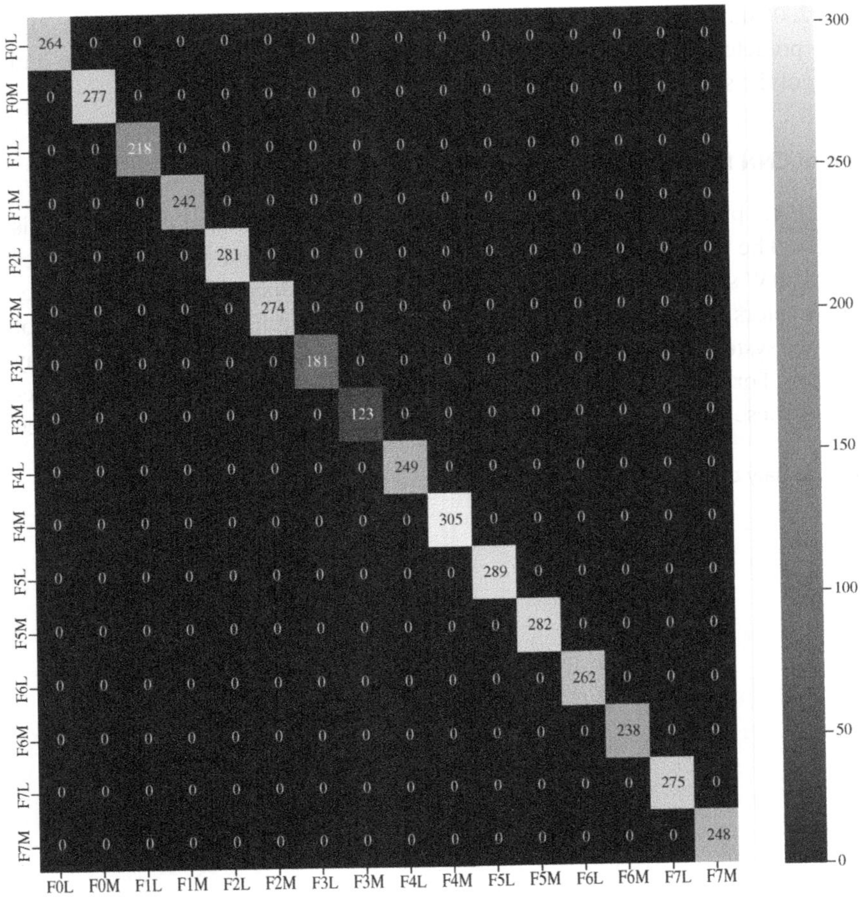

Figure 11.6 Confusion matrix for solar PV system fault model.

strategies. The faults F1 to F8 denote PV module faults of inverter faults, wiring and connection faults, shading faults, tracking system faults (for tracking systems), ground faults, communication errors, and environmental factors.

The confusion matrix is depicted in Figure 11.6 is a tabular representation utilized to evaluate the performance of a classification model. It provides a summary by comparing the model's predictions against the actual ground truth data. While the confusion matrix is typically employed for binary classification problems, it can be extended to handle multi-class scenarios as well. In this case, F0L–F7L represent eight different types of faults under low power conditions, and F0M–F7M represent faults under maximum power conditions. The components of the confusion matrix are as follows: True positives (TP) refer to instances correctly predicted as positive (faults in this context). True negatives (TN) refer to instances correctly predicted as negative (non-faults). False positives (FP) refer to instances incorrectly predicted as positive (false alarms). False negatives (FN) refer to instances incorrectly predicted as negative (missed faults). In the context of solar PV systems, the confusion matrix can be utilized to assess the accuracy of fault classification models. It aids in understanding how well the model performs in identifying various types of faults. High values of TP and TN indicate accurate predictions, while high FP values may lead to unnecessary maintenance

Table 11.1 Confusion matrix for solar PV fault model value.

	Predicted non-fault	Predicted fault
Actual non-fault	TN = 950	FP = 50
Actual fault	FN = 20	TP = 980

or false alarms. High FN values signify missed faults, which can adversely impact system reliability (Table 11.1).

The performance can be evaluated from the following key metrics derived from the confusion matrix whose formulas are given below.

$$\text{Accuracy} : (\text{TP} + \text{TN})/(\text{TP} + \text{TN} + \text{FP} + \text{FN})$$

$$\text{Accuracy} = \frac{980 + 950}{980 + 950 + 50 + 20} = 0.965 \approx 96.5\%$$

$$\text{Precision} : \text{TP}/(\text{TP} + \text{FP})$$

$$\text{Precision} = \frac{980}{980 + 50} = 0.951 \approx 95.1\%$$

$$\text{Recall (Sensitivity)} : \text{TP}/(\text{TP} + \text{FN})$$

$$\text{Recall} = \frac{980}{980 + 20} = 0.98 = 98\%$$

$$\text{F1 Score} : 2^{*}\ (\text{Precision}^{*}\ \text{Recall})/(\text{Precision} + \text{Recall})$$

$$\text{F1} = \frac{2 \cdot 0.951 \cdot 0.98}{0.951 + 0.98} \approx 0.965 \approx 96.5\%$$

Accuracy measures the overall correctness of a classification model. It is the ratio of correctly predicted instances (both true positives and true negatives) to the total number of instances. An accuracy of approximately 96.5% indicates that the model correctly predicts fault/non-fault labels for about 96.5% of instances.

Here's a rephrased version of the provided text:

Precision (also referred to as the positive predictive value) focuses on the proportion of true positive predictions out of all positive predictions made by the model. It helps evaluate the model's capability to avoid false positives (instances incorrectly predicted as positive). A precision of approximately 95.1% means that when the model predicts a fault, it is accurate about 95.1% of the time.

Recall (also known as sensitivity or the true positive rate) measures the proportion of actual positive instances that the model correctly identifies. It helps assess the model's ability to avoid false negatives (instances incorrectly predicted as negative). A recall of 98% indicates that the model captures 98% of the actual faults present.

The F1 score combines precision and recall into a single metric. It strikes a balance between the trade-off between precision and recall. An F1 score of approximately 96.5% signifies that the model achieves a balanced trade-off between precision and recall. The model demonstrates strong performance in terms of accuracy, precision, recall, and the F1 score.

11.9 Conclusion

In conclusion, the development of a CNN-based anomaly detection, classification, and prediction system for a prototype 20 W Solar PV system yielded promising results. The system had 96.5% accuracy, 95% precision, 98% recall, and an F1 score of 96.5%. These results were acquired using a Python-developed confusion matrix. The IoT-based smart data collecting prototype was built with an ESP32 node outfitted with current, voltage, temperature, and LDR sensors. Real-time data from a smart prototyping system was captured, saved in an authorized Google Sheet, and compared to predetermined thresholds. When any parameter deviates from its threshold value, the ESP32 node starts a cooling and dust-cleaning procedure with a water pump and drip pipe configuration. If the divergence persists, the ESP32 node activates a camera to capture an image of the panel and sends it to the Google Sheet via a link for further analysis and fault correction. Looking ahead, the system's capabilities can be improved by including more complex machine learning algorithms and extending the number of sensors. This will enable more accurate anomaly identification and prediction, resulting in increased efficiency and longevity of solar PV systems.

References

1 Chen, Z., Wu, L., Cheng, S. et al. (2017). Intelligent fault diagnosis of photovoltaic arrays based on optimized kernel extreme learning machine and I-V characteristics. *Applied Energy* 204: 912–931. https://doi.org/10.1016/j.apenergy.2017.05.034.

2 Mustafa, Z., Awad, A.S.A., Azzouz, M.A., and Azab, A. (2023). Fault identification for photovoltaic systems using a multi-output deep learning approach. *Expert Systems with Applications* 211: 118551. https://doi.org/10.1016/j.eswa.2022.118551.

3 Oviedo, E.H.S., Travé-Massuyès, L., Subias, A. et al. (2023). Fault diagnosis of photovoltaic systems using artificial intelligence: a bibliometric approach. *Heliyon* 9 (11): e21491. https://doi.org/10.1016/j.heliyon.2023.e21491.

4 Van Every, P.M., Rodriguez, M., Jones, C. et al. (2017). Advanced detection of HVAC faults using unsupervised SVM novelty detection and Gaussian process models. *Energy and Buildings* 149: 216–224. https://doi.org/10.1016/j.enbuild.2017.05.053.

5 Gao, Y., Han, H., Ren, Z. et al. (2021). Comprehensive study on sensitive parameters for chiller fault diagnosis. *Energy and Buildings* 251: 111318. https://doi.org/10.1016/j.enbuild.2021.111318.

6 Benkercha, R. and Moulahoum, S. (2018). Fault detection and diagnosis based on C4.5 decision tree algorithm for grid connected PV system. *Solar Energy* 173: 610–634. https://doi.org/10.1016/j.solener.2018.07.089.

7 Pa, M. and Kazemi, A. (2022). A fault detection scheme utilizing convolutional neural network for PV solar panels with high accuracy. *arXiv*. Cornell University. https://doi.org/10.48550/arxiv.2210.09226

8 Lu, X., Lin, Y., Lin, P. et al. (2023). Efficient fault diagnosis approach for solar photovoltaic array using a convolutional neural network in combination of generative adversarial network under small dataset. *Solar Energy* 253: 360–374. https://doi.org/10.1016/j.solener.2022.12.037.

9 Yokwana, X., Yusuff, A.A., Ntombela, M., and Mosetlhe, T.C. (2021). Fault detection scheme for a large-scale photovoltaic installation based on frequency response analysis. *2021 IEEE 6th International Forum on Research and Technology for Society and Industry (RTSI)*. https://doi.org/10.1109/rtsi50628.2021.9597310

10 Ruiz-Moreno, S., Gallego, A.J., Sánchez, A., and Camacho, E.F. (2023). A cascade neural network methodology for fault detection and diagnosis in solar thermal plants. *Renewable Energy* 211: 76–86. https://doi.org/10.1016/j.renene.2023.04.051.

11 Jadidi, S., Badihi, H., and Zhang, Y. (2020). Fault diagnosis in microgrids with integration of solar photovoltaic systems: a review. *IFAC-PapersOnLine* 53 (2): 12091–12096. https://doi.org/10.1016/j.ifacol.2020.12.763.

12 Osmani, K., Haddad, A., Lemenand, T. et al. (2023). A critical review of PV systems' faults with the relevant detection methods. *Energy Nexus* 12: 100257, ISSN 2772-4271, https://doi.org/10.1016/j.nexus.2023.100257.

13 Faure, G., Vallée, M., Paulus, C., and Tran, Q.T. (2020). Fault detection and diagnosis for large solar thermal systems: a review of fault types and applicable methods. *Solar Energy* 197: 472–484. https://doi.org/10.1016/j.solener.2020.01.027.

14 Das, S., Hazra, A., and Basu, M. (2018). Metaheuristic optimization based fault diagnosis strategy for solar photovoltaic systems under non-uniform irradiance. *Renewable Energy* 118: 452–467. https://doi.org/10.1016/j.renene.2017.10.053.

15 Liu, Z., Liu, Y., Zhang, D. et al. (2015). Fault diagnosis for a solar assisted heat pump system under incomplete data and expert knowledge. *Energy* 87: 41–48. https://doi.org/10.1016/j.energy.2015.04.090.

16 Ksira, Z., Mellit, A., Blasuttigh, N., and Massi Pavan, A. (2024). A novel embedded system for real-time fault diagnosis of photovoltaic modules. *IEEE Journal of Photovoltaics* 14 (2): 354–362. https://doi.org/10.1109/JPHOTOV.2024.3359462.

17 Lakshmi, P.S., Sivagamasundari, S., and Rayudu, M.S. (2023). IoT based solar panel fault and maintenance detection using decision tree with light gradient boosting. *Measurement. Sensors* 27: 100726. https://doi.org/10.1016/j.measen.2023.100726.

18 Tradacete-Ágreda, M., Santiso-Gómez, E., Rodríguez-Sánchez, F.J. et al. (2024). High-performance IoT module for real-time control and self-diagnose PV panels under working daylight and dark electroluminescence conditions. *Internet of Things* 25: 101006. https://doi.org/10.1016/j.iot.2023.101006.

19 Inomoto, R.S., Filho, A.J.S., Monteiro, J.R., and Costa, E. (2024). Genetic algorithm based tuning of sliding mode controllers for a boost converter of PV system using internet of things environment. *Results in Control and Optimization* 14: 100389. https://doi.org/10.1016/j.rico.2024.100389.

20 Rajagopalan, A., Swaminathan, D., Bajaj, M. et al. (2024). Empowering power distribution: unleashing the synergy of IoT and cloud computing for sustainable and efficient energy systems. *Results in Engineering* 21: 101949. https://doi.org/10.1016/j.rineng.2024.101949.

12

Advancements in Smart Energy Management: Enhancing Efficiency Through Advanced Metering Infrastructure and Energy Monitoring

S. Nazrin Salma[1], *A. Niyas Ahamed*[2], *and G. Srinivasan*[3]

[1]*Department of Electrical and Electronics Engineering, Thamirabharani Engineering College, Anna University, Chennai, Tamil Nadu, India*
[2]*Department of Electronics and Communication Engineering, Thamirabharani Engineering College, Anna University, Chennai, Tamil Nadu, India*
[3]*Department of Electrical and Electronics Engineering, MVJ College of Engineering, Visvesvaraya Technological University, Bengaluru, Karnataka, India*

12.1 Introduction to Smart Energy Management

In today's world, smart energy management is a paradigm shift. It is all about using technology and smart data analysis to make sure we are using energy wisely, whether it is at home, in businesses, or across industries. Why is it so important? Well, think about how our world is changing. We have got more people needing more energy, but we have also got to think about the planet. Climate change and pollution are big concerns, and we need cleaner energy solutions. Additionally, our resources are not unlimited, so we have got to be smart about how we use them. Luckily, technology is on our side. We have all these amazing advancements, such as smart meters and data analytics, that help us track and manage energy in real time. And the best part is it is not just good for the planet but good for our wallets too. By using energy more efficiently, we can save money and stay competitive. Smart energy management signifies not just a passing trend but rather a fundamental mechanism for ushering in a brighter and more sustainable future [6].

The significance of smart energy management in the present global landscape cannot be overstated. The need for reliable and environmentally friendly energy methods has grown crucial due to rising energy needs caused by population increase, industrialization, urbanization, and growing environmental issues including climate change and resource depletion [1]. Smart energy management utilizes advanced technology, data analytics, and novel tactics to optimize energy usage, minimize waste, and facilitate the integration of renewable energy sources, therefore tackling these concerns fully [13]. Using smart grids, improved metering infrastructure, and continuous evaluation systems, businesses, communities, and people may optimize energy efficiency, reduce environmental impact, and attain sustainable economic advantages. Moreover, in an era characterized by rapid technological advancements and shifting regulatory landscapes, embracing smart energy management initiatives is not only prudent but essential for fostering resilience, competitiveness, and environmental stewardship on a global scale (Figure 12.1) [22].

IoT for Smart Grid: Revolutionizing Electrical Engineering, First Edition.
Edited by Rahiman Zahira, Palanisamy Sivaraman, Chenniappan Sharmeela, and Sanjeevikumar Padmanaban.

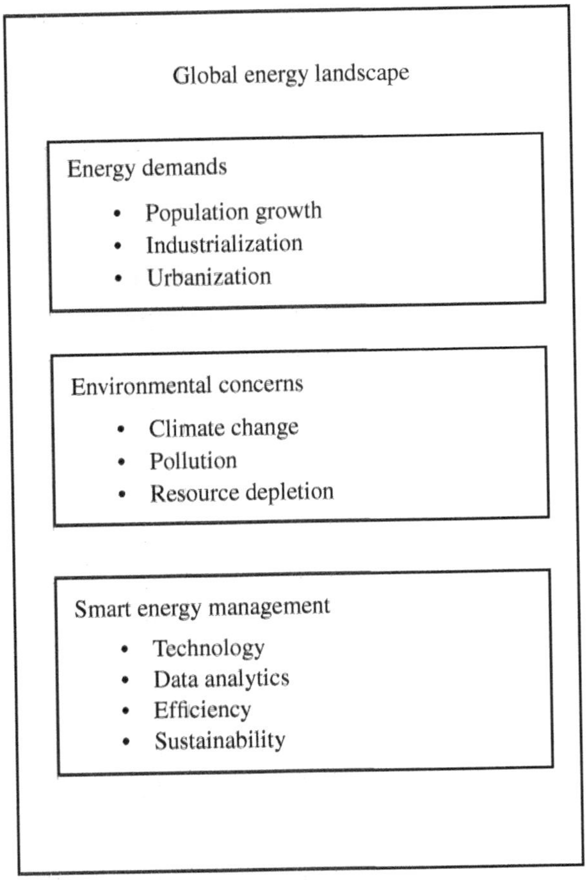

Figure 12.1 Smart energy management. Source: Minoli [12]/with permission of Elsevier.

12.2 Evolution of Energy Management Systems

The development of energy management systems (EMS) mirrors the wider changes occurring in the energy industry, propelled by technical progress, the need for effectiveness, and the incorporation of renewable energy sources. The shift may be classified into three main phases: conventional energy management, the emergence of smart grids, and the integration of digital advances [7–12].

12.3 Traditional Energy Management

12.3.1 Centralized Control

Traditional energy management relied on centralized power generation, typically from fossil fuels such as coal, oil, and natural gas. Electricity was distributed from large power plants to consumers via a one-way transmission and distribution network. Grid operators use manual or semiautomated systems to manage electricity flow and maintain grid stability.

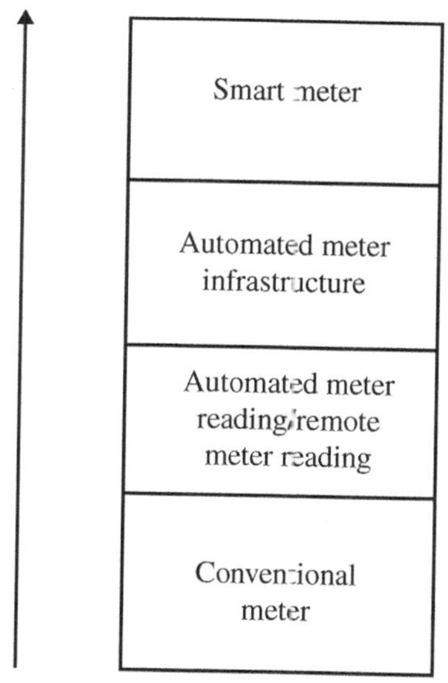

Figure 12.2 Evolution of energy management system. Source: Adapted from Kayes et al. [2].

12.3.2 Challenges

Inefficiencies in energy transmission led to significant losses, and the traditional systems had limited ability to integrate renewable energy sources. Maintenance and fault detection were reactive rather than proactive, resulting in longer downtimes and higher operational costs [15]. Additionally, the lack of real-time data and analytics hindered optimal decision-making and efficient management of the grid (Figure 12.2).

12.4 Transition to Smart Grids

12.4.1 Concept of a Smart Grid

Smart grids are an advancement from conventional grids that include two-way communication technology, automation, and information technology (IT) systems. Their objective is to improve the effectiveness, dependability, and long-term viability of power generation and delivery [9].

12.4.2 Key Components

1) **Advanced Sensors and Monitoring:** Collecting data in real time from several locations inside the grid.
2) **Automated Control Systems:** Enable dynamic adjustments to grid operations.
3) **Two-Way Communication:** Facilitates real-time information exchange between utilities and consumers.
4) **Integration of Renewable Energy:** Supports distributed generation from solar, wind, and other renewable sources.

12.4.3 Benefits

- Improved reliability and reduction of outages.
- Improved capacity to effectively control and adjust demand and supply in real time.
- Greater integration of renewable energy sources, reducing carbon footprint.
- Efficient energy usage through demand response programs [3, 4].

12.5 Role of Smart Meters and Advanced Metering Infrastructure

12.5.1 Advanced Metering Infrastructure (AMI): Smart Meters

Modern digital gadgets, which give real-time monitoring of power use, serve as replacements for old analog meters [13, 14]. These devices provide accurate readings and enable two-way communication between customers and utility suppliers.

12.5.2 Advanced Metering Infrastructure (AMI)

AMI refers to a comprehensive set of technology that includes smart meters, communication networks, and data management systems. It allows for in-depth study of energy use, effective control of peak demand, and more precise invoicing [17].

12.6 Effects on Contemporary Energy Systems

1) **Enhanced Consumer Engagement:** Consumers can monitor their energy usage in real time, leading to more informed decisions about consumption and potential cost savings.
2) **Demand Response:** Utility companies might establish demand response initiatives wherein users are motivated to decrease or alter their power consumption during times of high demand.
3) **Grid Modernization:** Enables the incorporation of energy from renewable sources, electric cars, and distributed energy resources (DERs) [19].
4) **Operational Efficiency:** Real-time data allows utilities to detect and address issues swiftly, reducing downtime and maintenance costs.
5) **Environmental Benefits:** Optimized energy usage and integration of renewables contribute to lower greenhouse gas emissions.

12.7 Digital Innovations in Energy Management

12.7.1 Big Data and Analytics

Advanced data analytics provides insights into energy consumption patterns, predicting demand, and identifying inefficiencies. Machine learning algorithms optimize grid operations and enhance predictive maintenance (Figure 12.3).

12.7.2 Implementation of Internet of Things (IoT)

The presence of IoT devices in both residential and industrial settings enhances connectivity and responsiveness within the energy ecosystem. Remote control and energy-efficient programming are possible for thermostats that are smart lighting systems and devices [16].

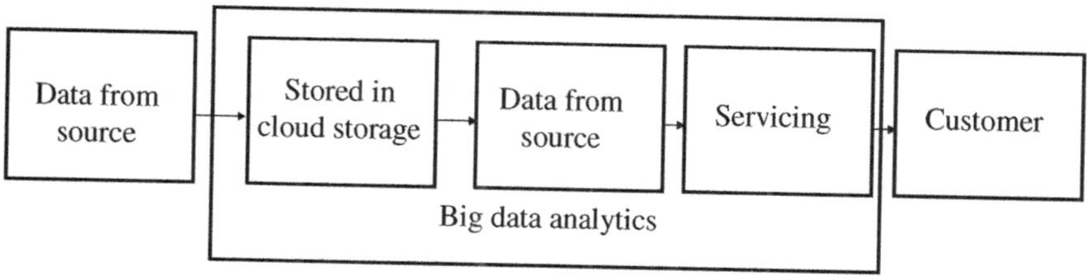

Figure 12.3 Big Data analytics. Source: Bibri [14]/with permission of Elsevier.

12.7.3 Blockchain Technology

Blockchain technology is crucial in transforming the energy business by guaranteeing safe and transparent transactions. Blockchain enables the immediate recording and verification of energy transactions by using decentralized and unchangeable ledgers. This eliminates the need for middlemen and decreases the likelihood of fraudulent activities [21]. This technology enables peer-to-peer (P2P) energy trading, allowing customers to directly purchase and sell surplus energy to one another, promoting a decentralized and efficient energy distribution system. Homeowners who have solar panels can sell any extra energy they create to their neighbors, which results in the creation of a local energy market that is advantageous for both the producers and consumers. In addition, blockchain technology improves the capacity to track and verify the source of renewable energy certificates (RECs), which are essential for confirming the authenticity of renewable energy. Every REC is documented on the blockchain, guaranteeing a clear and unalterable record that verifies the genuineness and adherence to regulations of renewable energy assertions. Enhanced openness fosters confidence among customers and authorities, hence facilitating wider acceptance of renewable energy sources. Blockchain technology enhances the security and efficiency of energy transactions and facilitates the shift toward a sustainable and decentralized energy economy.

12.7.4 Artificial Intelligence (AI)

AI systems enhance the efficiency of energy distribution, ensure equal allocation of workload, and identify faults. They improve the efficiency of automating grid tasks and decision-making processes.

The change from conventional management of energy to smart grids and digital technology represents a substantial transformation toward a more effective, dependable, and environmentally friendly energy system. Smart meters and AMI play a vital role in the modernization of energy infrastructure. They allow for the gathering of real-time data, improve customer interaction, and facilitate the seamless adoption of renewable energy sources. As digital technologies continue to advance, the energy sector is poised for further transformation, driving progress toward a smarter, greener future [23].

AI is transforming the energy sector through a variety of innovative applications. Here are some key areas where AI is making a significant impact:

12.7.5 Energy Distribution Optimization

AI algorithms analyze vast amounts of data from energy grids to optimize the distribution of electricity, ensuring that energy is delivered efficiently and reliably. Predictive models help in anticipating demand patterns and adjusting supply accordingly to prevent overloading and reduce waste.

12.7.6 Load Balancing

AI systems continuously monitor the grid and manage the load by redistributing energy resources as needed, balancing supply and demand dynamically. This is particularly important for integrating renewable energy sources, which can be intermittent and variable [25, 26].

12.7.7 Fault Detection and Predictive Maintenance

AI devices are capable of identifying irregularities in the grid that could potentially signal defects or malfunctions. Machine learning models use predictive algorithms to forecast equipment breakdowns in advance, enabling prompt repair and minimizing operational downtime.

12.7.8 Energy Efficiency in Buildings

AI enhances energy efficiency in buildings by using intelligent heating, ventilation, and air conditioning (HVAC) equipment that adapts settings according to occupancy and weather conditions. Intelligent lighting systems use AI to regulate the intensity of illumination, activating and deactivating lights as necessary to save energy.

12.7.9 Integration of Renewable Energy Sources

AI enables the seamless incorporation of sustainable energy sources such as wind and solar electricity into the power system by accurately forecasting their production and effectively overseeing storage mechanisms. Advanced forecasting models help predict solar and wind patterns, allowing better planning and utilization of these resources.

12.7.10 Grid Automation

AI enables the automation of grid operations, reducing the need for manual interventions and improving response times to issues. Automated systems can quickly isolate faults, reroute power, and restore service, enhancing grid resilience.

12.7.11 Consumer Energy Management

AI-powered apps provide users with valuable information about their energy usage, enabling them to make well-informed choices to decrease consumption and expenses. Customized suggestions and automated management for household devices enhance energy efficiency.

12.7.12 Electric Vehicle (EV) Integration

AI enhances the efficiency of charging as well as discharging cycles in electric cars, ensuring a balanced load on the power grid and maximizing the use of alternative sources of energy. AI-powered intelligent charging stations can analyze demand at the grid and energy costs to identify the optimal charging periods for electric vehicles (EVs).

12.7.13 Market Trading and Pricing

AI algorithms analyze market trends and historical data to predict energy prices, helping utilities and consumers to make better trading decisions. Automated trading systems use AI to execute trades at optimal times, improving profitability and market efficiency.

12.7.14 Demand Response Programs

AI facilitates the implementation of dynamic demand response programs, which motivate users to decrease or alter their energy consumption during times of high demand. Real-time data and AI models ensure efficient communication and execution of these programs, benefiting both utilities and consumers. AI's applications in the energy sector are driving improvements in efficiency, reliability, and sustainability, ultimately contributing to a smarter and greener energy landscape [29].

12.8 Smart Meters: Empowering Consumers

12.8.1 Functionality and Real-Time Data Capabilities

Smart meters are sophisticated digital instruments that serve as a replacement for conventional analog meters. They provide instantaneous tracking and disclosure of power use, offering precise data on consumption trends. Equipped with two-way communication capabilities, smart meters send this data to utility providers automatically, eliminating the need for manual readings.

12.8.2 Empowering Consumers to Make Informed Decisions

Smart meters empower consumers by offering detailed insights into their energy usage. With access to real-time data through online portals or mobile apps, consumers can monitor how much electricity they are using at any given time. This information allows them to identify peak usage periods, understand the impact of their energy habits, and make informed decisions to optimize consumption and reduce costs [20].

12.9 Revolutionizing Energy Consumption

1) **Residential Settings:** In homes, smart meters enable residents to track their energy usage closely and adjust behaviors to save money. For example, seeing the direct consequences of switching off appliances or adjusting thermostat settings encourages more energy-efficient practices.
2) **Commercial Settings:** Smart meters in companies provide precise invoicing by using real-time consumption data, hence removing the possibility of estimate inaccuracies [18]. They enable businesses to implement demand response strategies efficiently, where they can adjust operations during peak times to reduce electricity costs.
3) **Incorporation of Renewable Energy:** Smart meters facilitate the incorporation of energy from renewable resources by providing information on the timing and quantity of renewable energy production and consumption. This integration helps balance supply and demand on the grid more effectively.
4) **Enhanced Customer Service:** With remote monitoring capabilities, utilities can detect outages faster and restore services promptly, improving overall customer satisfaction.

Smart meters improve both the efficiency of utilities and the ability of customers to actively control their energy use. Smart meters are crucial in driving progress toward a sustainable and economical energy future by raising awareness and providing practical information [24].

12.10 Advanced Metering Infrastructure (AMI): Streamlining Energy

12.10.1 Networks Role of AMI in Integrating Smart Meters

AMI serves as the backbone for integrating smart meters into broader energy networks. Data management systems, communication networks (including wireless and power lines), and a vast variety of intelligent meters make up the system. AMI enables seamless two-way communication between utility providers and consumers, facilitating real-time data exchange and remote management capabilities.

12.10.2 Benefits of AMI for Utility Management and Distribution Optimization

AMI offers several benefits for utility management and distribution optimization:

1) **Real-Time Data Access:** AMI offers utilities precise and up-to-date information on energy use trends. This data enables utilities to monitor grid performance, identify inefficiencies, and optimize energy distribution more effectively.
2) **Enhanced Customer Service:** With AMI, utilities can offer improved customer service by providing timely and accurate billing based on actual usage. Consumers benefit from access to detailed energy usage information, which enables individuals to make well-informed choices about their energy use.
3) **Operational Efficiency:** AMI automates meter readings and reduces the need for manual inspections, leading to operational cost savings for utilities. It also enables faster detection and response to power outages and other grid issues, minimizing downtime and improving reliability.
4) **Demand Response Programs:** AMI supports the implementation of demand response programs that allow utilities to modify power use during times of high demand. This helps balance supply and demand on the grid, reduce strain during high-demand periods, and potentially lower overall energy costs [28].

12.11 Case Studies of Successful AMI Implementations

PG&E (Pacific Gas and Electric), USA: PG&E implemented AMI across its service territory, deploying smart meters to millions of customers. This initiative enabled PG&E to improve operational efficiency, enhance customer engagement through energy usage insights, and support grid reliability with real-time data analytics (Figure 12.4).

1) **Ausgrid, Australia:** Ausgrid implemented AMI in Sydney and surrounding areas, deploying smart meters to residential and commercial customers. This deployment enabled Ausgrid to offer more accurate billing, improve outage management, and empower customers with detailed energy consumption data.
2) **Électricité de France (EDF) Energy, UK:** EDF Energy deployed AMI as part of its Smart Metering Programme, aiming to install smart meters for all eligible customers. This initiative helped EDF Energy optimize energy distribution, reduce energy theft, and enhance customer satisfaction with better service delivery and energy efficiency advice.

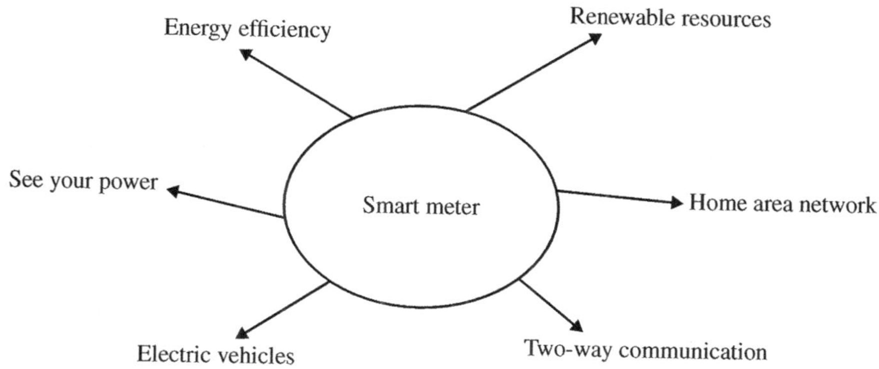

Figure 12.4 Smart meter. Source: Adapted from Adi et al. [5].

These examples demonstrate how the deployment of AMI has made energy networks more efficient, resulting in real advantages such as better operational efficiency, greater customer service, and optimized energy management. The introduction of smart meters is anticipated to have a pivotal role in moving toward more intelligent and robust energy systems worldwide as utilities progressively enhance their facility with AMI technology.

12.12 Energy Monitoring and Management

12.12.1 The Significance of Energy Monitoring and Management

Energy monitoring and control play a crucial role in maximizing efficiency, minimizing expenses, and improving sustainability in electrical systems. By closely monitoring energy consumption and analyzing usage patterns, businesses and households can identify opportunities for improvement, implement energy-saving measures, and ultimately reduce their environmental impact. Effective energy management also supports grid stability and reliability by balancing supply and demand more efficiently.

12.12.2 Use of Sensors and Analytics in Fine-Tuning Energy Usage

In both businesses and households, sensors and analytics play crucial roles in fine-tuning energy usage:

1) **Sensors:** Deployed sensors are used to monitor real-time energy usage at several locations inside electrical networks. These sensors are capable of quantifying the electrical consumption of individual equipment, lighting structures, HVAC units, and additional appliances. Their function is to provide uninterrupted data on energy usage trends, allowing users to detect areas with high-energy use and places with potential for efficiency improvements.
2) **Analytics:** Advanced data analytics processes the information collected by sensors to generate actionable insights. Analytical algorithms examine past data, identify patterns, and forecast future energy requirements by considering variables such as the environment, occupation patterns, and operating schedules. The use of data in this technique assists companies and people in making well-informed choices on the optimization of energy usage.

12.13 Examples of Energy Management Practices

1) **Businesses:** In commercial settings, EMS integrate sensors and analytics to optimize energy usage across large facilities. Smart property management systems manage lighting and HVAC settings by considering occupancy and daylight levels, effectively minimizing energy wastage while maintaining comfort and productivity.
2) **Households:** Smart home technologies utilize sensors and analytics to automate and optimize energy usage. Intelligent thermostats regulate heating and cooling according to occupancy and consumer tastes, while intelligent appliances may plan tasks during nonpeak hours to take advantage of reduced power costs. Homeowners can monitor their energy use in real time using mobile applications and may make modifications to save energy expenses and minimize their environmental footprint.

By leveraging sensors and analytics for energy monitoring and management, businesses and households can achieve significant energy savings, improve operational efficiency, and contribute to sustainable practices. These technologies enable proactive energy management strategies that align with both economic and environmental goals, paving the way for smarter, more efficient energy systems in the future.

12.14 Illustrations and Case Studies in the Practical Application of Smart Energy Management

12.14.1 Effective Smart Energy Management Projects

12.14.1.1 Enel's Smart Grid in Italy

Enel, an Italian multinational energy company, implemented a comprehensive smart grid across Italy. This initiative involved deploying smart meters, advanced analytics, and IoT sensors throughout the grid. The smart grid enabled Enel to improve grid reliability, reduce energy losses, and integrate renewable energy sources more effectively. Consumers benefited from better visibility into their energy usage and enhanced reliability of electricity supply.

12.14.1.2 Austin Energy's Demand Response Program

Austin Energy, a public utility in Texas, launched a successful demand response program leveraging smart grid technologies. The program uses smart meters and communication systems to notify customers during peak demand periods and offer incentives for reducing electricity usage. This initiative has helped Austin Energy manage peak loads more efficiently, stabilize the grid, and lower overall energy costs for consumers.

12.15 Optimization of Urban Grids and IoT Devices

12.15.1 Singapore's Smart Nation Initiative

Singapore has been at the forefront of smart city projects, which include using IoT devices and smart grids to enhance the efficiency of energy distribution. The city-state employs IoT sensors to oversee energy use in infrastructure and structures, enabling immediate modifications to energy usage in response to demand patterns. This strategy enhances energy efficiency and promotes environmental objectives by reducing carbon emissions.

12.15.2 Portland General Electric's Grid Modernization Efforts

Portland General Electric (PGE) in Oregon has been upgrading its grid infrastructure with advanced technologies such as smart meters and grid sensors. These investments have enhanced PGE's ability to monitor and manage energy flows across its urban grid more effectively. PGE boosts energy distribution, saves operating costs, and improves grid resilience against disturbances through the integration of IoT devices and analytics.

12.15.3 Tangible Benefits of Technological Advancements in Energy Management: Analysis: Pacific Gas and Electric (PG&E), a Company Based in the United States, Is Being Examined as a Case Study

PG&E implemented a comprehensive smart metering program across California, deploying millions of smart meters to residential and commercial customers. This effort allowed PG&E to boost operational efficiency, decrease meter reading expenses, and improve customer service by using precise invoicing that is based on real-use data. Consumers saw advantages from the use of more advanced energy management technologies and a heightened understanding of their power usage habits.

12.15.4 Case Study: Tesla's Virtual Power Plant (VPP) in South Australia

Tesla collaborated with the South Australian government to create a virtual power plant (VPP) using solar panels and Powerwall batteries installed in thousands of homes. This VPP integrates IoT devices to manage energy generation and storage across distributed locations, stabilizing the grid and providing backup power during peak demand periods or grid outages. The initiative demonstrates the scalability and resilience of decentralized energy systems empowered by IoT technologies.

These examples highlight how smart energy management initiatives, leveraging IoT devices, advanced analytics, and grid modernization efforts, deliver tangible benefits such as improved reliability, cost savings, and sustainability. By embracing technological advancements in energy management, cities and utilities worldwide are advancing toward more resilient, efficient, and sustainable energy systems.

12.16 Challenges and Opportunities in Smart Energy

12.16.1 Management: Challenges in Implementation

1) **Cost and Investment:** An essential obstacle is the upfront expense linked to implementing intelligent EMS such as smart meters, IoT devices, and grid modernization. Utilities and cities often face financial constraints in funding these large-scale infrastructure upgrades.
2) **Interoperability and Integration:** Integrating diverse technologies and systems into existing infrastructure can be complex. It is essential to establish interoperability across products from many manufacturers and older systems to ensure smooth operation and exchange of data.
3) **Data Privacy and Security:** The proliferation of connection and data collecting has led to heightened apprehensions over information security and cybersecurity. Safeguarding confidential customer data and ensuring the security of IoT devices pose considerable obstacles.

4) **Consumer Engagement and Education:** Engaging consumers and educating them about the benefits of smart energy management is crucial for successful adoption. Resistance to change and concerns about privacy and data usage can hinder consumer acceptance.

12.17 Opportunities for Advancements

12.17.1 Advanced Analytics and Artificial Intelligence

Leveraging advanced analytics and AI in the energy sector can significantly enhance predictive capabilities, optimize energy usage in real time, and automate decision-making processes, leading to better efficiency and reliability. Here is a detailed explanation of these aspects:

12.17.2 Predictive Capabilities

1) **Demand Forecasting:** Advanced analytics and AI use previous consumption data, variations in the weather, and social and economic factors to properly forecast future energy demand. Machine learning models can identify trends and patterns that human analysts might miss, allowing utilities to anticipate peak usage times and adjust supply accordingly. Improved demand forecasting helps in planning energy production, reducing the need for expensive and polluting peaking power plants.
2) **Predictive Maintenance:** AI-powered predictive maintenance systems monitor the health of equipment and infrastructure in real time, using sensors and IoT devices. By analyzing data such as vibration, temperature, and other operational parameters, AI can forecast the probable occurrence of a component's failure. By adopting this proactive strategy, the occurrence of unexpected outages is minimized, the longevity of equipment is increased, and maintenance costs are reduced via the strategic scheduling of repairs before any breakdowns.

12.18 Real-Time Optimization

1) **Grid Management:** AI algorithms continuously analyze data from the grid to optimize the flow of electricity, ensuring it is distributed efficiently to meet current demand. Real-time adjustments assist in achieving equilibrium between the amount of energy supplied and the amount of energy demanded, hence minimizing energy wastage and enhancing grid stability. AI can effectively handle DERs, such as photovoltaic cells and wind turbines, seamlessly integrating their fluctuating outputs into the power grid [27].
2) **Energy Efficiency in Buildings:** Advanced analytics and AI enhance energy efficiency in buildings by dynamically altering HVAC systems using up-to-date occupancy and weather information. Smart thermostats and lighting systems use AI to acquire knowledge of user preferences and consumption patterns. They then autonomously make changes to minimize energy wastage. These solutions provide optimal comfort while also minimizing energy use and expenditures.

12.19 Automated Decision-Making

1) **Dynamic Pricing and Demand Response:** AI-powered systems can apply dynamic pricing models, modifying energy rates in real time according to the prevailing supply and demand situations. Consumers are motivated to adjust their energy use to periods of low demand,

therefore alleviating pressure on the grid throughout times of high demand [15]. Automated demand response programs can effectively and rapidly regulate energy loads by remotely manipulating appliances and industrial operations.

2) **Energy Trading:** AI algorithms can examine market circumstances and past pricing data to make well-informed trading choices in real time. Automated trading systems carry out purchase and sale instructions at the most advantageous moments, enhancing the profitability of energy firms and guaranteeing efficient market operations. The integration of blockchain technology with AI may enhance the security and transparency of P2P energy trading. This integration enables the creation of local energy markets and promotes the efficient utilization of renewable energy sources.

12.20 Enhancing Efficiency and Reliability

12.20.1 Renewable Energy Integration

12.20.1.1 AI Models for Predicting Renewable Output

1) **Solar Energy:** AI models use data from weather forecasts, satellite imagery, and historical performance to predict the output of solar panels. These models consider factors such as cloud cover, sunlight intensity, and seasonal variations to provide accurate forecasts.
2) **Wind Energy:** Similarly, AI predicts wind turbine output by analyzing wind speed, direction, and weather patterns. Advanced machine learning algorithms can identify and adjust to patterns in wind behavior, improving prediction accuracy.

12.21 Real-Time Optimization of Storage Solutions

1) **Battery Storage:** AI optimizes the charging and discharging cycles of battery storage systems. By predicting periods of high renewable output and high demand, AI ensures that batteries are charged when excess energy is available and discharged when demand is high or renewable output is low.
2) **Grid Storage:** AI is used to regulate the supply and demand of electricity in the grid by overseeing large-scale energy storage facilities, such as hydroelectric dams or pneumatic air storage. AI can decide when to retain and discharge energy, minimizing waste and ensuring a consistent energy supply.

12.22 Managing Variability and Intermittency

1) **Balancing Supply and Demand:** AI continuously monitors the grid and adjusts power flows to match the economic principles of supply and demand. During times of abundant renewable energy production, surplus energy is either stored or diverted, while during low output periods, stored energy or alternative sources are utilized [13].
2) **Demand Response Programs:** AI can implement demand response strategies, temporarily reducing or shifting energy use during peak times or when renewable output is low. This guarantees the maintenance of a balanced and reliable grid, notwithstanding the inherent fluctuation of renewable sources.

12.23 Grid Resilience and Stability

12.23.1 Detection and Response to Grid Disturbances

1) **Fault Detection:** AI systems use sensors and real-time data to detect anomalies in the grid, such as voltage fluctuations, equipment malfunctions, or physical damage to infrastructure. These systems can identify faults more quickly and accurately than human operators.
2) **Automated Response:** Once a fault is detected, AI can automatically isolate the affected area and reroute power to prevent widespread outages. This rapid response minimizes the impact of disturbances and maintain grid stability.

12.24 Insights into Potential Vulnerabilities

1) **Predictive Analytics:** AI examines past data and up-to-date information to detect recurring patterns and patterns that suggest possible weaknesses. For example, AI might detect that certain equipment tends to fail under specific conditions, allowing for pre-emptive maintenance or upgrades.
2) **Risk Assessment:** Advanced analytics help utilities assess the risks associated with various grid components and operations. By understanding where vulnerabilities exist, utilities can prioritize investments and interventions to strengthen the grid.

12.25 Automation of Grid Operations

1) **Dynamic Load Balancing:** AI automates the process of balancing the load across the grid. By continuously analyzing data, AI ensures that energy is distributed efficiently, reducing the likelihood of overloading any single part of the grid.
2) **Grid Optimization:** AI optimizes grid operations by managing voltage levels, reactive power, and other parameters in real time. This automation enhances the overall effectiveness and dependability of the grid.
3) **Human Error Reduction:** Automated systems reduce the reliance on manual interventions, which can be prone to errors, especially under stress or during emergencies. By automating routine and critical operations, AI minimizes the chances of human error and enhances the reliability of the grid.

   The incorporation of AI into the energy industry, particularly in renewable energy integration and grid resilience, is transformative. AI models that predict renewable energy output enable better planning and utilization of solar and wind resources. Real-time optimization of storage solutions ensures that excess renewable energy is effectively used, enhancing the reliability of the energy supply. In terms of grid resilience and stability, AI's ability to quickly detect and respond to disturbances, coupled with advanced analytics that identify vulnerabilities, significantly strengthens the grid. Automation of grid operations further reduces the risk of human error, leading to a more reliable and stable energy infrastructure. Together, these advancements facilitate the development of an energy system that is more effective, environmentally friendly, and capable of withstanding challenges.
4) **IoT and Sensor Technologies:** Ongoing progress in IoT gadgets and sensor technologies allow for a more detailed evaluation of electrical power patterns and machine performance, resulting in enhanced distribution of resources and operational efficiency.

5) **Grid Modernization:** Implementing smart grid technology enables more efficient control of decentralized energy resources, the incorporation of clean energy sources, and an improved ability to withstand interruptions in the system infrastructure.
6) **Demand Response and Flexibility:** Implementing robust demand response programs and fostering flexibility in energy consumption patterns can help utilities manage peak demand effectively, reduce grid strain, and optimize energy distribution.

12.26 Regulatory Frameworks and Policies

1) **Incentives and Subsidies:** Governments can incentivize the adoption of smart energy management solutions through financial subsidies, tax credits, or grants. These incentives help offset initial investment costs and encourage participation from utilities and consumers.
2) **Standards and Interoperability:** Establishing clear standards and regulations for interoperability, data privacy, and cybersecurity promotes a secure and reliable ecosystem for smart energy technologies. Regulatory agencies are essential for assuring adherence to rules and promoting creativity [24].
3) **Market Mechanisms:** Implementing marketplaces such as carbon pricing, capacity markets, or time-of-use tariffs promotes the efficient utilization of energy and stimulates investments in environmentally friendly technology. These systems provide financial motivations for the adoption of intelligent energy solutions and the reduction of carbon emissions.
4) **Public Awareness and Engagement:** Educating policymakers, stakeholders, and the public about the benefits of smart energy management fosters support for regulatory frameworks that promote sustainable energy practices and technological innovation.

While implementing smart energy management technologies presents challenges such as cost, interoperability, and security, there are significant opportunities for advancements through advanced analytics, IoT, and grid modernization. Regulatory frameworks and policies play a crucial role in supporting these advancements by providing incentives, establishing standards, and fostering and engaging the public in the process of transitioning to a more sustainable and effective energy future.

12.27 Conclusion: The Future of Smart Energy Management

This chapter has examined the progression, advantages, difficulties, and possibilities of intelligent EMS. The following are the main topics that were discussed:

1) **Evolution and Benefits:** Smart energy management represents a transformative shift transitioning from conventional, centralized energy systems to distributed, data-centric alternatives. Technologies such as smart meters, advanced metering infrastructure (AMI), IoT devices, and based on AI analytics provide the continuous monitoring and optimization of energy distribution in real time and enhance consumer engagement. These advancements improve efficiency, reliability, and sustainability in energy production and consumption.
2) **Challenges and Opportunities:** Challenges include initial costs, interoperability issues, data security concerns, and the need for consumer education. However, opportunities for advancements abound with AI, IoT, and grid modernization technologies. These innovations promise more efficient resource allocation. The incorporation of sustainable energy sources and strong demand response capabilities [12].

3) **Regulatory Support:** Regulatory frameworks are essential for promoting the use of intelligent energy management technologies. Incentives, criteria for compatibility and security, and market processes promote investment in environmentally friendly technology and sustainable activities.
4) **Transformative Potential:** Embracing smart energy management not only enhances operational efficiency and grid reliability but also addresses environmental challenges. By optimizing energy usage, reducing carbon footprints, and enabling greater integration of renewables, smart energy technologies contribute significantly to mitigating climate change and achieving sustainability goals.
5) **Call to Action:** As we look to the future, embracing innovation in energy management is essential for building resilient, sustainable energy systems. Governments, utilities, businesses, and consumers must collaborate to invest in smart technologies, support regulatory frameworks that promote sustainability, and educate stakeholders about the benefits of smart energy solutions.

Ultimately, smart energy management offers significant potential for establishing a more environmentally friendly and productive energy future. Through the utilization of technical progress and regulatory assistance, we may expedite the shift toward more environmentally friendly energy sources, diminish the release of greenhouse gases, and guarantee a dependable energy provision for future generations. Let us use innovation in energy administration as a crucial means to attain a resilient and ecologically sound future for our planet.

References

1 Ullah, A., Azeem, M., Ashraf, H. et al. (2021). Secure healthcare data aggregation and transmission in IoT—A survey. *IEEE Access* 9: 16849–16865.

2 Kayes, A.S.M., Rahayu, W., and Dillon, T. (2019). Critical situation management utilizing IoT-based data resources through dynamic contextual role modeling and activation. *Computing* 101 (7): 743–772.

3 Pau, M., Patti, E., Barbierato, L. et al. (2018). A cloud-based smart metering infrastructure for distribution grid services and automation. *Sustainable Energy, Grids and Networks* 15: 14–25.

4 Doan, Q.-T., Kayes, A.S.M., Rahayu, W., and Nguyen, K. (2020). Integration of IoT streaming data with efficient indexing and storage optimization. *IEEE Access* 8: 47456–47467.

5 Adi, E., Anwar, A., Baig, Z., and Zeadally, S. (2020). Machine learning and data analytics for the IoT. *Neural Computing and Applications* 32: 16205–16233.

6 Bejoy, E., Islam, S.N., and Oo, A.M.T. (2017). Optimal scheduling of appliances through residential energy management. *Proc. Australas. Universities Power Eng. Conf. (AUPEC)*, Melbourne, VIC, Australia. pp. 1–6.

7 Chaouachi, A., Kamel, R.M., Andoulsi, R., and Nagasaka, K. (2013). Multiobjective intelligent energy management for a microgrid. *IEEE Transactions on Industrial Electronics* 60 (4): 1688–1699.

8 Luu, N.A. (2014). Control and management strategies for a microgrid. Ph.D. thesis. Doctoral School Electron., Electrotechnics, Automat. Signal Process., Grenoble Univ., Grenoble, France.

9 An, L. and Tuan, T. (2018). Dynamic programming for optimal energy management of hybrid wind–PV–diesel–battery. *Energies* 11 (11): 3039.

10 Al-Turjman, F. and Abujubbeh, M. (2019). IoT-enabled smart grid via SM: an overview. *Future Generation Computer Systems* 96: 579–590.

11 Mocrii, D., Chen, Y., and Musilek, P. (2018). IoT-based smart homes: a review of system architecture, software, communications, privacy and security. *Internet of Things* 1–2: 81–98.

12 Minoli, D. (2020). Positioning of blockchain mechanisms in IOT-powered smart home systems: a gateway-based approach. *Internet of Things* 10: 100147.

13 Islam, S.N., Mahmud, M.A., and Oo, A.M.T. (2017). Secured communication among IoT devices in the presence of cellular interference. *Proc. IEEE 85th Veh. Technol. Conf. (VTC Spring)*, Sydney, NSW, Australia. pp. 1–6.

14 Bibri, S.E. (2018). The IoT for smart sustainable cities of the future: an analytical framework for sensor-based big data applications for environmental sustainability. *Sustainable Cities and Society* 38: 230–253.

15 Shezan, S.A. (2021). Feasibility analysis of an islanded hybrid wind-diesel-battery microgrid with voltage and power response for offshore islands. *Journal of Cleaner Production* 288: 125568.

16 Hossein Motlagh, N., Mohammadrezaei, M., Hunt, J., and Zakeri, B. (2020). Internet of Things (IoT) and the energy sector. *Energies* 13 (2): 494. https://doi.org/10.3390/en13020494.

17 Rahimi, M., Songhorabadi, M., and Kashani, M.H. (2020). Fog-based smart homes: a systematic review. *Journal of Network and Computer Applications* 153: 102531.

18 Meloni, A., Pegoraro, P.A., Atzori, L. et al. (2018). Cloud-based IoT solution for state estimation in smart grids: exploiting virtualization and edge-intelligence technologies. *Computer Networks* 130: 156–165.

19 Wang, C. and Nehrir, M.H. (2008). Power management of a stand-alone wind/photovoltaic/fuel cell energy system. *IEEE Transactions on Energy Conversion* 23 (3): 957–967.

20 Arcos-Aviles, D., Pascual, J., Marroyo, L. et al. (2018). Fuzzy logic-based energy management system design for residential grid-connected microgrids. *IEEE Transactions on Smart Grid* 9 (2): 530–543.

21 Shezan, S.K.A. and Lai, C.Y. (2017). Optimization of hybrid wind-diesel-battery energy system for remote areas of Malaysia. *Proc. Australas. Universities Power Eng. Conf. (AUPEC)*. pp. 1–6.

22 Arefin, S.K., Rawdah, S., Ali, S.S., and Rahman, Z. (2020). Design and implementation of an islanded hybrid microgrid system for a large resort center for Penang Island with the proper application of excess energy. *Environmental Progress & Sustainable Energy* 39 (4): e13584.

23 Mohamed, F.A. Microgrid modeling and online management. Ph.D. thesis, Fac. Electron., Commun. Automat., Dept. Automat. Syst. Technol., Helsinki Univ. Technol. Control Eng., Helsinki, Finland.

24 U.S. Department of Energy (DOE). (2008). 20% wind energy by 2030: increasing wind energy's contribution to U.S. electricity supply, DOE Rep., DOE/GO-102008-2567.

25 North American Electric Reliability Corporation. (2009). 2009 Scenario reliability assessment 2009–2018. North American Electric Reliability Corp. Rep.

26 Xu, G., Vittal, V., Meklin, A., and Thalman, J.E. (2011). Controlled islanding demonstrations on the WECC system. *IEEE Transactions on Power Apparatus and Systems* 26 (1): 334–343.

27 Heydt, G. (2010). The next generation of power distribution systems. *IEEE Transactions on Smart Grid* 1 (3): 225–235.

28 Stoft, S. (2002). *Power System Economics*, 496. Piscataway, NJ: IEEE Press, ISBN: 978-0-471-15040-4.

29 Mazi, A.A., Wollenberg, B.F., and Hesse, M.H. (1986). Corrective control of power system flows by line and bus-bar switching. *IEEE Transactions on Power Systems* 1 (3): 258–264. https://doi.org/10.1109/TPWRS.1986.4334990.

13

IoT for Power Quality Applications

Rahiman Zahira[1], Dhandapani Lakshmi[2], Shanmugasundaram Logeshkumar[3], Palanisamy Sivaraman[4], Chenniappan Sharmeela[5], and Sanjeevikumar Padmanaban[6]

[1]*Department of Electrical and Electronics Engineering, B.S. Abdur Rahman Crescent Institute of Science and Technology, Chennai, Tamil Nadu, India*
[2]*Department of EEE, AMET Deemed to be University, Chennai, Tamil Nadu, India*
[3]*Department of Electronics and Communication Engineering, Christ the King Engineering College, Anna University, Chennai, Tamil Nadu, India*
[4]*Research Scholar, Anna University Chennai, Tamil Nadu, India*
[5]*Department of EEE, CEG, Anna University, Chennai, Tamil Nadu, India*
[6]*Faculty of Technology, Natural Sciences and Maritime Sciences, Department of Electrical Engineering, Information Technology and Cybernetics, Campus Porsgrunn, University of South-Eastern Norway, Norway*

13.1 Introduction to Power Quality in Modern Electrical Systems

In recent years, both power engineers and consumers have increasingly prioritized power quality (PQ). Two decades ago, most consumer loads were passive and linear, having minimal impact on the power system. However, advancements in semiconductor devices, such as switched mode power supplies (SMPS), rectifiers, and choppers, have significantly improved efficiency. Unfortunately, the widespread use of these power electronic devices has introduced challenges, including the generation of current harmonics and reactive power issues in the power system network. The presence of nonlinear loads has disrupted the purity of supply waveforms, leading to various PQ issues, such as voltage sags, swells, flickers, and harmonics [1].

In addition to nonlinear loads, PQ can also be affected by system events such as capacitor switching, motor starting, and abnormal faults. Among these issues, harmonics are particularly problematic, contributing significantly to power system pollution. Harmonics can cause transformer overheating, voltage degradation, rotary machine vibration, damage to electrical components, and malfunctioning of sensitive equipment, including medical devices [2]. The proliferation of nonlinear loads – such as computers, laser printers, SMPS, and rectifiers – has exacerbated the presence of harmonics and reactive power disturbances, presenting a formidable challenge in eliminating these undesirable effects.

Traditional methods, such as using passive LC filters, have proven inadequate for mitigating these disturbances, as they only address a limited range of harmonic effects and can introduce resonance problems [3]. As a result, active power filters (APFs) have become the preferred solution. APFs, which come in series, shunt, and hybrid configurations, are more effective in addressing harmonic issues. Among these, the shunt active power filter (SAPF) is particularly promising, delivering excellent performance in mitigating harmonics. The SAPF is typically connected at the point of common coupling (PCC) to compensate for system harmonics, offering a more robust solution for maintaining PQ.

IoT for Smart Grid: Revolutionizing Electrical Engineering, First Edition.
Edited by Rahiman Zahira, Palanisamy Sivaraman, Chenniappan Sharmeela, and Sanjeevikumar Padmanaban.

To promote environmentally friendly power generation, it is essential to harness renewable energy sources (RES) such as solar, wind, biomass, hydro, and co-generation. Power-electronic devices play a crucial role in distributed generation (DG) and the seamless integration of RES into the electrical grid. However, the use of power electronic converters introduces harmonics into the system. Consequently, controllers initially designed for nonrenewable sources have been adapted for wind energy systems and subsequently applied to multiple DG microgrid systems, with their total harmonic distortion (THD) values being verified. Simulations of wind energy systems connected to nonlinear loads have been conducted using MATLAB/Simulink.

The integration of the Internet of Things (IoT) into PQ applications offers a modern solution to these challenges. IoT sensors can be deployed throughout the power system to continuously monitor parameters such as voltage, current, and harmonic distortion in real time. This data is then transmitted to centralized platforms for analysis, enabling proactive management and quick identification of PQ issues. By leveraging IoT technology, it becomes possible to automatically detect and correct deviations from established standards, ensuring a reliable and high-quality power supply while minimizing the risks associated with PQ disturbances.

Reid [4] highlighted key PQ issues such as overvoltage, undervoltage, and harmonic distortion, providing guidelines and case studies across different systems. Dugan et al. [5] identified sources of PQ problems and offered solutions. Subjak and McQuilkin [6] developed a portable computer-based system for online harmonic measurement, enabling efficient data acquisition and analysis with minimal service interruption. Gopalakrishnan et al. [7] conducted field measurements in manufacturing plants and designed a shunt active filter, reducing current harmonic distortions from 15.58% to 2.11%. Patel et al. [8] explored the effects of harmonics on PQ and suggested various pulse width modulation (PWM) techniques for improvement.

Akagi [1] reviewed the state of active filters and their future prospects in power electronics. Aredes et al. [2] introduced a three-phase four-wire shunt APF using a conventional three-leg converter, developing control strategies to address harmonics and zero-sequence components. Singh et al. [9] provided a comprehensive review of active filters, covering harmonics, filter configurations, control methodologies, and application-specific considerations. Valouch [10] compared different power theories based on performance criteria such as supply currents and power losses. Al-Zamil and Torrey [3] implemented a hybrid series passive/shunt active filter, reducing THD from 28.16% to 3.9%. Aredes and Monteiro [11] discussed a simplified control strategy using sinusoidal Fryze currents and a phase-locked loop (PLL) circuit. Dell'Aquila and Lecci [12] proposed a shunt active filter, demonstrating its performance under polluting load conditions. Omeiri et al. [13] developed a three-phase shunt power filter for harmonic suppression and reactive power compensation, achieving a THD reduction from 23.53% to 4.16%. Chelladurai et al. [14] investigated PWM techniques for SAPF, finding only a slight improvement in THD with space vector pulse width modulation (SVPWM) compared to sinusoidal pulse width modulation (SPWM). Herrera et al. [15] analyzed various control strategies for unbalanced nonlinear loads under different conditions.

13.2 Power Quality Standards

PQ is governed by various standards, including the Computer and Business Equipment Manufacturers Association (CBEMA) curve, Information Technology Industry Council (ITIC) curve, IEC 61000, EN 50160:2001, and Institute of Electrical and Electronics Engineers (IEEE) standards. Among these, the most widely recognized standards for harmonic control are those set by the IEEE in the United States and the International Electrotechnical Commission (IEC) in the European Union.

In 1981, the IEEE introduced Standard 519-1981, which provided guidelines and recommended practices for managing issues such as commutation notching, voltage distortion, telephone influence, and flicker limits caused by power converters. The revised IEEE 519-1992 standard placed a stronger emphasis on limiting the maximum amount of harmonic currents at the connection point with the power utility to prevent excessive voltage distortion. This standard has continued to evolve, with the most recent revision, IEEE 519-2014, providing updated guidelines on the permissible harmonic levels in electrical power systems, focusing on both voltage and current distortion limits at various voltage levels. The standard is now more comprehensive in addressing the needs of modern power systems that incorporate a high number of nonlinear loads.

Additional IEEE standards relevant to PQ include:

- **IEEE 1100-2005 (Emerald Book):** Guidelines for powering and grounding sensitive electronic equipment.
- **IEEE 1159-2019:** Recommended practice for monitoring electric PQ, providing updates on the monitoring and characterization of PQ events.
- **IEEE 1250-2018:** Guide for service to sensitive electronic equipment, focusing on the impact of PQ disturbances on sensitive equipment and methods to mitigate these impacts.
- **IEEE P2413:** An overarching architecture for IoT standards that provides a unified framework for interoperability among different IoT devices, networks, and platforms. This standard is crucial for ensuring that IoT-enabled PQ systems can effectively communicate and operate within the broader smart grid ecosystem.
- **IEEE 1451:** A standard for smart transducers, this specification outlines how sensors and actuators should communicate over networks. This is particularly relevant for IoT sensors used in monitoring PQ, ensuring that they can effectively transmit accurate data to control systems.

The IEC standards are also crucial in specifying harmonic levels aimed at protecting low-voltage systems at the customer end and utility installations. **IEC 61000-3-2:2018** is a key standard that establishes harmonic current limits for different classes of equipment:

- **Class A:** Balanced three-phase equipment; household appliances (excluding equipment identified as Class D); tools (except portable); dimmers for incandescent lamps (but not other lighting equipment); audio equipment; anything not otherwise classified.
- **Class B:** Portable power tools.
- **Class C:** All lighting equipment except incandescent lamp dimmers.
- **Class D:** Single-phase devices under 600 W, including personal computers, personal computer (PC) monitors, and television (TV) receivers.
- **IEC 62443:** A series of standards for industrial communication networks, focusing on cybersecurity. As IoT devices are integrated into power systems, ensuring the security of these devices and the data they transmit is vital to maintaining PQ and preventing cyber threats.
- **IEC 61850:** Although primarily for communication networks and systems in substations, this standard is increasingly relevant for integrating IoT devices within the smart grid, enabling real-time monitoring and control of PQ across the grid.

13.3 Power Quality Solutions

To maintain high PQ standards, it is crucial to control the levels of harmonics in the electrical system. If these harmonic levels exceed the specified limits, consumer equipment may suffer damage or malfunction. Various filtering techniques are employed to mitigate waveform distortions and maintain PQ, including passive, active, and hybrid filters. With the advent of the IoT and smart

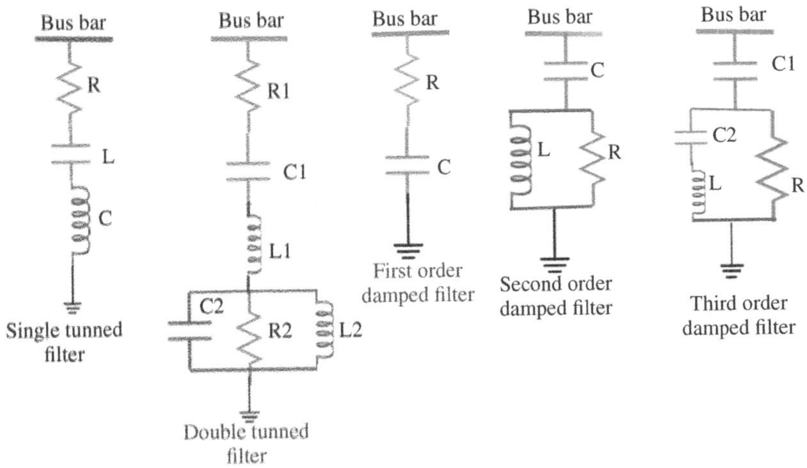

Figure 13.1 Types of passive filters.

grids, these solutions are becoming increasingly intelligent, allowing for real-time monitoring and dynamic adjustment to optimize PQ.

13.3.1 Passive Filter

The traditional approach to mitigating harmonics is using passive filters, which are made up of dampening resistors, inductors, and capacitors. Based on their characteristics, they are categorized as low-pass, band-pass, high-pass, and tuned filters. They are reasonably priced when compared to other techniques. Band-pass filters (BPFs) target a certain frequency band, whereas low-pass filters (LPFs) and high-pass filters (HPFs) are used to cancel high-order and low-order harmonics, respectively [3]. The different kinds of passive filters are depicted in Figure 13.1. Tuned filters are designed to eliminate the specific frequencies; their optimization involves examining the properties of harmonic-producing loads as well as system interactions. However, their efficiency is dependent on proximity to harmonic generators. As the power system size increases, the design complexity and cost of these filters also increase. Additionally, passive filters can cause resonance problems within the load and network. The integration of IoT sensors within smart grids allows for real-time monitoring of these filters, enabling adjustments that minimize resonance issues and optimize filter performance [16].

13.3.2 Active Filter

Traditionally, passive inductor-capacitor (LC) filters and capacitors have been used to eliminate line current harmonics and improve power factor. However, these conventional solutions can become ineffective due to variations in magnitude and frequency, leading to unpredictable harmonic distortion. Active filters offer a more adaptive solution, particularly when integrated with IoT and smart grid technologies.

Active filters dynamically adjust to changing load conditions by injecting harmonic currents that counteract existing distortions. Unlike passive filters, they are not limited to filtering specific frequencies and can respond in real time to variations in harmonic levels [17]. By utilizing IoT sensors and smart grid technologies, active filters can continuously monitor the electrical network, optimizing their operation based on real-time data. Active filters are classified into series, shunt, and hybrid filters.

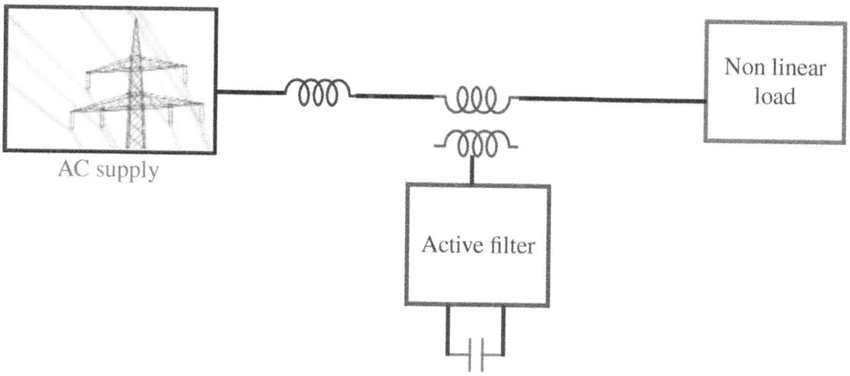

Figure 13.2 Series active filter.

13.3.3 Series Active Filter

A series active filter is connected to the PCC in series with the mains using a matching transformer, as shown in Figure 13.2. This filter type is effective in eliminating voltage harmonics, balancing terminal voltage, and reducing negative-sequence voltage in a three-phase system [18]. Electric utilities can install a series of active filters to compensate for voltage harmonics and dampen harmonic propagation caused by resonance with line impedances and passive shunt compensators. However, series active filters primarily address voltage harmonics and are not designed to compensate for load current harmonics. They act as a high impedance to current harmonics from the power source.

The inclusion of IoT sensors allows for real-time detection and response to harmonic distortions, enhancing the filter's ability to maintain voltage stability and protect sensitive equipment in smart grid environments.

13.3.4 Shunt Active Power Filter

SAPFs are primarily used for mitigating current harmonics, compensating reactive power, and balancing unbalanced currents. The SAPF injects compensating currents that are equal in magnitude but opposite in phase to the harmonics produced by nonlinear loads, effectively canceling them out [11, 19]. Additionally, SAPFs can stabilize and improve the voltage profile within the system, as shown in Figure 13.3.

With the integration of IoT and smart grid technologies, SAPFs are now equipped with microprocessor or microcontroller-based sensors that estimate harmonic content and determine control logic in real time. This allows for dynamic adjustment of the filter's operation based on current system conditions, ensuring optimal performance and improved PQ.

13.3.5 Hybrid Filter

Hybrid filters combine the benefits of both active and passive filters, offering a cost-effective solution for harmonic mitigation. Figure 13.4 shows a typical hybrid filter configuration that includes a passive shunt L-C filter that targets lower-order harmonics, combined with an active filter that addresses higher-order harmonics and dynamic conditions. This hybrid approach reduces the size

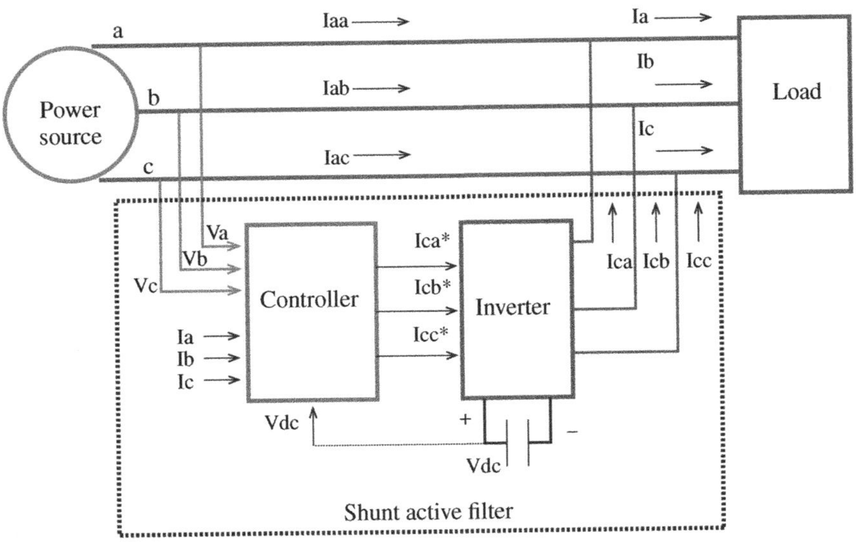

Figure 13.3 Shunt active filter.

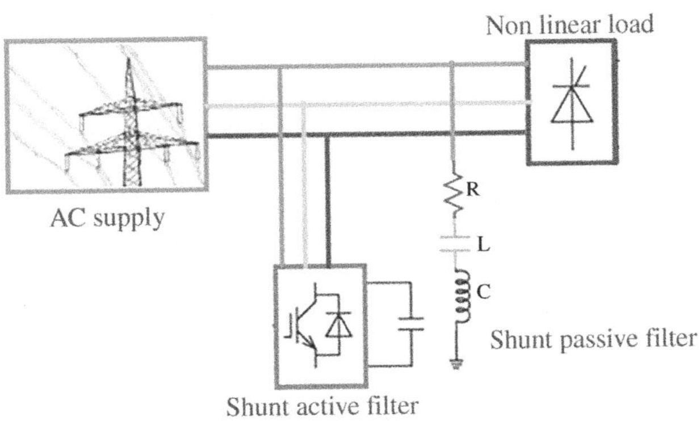

Figure 13.4 Hybrid filter with shunt active and shunt passive filter.

and cost of both active and passive filters while maintaining the capability to reduce voltage and current harmonics effectively [20].

In smart grid environments, hybrid filters can leverage IoT-enabled sensors to monitor system performance in real time, adjusting their operation to maintain PQ across a wide range of conditions. This ensures that both voltage and current harmonics are minimized, protecting equipment and improving overall system reliability.

13.4 IOT for Power Quality

The integration of IoT sensors and devices in PQ monitoring has revolutionized how electrical systems are managed and maintained. IoT sensors enable real-time data collection on critical parameters such as voltage, current, frequency, and harmonic distortion across the power network [21]. These sensors, deployed throughout the grid, transmit data continuously to centralized systems, allowing for instantaneous analysis and response to PQ issues. The deployment of IoT

Figure 13.5 IoT-enabled power quality monitoring system architecture.

devices facilitates proactive maintenance, early detection of anomalies, and automatic adjustment of control systems to mitigate disturbances like voltage sags, swells, and harmonics. By leveraging IoT technology, PQ monitoring becomes more efficient and accurate, leading to improved system reliability, reduced downtime, and enhanced protection of sensitive equipment [22]. Figure 13.5 shows IoT-enabled PQ Monitoring System Architecture.

Real-time PQ data analytics using IoT involves leveraging IoT technologies to monitor and analyze the quality of electrical power in real time. This approach integrates IoT sensors to collect comprehensive data on various PQ parameters such as voltage, current, and harmonics. The data is then analyzed using advanced analytics techniques, including machine learning and statistical methods, to identify and address PQ issues promptly. This real-time monitoring and analysis enhance the stability and efficiency of electrical systems, allowing for immediate corrective actions and improved overall grid performance [21, 22]. Additionally, considerations for scalability and security are essential to handle large volumes of data and protect against cyber threat.

13.5 The Role of IoT in Enhancing Power Quality

The integration of IoT technology significantly advances PQ management through several key functions:

- **Real-Time Monitoring:** IoT devices facilitate the continuous monitoring of PQ parameters such as voltage, current, and frequency. This capability ensures the swift detection of disturbances, enabling quick responses to prevent issues.

- **Remote Monitoring:** By enabling remote access to PQ data, IoT reduces the need for on-site inspections. This allows for timely intervention and management of PQ issues from a distance.
- **Predictive Maintenance:** IoT data analytics tools analyze trends and patterns to forecast potential PQ problems. This predictive capability allows for proactive maintenance, thereby minimizing downtime and avoiding unexpected failures.
- **Automated Control:** IoT systems can automatically implement control measures to address PQ disturbances. For example, they can switch to backup power sources or adjust load levels to maintain stable PQ.
- **Advanced Data Analytics:** The vast data collected by IoT devices supports sophisticated analytics and machine learning algorithms. These tools uncover trends, patterns, and correlations within PQ data, leading to more informed decision-making.
- **Smart Grid Integration:** IoT plays a crucial role in integrating distributed energy resources (DERs) and microgrids. This integration enhances overall PQ and strengthens grid resilience.
- **Energy Efficiency:** IoT optimizes energy consumption by identifying inefficiencies and opportunities for savings. This leads to more efficient energy use and reduced waste.
- **Asset Management:** Through continuous tracking of equipment performance, IoT facilitates proactive maintenance and extends the lifespan of assets. This ensures that infrastructure remains in optimal condition.
- **Grid Modernization:** IoT contributes to the advancement of smart grids, enabling real-time monitoring, control, and optimization of PQ. This modernization supports more reliable and efficient grid operations.
- **Customer Engagement:** IoT empowers consumers with real-time information about PQ and energy usage. This transparency allows for personalized recommendations and better management of energy consumption.

13.6 Architecture for Power Quality Management Using IoT

Figure 13.6 indicates a smart energy management strategy that can help manufacturers reduce energy consumption. This block diagram depicts an architecture of the integration of intelligent energy management system modules in industrial settings within a sensor area network (SAN) for monitoring and managing diverse loads in the downstream zone, resulting in more efficient energy use. This system is designed to automatically detect transient behaviors and anomalies associated with PQ issues, such as voltage sags, poor power factor, and high-order harmonics, which can develop as a result of nonlinear grid load behavior. Unlike typical reactive techniques, which rely on sample measurements after problems occur, this methodology actively monitors and identifies possible problems, allowing for immediate action. The primary objective of this investigation is to utilize transit-time domain analysis to transform the grid system into an autonomous one that may identify problems on its own without human intervention [23].

The approach employs the use of intra-load communication, source-load communication, and onboard load processing to integrate artificial intelligence to automatically recognize abnormal events, even during occasional occurrences. The concept also emphasizes the significance of machine-to-machine communication, which is essential to the widely recognized IoT technology [24]. In response to load dynamics, the central processing server (CPS) manages the correction nodes on the correction plane. Based on measurements from the IoT layer, it computes and sends configuration settings to the relevant corrective nodes. The suggested technique stands out from other mitigation techniques since it provides a thorough approach to restoration management.

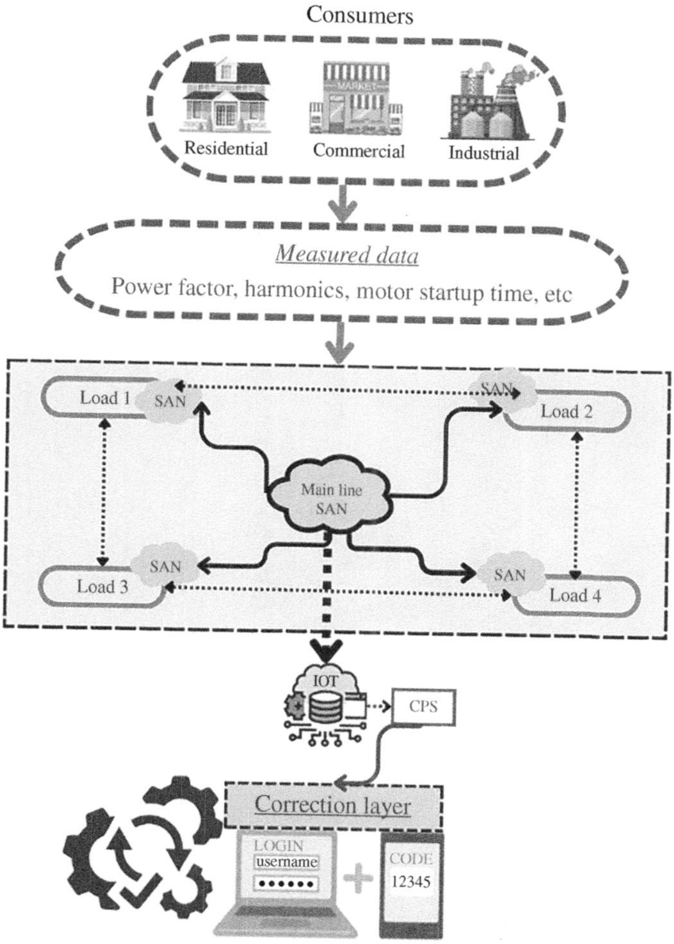

Figure 13.6 The integration of intelligent energy management system modules in industrial settings.

This approach combines multi-factor causal analysis to handle identified problems in an integrated way. Since traditional capacitor banks often fail due to their slow response time, a more intelligent approach is necessary. Active cancellation becomes more effective in the presence of multiple harmonic orders compared to passive filters. In this context, the IoT layer excels by delivering an integrated view of potential problem sources, enabling informed decisions and the application of optimal remediation techniques [25].

13.7 IoT Architecture for Smart Grid and Power Quality Applications

Figure 13.7 shows the overview of IoT enabled smart grid and Figure 13.8 presents the concept of an IoT smart grid, which incorporates intelligence at each stage to improve power consumption and disaster prevention, and eventually reduce total energy consumption in compliance with industrial practices [26]. There is a flexible framework for implementing different strategies owing to this three-layer architecture. A majority of the electricity used in the manufacturing sector is utilized

Figure 13.7 Overview of IOT-enabled smart grid.

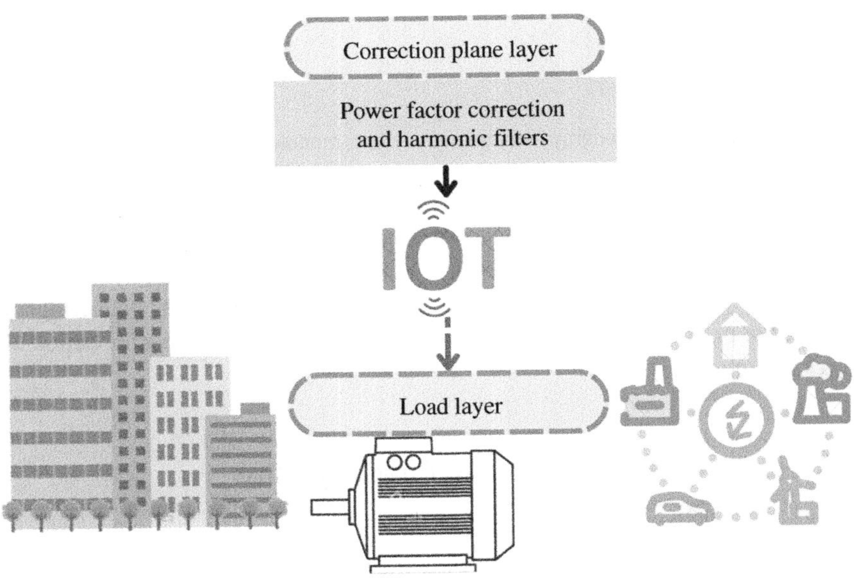

Figure 13.8 IoT architecture for smart grid.

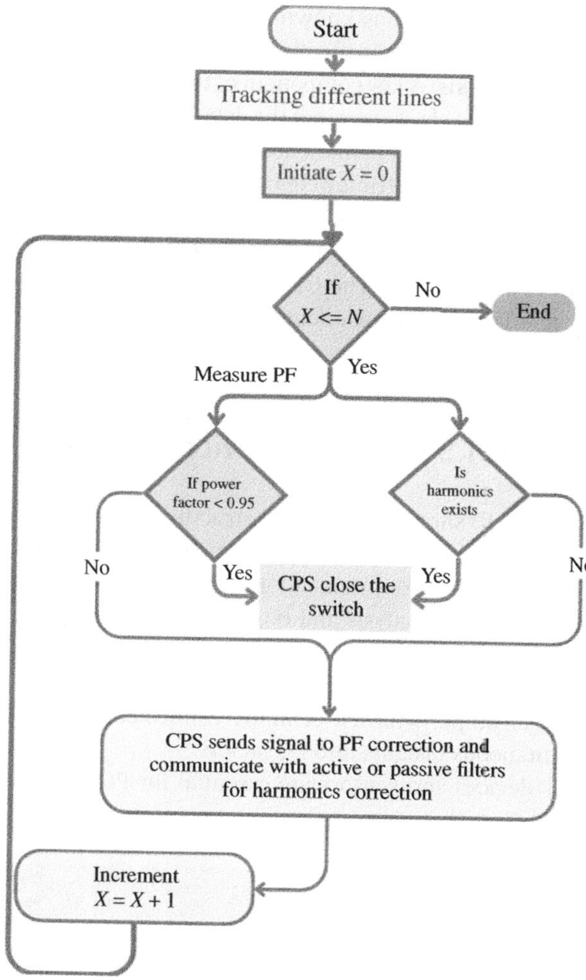

Figure 13.9 Energy savings analysis flowchart.

by machinery, which includes electric motors, pumps, and fans [27]. For comprehensive tracking, IoT sensors in the second layer, known as the IoT layer, act as connecting links. Perturbative plane layer: This layer contains the harmonic filters and power factor correction banks.

The flow chart in Figure 13.9 highlights the energy efficiency arrangement. It begins with the monitoring of multiple processes, each with a different load number (X denoting the load order and N the total number of loads) [22]. Each load is evaluated in terms of two crucial factors: power factor and harmonics. These parameters' inefficiencies can be found by the CPS [28]. In the event that problems are found, the CPS is alerted and takes appropriate action. The flowchart provides a detailed overview of the energy efficiency process:

- **Multiple Process Monitoring:** The flowchart begins with the continuous monitoring of multiple production processes. Different loads are linked to each process; "X" denotes the particular order or kind of each load, while "N" denotes the total number of loads under monitoring.
- **Evaluation of Critical Parameters:** Two crucial parameters that are evaluated for every load are power factor and harmonics.

  Power factor determines how efficiently the load uses energy. A low power factor indicates insufficient use of considerable quantities of energy, which could result in energy losses. The electrical waveform distortions known as harmonics are caused by nonlinear loads. Elevated harmonic levels may result in ineffectiveness and possible harm to electrical apparatus.

- **CPS Analysis:** A key component of the system is the CPS. With an emphasis on power factor and harmonic levels, it examines the data gathered during the monitoring phase. In these parameters, the CPS can detect inefficiencies or performance deviations from optimal performance.
- **Detection of Inefficiencies:** The CPS sends out an alarm if it detects complications with the power factor or harmonic levels. In this step, the measured values are compared to expected performance metrics or predefined criteria.
- **Alert and Action:** The CPS is alerted when abnormalities arise and initiates corrective measures. This might include alerting operations for actions taken by individuals, implementing system settings adjustments, or turning on corrective mechanisms [29].

13.8 IoT Sensors and Devices for Power Quality Monitoring

Figure 13.10 shows the physical structure of the IoT-based PQ monitoring system. It consists of integrating sensors and devices to gather data in real time, a central processing unit to analyze PQ metrics, and communication channels to send alerts and information. The figure illustrates how intelligent analysis and ongoing monitoring provided by IoT technology improve PQ issue detection and management [30]. The incorporation of IoT technology into PQ monitoring is a significant advancement in the maintenance and improvements of electrical networks. This type of approach employs the use of an IoT-connected network of sensors and devices to deliver thorough, instantaneous insights into PQ factors.

IoT devices and sensors are essential for PQ monitoring because they offer data and insights in real time that support the maintenance of reliable and efficient electrical networks [31, 32].

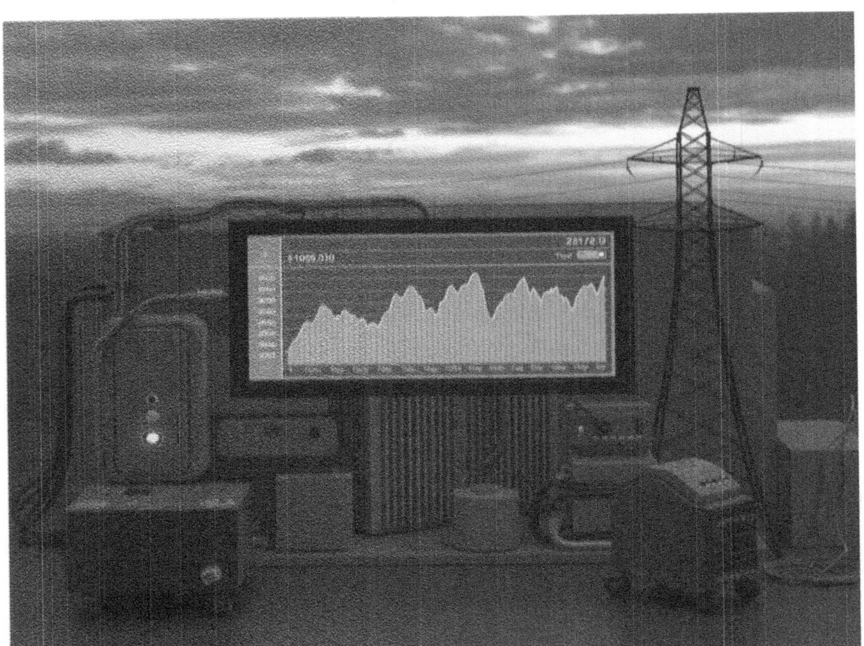

Figure 13.10 IoT for power quality monitoring.

An extensive summary of the many different types of sensors and devices, together with a description of their uses and contributions to PQ monitoring, is provided below:

- **Voltage Sensors:** It is used for detecting fluctuations in voltage, such as sags, and swells, which are crucial to assess PQ. They keep voltage levels within permissible limits to protect equipment and ensure effective operation.
- **Current Sensors:** These sensors detect current magnitudes and fluctuations. They are helpful in identifying overcurrent scenarios, which might signal potential issues such as short circuits or overloads. They also help in calculating the power factor.
- **Power Factor Meters:** They measure the effectiveness with which electrical power is turned into efficient operation. The power factor, which calculates the ratio of real to perceived power, indicates the efficiency of power use. A poor power factor can cause higher energy expenses and inefficiencies. Power factor meters assist in determining the requirement for corrective devices, such as capacitors, to increase energy efficiency.
- **Harmonic Analyzers:** This analyzer detects and evaluates the harmonic distortions in electrical waveforms. Nonlinear loads produce harmonics, which can cause inefficiency and equipment damage. By detecting harmonic levels, these analyzers aid in the diagnosis of problems and the development of appropriate filtering solutions to lessen their impacts.
- **Frequency Sensors:** This sensor is used to monitor the electrical supply's frequency to ensure that it maintains within the specified range of frequencies. Frequency variations can indicate underlying issues and restrict system performance. These sensors help to the robustness of the power supply by alerting operators to any abnormalities.
- **Temperature Sensors:** It measures the temperature of electrical components and systems. Overheating could indicate inefficiencies or possible problems. These sensors aid in the prevention of thermal damage and the maintenance of dependable system operation [33].

IoT Gateways act as intermediates, collecting data from multiple sensors and sending it to centralized systems or cloud platforms [34]. They help integrate IoT devices into the monitoring system by handling data aggregation, preprocessing, and communication. Energy meters measure the overall consumption of energy in different loads and operations. Sensors give information on patterns of usage, which is critical for figuring out ways that will decrease energy consumption and increase efficiency. This information assists in making intelligent choices regarding energy management and conservation techniques. Data loggers gather and store data from sensors periodically, enabling past information on PQ. This information is useful for analyzing changes, recognizing recurring issues, and assessing system performance over time. Cloud-based analytics platforms collect and analyze data from IoT sensors to provide real-time dashboards, alerts, and reports [35]. These platforms use analytical algorithms to offer actionable data, allowing for proactive monitoring and timely response to PQ issues.

13.9 Conclusions

This chapter explored the revolutionary effects of incorporating IoT technology into PQ monitoring and data analytics. By employing real-time data collection and powerful analytics, IoT improves the capacity to continuously monitor and regulate PQ, resolving issues with unparalleled precision and efficiency. The use of numerous IoT sensors, such as those for voltage, current, power factor, harmonics, frequency, and temperature, gives full information on the electrical system's performance. These sensors transmit important data to central processing systems via IoT gateways, where it is processed to detect anomalies, predict potential problems, and optimize energy consumption. IoT systems' real-time capabilities enable efficient detection and correction of PQ issues, hence improving system reliability, efficiency, and overall operational efficiency. Advanced analytics technologies provide precise information via real-time reports and automatic alerts, enabling data-driven choices and preventive maintenance. The aforementioned subsequently decreases operational costs, and additionally increases the lifespan of equipment and energy efficiency. Overall, the integration of IoT into PQ monitoring is a huge step forward in electrical system management. It encourages a more dynamic, responsive, and efficient approach to PQ, providing the platform for more sustainable and intelligent energy management techniques. The insights and strategies presented in this chapter highlight IoT's vital role in PQ optimization, as well as its potential to transform electrical engineering and energy management.

References

1 Akagi, H. (1996). New trends in active filters for improving power quality. *International Conference on Power Electronics Drives and Energy Systems for Industrial Growth*. pp. 417–425. IEEE.

2 Afonso, J.L., Aredes, M., Watanabe, E., and Martins, J.S. (2000). Shunt active filter for power quality improvement: electricity for a sustainable urban development. pp. 683–691.

3 Al-Zamil, A. and Torrey, D. (2001). A passive series, active shunt filter for high power applications. *IEEE Transactions on Power Electronics* 16 (1): 101–109.

4 Reid, W.E. (1996). Power quality issues-standards and guidelines. *IEEE Transactions on Industry Applications* 32 (3): 625–632.

5 Dugan, R.C., McGranaghan, M.F., and Beaty, H.W. (1996). *Electrical Power Systems Quality*. New York: McGraw Hill.

6 Subjak, J.S. and Mcquilkin, J.S. (1990). Harmonics-causes, effects, measurements, and analysis: an update. *IEEE Transactions on Industry Applications* 26 (6): 1034–1042.

7 Gopalakrishnan, C., Udayakumar, K., and Raghavendiran, T.A. (2002). Survey of harmonic distortion from power quality measurements and the application of standards including simulation. In: *IEEE/PES Transmission and Distribution Conference and Exhibition*, vol. 2, 1054–1058. IEEE.

8 Patel, M.A., Patel, A.R., Vyas, D.R., and Patel, K.M. (2009). Use of PWM techniques for power quality improvement. *International Journal of Recent Trends in Engineering* 1 (4): 99.

9 Singh, B., Al-Haddad, K., and Chandra, A. (1999). A review of active filters for power quality improvement. *IEEE Transactions on Industrial Electronics* 46 (5): 960–971.

10 Valouch, V. (1999). Active filter control methods based on different power theories. *ISIE'99. Proceedings of the IEEE International Symposium on Industrial Electronics (Cat. No. 99TH8465)*, vol. 2, 521–526. IEEE.

11 Aredes, M. and Monteiro, L.F. (2002). A control strategy for shunt active filter. *International Conference on Harmonics and Quality of Power*. pp. 472–477.

12 Dell'Aquila, A. and Lecci, A. (2003). A current control for three-phase four-wire shunt active filters. *Automatika* 44 (3–4): 129–135.

13 Omeiri, A., Haddouche, A., Zellouma, L., and Saad, S. (2006). Suppression and reactive power compensation. *Asian Journal of Information Technology* 5 (12): 1454–1457.

14 Chelladurai, J., Ilango, G.S., Nagamani, C., and Kumar, S.S. (2008). Investigation of various PWM techniques for shunt active filter. *World Academy of Science, Engineering and Technology* 39: 192–198.

15 Herrera, R.S., Salmerón, P., and Kim, H. (2008). Instantaneous reactive power theory applied to active power filter compensation: different approaches, assessment, and experimental results. *IEEE Transactions on Industrial Electronics* 55 (1): 184–196.

16 Rüstemli, S., Okuducu, E., Almalı, M.N., and Efe, S.B. (2015). Reducing the effects of harmonics on the electrical power systems with passive filters. *Bitlis Eren University Journal of Science and Technology* 5 (1): 1–10. https://doi.org/10.17678/beujst.57339.

17 Aredes, M., Hafner, J., and Heumann, K. (1997). Three-phase four-wire shunt active filter control strategies. *IEEE Transactions on Power Electronics* 12 (2): 311–318.

18 Watanabe, E.H. and Aredes, M. (2002). Power quality considerations on shunt/series current and voltage conditioners. *10th International Conference on Harmonics and Quality of Power. Proceedings (Cat. No. 02EX630).* Vol. 2, pp. 595–600. IEEE.

19 Aredes, M. and Watanabe, E.H. (1995). New control algorithms for series and shunt three-phase four-wire active power filters. *IEEE Transactions on Power Delivery* 10 (3): 1649–1656.

20 Singh, B., Verma, V., Chandra, A., and Al-Haddad, K. (2005). Hybrid filters for power quality improvement. *IEE Proceedings-Generation, Transmission and Distribution* 152 (3): 365–378.

21 Bagdadee, A.H., Hoque, M.Z., and Zhang, L. (2020). IoT based wireless sensor network for power quality control in smart grid. *Procedia Computer Science* 167: 1148–1160.

22 Panjaitan, S.D., Tjen, J., Sanjaya, B.W. et al. (2022). A forecasting approach for IoT-based energy and power quality monitoring in buildings. *IEEE Transactions on Automation Science and Engineering.* 20 (2): 892–900.

23 Hafidz, I., Priyadi, A., Pujiantara, M. et al. (2023). Development of IoT-based portable power quality monitoring on microgrids by enhancing protection features. *IEEE Access* 11: 49481–49492.

24 Gomaa, N.N., Youssef, K.Y., and Abouelatta, M. (2020). On design of IoT-based power quality oriented grids for industrial sector. *Advances in Science Technology and Engineering Systems* 5 (6): 1634–1642.

25 Bagdadee, A.H. and Zhang, L. (2019). A review of the smart grid concept for electrical power system. *International Journal of Energy Optimization and Engineering (IJEOE)* 8 (4): 105–126. https://doi.org/10.4018/IJEOE.2019100105.

26 Zahira, R., Lakshmi, D., Ravi, C.N., and Sasikala, S. (2018). Power quality enhancement using grid connected PV inverter. *Journal of Advanced Research in Dynamical and Control Systems* 10 (05): 309–317.

27 Rahiman, Z., Dhandapani, L., Chengalvarayan Natarajan, R. et al. (2023). Power quality conditioners in smart power system. In: *Artificial Intelligence-based Smart Power Systems* (ed. P. Sanjeevikumar, S. Palanisamy, S. Chenniappan, and J.B. Holm-Nielsen), 233–258. Wiley.

28 Cao, Z., Chung, Y.W., Xiong, Y. et al. (2016). IoT based manufacturing system with a focus on energy efficiency. *2016 IEEE Innovative Smart Grid Technologies-Asia (ISGT-Asia).* pp. 545–552. IEEE.

29 Aburukba, R.O., Al-Ali, A.R., Landolsi, T. et al. (2016). IoT based energy management for residential area. *2016 IEEE International Conference on Consumer Electronics-Taiwan (ICCE-TW).* pp. 1–2. IEEE.

30 Indira, G., Bhavani, M., Brinda, R., and Zahira, R. (2024). Electricity load demand prediction for microgrid energy management system using hybrid adaptive barnacle-mating optimizer with artificial neural network algorithm. *Energy Technology* 12 (5): 2301091.

31 Ildarabadi, R. and Zadehbagheri, M. (2023). New technology and method for monitoring the status of power systems to improve power quality—a case study. *Processes* 11 (8): 2468.

32 Palanisamy, S., Rahiman, Z., and Chenniappan, S. (2023). Introduction to smart power systems. In: *Artificial Intelligence-based Smart Power Systems* (ed. S. Padmanaban, S. Palanisamy, S. Chenniappan, and J.B. Holm-Nielsen), 1–13. Wiley.

33 Lee, C.H. and Lai, Y.H. (2016). Design and implementation of a universal smart energy management gateway based on the Internet of Things platform. *2016 IEEE International Conference on Consumer Electronics (ICCE)*. pp. 67–68. IEEE.

34 Rokonuzzaman, M., Akash, M.I., Mishu, M.K. et al. (2022). IoT-based distribution and control system for smart home applications. *2022 IEEE 12th Symposium on Computer Applications & Industrial Electronics (ISCAIE)*. pp. 95–98. IEEE.

35 Singh, R.R., Yash, S.M., Shubham, S.C. et al. (2020). IoT embedded cloud-based intelligent power quality monitoring system for industrial drive application. *Future Generation Computer Systems* 112: 884–898.

14

An IoT and 1D Convolutional Neural Network-Based Method for Smart Building Energy Management

Aleena Swetapadma, Nalini P. Behera, Harsh Saran, and Saurav Kumar

School of Computer Engineering, KIIT Deemed to be University, Bhubaneswar, Odisha, India

14.1 Introduction

Energy consumption in buildings is nearly about 20.1% of the total. To make the buildings energy efficient, various measures can be implemented. One of the measures that can be implemented to reduce energy consumption is to determine the occupancy of the building. Different researchers have suggested different methods to detect the occupancy of a building. Some of the existing methods have been discussed in the section below.

A method has been suggested to detect occupancy based on blind system identification and a prediction model of electricity consumption by an air-conditioning system [1]. It uses the mass-conservation law, venting levels, frequentist maximum-likelihood algorithm, and Bayesian estimation. To detect building occupancy, advanced metering infrastructure data and deep learning have been used. The deep learning models used are convolutional neural network (CNN) and long short-term memory (LSTM) network. The deep learning method for occupancy detection has nearly 90% accuracy. A method has been proposed using received signal strength indicator from user equipment and machine learning in ref. [2]. The presence of users in specific sections of the building is also determined.

A deep learning model has been used to count occupancy in ref. [3]. The energy consumption based on occupancy has been predicted. The root mean square error (RMSE) performance of occupancy estimation was 1.9. A method has been proposed for occupancy detection based on temperature and motion data in ref. [4]. The data was given to machine learning models for occupancy detection. The method has an accuracy of up to 95% and an F1 score of 95%. In ref. [5], a study has been carried out that provides insight into the machine learning-based occupancy prediction model, including data collection, prediction, and validation. A method has been suggested using decision trees to explore the relationship between different types of sensor data and occupancy in ref. [6]. The accuracy of the sensor features with the decision tree is found to be 98.4%.

In ref. [7], a method has been suggested for occupancy detection in buildings from electricity consumption data. A method has been suggested for occupancy modeling using 12 ambient sensor variables and machine-learning techniques in ref. [8]. The best accuracy using the decision-tree technique is between 96.0% and 98.2%. In ref. [9], a neural-network-based method has been proposed for building occupancy detection using temperature, light, CO_2, humidity, etc. The accuracy of the occupancy detection method is found to be 95.6%. A method has been proposed using

IoT for Smart Grid: Revolutionizing Electrical Engineering, First Edition.
Edited by Rahiman Zahira, Palanisamy Sivaraman, Chenniappan Sharmeela, and Sanjeevikumar Padmanaban.

the Internet of Things (IoT) framework and LSTM network for occupancy detection in ref. [10]. The method has been compared with support vector machine (SVM), Naive Bayes network, and neural network.

In ref. [11], machine learning methods are used to predict occupants' thermal comfort votes and thermal sensation votes. A vision-based deep learning framework for occupancy detection has been proposed in ref. [12]. The region-based CNN model has an accuracy of 85.63% occupancy detection. A method has been suggested for predicting occupancy using heat sensors and CNN in ref. [13]. In ref. [14], an explicit duration hidden Markov model-based occupancy prediction method has been proposed.

Considering all the above-discussed methods, there is still a scope of improvement in detecting and predicting occupancy. In this work, a method has been proposed using sensor data and one-dimensional convolutional neural network (1D CNN). The paper has been organized as follows. Section 14.2 describes the architecture of 1D CNN. Section 14.3 describes the proposed occupancy detection method. Section 14.4 contains the result of the occupancy detection method. Section 14.5 contains the discussion and Section 14.6 contains the conclusion.

14.2 One-Dimensional Convolutional Neural Network

The classification and prediction work can be done using different techniques. Among all the techniques that exist, machine learning has been used more commonly. Machine learning methods can be of shallow learning and deep learning. The deep learning methods have more advantages than shallow learning for classification. The deep learning methods are mostly used for image processing and analysis. Some deep learning methods are also used for signal processing and analysis. One such deep learning method is 1D-CNN. Figure 14.1 shows the architecture of the 1D-CNN used in this work. It has two convolutional layers, two max-pooling layers, one flattened layer, and one dense layer.

14.3 Proposed Method

The proposed 1D CNN method has two objectives: first, it detects if occupied or not, and second is to identify the number of occupants. Figure 14.2 shows the flowchart of the proposed method.

14.3.1 Inputs Used

The inputs used in this work are different for the occupancy detection module and occupancy prediction module. Details of the two data sets used have been given in the section below.

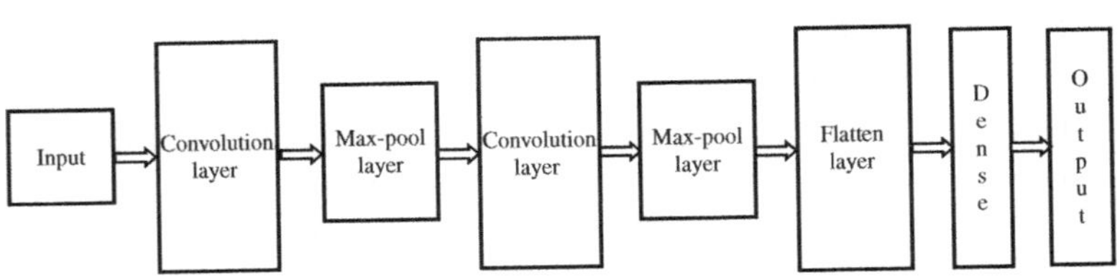

Figure 14.1 1D CNN architecture.

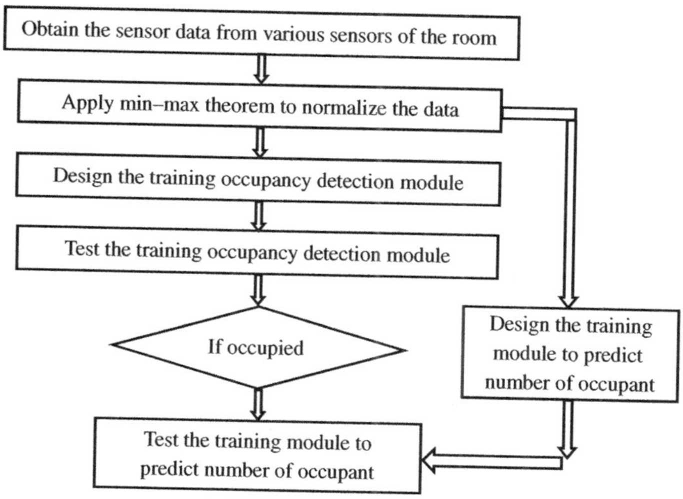

Figure 14.2 Flowchart of the proposed method.

14.3.1.1 Data for Occupancy Detection

The occupancy detection method is validated using a data set from the University of California Irvene (UCI) machine learning repository. The data from an office room such as temperature, humidity, light, and CO_2 levels has been collected [15]. The occupancy of the room is recorded with a camera and the humidity ratio is also included [16]. Figure 14.3 shows the different input features recorded from different sensors and the corresponding output. It contains the data recorded from the temperature sensor, light sensor, and CO_2 sensor along with the occupancy of the room.

14.3.1.2 Data for Occupancy Prediction

The data set used for occupancy prediction was collected from seven sensor nodes and one edge node in a star configuration [17]. Five nonintrusive sensors were used for temperature, light, sound, CO_2, and digital passive infrared. The data was collected for a period of four days with the occupancy in the room varying between 0 and 3 people. Figure 14.4 shows the different input features recorded from different sensors and the corresponding output.

14.3.2 Preprocessing

The recorded data from the sensors has been taken for processing. The goal of normalizing data is to ensure that each dataset is on the same scale, making each feature equally important. The data has been normalized using min–max theorem. The min–max theorem has been given in equation (14.1),

$$X = \frac{\text{Value} - \text{Minimum value}}{\text{Maximum value} - \text{Minimum value}} \tag{14.1}$$

14.3.3 Occupancy Detection Method

The proposed method first detects if the room is occupied or not. The occupancy detection method first trains the network with known output and then tests the train network with unknown output. The normalized features are taken as input and the corresponding targets are designed for the training network. The desired targets of the method are "0" for a room not occupied and "1" for

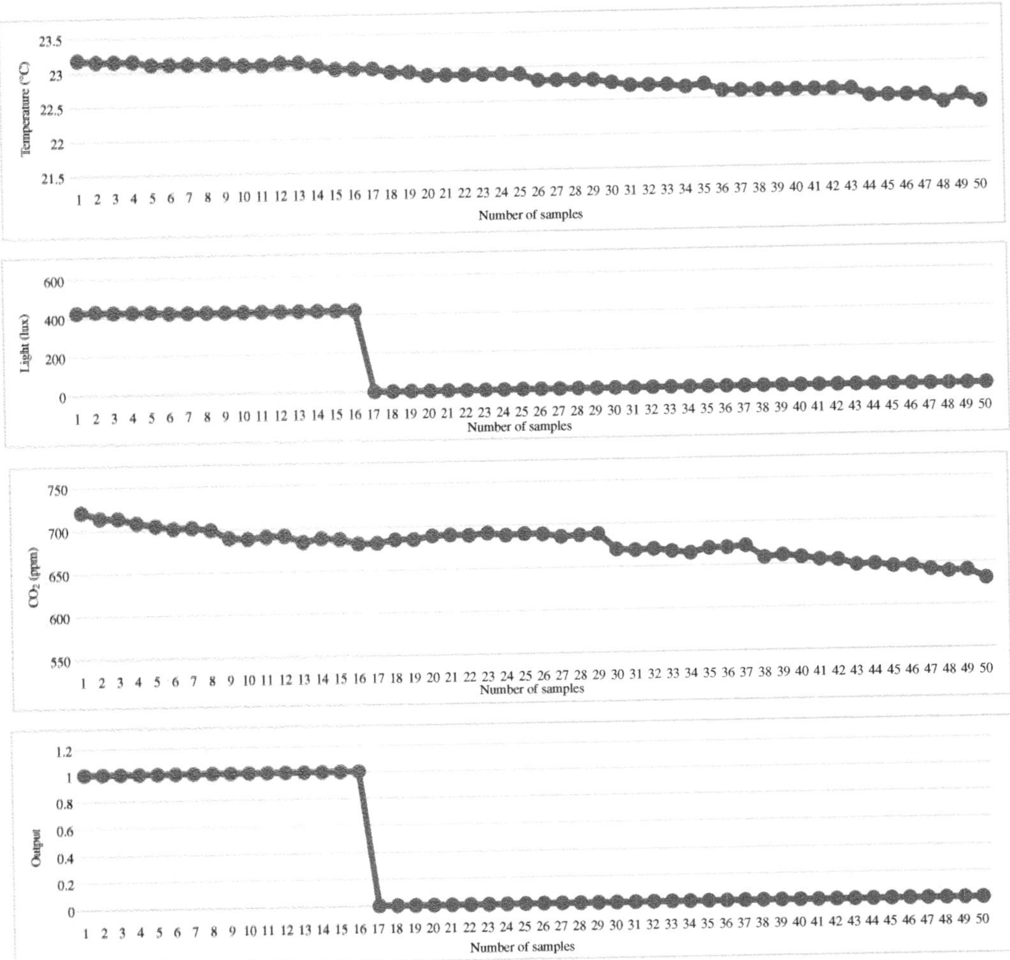

Figure 14.3 Inputs and output of occupancy detection.

a room occupied. The input and target are given to the 1D CNN method. The 1D CNN method has been designed with various hyper parameters. The optimal value of the hyper parameters has been chosen after various trials. Table 14.1 shows the optimal value of parameters for training. Figure 14.5 shows the training performance in terms of accuracy and loss.

14.3.4 Occupancy Prediction Method

The proposed occupancy prediction method predicts the number of occupants in the room. The normalized features are taken as input and the corresponding targets are designed for the training network. The desired targets of the method are "0" for not occupied, "1" for one person occupied, and so on. The 1D CNN method has been designed with various hyper parameters. The optimal value of the hyper parameters has been chosen after various trials. Table 14.2 shows the optimal value of parameters for training. Figure 14.6 shows the training performance in terms of accuracy and loss.

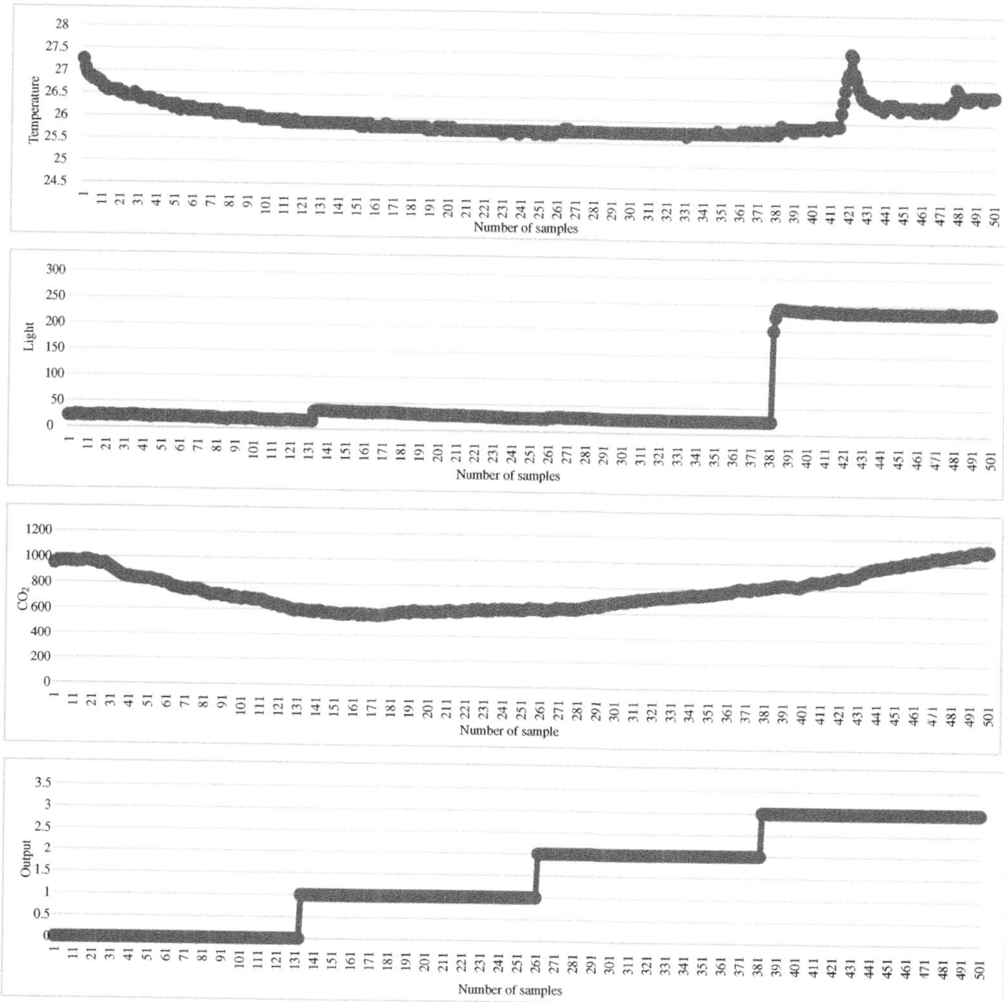

Figure 14.4 Inputs and output of occupancy prediction.

Table 14.1 Parameters used for training occupancy detection module.

Layers with neurons	Epochs	Optimizer	Activation function	Training accuracy (%)
Conv1D(16), Dense(32)	10	Adam	Sigmoid	95.45
Conv1D(16), Dense(32)	50	Adam	Sigmoid	98.83
Conv1D(16), Dense(32)	100	Adam	Sigmoid	99.18
Conv1D(16), Conv1D(16), Dense(32)	150	Adam	Sigmoid	99.87
Conv1D(16), Conv1D(32), Dense(32)	200	Adam	Sigmoid	99.97
Conv1D(16), Conv1D(32), Dense(32)	250	Adam	Sigmoid	99.98
Conv1D(16), Conv1D(16), Dense(32)	300	Adam	Sigmoid	99.99

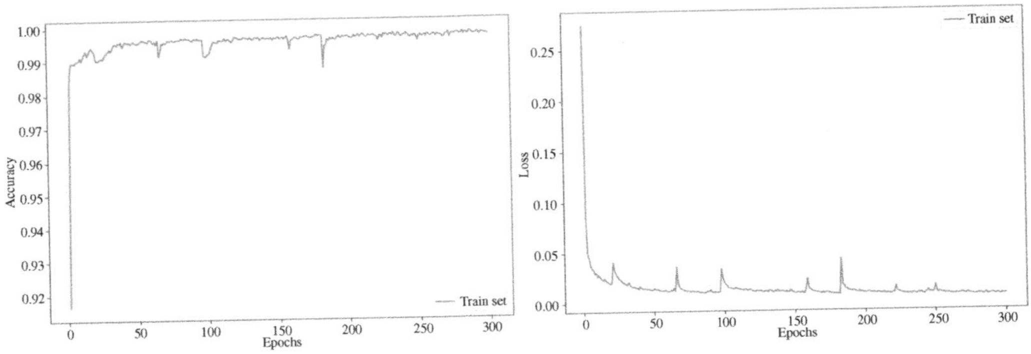

Figure 14.5 Training performance of occupancy detection module.

Table 14.2 Parameters used for training number of occupant prediction module.

Layers with neurons	Epochs	Optimizer	Activation function	Training accuracy (%)
Conv1D(16), Dense(32)	10	Adam	Sigmoid	98.827
Conv1D(16), Dense(32)	40	Adam	Sigmoid	99.382
Conv1D(16), Dense(32)	100	Adam	Sigmoid	99.568
Conv1D(32), Dense(32)	100	Adam	Sigmoid	99.407
Conv1D(16), Conv1D(16), Dense(32)	100	Adam	Sigmoid	99.592
Conv1D(16), Conv1D(32), Dense(32)	150	Adam	Sigmoid	99.617
Conv1D(16), Conv1D(32), Dense(32)	200	Adam	Sigmoid	99.716
Conv1D(16), Conv1D(32), Dense(32)	500	Adam	Sigmoid	99.925

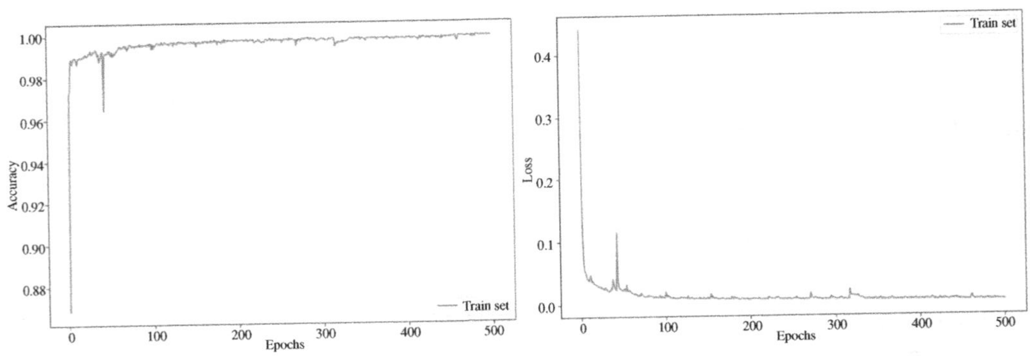

Figure 14.6 Training performance of a number of occupant prediction modules.

14.4 Result

The proposed building occupancy detection and prediction method has been tested. The performance of the method has been analyzed using various matrices. The test results of the detection and prediction method have been discussed below.

Table 14.3 Confusion matrix.

Confusion matrix		Target	
		0	1
Output	0	1639	54
	1	3	968

14.4.1 Performance of Occupancy Detection Method

The occupancy detection method has been tested with doors open data and doors closed data. It has been discussed in detail below.

14.4.1.1 With Door Closed

The proposed occupancy detection method is tested with the data when door is closed. The results of the method have been discussed here after testing the samples. Table 14.3 shows the confusion matrix obtained. The F1-score in detecting occupancy is 0.97 while F1-score in detecting non-occupancy is 0.98. It can be observed from Table 14.4 that the overall accuracy of the method is 97.86%.

14.4.1.2 With Door Open

The proposed occupancy detection method is tested with the data when the door is open. The results of the method have been given in Tables 14.5 and 14.6 after testing the samples. Table 14.5 shows the confusion matrix obtained. The F1-score in detecting occupancy is 0.98, while the F1-score in detecting non-occupancy is 1.00. It can be observed from Table 14.6 that the overall accuracy of the method is 99.25%.

Table 14.4 Performance matrices.

Precision	Recall	F1-score	Accuracy (%)
0.97	0.98	0.98	97.86

Table 14.5 Confusion matrix.

Confusion matrix		Target	
		0	1
Output	0	7644	59
	1	14	2034

Table 14.6 Performance matrices.

Precision	Recall	F1-score	Accuracy (%)
0.98	0.99	0.99	99.25

Table 14.7 Performance of the occupancy prediction.

Accuracy (%)	Loss
99.654	0.013

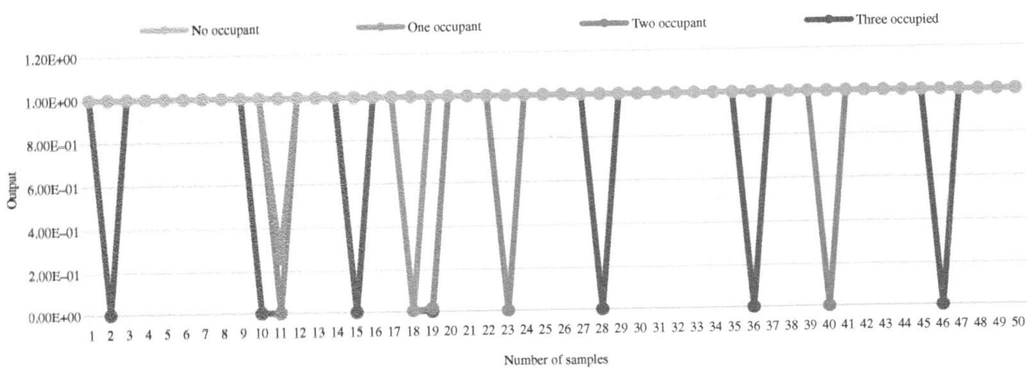

Figure 14.7 Output of the occupancy prediction method.

14.4.2 Performance of Number of Occupant Prediction

The proposed occupancy prediction method has been tested and the results are discussed here. Table 14.7 shows the performance of the occupancy prediction method with accuracy and loss. Figure 14.7 shows the output obtained for the occupancy prediction method for different test cases. It can be observed that the proposed 1D CNN method can accurately predict the number of occupants in the building.

14.5 Discussion

The proposed occupancy detection method has been compared with various methods suggested by different researchers. The method has been compared in terms of the algorithms used and their accuracy, as shown in Table 14.8. From Table 14.8, it can be observed that the proposed method occupancy detection method has an accuracy of 99.86% and the occupancy prediction method has 99.65%. Most of the methods detect the occupancy but there are very few methods that predict the number of occupants.

Table 14.8 Comparison of occupancy detection method with other methods.

Authors	Method used	Task	Performance
Feng et al. [18]	CNN and LSTM	Occupancy detection	Accuracy – 90%
Wang et al. [4]	Random forest, decision tree, K-nearest neighbor, and support vector machine	Occupancy detection	Accuracy – 95%
Hailemariam et al. [6]	Decision tree	Occupancy detection	Accuracy – 81–98%
Kleiminger et al. [7]	SVM, K-NN, thresholding	Occupancy detection	Accuracy – 59–90%
Yang et al. [8]	SVM, K-NN, NB, DT	Occupancy detection	Accuracy – 88–98%
Das et al. [9]	Feed-forward back-propagation neural network	Occupancy detection	Accuracy – 95.6%
This work	1D CNN	Occupancy detection	Accuracy – 97.86%
		Occupancy prediction	Accuracy – 99.65%

14.6 Conclusion

In this work, an occupancy detection method and an occupancy prediction method have been suggested using 1D CNN method using sensor data. The inputs used are CO_2, temperature, light, humidity, etc. The highest accuracy obtained for the occupancy detection method is 97.86%. The highest accuracy obtained for the occupancy prediction method is 99.65%. Occupancy detection is an important factor in the estimation of building energy consumption. It can be used to take measures to reduce the energy of the buildings. The future scope of the work can be adding more features to predict the number of occupants correctly.

References

1 Wei, Y., Xia, L., Pan, S. et al. (2019). Prediction of occupancy level and energy consumption in office building using blind system identification and neural networks. *Applied Energy* 240: 276–294.

2 Al-Habashna, A., Wainer, G., and Aloqaily, M. (2022). Machine learning-based indoor localization and occupancy estimation using 5G ultra-dense networks. *Simulation Modelling Practice and Theory* 118: 102543.

3 Kim, M.-L., Park, K.-J., and Son, S.-Y. (2023). Occupancy-based energy consumption estimation improvement through deep learning. *Sensors* 23: 2127.

4 Wang, C., Jiang, J., Roth, T. et al. (2021). Integrated sensor data processing for occupancy detection in residential buildings. *Energy and Buildings* 237: 110810.

5 Zhang, W., Wu, Y., and Calautit, J.K. (2022). A review on occupancy prediction through machine learning for enhancing energy efficiency, air quality and thermal comfort in the built environment. *Renewable and Sustainable Energy Reviews* 167: 112704.

6 Hailemariam, E., Goldstein, R., Attar, R., and Khan, A. (2011). Real-time occupancy detection using decision trees with multiple sensor types. *Proceedings of the 2011 Symposium on*

Simulation for Architecture and Urban Design. pp. 141–148. San Diego, CA: Society for Computer Simulation International.

7 Kleiminger, C., Beckel, T., and Staake, S.S. (2013). Occupancy detection from electricity consumption data. *Proceedings of the 5th ACM Workshop on Embedded Systems for Energy-Efficient Buildings*. pp. 1–8. Rome, Italy: ACM.

8 Yang, Z., Li, N., Becerik-Gerber, B., and Orosz, M. (2014). A systematic approach to occupancy modeling in ambient sensor-rich buildings. *Simulation* 90 (8): 960–977.

9 Das, S., Swetapadma, A., and Panigrahi, C. (2017). Building occupancy detection using feed forward back-propagation neural networks. *2017 3rd International Conference on Computational Intelligence and Networks (CINE)*, Odisha, India. pp. 63–67.

10 Hitimana, E., Bajpai, G., Musabe, R. et al. (2021). Implementation of IoT framework with data analysis using deep learning methods for occupancy prediction in a building. *Future Internet* 13: 67.

11 Chai, Q., Wang, H., Zhai, Y., and Yang, L. (2020). Machine learning algorithms to predict occupants' thermal comfort in naturally ventilated residential buildings. *Energy and Buildings* 217: 109937.

12 Tien, P.W., Wei, S., Calautit, J.K. et al. (2022). Real-time monitoring of occupancy activities and window opening within buildings using an integrated deep learning-based approach for reducing energy demand. *Applied Energy* 308: 118336.

13 Arvidsson, S., Gullstrand, M., Sirmacek, B., and Riveiro, M. (2021). Sensor fusion and convolutional neural networks for indoor occupancy prediction using multiple low-cost low-resolution heat sensor data. *Sensors* 21 (4): 1036.

14 Rueda, L., Agbossou, K., Henao, N. et al. (2021). Online unsupervised occupancy anticipation system Applied to residential heat load management. *IEEE Access* 9: 109806–109821.

15 Candanedo, L.M. and Feldheim, V. (2016). Accurate occupancy detection of an office room from light, temperature, humidity and CO_2 measurements using statistical learning models. *Energy and Buildings* 112: 28–39.

16 UCI Machine Learning Repository.

17 Singh, A.P., Jain, V., Chaudhari, S. et al. (2018). Machine learning-based occupancy estimation using multivariate sensor nodes. *2018 IEEE Globecom Workshops*, Abu Dhabi, United Arab Emirates. pp. 1–6.

18 Feng, C., Mehmani, A., and Zhang, J. (2020). Deep learning-based real-time building occupancy detection using AMI data. *IEEE Transactions on Smart Grid* 11: 4490–4501.

15

IoT for E-Mobility

Shanmugasundaram Logeshkumar[1], *Krishnakumar Shanmugasundaram*[2], *Rahiman Zahira*[3], *Palanisamy Sivaraman*[4], *and Chenniappan Sharmeela*[5]

[1] *Department of Electronics and Communication Engineering, Christ the King Engineering College, Anna University, Chennai, Tamil Nadu, India*
[2] *Capgemini US Corp, Charlotte, NC, USA*
[3] *Department of Electrical and Electronics Engineering, B.S. Abdur Rahman Crescent Institute of Science and Technology, Chennai, Tamil Nadu, India*
[4] *Research Scholar, Anna University, Chennai, Tamil Nadu, India*
[5] *Department of EEE, CEG, Anna University, Chennai, Tamil Nadu, India*

Introduction

The Internet of Things (IoT) is revolutionizing many aspects of our lives, and the transportation sector is no exception. In the realm of e-mobility, IoT is playing a pivotal role in transforming how electric vehicles (EVs) are designed, operated, and maintained. By connecting EVs to a network of sensors, devices, and software applications, IoT is enabling a new era of intelligent and sustainable transportation.

15.1 What Is IoT for E-Mobility?

IoT for e-mobility refers to the use of a network of interconnected devices to collect, transmit, and analyze data related to EVs. These devices can be embedded within the EV itself, such as battery management systems (BMS) and motor controllers, or they can be part of the charging infrastructure and surrounding environment. The data sets received will be used to improve various aspects of e-mobility, including:

The magic of connected e-mobility lies in its ability to transform EVs into intelligent machines Figure 15.1. A network of IoT sensors continuously monitors your vehicle's health, providing real-time data on battery health, motor performance, and energy consumption. This data becomes the key to optimizing your EV's performance. Imagine the system suggesting adjustments to maximize range or extending battery life through personalized charging recommendations. The benefits extend beyond the vehicle itself. Charging infrastructure management gets a boost with IoT-enabled stations. Real-time information on availability, power output, and usage patterns allows for smarter deployment of charging networks, improving efficiency and reducing wait times for drivers. Predictive maintenance takes center stage as well. By constantly monitoring various vehicle parameters, the system can identify potential issues early on, alerting you or service centers for preventive maintenance. This proactive approach can prevent breakdowns and extend

IoT for Smart Grid: Revolutionizing Electrical Engineering, First Edition.
Edited by Rahiman Zahira, Palanisamy Sivaraman, Chenniappan Sharmeela, and Sanjeevikumar Padmanaban.

Figure 15.1 Portrays the Internet of Things in electric mobility. Source: Solveig/Adobe Stock Photos.

the lifespan of your EV. Finally, connected e-mobility paves the way for a smarter grid. IoT helps integrate EVs seamlessly with the power grid. Utilities can leverage data on charging patterns and energy consumption to optimize grid management, ensuring a stable and reliable power supply to support the growing number of EVs on the road. In essence, connected e-mobility, empowered by IoT, creates a win–win situation for drivers, utilities, and the environment [1, 2].

15.2 Benefits of IoT for E-Mobility

The adoption of IoT in e-mobility offers a multitude of benefits for various stakeholders, including:

- **EV Manufacturers:** IoT data can provide valuable insights into vehicle performance and user behavior. This information can be used to improve future EV designs, develop new features, and optimize manufacturing processes.
- **EV Drivers:** IoT-based applications can provide EV drivers with real-time information on battery health, charging availability, and nearby amenities. This can improve the driving experience, reduce range anxiety, and make EV ownership more convenient.
- **Energy Providers:** IoT data can help energy providers understand charging patterns and optimize grid management. This can lead to a more efficient and stable power grid, better prepared to handle the increasing demand from EVs.
- **Cities and Governments:** IoT-enabled e-mobility solutions can help cities reduce traffic congestion, improve air quality, and promote sustainable transportation practices [1, 2].

15.3 Challenges of IoT for E-Mobility

Despite its many benefits, the integration of IoT into e-mobility also presents several challenges that need to be addressed:

- **Security and Privacy:** The vast amount of data collected by IoT devices creates new security and privacy concerns. Robust security measures are needed to protect sensitive data from cyber-attacks.
- **Data Standardization:** The lack of standardized data formats across different IoT devices and platforms can hinder data integration and interoperability.
- **Infrastructure Development:** The widespread adoption of IoT for e-mobility requires significant investment in infrastructure development, including charging stations, communication networks, and data management systems.
- **User Acceptance:** Gaining user acceptance for new IoT-based technologies is crucial. Consumers need to be assured of the benefits and security of these technologies before they are widely adopted.

15.4 The Future of IoT for E-Mobility

The future of IoT for e-mobility is bright. As technology continues to evolve, we can expect to see even more innovative applications emerge. Here are some potential future directions for IoT in e-mobility.

The road to autonomous EVs is paved with data, and IoT sensors and communication networks are key ingredients. Imagine a future where EVs leverage IoT to navigate their environment safely and efficiently (Figure 15.2). This is achieved through **Vehicle-to-Everything (V2X) communication**. These networks allow EVs to talk to each other and surrounding infrastructure, like traffic lights and smart roads. This paves the way for groundbreaking safety features, smoother traffic flow through real-time adjustments, and even optimized energy consumption for a greener journey.

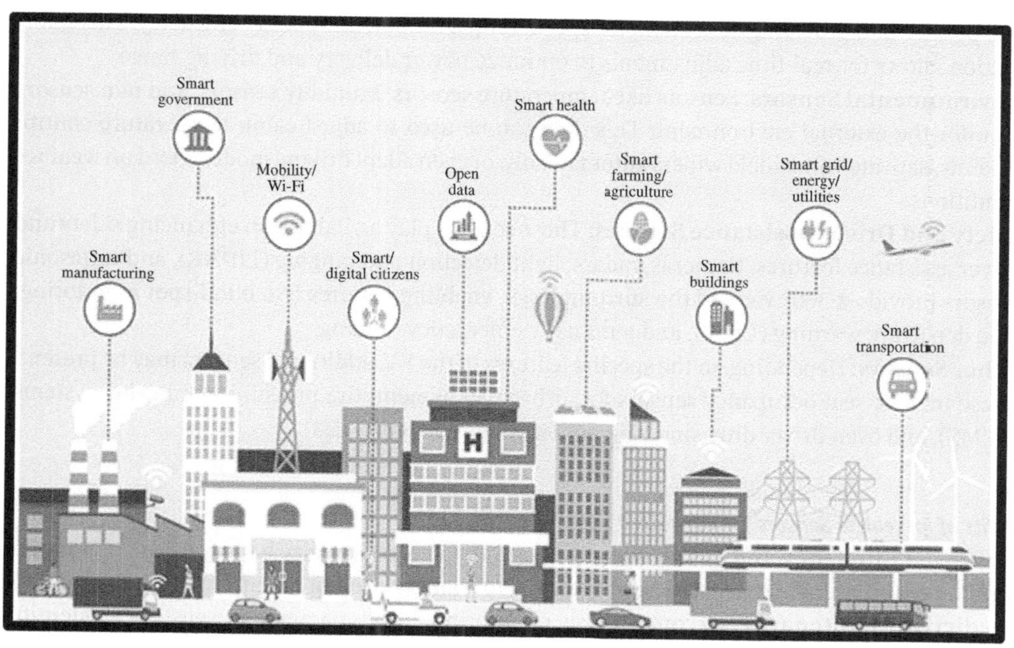

Figure 15.2 IoT in everything (in all fields).

But the benefits go beyond the road itself. The data collected by IoT can be used to develop **personalized e-mobility services**. Imagine customized charging plans based on your driving habits, routes optimized for current traffic conditions, or real-time recommendations for nearby amenities during your trip. In essence, IoT will be the central nervous system of autonomous EVs, creating a smarter, safer, and more personalized transportation experience for everyone [1].

15.5 Various Considerations and Possibilities of IoT for E-Mobility

15.5.1 Connected Vehicle Technologies

15.5.1.1 In-Vehicle Sensors (Battery Health, Motor Performance, Energy Consumption) – The Eyes and Ears of Connected Electric Vehicles

In-vehicle sensors are the foundation of connected EVs [3, 5]. These tiny powerhouses act as the eyes and ears of the car, constantly gathering data about the vehicle's internal workings and its surrounding environment. This data is then fed into the car's computer system and transmitted to the cloud via an internet connection, enabling a whole new world of possibilities for e-mobility.

Here is a deeper dive into the world of in-vehicle sensors for connected EVs.

Types of In-Vehicle Sensors A wide range of sensors are used in connected EVs [6], each playing a specific role:

- **BMS Sensors** [10]**:** These sensors monitor the health and performance of the EV's battery pack, tracking parameters like voltage, current, temperature, and remaining capacity. This data is crucial for optimizing battery life, preventing overcharging, and ensuring safe operation.
- **Motor and Powertrain Sensors:** These sensors monitor the performance of the electric motor and powertrain, providing data on motor speed, torque output, and energy efficiency. This information allows for real-time adjustments to optimize power delivery and driving range.
- **Environmental Sensors:** Sensors like temperature sensors, humidity sensors, and rain sensors monitor the external environment. This data can be used to adjust cabin temperature control systems, activate windshield wipers automatically, or even adapt driving modes based on weather conditions.
- **Safety and Driver Assistance Sensors:** These sensors play a vital role in enhancing safety and driver assistance features. Cameras, radars, light detection and ranging (LiDAR), and ultrasonic sensors provide a 360° view of the surroundings, enabling features like blind-spot monitoring, lane departure warning (LDW), and automatic emergency braking.
- **Other Sensors:** Depending on the specific features of the EV, additional sensors may be present. These include seat occupancy sensors for airbag deployment, tire pressure monitoring systems (TPMS), and even driver drowsiness detection systems (Figure 15.3).

Benefits of In-Vehicle Sensors for Connected EVs

- **Improved Efficiency and Performance:** Sensor data helps optimize battery usage, motor performance, and overall vehicle efficiency, leading to a longer driving range.
- **Predictive Maintenance:** By continuously monitoring various parameters, sensors can identify potential problems early on, allowing for preventive maintenance and avoiding breakdowns.
- **Enhanced Safety Features:** Sensor data enables advanced driver-assistance systems (ADAS) to provide real-time warnings and intervene in critical situations, improving overall road safety.

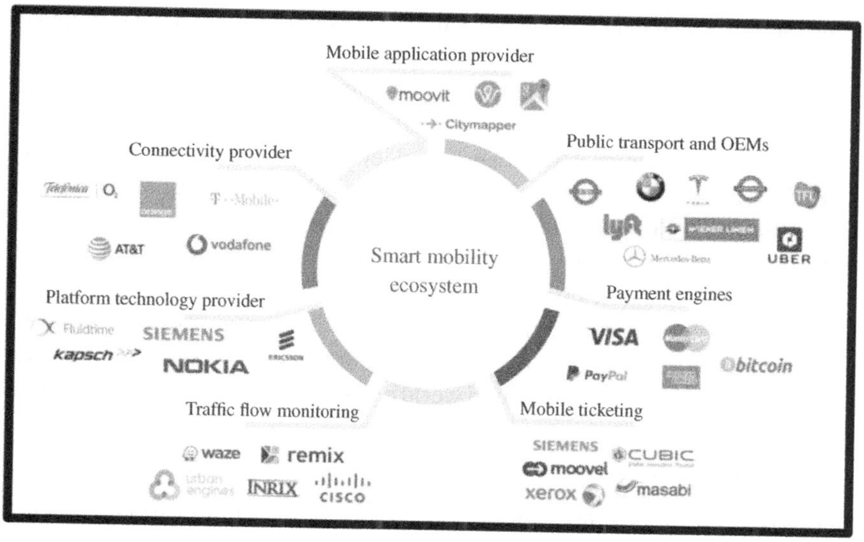

Figure 15.3 Integrated mobility platform.

- **Personalized Driving Experience:** Sensor data can be used to personalize the driving experience, such as adjusting cabin temperature based on occupancy or optimizing route navigation based on traffic conditions.
- **Connected Services:** Sensor data can be used to enable connected services like remote diagnostics, charging station recommendations, and integration with smart city infrastructure.

Challenges and Considerations

- **Data Security and Privacy:** The vast amount of data collected by in-vehicle sensors raises concerns about security and privacy. Robust measures are needed to protect this data from cyberattacks and ensure user privacy.
- **Sensor Fusion and Data Management:** With so many sensors generating data, efficient data fusion and management techniques are crucial to extract meaningful insights and avoid information overload.
- **Standardization:** Standardization of sensor protocols and data formats is essential to ensure seamless communication and interoperability between different EVs and infrastructure.

The Future of In-Vehicle Sensors The future of in-vehicle sensors is bright. As technology advances, we can expect even more sophisticated sensors to emerge, with higher resolution, wider range, and enhanced capabilities. This will pave the way for even more advanced connected car features, ultimately leading to a safer, more efficient, stable batteries with nanomaterials [12] and sustainable future for e-mobility.

15.5.1.2 Connected Vehicle Technologies: Telematics and Data Collection Units – The Unsung Heroes of E-Mobility

In the world of connected EVs, in-vehicle sensors play a starring role, constantly gathering data. But this data would not be very useful without the behind-the-scenes heroes: telematics and data collection units. These units act as the brains and brawn of data transmission, transforming raw sensor data into actionable information for a smarter and more connected e-mobility experience.

What Are Telematics and Data Collection Units? Telematics units are essential components equipped onboard computers.

The beating heart of a telematics unit is a symphony of components working together. **Sensors,** either built-in or leveraging existing ones in your vehicle, collect raw data on various parameters. This might include information like engine performance, fuel levels, or even location. A **processing unit** then takes center stage, transforming this raw data into a usable format The processed data is then ready for transmission. A **cellular modem** acts as the communication hub, securely transmitting the data wirelessly to a cloud platform via cellular networks. Finally, a **Global Positioning System (GPS) module** plays a crucial role by providing precise location data. This facilitates features like remote tracking of your vehicle or setting up geofences to receive alerts when your car enters or exits a designated area. These components working in harmony allow telematics units to collect, process, and transmit valuable data, unlocking a world of possibilities for fleet management, personal safety, and improved driving efficiency.

What Data Do They Collect? Telematics units collect a wide range of data from various sources, including: Vehicle Health, Driving behaviour, Location data, Charging information.

Telematics units turn your vehicle into a data powerhouse, collecting a wealth of information that can be categorized into four key areas. Firstly, **vehicle health** takes center stage. Data on battery health, motor performance, and even diagnostic trouble codes provides valuable insights for preventative maintenance, helping you avoid unexpected breakdowns. Secondly, the system acts as your personal driving coach. By monitoring **driving behavior** like speed, acceleration, and braking patterns, it can offer suggestions to improve fuel efficiency and make you a more eco-conscious driver. **Location data** plays a key role as well. GPS data unlocks features like real-time traffic updates and route optimization, making every journey smoother and faster. This data can also be used for remote vehicle tracking, providing peace of mind and an extra layer of security. Finally, **charging information** is collected, including details on charging stations used, duration, and energy consumption. This data helps optimize charging strategies for individual needs and improve infrastructure planning for a more robust charging network in the future. In essence, telematics units transform your car from a simple mode of transportation into a data-driven machine, providing valuable insights to improve efficiency, safety, and the overall driving experience.

Benefits of Telematics and Data Collection Units

- **Improved Fleet Management:** For companies with electric fleets, telematics data provides valuable insights for optimizing vehicle usage, reducing operating costs, and scheduling maintenance.
- **Enhanced Safety and Security:** Real-time location data and crash detection features can aid emergency response and stolen vehicle recovery.
- **Personalized Driver Experience:** Telematics data can be used to provide drivers with personalized feedback on their driving habits, helping them improve efficiency and potentially qualify for insurance discounts.
- **Remote Diagnostics and Maintenance:** By analyzing data remotely, service centers can proactively identify potential issues and schedule preventive maintenance, minimizing downtime for EV owners.
- **Integration with Connected Services:** Telematics data is the backbone of many connected car services, such as remote charging management, over-the-air (OTA) software updates, and integration with smart city infrastructure.

Challenges and Considerations

- **Data Privacy Concerns:** As with in-vehicle sensors, robust security measures are essential to protect user privacy and prevent unauthorized access to sensitive data.
- **Data Ownership and Usage:** Clear regulations are needed regarding data ownership and how it can be used by manufacturers, service providers, and third parties.
- **Cost and Integration:** The cost of telematics units and the complexity of data integration can be a barrier to wider adoption, particularly in personal EVs.

The Future of Telematics and Data Collection Units As connected car technology evolves, telematics units will become even more sophisticated. We can expect the following:

- **Advanced Data Analytics:** Machine learning and artificial intelligence (AI) will be used to extract deeper insights from data, enabling predictive maintenance and personalized recommendations.
- **V2X Communication:** Telematics units will play a crucial role in facilitating communication between EVs and other vehicles and infrastructure, enhancing safety and improving traffic flow.
- **Standardization and Interoperability:** Standardized data formats and protocols will ensure seamless communication between different telematics units and cloud platforms.

By acting as the silent workhorses of data collection and transmission, telematics and data collection units are the backbone of a connected e-mobility future. As these technologies continue to develop, they will play an increasingly critical role in creating a safer, more efficient, and sustainable transportation system.

15.5.1.3 Connected Vehicle Technologies: Vehicle-to-Infrastructure (V2I) Communication – The Language of the Smart Road

Imagine a world where your EV seamlessly talks to the infrastructure around it, receiving real-time updates on traffic congestion, upcoming hazards, and even optimized traffic light timings [7]. This futuristic vision is becoming a reality with vehicle-to-infrastructure (V2I) communication, the language of the smart road for connected e-mobility.

What Is V2I Communication? V2I communication is a two-way exchange of data between vehicles and roadside infrastructure. Vehicles equipped with V2I technology can "talk" to traffic lights, road signs, and other roadside units (RSUs) using dedicated short-range wireless communication protocols. This enables real-time data exchange, creating a network of intelligent transportation systems.

How Does V2I Work? The core components of V2I communication include onboard units, Infrastructure side unit, communication channels.

The magic of V2X communication lies in its interconnected infrastructure. Imagine a network composed of onboard units (OBUs) installed within EVs and RSUs strategically placed along roads. These OBUs act as transceivers, equipped with communication modules and software to not only process data from the vehicle but also transmit it securely. On the other side of the conversation are RSUs. These RSUs mirror the functionality of OBUs, boasting similar communication modules and software. Their role is to receive data from passing vehicles and transmit relevant information back. The beauty lies in the central network that ties everything together. Data collected from RSUs is aggregated and processed within this centralized hub. This processed information can then be sent back to vehicles in real-time, providing crucial updates on traffic conditions or potential hazards. Additionally, the network can integrate with traffic management systems, allowing for real-time adjustments and optimized traffic flow for everyone on the road. V2X communication paves the way for a safer, more efficient, and ultimately smarter transportation ecosystem.

Benefits of V2I Communication for E-Mobility

- **Enhanced Safety:** V2I can warn drivers of upcoming hazards like accidents, red lights, or slippery roads, allowing for preventive action and improving overall road safety.
- **Improved Traffic Flow:** Real-time traffic data from vehicles can be used to optimize traffic light timings, reducing congestion and travel times for EVs.
- **Reduced Emissions:** Smoother traffic flow due to V2I communication can lead to fewer stop-and-go situations, resulting in lower emissions from EVs.
- **Priority for EVs:** V2I can be used to prioritize EVs at traffic lights, further improving their efficiency and encouraging wider adoption.
- **Enhanced Infrastructure Management:** Data collected from EVs can be used to optimize infrastructure planning, and maintenance scheduling, as well as to improve overall traffic management.

Challenges and Considerations

- **Standardization:** Standardization of communication protocols and data formats is crucial for seamless V2I communication across different vehicle and infrastructure manufacturers.
- **Deployment Cost:** The large-scale deployment of RSUs requires significant investment in infrastructure development.
- **Cybersecurity:** Robust security measures are needed to protect V2I communication networks from cyberattacks and ensure data integrity.

The Future of V2I Communication V2I communication is on the cusp of a revolution. As technology advances, we can expect to see the following.

V2X communication is the nervous system of the future transportation landscape. While V2I takes center stage for autonomous vehicles, it's just one branch of a larger ecosystem. **V2X communication**, encompassing both V2I and vehicle-to-vehicle (V2V) communication, creates a comprehensive network for connected vehicles and infrastructure. Imagine a future where self-driving EVs seamlessly communicate with traffic lights, road signs, and even other vehicles. **Integration with autonomous vehicles** becomes paramount. V2I data will be the lifeblood of safer and more efficient self-driving operations, allowing autonomous vehicles to navigate their environment with real-time awareness. Furthermore, V2I data will be integrated with **advanced traffic management systems**. This translates to real-time route optimization and dynamic traffic control, ushering in an era of smoother traffic flow and a more efficient transportation network for everyone. By enabling communication between EVs and the smart road, V2I technology is paving the way for a future of safer, more efficient, and sustainable transportation. As V2I continues to evolve, it will play a key role in transforming our cities and revolutionizing the way we travel.

15.5.2 Smart Charging Infrastructure

15.5.2.1 Smart Charging Infrastructure: IoT-Enabled Charging Stations – Powering the Future of E-Mobility

The rise of EVs has brought a new challenge: ensuring a robust and efficient charging infrastructure [8, 9, 16]. Here's where IoT-enabled charging stations step in, transforming the way EVs are charged and paving the way for a smarter and more sustainable future.

What Are IoT-Enabled Charging Stations? These charging stations are equipped with sensors, processors, and internet connectivity, allowing them to communicate and exchange data with a central network. This "smartness" unlocks a range of features and benefits.

The magic of connected charging stations lies in their ability to transform the entire EV experience. Imagine a world where frustration is eliminated before you even leave home. **Real-time availability and status** via mobile apps allow users to locate nearby charging stations and see if they are available in real time. No more arriving at a station only to find it occupied! The benefits extend beyond user convenience. **Remote monitoring and diagnostics**, powered by IoT, empower station operators to keep a watchful eye on their infrastructure. The system can identify potential issues with the station's health or performance remotely, allowing for proactive maintenance and minimizing downtime. But connected charging stations are not just about efficiency; they are also environmentally conscious. **Dynamic pricing and load management** utilize IoT to adjust prices based on electricity costs and demand. This incentivizes off-peak charging, which in turn helps utilities manage the strain on the power grid. Furthermore, **smart charging protocols** establish communication between the station and the EV, enabling optimized charging profiles that maximize battery life and efficiency. The cherry on top? Integration with renewable energy sources. Imagine charging stations that prioritize clean energy whenever possible, seamlessly integrating with solar panels or other renewable sources to create a greener transportation ecosystem. This is the future of connected charging stations – a world of convenience, efficiency, and environmental responsibility.

Benefits of IoT-Enabled Charging Stations

- **Improved User Experience:** Convenient station location, real-time availability, and potential for faster charging due to optimized protocols.
- **Reduced Infrastructure Costs:** Predictive maintenance and remote monitoring can minimize downtime and extend the lifespan of charging stations.
- **Enhanced Grid Stability:** Dynamic load management and integration with renewables contribute to a more stable and sustainable power grid.
- **Promotes Wider EV Adoption:** A reliable and efficient charging network assuages range anxiety, encouraging more people to switch to EVs.
- **Data-Driven Insights:** Data collected from charging stations can inform future infrastructure planning and development in line with EV adoption rates.

Challenges and Considerations

- **Standardization:** Standardization of communication protocols and data formats is essential for seamless interoperability between different charging stations and EVs.
- **Cybersecurity:** Robust security measures are needed to protect charging station networks from cyberattacks and ensure secure user data transmission.
- **Grid Modernization:** Large-scale adoption of EVs may necessitate grid upgrades to handle increased demand for electricity.
- **Deployment Cost:** The initial investment in setting up a network of IoT-enabled charging stations can be significant.

The Future of IoT-Enabled Charging Stations The future of charging infrastructure is bright and brimming with innovation:

- **Wireless Charging:** Wireless charging pads at parking spots or even while driving could revolutionize the charging experience, offering ultimate convenience.

- **Bidirectional Charging (Vehicle-to-Grid [V2G]):** EVs could potentially act as power sources, feeding excess energy back into the grid during peak demand periods.
- **Integration with Smart Cities:** Charging stations will seamlessly integrate with smart city infrastructure, providing real-time data for traffic management and energy optimization.

By harnessing the power of IoT, charging stations are evolving beyond mere power points. They are becoming intelligent hubs that will be instrumental in accelerating the transition toward a sustainable and electrified future of transportation.

15.5.2.2 Smart Charging Infrastructure: Smart Grid Integration and Load Balancing – Orchestrating the Flow of Electrons

The widespread adoption of EVs presents both opportunities and challenges for the power grid [11]. While EVs offer a cleaner and more sustainable transportation option, managing the increased demand for electricity during charging periods requires a smarter approach. This is where smart grid integration and load balance come into play, working in concert with IoT-enabled charging stations to orchestrate a smooth flow of electrons and ensure grid stability for the future of e-mobility.

Smart Grid Integration: A Connected Ecosystem The traditional power grid is undergoing a transformation toward a "smart grid." This intelligent network utilizes two-way communication and advanced technologies to optimize energy use and distribution. Here is how EVs and smart grids interact:

- **Real-Time Data Exchange:** IoT-enabled charging stations and EVs can communicate with the smart grid, providing real-time data on charging demand and energy consumption.
- **Dynamic Pricing and Demand Response:** Based on real-time data, the smart grid can adjust electricity prices to incentivize off-peak charging and discourage peak-hour charging. This "demand response" helps manage peak loads and grid stability.
- **Integration with Renewable Energy:** The smart grid can prioritize charging EVs with renewable energy sources like solar or wind when available, further reducing reliance on fossil fuels.

Load Balancing: Keeping the Grid in Equilibrium Load balancing refers to the process of distributing the electrical load evenly across the grid to prevent overloading and potential blackouts. Here is how EVs and smart grids work together for load balancing:

- **Smart Charging Protocols:** Communication between the smart grid and EVs allows for optimized charging profiles. EVs can be charged slowly during peak hours and at a faster rate during off-peak hours, reducing the overall demand on the grid at any given time.
- **V2G Technology:** Emerging V2G technology allows EVs to act as mobile energy storage units. During periods of high demand, EVs can potentially feed excess stored energy back into the grid, alleviating strain and contributing to grid stability.
- **Distributed Energy Resources:** The smart grid can integrate with other distributed energy resources, such as rooftop solar panels or home energy storage systems, to further diversify the energy mix and reduce reliance on centralized power plants.

Benefits of Smart Grid Integration and Load Balancing

- **Enhanced Grid Stability:** By managing charging demand and integrating EVs with renewable energy sources, the grid can become more resilient and efficient.
- **Reduced Reliance on Fossil Fuels:** Prioritizing off-peak charging and utilizing renewable energy sources helps minimize reliance on fossil fuels and contributes to a cleaner environment.

- **Lower Electricity Costs:** Efficient load balancing can lead to lower electricity costs for both EV owners and the overall grid system.
- **Promotes Wider EV Adoption:** A stable and efficient grid with managed charging infrastructure addresses concerns about grid overload and encourages wider EV adoption.

Challenges and Considerations

- **Grid Modernization:** Large-scale EV adoption may necessitate significant investments in grid modernization to accommodate the increased demand and enable two-way communication.
- **Consumer Behavior:** Encouraging off-peak charging habits and educating users about smart charging practices is crucial for optimizing grid management.
- **Data Security and Privacy:** Robust security measures are essential to protect sensitive data exchanged between EVs, charging stations, and the smart grid.

The Future of Smart Grid Integration and Load Balancing The future of smart grid integration and load balancing is full of potential:

- **Advanced Algorithms and Machine Learning:** Machine learning can be used to predict charging patterns and optimize load-balancing strategies in real time.
- **Advanced Energy Storage Solutions:** Development of more efficient and cost-effective energy storage solutions will further enhance the grid's ability to integrate EVs effectively.
- **Decentralized Grid Models:** The future may see a shift toward more decentralized grid models with local energy generation and consumption, further improving grid resilience and sustainability.

By fostering a collaborative relationship between the power grid, EVs, and renewable energy sources, smart grid integration and load balancing offer a pathway toward a sustainable and efficient future for e-mobility. As technology continues to evolve, this intelligent orchestration of electrons will ensure a smooth and reliable flow of energy for a cleaner and more electrified transportation landscape.

15.5.2.3 Smart Charging Infrastructure: Wireless Charging Technologies – The Future of Convenience

Imagine a world where charging your EV becomes as effortless as parking your car [11]. Wireless charging technologies are on the horizon, promising a future of convenience and potentially revolutionizing the way we charge EVs.

What Is Wireless Charging for EVs? Wireless charging eliminates the need for physical cables. Instead, electromagnetic energy is transferred wirelessly between a charging pad embedded in the ground and a receiver coil mounted on the underside of the EV.

Types of Wireless Charging Technologies There are two main approaches to wireless charging for EVs:

- **Inductive Charging:** This widely used technology relies on a primary coil in the charging pad and a secondary coil in the EV's receiver. When these coils are close together, an alternating magnetic field is created, inducing a current in the secondary coil, which charges the EV battery.
- **Resonant Charging:** This technology uses a specific resonant frequency to transfer energy wirelessly. The charging pad and the EV's receiver coil are tuned to the same frequency, enabling efficient energy transfer even with some misalignment between the pad and the vehicle.

Benefits of Wireless Charging

- **Enhanced Convenience:** Wireless charging eliminates the need to fumble with cables, significantly improving the user experience.
- **Reduced Wear and Tear:** By eliminating the need to constantly plug and unplug cables, wireless charging can minimize wear and tear on charging connectors and ports.
- **Weatherproof Operation:** Wireless charging systems are not susceptible to weather elements like rain or snow, unlike traditional cable-based charging.
- **Potential for In-Motion Charging:** Future advancements may enable wireless charging while EVs are parked or even while driving on specially equipped roads, maximizing charging efficiency.
- **Integration with Smart Grids:** Wireless charging systems can be integrated with smart grids to optimize charging based on real-time electricity availability and cost.

Challenges and Considerations

- **Efficiency:** Currently, wireless charging systems have lower efficiency compared to traditional cable-based charging due to energy losses during the wireless transfer.
- **Cost and Infrastructure Development:** The upfront cost of installing wireless charging pads can be higher than traditional charging stations.
- **Standardization:** Standardization of charging pad formats and power levels is crucial for interoperability between different EVs and charging systems.
- **Foreign Object Detection:** Robust foreign object detection systems are necessary to prevent the charging pad from activating when metallic objects are present on the ground.

The Future of Wireless Charging Technologies Wireless charging for EVs is still in its early stages, but advancements are happening rapidly:

- **Improved Efficiency:** Research and development are focused on improving the efficiency of wireless charging systems, minimizing energy losses, and maximizing charging speed.
- **Higher Power Levels:** Future wireless charging systems are expected to handle higher power levels, enabling faster charging times for EVs.
- **Dynamic Charging:** The concept of dynamic wireless charging, where EVs can be charged while in motion, is being explored for potential future implementation on highways or designated lanes.

Wireless charging technologies hold immense promise for the future of e-mobility. As these technologies evolve, they have the potential to transform the charging experience, improve convenience, and encourage wider EV adoption, paving the way for a cleaner and more sustainable transportation landscape.

15.5.3 Predictive Maintenance and Diagnostics

15.5.3.1 Predictive Maintenance and Diagnostics: Remote Monitoring of Vehicle Health – Crystal Balls for Connected EVs

EVs are brimming with technology, but what if they could predict their own problems? This is the magic of predictive maintenance and diagnostics, utilizing remote monitoring of vehicle health to ensure a smooth ride and prevent unexpected breakdowns.

Remote Monitoring: Keeping an Eye on Your EV Imagine a system that constantly monitors your EV's health, even when you are not behind the wheel. This is what remote monitoring achieves:

- **In-Vehicle Sensors:** A network of sensors collects data on various aspects of the EV, including battery health, motor performance, temperature readings, and other vital parameters.
- **Telematics Unit:** This unit acts as the brain, processing sensor data and transmitting it securely to a cloud platform via cellular networks.
- **Cloud-Based Analytics:** Powerful analytics tools analyze the collected data, identifying trends and potential issues.

Predictive Maintenance: Addressing Problems Before They Arise By analyzing historical data and identifying patterns, the system can predict potential problems before they escalate into major breakdowns. This allows for the following:

- **Proactive Maintenance Alerts:** EV owners receive alerts recommending maintenance actions, such as battery cooling system checks or software updates, preventing future failures.
- **Optimized Service Scheduling:** Service centers can schedule maintenance based on the predicted needs of the EV, improving service efficiency and minimizing downtime for owners.
- **Extended Vehicle Lifespan:** By addressing potential issues early on, predictive maintenance helps extend the lifespan of the EV's components and battery.

Benefits of Remote Monitoring and Predictive Maintenance

- **Reduced Downtime:** Early detection and prevention of problems minimize the risk of unexpected breakdowns, keeping your EV on the road.
- **Lower Maintenance Costs:** Addressing issues proactively can prevent costly repairs down the line.
- **Improved Safety:** By identifying potential safety hazards early on, predictive maintenance contributes to a safer driving experience.
- **Enhanced Peace of Mind:** EV owners can enjoy peace of mind knowing their vehicles are constantly monitored and potential problems are identified before they become major issues.
- **Data-Driven Insights:** The data collected from remote monitoring can be used by manufacturers to improve future EV designs and enhance overall vehicle reliability.

Challenges and Considerations

- **Data Security and Privacy:** Robust security measures are essential to protect sensitive vehicle data transmitted during remote monitoring.
- **User Acceptance:** Some EV owners may have concerns about data privacy or the potential for over-reliance on predictive systems. Clear communication and user education are crucial.
- **Cost of Implementation:** The widespread adoption of predictive maintenance systems may require investment in infrastructure and technology.

The Future of Predictive Maintenance and Diagnostics The future of predictive maintenance is brimming with possibilities [13]:

- **Advanced Machine Learning:** Machine learning algorithms will become even more sophisticated, enabling more accurate predictions and earlier detection of potential problems.
- **Integration with Smart Grids:** Data from remote monitoring systems could be integrated with smart grids to optimize charging strategies and grid stability, particularly for fleets of EVs.

- **OTA Updates:** Predictive diagnostics could trigger OTA software updates for the EV's control systems, addressing potential issues remotely without the need for a service visit.

By offering a glimpse into the future of your EV's health, predictive maintenance and diagnostics are a game-changer for e-mobility. As these technologies evolve, they will play a critical role in ensuring a reliable, safe, and cost-effective driving experience, paving the way for a more sustainable transportation future.

15.5.3.2 Predictive Fault Detection and Alerts: Spotting Trouble Before It Starts

Within the realm of predictive maintenance and diagnostics for EVs, predictive fault detection and alerts play a starring role. Imagine your EV acting as its own fortune teller, identifying potential problems before they morph into major breakdowns. This is exactly what this technology strives for, keeping your EV running smoothly and preventing unexpected detours.

How Does Predictive Fault Detection Work? Predictive fault detection relies on a powerful combination of:

- **In-Vehicle Sensors:** A network of sensors strategically placed throughout the EV constantly monitor critical parameters like battery health, motor performance, temperature readings, vibration levels, and other vital data points.
- **Telematics Unit:** This unit acts as the data hub, collecting sensor data and transmitting it securely to a cloud platform via cellular networks.
- **Cloud-Based Analytics:** Powerful analytics engines in the cloud analyze the incoming data stream in real time. Machine learning algorithms are trained on historical data to identify patterns and anomalies that could signal potential faults.

Predictive Alerts: Taking Action Before Trouble Arrives When the cloud-based analytics detect a potential issue, the system swings into action:

- **Alerts and Notifications:** EV owners receive timely alerts via mobile apps or email, informing them about the potential fault and recommending appropriate action. These could be suggestions like scheduling a service appointment for a battery check or visiting a service center for a software update.
- **Severity Levels:** The alerts can be categorized based on severity, allowing owners to prioritize critical issues that require immediate attention.
- **Information and Recommendations:** The alerts may also provide additional information about the potential fault and suggest actions to take to minimize further damage or maintain safe operation until a service appointment is scheduled.

Benefits of Predictive Fault Detection and Alerts

- **Reduced Downtime:** Early detection of potential faults allows for proactive maintenance, preventing breakdowns and keeping your EV on the road.
- **Lower Repair Costs:** Addressing potential issues early on can prevent costly repairs down the line by identifying problems before they escalate.
- **Enhanced Safety:** By identifying potential safety hazards like a failing battery or a faulty motor, predictive fault detection contributes to a safer driving experience.
- **Improved Peace of Mind:** EV owners can enjoy peace of mind knowing their vehicles are constantly monitored, and potential issues are flagged before they become major concerns.
- **Data-Driven Optimization:** The data collected from fault detection systems can be used by manufacturers to improve future EV designs and enhance overall component reliability.

Challenges and Considerations

- **Data Security and Privacy:** Robust security measures are essential to protect sensitive vehicle data transmitted during the fault detection process.
- **False Positives:** The system needs to be fine-tuned to minimize false positives, preventing unnecessary service visits due to misinterpreted data.
- **User Education:** EV owners need to understand the system's capabilities and limitations, along with the importance of acting upon the received alerts.

The Future of Predictive Fault Detection and Alerts The future of predictive fault detection holds immense promise in the following areas.

- **Advanced Machine Learning:** Machine learning algorithms will become even more sophisticated, enabling more accurate fault detection and earlier identification of potential problems.
- **Integration with Augmented Reality (AR):** AR technology could be used to provide technicians with real-time information about potential faults, simplifying and speeding up the repair process.
- **Proactive Part Replacement:** Based on predicted faults, critical parts nearing failure could be preemptively replaced during scheduled maintenance, minimizing downtime and maximizing vehicle uptime.

By acting as a proactive guardian for your EV's health, predictive fault detection and alerts empower you to take control and prevent problems before they arise. As technology advances, this system will become even more refined, ensuring a smoother, safer, and more reliable driving experience for the future of e-mobility.

15.5.3.3 Predictive Maintenance and Diagnostics: Proactive Maintenance Scheduling – Scheduling Service Based on Needs, Not Mileage

In the world of EVs, predictive maintenance and diagnostics are revolutionizing how we care for these vehicles. One key aspect of this transformation is proactive maintenance scheduling. Gone are the days of blindly following a manufacturer's service schedule based solely on mileage. Proactive maintenance scheduling leverages the power of data and analytics to tailor service needs to the specific health of your EV, optimizing maintenance efficiency and maximizing vehicle lifespan.

How Does Proactive Maintenance Scheduling Work? This innovative approach relies on the following key elements:

- **In-Vehicle Sensors:** A network of sensors embedded throughout the EV continuously monitors critical parameters like battery health, motor performance, cooling system efficiency, and various fluid levels.
- **Telematics Unit:** This unit acts as the data collection hub, gathering sensor data and securely transmitting it to a cloud platform via cellular networks.
- **Cloud-Based Analytics:** Powerful analytics engines in the cloud analyze the incoming data stream. Machine learning algorithms, trained on vast datasets, identify trends, predict potential issues, and assess the overall health of the EV.

Shifting from Mileage-Based to Condition-Based Service Traditionally, car maintenance schedules are based on mileage intervals. Proactive maintenance scheduling flips this script:

- **Customized Service Plans:** Instead of a one-size-fits-all approach, service schedules are tailored to the specific conditions and usage patterns of each EV. Factors like driving habits, climate conditions, and rapid charging frequency are all considered.
- **Early Detection and Intervention:** By identifying potential issues before they escalate, proactive scheduling allows for timely maintenance, preventing breakdowns and minimizing repair costs.
- **Optimized Service Intervals:** EVs with efficient driving habits and optimal operating conditions may require service less frequently than those driven in harsh environments or subjected to frequent fast charging.

Benefits of Proactive Maintenance Scheduling

- **Reduced Downtime:** Early detection and prevention of problems minimizes the risk of unexpected breakdowns, keeping your EV on the road.
- **Lower Maintenance Costs:** Addressing issues early on can prevent costly repairs down the line.
- **Extended Vehicle Lifespan:** By providing targeted maintenance based on the EV's actual needs, proactive scheduling helps extend the lifespan of the battery and other critical components.
- **Improved Efficiency:** Regular maintenance of cooling systems and other performance-related components optimizes energy efficiency, potentially leading to a longer driving range.
- **Data-Driven Insights:** The data collected from proactive maintenance systems can be used by manufacturers to improve future EV designs and enhance overall component durability.

Challenges and Considerations

- **Data Security and Privacy:** Robust security measures are essential to protect sensitive vehicle data transmitted during the maintenance scheduling process.
- **User Acceptance:** Some EV owners may be accustomed to traditional mileage-based schedules. Clear communication and education are crucial to establish trust in the proactive approach.
- **System Integration:** Ensuring seamless integration between in-vehicle sensors, telematics units, and cloud-based analytics platforms is vital for effective proactive maintenance scheduling.

The Future of Proactive Maintenance Scheduling The future of proactive maintenance scheduling is bright and brimming with possibilities:

- **Advanced Diagnostics:** As diagnostic capabilities evolve, the system will be able to predict not just potential failures but also the remaining lifespan of critical components, enabling even more precise maintenance scheduling.
- **Integration with Service Providers:** Proactive maintenance systems could seamlessly connect with service providers, allowing for automated appointment scheduling based on the EV's detected needs.
- **Personalized Recommendations:** The system could provide personalized recommendations for optimizing EV performance and efficiency based on individual driving habits and usage patterns.

By taking a data-driven approach to service scheduling, proactive maintenance empowers EV owners to optimize their vehicle's health and performance. As technology advances, this system will become even more sophisticated, paving the way for a future of predictive and personalized EV care.

15.5.4 Advanced Driver-Assistance Systems (ADAS)

15.5.4.1 Advanced Driver-Assistance Systems (ADAS) – Sensor Fusion: Seeing the Bigger Picture for Enhanced Safety

Imagine driving down the road with a virtual co-pilot constantly monitoring your surroundings, keeping you informed and safe. This is the promise of ADAS, and sensor fusion plays a pivotal role in making it a reality. Let us delve into how sensor fusion elevates situational awareness for a safer driving experience.

What Are ADAS? ADAS [14] are a suite of technologies that work together to enhance driver safety and improve vehicle control. These systems utilize various sensors to gather data about the vehicle's surroundings and use this information to:

- **Warn Drivers of Potential Hazards:** Blind-spot detection, LDW, and forward collision warning are some examples (Figure 15.4).
- **Provide Semi-Autonomous Driving Assistance:** Features like adaptive cruise control and lane centering assist can take some of the burden off the driver.

What Is Sensor Fusion? Sensor fusion is the heart of ADAS, where data from multiple sensors is combined and analyzed to create a comprehensive picture of the driving environment. Here are the key sensors involved:

- **Cameras:** Provide visual information about the road, traffic lights, lane markings, and pedestrians.
- **Radar:** Uses radio waves to detect objects with high precision, particularly effective in low-light conditions.
- **LiDAR:** Emits laser pulses to create a highly detailed 3D map of the surroundings.
- **Ultrasonic Sensors:** Short-range sensors are ideal for detecting nearby obstacles during parking maneuvers.

Benefits of Sensor Fusion for Enhanced Situational Awareness

- **Improved Accuracy and Reliability:** Combining data from various sensors reduces the reliance on any single source, leading to more accurate hazard detection and system performance, especially in challenging weather conditions.

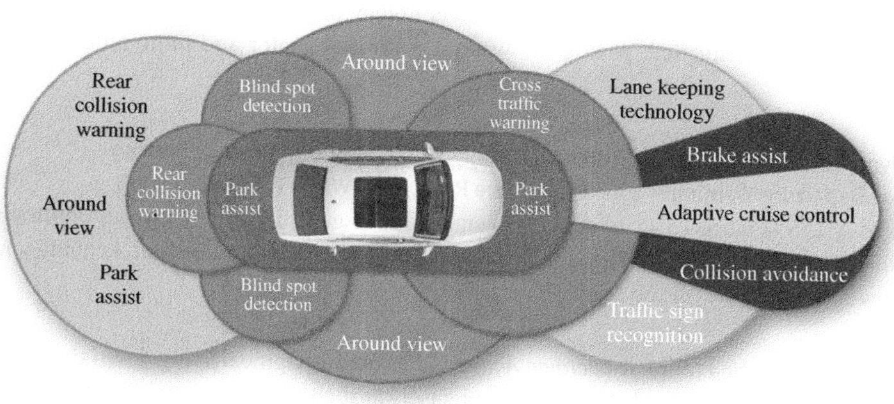

Figure 15.4 Advanced driver assistance system feature.

- **Reduced False Positives:** By cross-referencing data from multiple sensors, the system can differentiate between real threats and potential false alarms, minimizing unnecessary driver distraction.
- **Comprehensive Understanding of the Environment:** Sensor fusion provides a richer picture of the surroundings, allowing the ADAS system to anticipate potential hazards and react accordingly. This includes factors like the size, speed, and direction of nearby objects.
- **Enhanced Decision-Making:** With a more complete understanding of the situation, the ADAS system can provide more precise and timely warnings to driver, enabling them to make informed decisions and take corrective actions.

Challenges and Considerations

- **Sensor Limitations:** Each sensor has its own limitations. Cameras struggle in low-light conditions, while radar may have difficulty differentiating between objects. Sensor fusion helps mitigate these limitations, but it's not a perfect solution.
- **Data Processing Power:** Fusing data from multiple sensors requires significant processing power. This necessitates powerful on-board computers within the vehicle to handle the complex calculations in real time.
- **Cost and Complexity:** Implementing ADAS with sensor fusion can add to the vehicle's cost. Additionally, the complexity of these systems requires robust software development and rigorous testing to ensure reliable functionality.

The Future of Sensor Fusion for ADAS The future of sensor fusion for ADAS is on an exciting trajectory:

- **Advanced Sensor Technologies:** Advancements in sensor technology, like higher resolution cameras and longer-range LiDAR, will further improve the accuracy and detail of the perceived environment.
- **AI Integration:** AI algorithms will play a more prominent role in processing sensor data, enabling real-time object recognition, behavior prediction, and even hazard forecasting.
- **V2X Communication:** Communication with other vehicles and roadside infrastructure will provide ADAS systems with additional data points, creating an even more comprehensive picture of the driving landscape.

By leveraging the power of sensor fusion, ADAS systems are constantly evolving to provide drivers with an extra layer of safety and support. As technology progresses, we can expect ADAS to become even more sophisticated, paving the way for a future of safer and more autonomous driving experiences.

15.5.4.2 Advanced Driver-Assistance Systems (ADAS) – Autonomous Emergency Braking (AEB) and Lane Departure Warning – Your Eyes on the Road, Even When You Blink

In the realm of ADAS (Figure 15.4), two features stand out as guardians of safety: autonomous emergency braking (AEB) and LDW. These technologies act as your watchful copilot, keeping you and your fellow motorists safe on the road.

Autonomous Emergency Braking (AEB): A Guardian Against Collisions Imagine driving down the highway when a car suddenly stops ahead. AEB steps in precisely at such moments, applying automatic braking to prevent a collision or significantly reduce its impact. Here is how it works:

- **Sensor Network:** AEB relies on a combination of sensors like radar, cameras, or LiDAR to detect objects in front of the vehicle.
- **Collision Imminence Detection:** Sophisticated algorithms analyze the distance, speed, and relative position of the approaching object. If a collision is deemed imminent, the system prepares to intervene.
- **Automatic Braking:** If the driver does not react in time, AEB automatically applies brakes to slow down the vehicle or bring it to a complete stop, potentially mitigating the severity of the collision or even preventing it altogether.

Benefits of Autonomous Emergency Braking

- **Reduced Rear-End Collisions:** AEB is particularly effective in preventing or mitigating the impact of rear-end collisions, a common occurrence on busy roads.
- **Enhanced Safety for Vulnerable Road Users:** AEB can also help prevent collisions with pedestrians or cyclists who may suddenly appear in front of the vehicle.
- **Improved Driver Confidence:** The presence of AEB can provide drivers with a sense of security, knowing they have an extra layer of protection in case of emergencies.
- **Potential for Lower Insurance Costs:** Some insurance companies offer discounts for vehicles equipped with AEB due to the reduced risk of accidents.

Lane Departure Warning (LDW): Keeping You in Your Lane Ever drifted out of your lane unintentionally? LDW acts as a gentle nudge, reminding you to stay within the designated lane markings. Here is how it works:

- **Lane Detection System:** A forward-facing camera detects lane markings on the road.
- **Unintentional Lane Departure:** If the vehicle starts to drift out of its lane without the turn signal activated, the LDW system triggers an alert.
- **Audio or Visual Alerts:** The driver is typically warned through an audible beep, steering wheel vibration, or a visual indicator on the dashboard.

Benefits of Lane Departure Warning

- **Reduced Risk of Run-Off-Road Collisions:** LDW helps prevent drivers from unintentionally drifting off the road and colliding with oncoming traffic or objects on the shoulder.
- **Minimizes Sideswipe Accidents:** By alerting drivers about lane departure, LDW can help prevent accidents caused by drifting into neighboring lanes.
- **Combats Driver Drowsiness:** LDW can be especially beneficial on long journeys when driver fatigue can set in, and attention may wane.

Limitations of ADAS: Not a Replacement for Safe Driving It is important to remember that ADAS features like AEB and LDW are driver-assistance systems, not replacements for safe driving practices. Here are some key points to consider:

- **System Limitations:** Sensors may have blind spots or limitations in certain weather conditions. Always stay alert and prioritize your own judgment.
- **Driver Responsibility:** It is the driver's responsibility to remain focused on the road and be prepared to take corrective action whenever necessary.
- **False Alarms:** While rare, these systems can generate occasional false alarms. Do not become reliant solely on the ADAS features and maintain situational awareness.

The Future of ADAS The future of ADAS is bright, with continuous advancements in sensor technology, software algorithms, and computing power. We can expect to see the following:

- **More Sophisticated Features:** Future ADAS systems may incorporate features like automatic lane-changing assistance and emergency evasive steering.
- **Improved System Reliability:** As technology matures, false alarms will become less frequent, and ADAS systems will become even more reliable.
- **Greater Vehicle Autonomy:** The evolution of ADAS paves the way for increasing levels of vehicle autonomy, ultimately leading to a future of safer and more efficient transportation.

By providing an extra layer of safety on the road, ADAS features like AEB and LDW are transforming the driving experience. Remember, these are valuable tools, but safe driving practices will always be your most important asset on the road

15.5.4.3 ADAS Meets IoT: Integration with IoT Data for Route Optimization and Traffic Management – A Powerful Alliance for Smarter Roads and Safer Journeys

ADAS are revolutionizing driving by enhancing safety and offering a glimpse into the future of autonomous vehicles. However, ADAS can become even more powerful when it joins forces with the vast network of data collected by IoT devices. This potent combination unlocks exciting possibilities for route optimization, traffic management, and ultimately, a smarter transportation ecosystem.

How ADAS and IoT Data Work Together Imagine a scenario where your car is not just aware of its immediate surroundings but also has access to real-time information about the broader traffic landscape. This is the magic of ADAS and IoT data integration:

- **V2X Communication:** ADAS-equipped vehicles become active participants in a network, communicating with other vehicles (V2V) and roadside infrastructure (V2I) equipped with IoT sensors.
- **Real-Time Traffic Data:** Traffic lights, congestion sensors, and connected infrastructure devices collect and share real-time data on traffic flow, accidents, and road closures.
- **Cloud-Based Processing:** This vast amount of data is processed in the cloud, generating insights into traffic patterns, potential bottlenecks, and alternative routes.

Optimizing Routes for a Smoother Journey By leveraging this data, ADAS systems can offer a range of benefits to drivers:

- **Dynamic Route Guidance:** Navigation systems can consider real-time traffic conditions and suggest alternative routes to avoid congestion and save time and fuel.
- **Accident Alerts and Detours:** ADAS can receive immediate alerts about accidents or road closures, allowing drivers to reroute and avoid potential delays.
- **Improved Efficiency:** By optimizing routes based on real-time data, traffic flow can become smoother, leading to reduced emissions and fuel consumption.

Enhancing Traffic Management for a Smarter City The integration of ADAS and IoT data also benefits traffic management authorities:

- **Real-Time Traffic Monitoring:** City authorities gain a comprehensive view of traffic flow across the city, enabling them to dynamically adjust traffic light timings and deploy resources effectively.

- **Congestion Prediction and Prevention:** By analyzing traffic patterns, authorities can predict potential congestion points and take proactive measures to prevent gridlock.
- **Improved Infrastructure Planning:** Traffic data can inform future infrastructure development projects, ensuring roads cater to the evolving traffic patterns of a smart city.

Challenges and Considerations

- **Standardization:** Standardization of data formats and communication protocols is crucial for seamless V2X communication between vehicles and infrastructure.
- **Privacy Concerns:** Data security and privacy measures need to be robust to ensure the confidentiality of driver and traffic information.
- **Infrastructure Investment:** Upgrading existing infrastructure with IoT sensors and communication networks necessitates significant investment.

The Future of ADAS and IoT Integration The future of ADAS and IoT integration is brimming with potential:

- **Cooperative Maneuvers:** Vehicles may communicate and coordinate maneuvers, like lane changes or merging, to improve traffic flow and safety.
- **Integration with Autonomous Vehicles:** As autonomous vehicles become a reality, ADAS and IoT data will play a vital role in their safe navigation and decision-making processes.
- **Smart City Planning:** Traffic data will be a cornerstone of smart city planning, enabling the creation of more efficient and sustainable transportation networks.

By combining the power of ADAS with the vast intelligence of IoT data, we are paving the way for a future of smarter, safer, and more efficient transportation. This powerful alliance will not only enhance individual driving experiences but also transform our cities into connected ecosystems that prioritize smooth traffic flow, environmental sustainability, and overall well-being.

15.5.5 Connected E-Mobility Services

The world of EVs is undergoing an exciting transformation, driven by connected e-mobility services. These services leverage the power of data and connectivity to provide a more convenient, efficient, and personalized EV ownership experience. Here is how connected e-mobility services are revolutionizing EV charging:

15.5.5.1 Real-Time Charging Station Availability and Navigation

Imagine a world where finding an available charging station is as easy as hailing a ride. Connected e-mobility services make this a reality.

Comprehensive Charging Station Database: These services maintain a constantly updated database of charging stations, including location information, real-time availability status (available, occupied, or out of order), and charger type (alternating current [AC] or direct current [DC], kW rating).

Seamless Navigation Integration: The charging station information is seamlessly integrated with navigation apps, allowing drivers to easily locate available stations along their planned route or find the nearest station when needed.

Advanced Filtering Options: Users can filter search results based on factors like charger type, charging speed, and amenities offered at the charging station (e.g., restrooms and restaurants).

Benefits:

- **Reduced Range Anxiety:** Knowing exactly where to find an available charging station eliminates range anxiety, a major concern for many EV owners.
- **Optimized Trip Planning:** Drivers can factor charging stops into their journeys seamlessly, ensuring they reach their destination without any worries.
- **Improved Efficiency:** Real-time data helps drivers locate the most suitable charging station, minimizing wasted time searching for unoccupied stations.

15.5.5.2 Personalized Charging Plans Based on Usage Patterns

Connected e-mobility services go beyond just finding chargers; they can also help you optimize your charging habits.

Connected e-mobility puts you in the driver's seat (quite literally) when it comes to managing your EV's energy consumption. The system goes beyond simply providing information; it analyzes your driving data to offer **personalized charging recommendations**. This data includes details like trip lengths, how often you charge, and your preferred charging locations. By understanding your habits, the service can suggest optimal charging strategies. Imagine a condition like slower overnight charging at home for your daily commute, maximizing efficiency and potentially benefiting from lower electricity rates. For longer trips, it might suggest faster DC charging to get you back on the road quicker. The system can even help you save money. Through **usage pattern analysis**, it may identify cost-effective charging options. This could involve informing you about time-of-day tariffs, where electricity prices fluctuate, or recommending subscriptions to charging networks that offer better rates. By taking advantage of these personalized recommendations, you can optimize your charging routine and keep your EV running smoothly and cost-effectively.

Benefits

- **Reduced Charging Costs:** By optimizing charging habits, you can minimize reliance on expensive DC fast charging and leverage more cost-effective charging options whenever possible.
- **Improved Battery Health:** Personalized recommendations can help maintain optimal battery health by avoiding unnecessary frequent fast charging, which can degrade battery life over time.
- **Enhanced User Experience:** A personalized approach to charging makes EV ownership more convenient and aligns with your individual driving needs.

The Future of Connected E-Mobility Services The future of connected e-mobility services (Figure 15.5) is brimming with exciting possibilities:

- **Predictive Maintenance:** By analyzing charging data, the system could predict potential battery issues and recommend preventative maintenance to maximize battery lifespan.
- **Integration with Smart Grids:** Connected e-mobility services could interact with smart grids, enabling EVs to charge during off-peak hours when electricity demand is lower, contributing to grid stability and potentially reducing charging costs.
- **V2G Integration:** Future advancements may enable EVs to not only charge from the grid but also feed electricity back into the grid during peak hours, creating a more sustainable energy ecosystem.

By harnessing the power of connectivity, connected e-mobility services are transforming the EV charging experience. These services offer real-time information, personalized recommendations, and a glimpse into a future where EVs seamlessly integrate with the broader energy grid. As technology continues to evolve, connected e-mobility services will play a pivotal role in accelerating the adoption of EVs and paving the way for a cleaner and more sustainable transportation landscape.

Figure 15.5 Future of connected E-mobility.

15.5.5.3 Connected E-Mobility Services: A Match Made in Mobility Heaven

The convergence of connected e-mobility services [15], ride-hailing/car-sharing platforms, and eco-driving recommendations is creating a transportation revolution. Imagine a world where seamlessly connected services empower eco-conscious mobility, benefitting both users and the environment. Here is how this powerful integration is transforming the transportation landscape.

Connected E-Mobility Meets Ride-Hailing/Car-Sharing Imagine a seamless experience where you can not only hail an EV but also see its capabilities in real time. Connected e-mobility services can integrate with ride-hailing and car-sharing platforms, offering **streamlined EV access** (Figure 15.6). This means you will be able to easily identify and book EVs directly within the existing apps you already use. No more switching between apps – everything is conveniently consolidated.

But convenience goes beyond booking. Transparency is key. Passengers will have access to **real-time battery information** before booking a ride. This includes the estimated range of the EV and the health of its battery. Knowing this upfront ensures a smooth journey without any unexpected range anxiety.

The integration does not stop there. The platforms can also display nearby charging stations within the app through **charging infrastructure integration**. This equips drivers with the knowledge of the nearest charging options, allowing them to conveniently plan for mid-trip top-ups if needed. This connected ecosystem makes choosing an EV for your ride-hailing or car-sharing needs a more informed and convenient decision.

Benefits

- **Increased EV Adoption:** By making EVs readily accessible through familiar platforms, this integration can encourage the widespread adoption of EV.

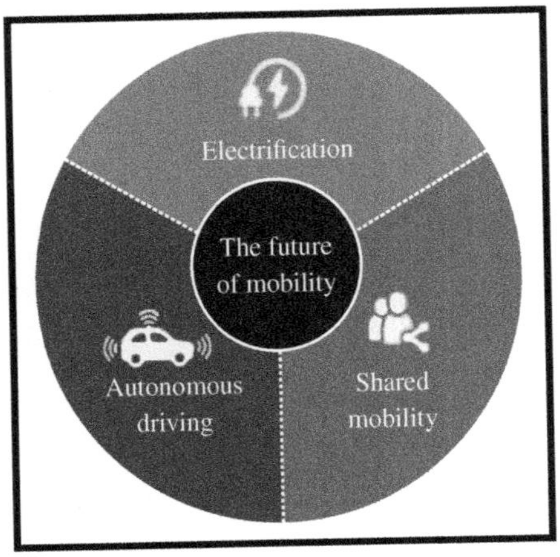

Figure 15.6 Future of mobility.

- **Reduced Emissions:** A shift toward EVs in ride-hailing and car-sharing significantly reduces overall fleet emissions, contributing to cleaner air and a healthier environment.
- **Enhanced User Experience:** A seamless booking and charging experience makes using EVs for ride-hailing and car-sharing convenient and user-friendly.

Eco-driving Recommendations Through Route Optimization Connected e-mobility services can leverage data to promote eco-friendly driving.

Connected e-mobility takes eco-driving to the next level by optimizing your journey for both efficiency and battery life. **Real-time traffic analysis** allows navigation systems to consider congestion and suggest routes that minimize stop-and-go situations, notorious for draining battery life in EVs. Imagine a system that anticipates traffic jams and reroutes you for a smoother ride, maximizing your battery range. The system's intelligence goes beyond traffic. **Terrain analysis** factors in elevation changes. By suggesting routes with minimal inclines, it can minimize energy consumption, as uphill driving demands more power from the battery. This data-driven approach ensures you get the most out of every charge. Finally, connected e-mobility becomes your personal eco-driving coach. **Driving style coaching** provides real-time feedback on your habits. Imagine gentle reminders to maintain constant speed or avoid harsh acceleration, both of which can significantly impact battery efficiency. By incorporating these suggestions into your driving, you can become a more eco-conscious driver and extend your EV's range, making every journey kinder to the environment.

Benefits

- **Extended EV Range:** Eco-friendly route optimization helps maximize EV range by minimizing unnecessary energy consumption.
- **Reduced Energy Costs:** By promoting fuel-efficient driving practices, users can potentially save money on charging costs.
- **Environmental Impact Reduction:** Eco-driving contributes to lower carbon emissions, leading to a cleaner and more sustainable transportation landscape.

Challenges and Considerations

- **Standardization:** Ensuring seamless data exchange between connected e-mobility services, ride-hailing/car-sharing platforms, and EV charging infrastructure is crucial.

- **User Education:** Raising awareness about eco-driving practices and the benefits of connected e-mobility services is essential for widespread adoption.
- **Privacy Concerns:** Data security and privacy measures need to be robust to protect user information and driving habits.

The Future Holds Promise The future of this integrated approach is bright.

The future of connected e-mobility promises a hyper-personalized experience that goes beyond just convenience. Imagine eco-routing that not only considers efficiency but also your preferences. **Personalized eco-routing** could allow you to choose between the fastest route or the most fuel-efficient one, tailoring the journey to your needs. This personalization extends beyond the individual. With **V2X integration**, EVs can communicate with infrastructure in real time. This allows for a more holistic approach to traffic management. Imagine your car receiving information about traffic lights, congestion levels, and even available charging stations. By integrating this data, connected e-mobility services can suggest eco-friendly routes that consider not just distance but also real-time conditions, leading to a smoother and greener driving experience for everyone. But eco-friendly choices should not feel like a chore. **Gamification and incentives** can play a role here. Imagine being rewarded with points or badges for adopting sustainable transportation choices like eco-driving. These playful nudges can encourage positive behavior and make opting for a cleaner commute more engaging and rewarding. By combining the power of connected e-mobility services, ride-hailing/car-sharing platforms, and eco-driving recommendations, we are creating a transportation ecosystem that prioritizes sustainability, efficiency, and user convenience. This collaborative approach paves the way for a future where electric mobility becomes the norm, leading to a cleaner and more sustainable transportation landscape for all.

15.5.6 Security and Privacy Considerations in Connected E-Mobility Services

Connected e-mobility services offer a plethora of benefits, but they also introduce new security and privacy challenges. Here is a breakdown of key considerations to ensure a safe and trustworthy experience.

15.5.6.1 Data Encryption and Authentication Protocols

- **Encryption:** All data transmitted between EVs, charging stations, cloud platforms, and mobile apps should be encrypted using robust algorithms like AES-256. This safeguards sensitive information like battery health, location data, and user profiles from unauthorized access.
- **Authentication Protocols:** Strong authentication protocols like mutual transport layer security (MTLS) ensure that only authorized devices and entities can communicate within the connected e-mobility ecosystem. This prevents unauthorized access attempts and protects against man-in-the-middle attacks.

15.5.6.2 Secure Communication Between Vehicles and Infrastructure

- **Secure Communication Channels:** Dedicated and secure communication channels are essential for data exchange between EVs and infrastructure like charging stations or RSUs. This minimizes the risk of data interception by malicious actors.
- **Vulnerability Management:** Regular vulnerability assessments and prompt patching of identified weaknesses in-vehicle software and charging station firmware are crucial to address potential security exploits.

15.5.6.3 User Privacy Controls and Data Ownership

The success of connected e-mobility hinges on building trust with users. This means prioritizing transparency and user consent. EV users deserve clear and concise information about what data is collected from their vehicles (battery health, driving patterns, etc.), how it is used (optimizing charging infrastructure, for example), and with whom it is shared (authorized entities within the connected e-mobility ecosystem). Furthermore, users should have the right to explicitly **consent** to this data collection and usage. This empowers them to make informed choices about their privacy. Beyond consent, **granular privacy controls** are crucial. Imagine users having the ability to opt-out of specific data collection practices, like location tracking while driving. They should also be able to set data retention periods, determining how long their data is stored, and request data deletion when desired.

Finally, the concept of **data ownership** needs to be addressed. Ideally, users should retain ownership of their data. This means they have the right to decide how it is used within the framework of connected e-mobility services. By establishing clear ownership and providing granular privacy controls, connected e-mobility can build a foundation of trust that will be essential for its long-term success.

15.5.6.4 Addressing Security and Privacy Concerns

- **Collaboration Among Stakeholders:** Governments, automotive manufacturers, technology providers, and service operators need to collaborate on developing robust security and privacy standards for connected e-mobility services.
- **Regulatory Frameworks:** Clear and comprehensive regulatory frameworks are needed to ensure compliance with data privacy laws like General Data Protection Regulation (GDPR) and to hold stakeholders accountable for data security breaches.

By prioritizing security and privacy, we can build trust in connected e-mobility services. This will encourage wider adoption of EVs and unlock the full potential of a sustainable and efficient transportation future. As technology evolves, so too must our commitment to robust cybersecurity practices and user privacy protections. This collaborative effort will pave the way for a secure and trustworthy connected e-mobility ecosystem that benefits everyone.

15.5.7 Standardization and Interoperability: The Backbone of Connected E-Mobility

The magic of connected e-mobility services lies in seamless communication and data exchange between various components – EVs, charging stations, apps, and cloud platforms. To achieve this, standardization and interoperability are paramount. Let us delve into why these aspects are crucial and how they can be achieved.

15.5.7.1 The Importance of Common Data Formats

Imagine different car manufacturers using their own unique language to communicate with gas stations. It would be chaos! Similarly, in connected e-mobility, a common language for data exchange is essential. The reason is presented here.

The smooth operation of connected e-mobility relies on a common language – standardized data formats. Imagine all devices and systems within the ecosystem speaking the same tongue. This **universal understanding** ensures seamless data exchange and eliminates compatibility issues that could otherwise hinder functionality. Furthermore, standardized formats act as a gateway for new players. **Simplified integration** makes it easier for innovative companies to enter the connected e-mobility space, fostering a competitive and dynamic industry. This not only benefits users with a wider range of options but also accelerates the overall development of the technology. Finally, common data formats lead to **enhanced efficiency**. Streamlined data processing and analysis allow for faster and more efficient implementation of connected e-mobility services. This translates to quicker rollouts of new features and a constantly improving user experience.

15.5.7.2 Open Platforms and APIs for Seamless Data Exchange

Think of Application Programming Interfaces (APIs) as bridges connecting different systems. Open platforms with well-defined APIs facilitate this connection.

The future of connected e-mobility hinges on open data platforms. These platforms act as a central hub where data can flow freely **between authorized entities**. This unrestricted flow is crucial for powering features like real-time charging station availability or personalized charging recommendations based on your individual needs. Furthermore, open platforms operate with **Open APIs**. These APIs essentially act as doorways for developers to access and utilize the data. This fosters a dynamic and competitive ecosystem, as developers are empowered to create innovative applications and services that leverage connected e-mobility data in unique ways. Ultimately, open platforms benefit everyone by enabling seamless data exchange between apps and services. This translates to a more convenient and user-friendly experience, making connected e-mobility a more attractive and accessible option for everyone.

15.5.7.3 Collaboration Between Industry Stakeholders

Achieving standardization and interoperability requires a collaborative effort from various stakeholders:

- **Automotive Manufacturers:** Car manufacturers need to agree on common data formats and communication protocols for V2I interaction.
- **Charging Network Operators:** Charging station providers need to ensure their infrastructure adheres to standardized data formats for seamless communication with EVs and other systems.
- **Tech Companies and Service Providers:** App developers and connected e-mobility service providers need to design their systems to work with open platforms and standardized APIs.
- **Government and Regulatory Bodies:** Governments can play a vital role by establishing clear regulations and promoting the adoption of standardized data formats and open platforms.

15.5.7.4 Benefits of Standardization and Interoperability

- **Faster EV Adoption:** A standardized and interoperable connected e-mobility ecosystem fosters faster EV adoption by ensuring compatibility and a smooth user experience.
- **Reduced Costs:** Standardization eliminates the need for custom solutions and simplifies integration, leading to reduced development and deployment costs for connected e-mobility services.
- **A Sustainable Future:** By enabling innovation and competition, standardization paves the way for a more efficient and sustainable transportation future powered by EV.

Standardization and interoperability are ongoing journeys in connected e-mobility. Industry collaboration, open communication, and continuous improvement are essential to ensure a future where connected e-mobility services are seamless, efficient, and accessible to all. As technology advances, so too must our commitment to developing and adhering to robust standards and open platforms. By working together, we can create a thriving connected e-mobility ecosystem that propels us toward a cleaner and more sustainable transportation landscape.

15.5.8 The Role of Big Data and Analytics: The Engine of Connected E-Mobility

Big data and analytics are the fuel that propels connected e-mobility services. By collecting and analyzing vast amounts of data from various sources, these technologies unlock a treasure trove of insights that optimize charging infrastructure, improve driving experiences, and pave the way for a smarter and more sustainable transportation ecosystem. Here is a closer look at how big data and analytics are driving the connected e-mobility revolution.

15.5.8.1 Data Collection and Aggregation from Various Sources

Connected e-mobility services generate a symphony of data from charging stations, mobile apps and user input and traffic management systems.

Connected e-mobility thrives on a rich tapestry of data collected from various sources. **EVs** themselves act as data hubs, recording information on battery health, driving patterns, location, and energy consumption. This data provides invaluable insights into how users interact with their vehicles and how efficiently EVs operate. **Charging stations** contribute their own set of data points, recording information on energy usage, station availability, and uptime. This allows for better management of the charging network and helps identify areas where expansion or maintenance might be needed. **Mobile apps and user input** are another crucial piece of the puzzle. User interactions with connected e-mobility apps provide insights into preferences and charging habits. This allows the system to personalize route planning and charging recommendations based on individual needs. Finally, **traffic management systems** add another layer of data by providing real-time information on congestion levels and road conditions. By integrating this data, connected e-mobility services can optimize route planning and suggest charging stops that consider not just distance but also traffic flow, making the entire experience more efficient and convenient.

15.5.8.2 Machine Learning for Predictive Insights and Optimization

Once this data is collected, machine learning algorithms step in:

- **Predictive Maintenance:** By analyzing battery health data, machine learning can predict potential issues and recommend preventative maintenance to avoid breakdowns and extend battery life.
- **Charging Infrastructure Optimization:** Data analysis helps identify areas with high demand for charging stations and optimize their placement for maximum user convenience.
- **Dynamic Route Planning with Charging Stops:** Real-time traffic data and energy consumption patterns allow for suggesting routes with optimal charging stops, minimizing travel time and range anxiety.
- **Personalized Charging Recommendations:** Machine learning can analyze individual driving habits and suggest personalized charging strategies for maximizing efficiency and minimizing reliance on expensive fast charging.

15.5.8.3 Real-Time Decision-Making Based on Data Analysis

The power of connected e-mobility extends beyond individual vehicles. This network of data can be harnessed to improve efficiency and user experience on a grand scale. **Grid management and demand response** come into play when utilities leverage EV charging data to manage electricity demands on the grid. Imagine a scenario where EVs act as distributed energy resources, potentially feeding excess power back into the grid during peak hours, contributing to a more stable and sustainable energy system. Traffic congestion, a major headache for drivers, can also be addressed with connected e-mobility. **Congestion management** empowers traffic authorities with real-time data to optimize traffic flow and minimize congestion, especially around charging stations experiencing high demand. This can lead to smoother commutes for everyone. Finally, the user experience is at the heart of connected e-mobility. Real-time data analysis allows these services to provide users with **up-to-date information on charging station availability, wait times, and alternative routes**. This translates to a seamless and efficient experience, whether you are planning a long trip or just looking for a quick top-up on the go.

15.5.8.4 Challenges and Considerations

- **Data Security and Privacy:** Robust measures are essential to protect sensitive user data and ensure compliance with data privacy regulations.
- **Data Quality and Standardization:** The effectiveness of data analytics relies on high-quality, standardized data collection practices across the connected e-mobility ecosystem.
- **Data Ownership and Governance:** Clear guidelines are needed to determine data ownership and establish responsible data governance frameworks.

Big data and analytics are transforming connected e-mobility services. As data collection becomes more sophisticated and machine learning algorithms evolve, we can expect even more innovative applications. By harnessing the power of data, we can unlock a future where connected e-mobility services are intelligent, proactive, and contribute significantly to a cleaner and more sustainable transportation landscape.

15.5.9 Connected E-Mobility: Powering Sustainable and Smart Cities

Connected e-mobility services [1], when integrated with smart city infrastructure, offer a powerful recipe for a greener and more efficient urban future. Here is how connected e-mobility can significantly impact cities and sustainability.

15.5.9.1 Smart Traffic Management and Congestion Reduction

Imagine a future where traffic lights adapt to the flow of EVs in real time. This is the potential of **real-time traffic data and routing**. Connected EVs can share their location and traffic data, allowing authorities to dynamically adjust traffic light timings and optimize traffic flow for everyone. This collaborative approach would not only reduce congestion but also improve overall efficiency. Furthermore, the convenience of connected e-mobility services could lead to a significant shift in how we travel. With **reduced reliance on personal vehicles**, services like ride-hailing and car-sharing become even more attractive. This could entice people to leave their cars at home, leading to fewer vehicles on the road and further alleviating congestion. To incentivize the switch to EVs and create a smoother flow for zero-emission vehicles, cities could implement **EV priority lanes**. These designated lanes would not only improve travel times for EV drivers but also create a visual cue for the environmental benefits of electric transportation, potentially accelerating EV adoption and a cleaner future for our cities.

15.5.9.2 Improved Air Quality Through Reduced Emissions

The large-scale shift from gas-powered vehicles to EVs powered by renewable energy sources is poised to bring a breath of fresh air to our cities. **EVs with zero tailpipe emissions** will significantly reduce air pollution, creating a healthier environment for everyone. But the benefits go beyond the tailpipe. **Efficient charging infrastructure**, optimized through connected e-mobility services, can minimize reliance on the grid during peak hours when power plants traditionally burn fossil fuels. This smarter approach to charging can further reduce emissions throughout the energy production cycle. Finally, rethinking how our cities are designed plays a crucial role. Urban planning that prioritizes walking, cycling, and public transport alongside connected e-mobility services can create a virtuous cycle. By making alternative transportation options more attractive and convenient, we can further **encourage reduced reliance on personal vehicles**, leading to cleaner air and a more breathable urban environment for all.

15.5.9.3 Promoting Sustainable Transportation Practices

Connected e-mobility is poised to revolutionize transportation not just for convenience, but also for sustainability. Imagine navigation systems that not only get you there but also suggest eco-friendly routes. These **eco-driving recommendations**, along with real-time feedback on your driving habits, can help you become a more fuel-efficient driver. The benefits extend beyond personal choices. Connected e-mobility services can seamlessly integrate with public transportation options through **multimodal transportation integration**. This would make it easy to combine EVs with buses, trains, or bike-sharing for shorter trips or congested areas, reducing overall reliance on cars. Additionally, the vast amount of data collected by these services can be a goldmine for policymakers. Through **data-driven policy decisions**, cities can optimize infrastructure development, implement smarter parking management solutions, and create incentives that encourage sustainable transportation choices, paving the way for a greener future.

15.5.9.4 Challenges and Considerations

The road to a future powered by EVs is paved with both promise and hurdles. One of the biggest challenges is the upfront investment required for a **robust charging infrastructure**. Building a network of charging stations necessitates collaboration between the public and private sectors. Governments can provide incentives and subsidies, while private companies can invest in building and operating charging stations. Another hurdle lies in our existing cities. **Urban planning and zoning regulations** need to adapt to accommodate this new technology. Cities will need to create space for charging stations in existing neighborhoods and integrate EV charging into future development plans. Finally, the transition to e-mobility should not exacerbate existing inequalities. **Equity and accessibility** are paramount. Efforts need to be made to ensure that everyone has access to connected e-mobility services, not just those who can afford EVs. This might involve subsidies for low-income residents or strategically placed charging stations in underserved communities. By addressing these challenges, we can create a future where everyone can benefit from a cleaner and more sustainable transportation system.

Connected e-mobility, when embraced as part of a comprehensive smart city strategy, has the potential to revolutionize urban transportation. By prioritizing sustainability, efficiency, and accessibility, cities can leverage connected e-mobility services to create cleaner, healthier, and more livable urban environments for all. As technology continues to develop, connected e-mobility will undoubtedly play a pivotal role in shaping a sustainable future for our cities.

15.5.10 The Future of IoT for E-Mobility: A Glimpse into a Connected Transportation Landscape

The future of e-mobility is brimming with exciting possibilities, powered by the ever-evolving world of IoT. Here is a roadmap to what lies ahead.

15.5.10.1 Autonomous Electric Vehicles and V2X Communication

Self-Driving EVs: Imagine a world where EVs navigate roads autonomously, powered by a network of sensors and communication technologies. V2X communication will be paramount:

- **V2V:** Enables real-time communication between autonomous EVs, allowing them to "talk" to each other and coordinate maneuvers for safer and more efficient traffic flow.
- **V2I:** Allows autonomous EVs to communicate with traffic lights, road signs, and smart infrastructure, enabling real-time adjustments to traffic flow, speed limits, and avoiding accidents.

15.5.10.2 Personalized E-Mobility Services Based on User Preferences

The future of e-mobility is getting smarter thanks to the power of AI. Imagine an e-mobility service that anticipates your needs. AI-powered recommendations will personalize your experience in amazing ways. No more wondering where to charge – the system will suggest stops based on your usual driving patterns and where you are headed. Need to save money? It can recommend optimal charging times based on electricity tariffs and flexibility. Planning a road trip? It would not just consider traffic but also your desire for scenic routes or specific charging stations, making every journey tailored to you.

15.5.10.3 Integration with Smart Cities and Intelligent Transportation Systems

The future of transportation is shaping up to be a connected and collaborative ecosystem Imagine a world where your EV seamlessly integrates with other transportation options. You could plan a trip that combines a subway ride with a dockless bike rental for the last mile, all orchestrated through a single app. This is the promise of seamless multimodal integration. But the benefits extend beyond personal convenience. EVs would not just be drawing power from the grid, they will become active participants. With smart grid integration, these vehicles could potentially feed excess electricity back into the grid during peak hours. This two-way flow would contribute to grid stability and pave the way for wider adoption of renewable energy sources like solar and wind power. The impact goes beyond energy too. Cities will be able to leverage data from connected e-mobility services for dynamic urban planning. Imagine traffic lights that adjust based on real-time EV usage, designated lanes for faster commutes, and charging stations strategically placed in high-demand areas. This data-driven approach would optimize traffic flow, reduce congestion, and create a more sustainable transportation system for everyone.

15.5.10.4 Challenges and Considerations

The challenges and considerations are listed under three headers: Cybersecurity and Data Privacy, Standardization and Interoperability, and Ethical Considerations. It explains that robust cybersecurity measures will be crucial to protect against hacking attempts and ensure the confidentiality of user data. Standardization and Interoperability claims that standardized communication protocols and data formats are essential for seamless integration between EVs, infrastructure, and various connected e-mobility services. Ethical Considerations explains that the development and deployment of autonomous vehicles raises ethical questions that need careful consideration, such as liability in case of accidents and ensuring equitable access to these technologies.

15.6 Conclusion

In conclusion, IoT transforms e-mobility by enabling a more connected, efficient, and sustainable transportation system. By addressing the challenges and embracing the opportunities, IoT has the potential to revolutionize the way we travel and power our vehicles. The future of IoT for e-mobility is a collaborative effort. Governments, automotive manufacturers, technology providers, and infrastructure developers need to work together to create a safe, sustainable, and accessible connected transportation ecosystem. By overcoming challenges and embracing innovation, we can unlock the full potential of IoT for e-mobility, paving the way for a cleaner, smarter, and more efficient transportation landscape for all. As technology advances exponentially, the possibilities for connected e-mobility seem limitless. This journey toward a future powered by sustainable EVs and intelligent transportation systems promises to transform how we travel and experience our cities.

References

1 Exner, J.-P., Bauer, S., Novikova, K. et al. (2020). Connected E-mobility, IoT and its emerging requirements for planning and infrastructures. *Real corp 2020* 175–181.

2 Urooj, S., Alrowais, F., Teekaraman, Y. et al. (2021). IoT based electric vehicle application using boosting algorithm for smart cities. *Energies* 14 (4): 1072.

3 Scrocca, M., Baroni, I., and Celino, I. (2021). Urban IoT ontologies for sharing and electric mobility. *Semantic Web* 14: 1–22. https://doi.org/10.3233/SW-210445.

4 Kabalci, Y., Kabalci, E., Padmanaban, S. et al. (2019). Internet of Things applications as energy internet in smart grids and smart environments. *Electronics* 8: 972.

5 Farmanbar, M., Parham, K., Arild, O., and Rong, C. (2019). A widespread review of smart grids towards smart cities. *Energies* 12: 4484.

6 Muralikrishnan, P., Kalaivani, M., and College, K.R. (2020). IOT based electric vehicle charging station using Arduino Uno. *International Journal of Advanced Science and Technology* 29: 4101–4106.

7 Ayob, A., Wan Mahmood, W.M.F., Mohamed, A. et al. (2014). Review on electric vehicle, battery charger, charging station and standards. *Research Journal of Applied Sciences, Engineering and Technology* 7: 364–372.

8 Kong, P.Y. and Karagiannidis, G.K. (2016). Charging schemes for plug-in hybrid electric vehicles in smart grid: a survey. *IEEE Access* 6846–6875.

9 Phadtare, K.S. (2020). A review on IoT based electric vehicle charging and parking system. *International Journal of Engine Research* 9: 831–835.

10 Sivaraman, P. and Sharmeela, C. (2020). IoT-based battery management system for hybrid electric vehicle. In: *Artificial Intelligent Techniques for Electric and Hybrid Electric Vehicles* (ed. A. Chitra, P. Sanjeevikumar, J.B. Holm-Nielsen, and S. Himavathi), 1–16. Wiley.

11 Sivaraman, P., Sharmeela, C., and Logeshkumar, S. (2021). Charging infrastructure layout and planning for plug-in electric vehicles. In: *Cable Based and Wireless Charging Systems for Electric Vehicles* (ed. P. Sanjeevikumar), 1–24. IET.

12 Logeshkumar, S. and Manoharan, R. (2014). Influence of some nanostructured materials additives on the performance of lead acid battery negative electrodes. *Electrochimica Acta* 144: 147–153.

13 Palanisamy, S., Shanmugasundaram, L., and Chenniappan, S. (2022). *Energy Storage Systems for Smart Power Systems*. Wiley.

14 Kukkala, V.K., Tunnell, J., Pasricha, S., and Bradley, T. (2018). Advanced driver-assistance systems: a path toward autonomous vehicles. *IEEE Consumer Electronics Magazine* 7: 18–25. https://doi.org/10.1109/MCE.2018.2828440.

15 Seuwou, P., Banissi, E., and Ubakanma, G. (2020). The future of mobility with connected and autonomous vehicles in smart cities. In: *Digital Twin Technologies and Smart Cities* (ed. M. Farsi, A. Daneshkhah, A. Hosseinian-Far, and H. Jahankhani). Springer Nature https://doi.org/10.1007/978-3-030-18732-3_3.

16 Zahira, R., Lakshmi, D., Ezhilarasi, G. et al. (2022). 6 - Stand-alone microgrid concept for rural electrification: a review. In: *Residential Microgrids and Rural Electrifications* (ed. S. Padmanaban, C. Sharmeela, P. Sivaraman, and J.B. Holm-Nielsen), 109–130. Elsevier.

16

Standards for Internet of Things (IoT)

Mohamed Mustafa Mohamed Iqbal[1], Balasubramanian Nandhan[1], Sakthivel Sruthi[1], Ravikumar Mithra[1], Rajagopal Logesh Krishna[1], Rahiman Zahira[2], Balan Gunapriya[3], and Veerasamy Balaji[4]

[1]*Department of Electrical and Electronics Engineering, PSG Institute of Technology and Applied Research, Coimbatore, Tamil Nadu, India*
[2]*Department of Electrical and Electronics Engineering, B.S. Abdur Rahman Crescent Institute of Science and Technology, Chennai, Tamil Nadu, India*
[3]*Department of Electrical and Electronics Engineering, New Horizon College of Engineering, Bangalore, Karnataka, India*
[4]*Department of Electrical and Electronics Engineering, PSG College of Technology, Coimbatore, Tamil Nadu, India*

16.1 Introduction

Internet of Things (IoT) refers to a network of physical objects, such as vehicles, buildings, and other items, that have sensors, software, and other technologies built into them [1, 2]. This allows the objects to exchange data and communicate with various devices and systems over the internet. In the context of the Internet of Things, "things" can be any natural or man-made objects that can be assigned an IP address and can send data over a network. Data flow over a network is made possible by IoT without the need for a human-to-human or human-to-computer interface [3, 4]. Due to the widespread availability of inexpensive, low-power sensor technology, IoT is used extensively. Sensors can now be easily connected to the cloud and other physical devices for efficient data transfer thanks to a range of internet network protocols. Improvements in neural networks have enabled natural language processing (NLP), which makes IoT devices appealing, reasonably priced, and useful for use at home. IoT provides enterprises with a real-time window into the inner workings of their systems, providing information on everything from supply chain and logistics operations to machine performance. IoT enables machines to complete laborious tasks without human assistance. IoT drives digital transformation in industries and organizations [5, 6].

IoT has many features such as interoperability, edge computing, real-time operation, and remote accessibility and control. Interoperability makes sure that gadgets, regardless of platform or vendor, can interact and cooperate [7]. Edge computing helps for processing and storing data close to the source whereas the cloud computing platform helps for data storage and processing and storage in a centrally located space for longer-term data storage and more thorough analysis. The real-time operation feature of IoT allows quick reactions to environmental changes by offering real-time monitoring and feedback. It also ensures timely updates and actions by transmitting and processing data quickly [8].

Remote accessibility and control ensures that the IoT devices are remotely controlled by users via web interfaces, allowing them to monitor and access the devices from any location. Automation and intelligence features of IoT systems make intelligent decisions based on data, increasing

IoT for Smart Grid: Revolutionizing Electrical Engineering, First Edition.
Edited by Rahiman Zahira, Palanisamy Sivaraman, Chenniappan Sharmeela, and Sanjeevikumar Padmanaban.

performance and efficiency, and they can automate repetitive processes, decreasing the need for human participation [9]. Moreover, the security and privacy feature of IoT helps to manage data privacy and makes sure that regulations are followed, safeguarding data transit and storage through encryption techniques, and limiting access to the IoT network to authorized people and devices only. Another important feature of the IoT system is that it enables an integration with other technologies through which the IoT data can be integrated with artificial intelligence, big data analytics, and blockchain easily for better performance [10].

IoT is being used extensively across many different sectors and industries, changing how businesses run and improving daily living. Sections are arranged in the following manner. Section 16.2 describes the importance and applications of IoT for smart grid and smart-cities-related domains. The need for standardization of IoT protocols and procedures has been presented in Section 16.3. Different standards both international and national level, pertaining to various applications, are elaborately presented in subsequent Sections 16.4–16.8. Section 16.4 describes the international and national standards of IoT for the healthcare sector. However, the international and national standards in the agriculture and food industry are being briefed in Section 16.5. Further, the IoT standards for smart-home and industrial automation applications are presented in Section 16.6. Various IEEE and ISO standards for the disaster management sector have been reviewed and presented in Section 16.7. Similarly, the IEEE, IEC/ISO standards for cybersecurity and data science domain are illustrated in Section 16.8. Section 16.9 describes the possible research scope and future work presented in Section 16.10. The overview and the summary of standards for various applications are presented in Section 16.10.

16.2 Smart Grid, Smart Transportation, and Smart Cities

In smart energy systems, IoT enables a wide array of applications spanning across the whole spectrum of energy systems, thus causing an influx of technology comprising quality variable solutions. One of the greatest benefits is rapid communication between subunits, maximizing the utilization of power as the impacts on ecology diminish while earnings from green power sources increase; hence, IoT has turned out to be a prospective breakthrough idea in this sector [11, 12]. In green energy, they support the implementation of smart grid applications using IOT technologies to enable India to move toward a clean electricity infrastructure that is also robust. To ensure harmony of operations within smart grid systems, as well as interoperability between them, and also safeguard against cyber-attacks while facilitating balanced use of resources among these two entities (people and industries), IoT solutions must conform to both global norms integrated with those of this country [13].

By combining sensors, cameras, and networked devices, IoT improves efficiency, sustainability, and the quality of life for residents in smart cities. Applications of smart cities include intelligent street lighting that adapts to pedestrian and traffic patterns, smart parking systems that manage availability, and real-time traffic monitoring that optimizes flow and reduces congestion. Waste management systems optimize collection routes, while environmental sensors check the quality of air and water. Real-time data and alarms from IoT-enabled emergency response systems enhance public safety. In general, smart cities use IoT to reduce administrative costs, preserve resources, and offer citizen-focused, responsive services for sustainable urban growth [14].

IoT applications in transportation are also possible with clever solutions that improve efficiency, safety, and traveler experience [15]. IoT is revolutionizing transportation. IoT-enabled connected cars optimize fleet management and save downtime with real-time tracking, diagnostics, and predictive maintenance. Efficient traffic control and monitoring systems lower emissions and traffic jams. IoT-enabled public transport systems provide occupancy tracking and real-time scheduling updates for increased dependability. Road safety and efficiency are improved by intelligent infrastructure, such as vehicle-to-infrastructure communication and smart intersections. With improved safety and efficiency measures, autonomous vehicles advance the future of transportation by integrating IoT for navigation, sensor data processing, and vehicle-to-vehicle communication [16].

16.3 Standardization of IoT Environment

IoT standardization, in general, encourages innovation, speeds up market acceptance, and increases stakeholder trust by defining best practices, simplifying processes, and guaranteeing uniformity in technological applications across borders [17]. Big amounts of data are produced by smart objects. It is necessary to handle, move, and store this data safely. It guarantees cost-effective and interoperable solutions, creates new opportunities, and lets the market realize its full potential. There is not just one widely recognized standard that defines IoT. Rather, it includes a broad spectrum of standards, protocols, and technologies that have been developed to serve different IoT application elements.

As standardization promotes interoperability, systems, and gadgets from many manufacturers can interact with one another without any problems. This encourages the development of coherent IoT ecosystems where various devices may cooperate well, increasing productivity and lowering integration costs. Standards are also crucial for security and privacy. To guard against cyber dangers and unauthorized access, standardized security protocols and encryption techniques are essential, given the billions of networked devices that are gathering and delivering data. Guidelines for data privacy are also defined by standardization frameworks, guaranteeing that private and sensitive data is managed safely and as per laws. By offering uniform frameworks for data formats, communication protocols, and device management, standards promote scalability. This makes it possible for IoT installations to grow seamlessly and effectively, supporting extensive networks and a wide range of applications without sacrificing dependability or speed [18].

16.3.1 International Standards

For the IoT to develop a single framework that supports interoperability, security, and reliability across worldwide deployments, international standards are essential. These standards guarantee flawless communication between IoT systems and devices from various manufacturers. They are created and maintained by international organizations, consortia, and standardization bodies. International standards make it easier to integrate and scale IoT solutions by defining common communication protocols such as message queuing telemetry transport (MQTT), constrained application protocol (CoAP), and standardizing data formats such as JavaScript Object Notation (JSON), and eXtensible Markup Language (XML) [19]. To build coherent IoT ecosystems where a variety of devices, from sophisticated industrial machinery and consumer electronics to sensors and actuators, can operate together effectively, interoperability is crucial.

When it comes to resolving security and privacy issues in IoT deployments, worldwide standards are essential. The best practices for safeguarding IoT devices, networks, and data are outlined in standards such as the ISO/IEC 27000 series. They set up mechanisms for user privacy protection and cyber threat defense, including authentication, encryption, access control, and secure data transmission. Adherence to these standards guarantees the implementation of strong security protocols, hence augmenting confidence among stakeholders and users [20].

International standards cover legal frameworks such as the United States' Health Insurance Portability and Accountability Act (HIPAA) [21]. Adherence to these standards guarantees that IoT implementations meet legal obligations for data security and confidentiality, an essential aspect of managing sensitive data in industries such as healthcare and finance.

16.3.2 Indian Standards

IoT standards are becoming important in India as a framework to promote security, dependability, and interoperability across different IoT applications. The Ministry of Consumer Affairs, Food, and Public Distribution delegates authority for developing and enforcing standards in a variety of domains, including IoT, to the Bureau of Indian Standards (BIS) [22]. To guarantee the stable implementation and uptake of IoT technology, Indian IoT standards concentrate on many important factors. By defining common protocols, communication frameworks, and data formats, these standards seek to advance interoperability. This facilitates cohesive IoT ecosystems by enabling easy integration and communication among IoT systems and devices from various manufacturers.

Indian IoT guidelines prioritize security and privacy as key considerations. Best practices for data encryption, safe communication protocols, access control systems, and data privacy frameworks are all included in the BIS standards [23]. These standards guarantee that sensitive data gathered by IoT devices is shielded from breaches and unwanted access, helping to reduce cybersecurity threats. BIS guidelines also guarantee that IoT solutions comply with Indian laws, including industry-specific mandates and data protection legislation, by aligning with national regulatory requirements and policies. Adherence to these standards fosters innovation, builds stakeholder trust, and speeds up IoT technology adoption throughout India's expanding digital economy. In general, Indian IoT standards are essential for promoting creativity, strengthening cybersecurity, and making it easier for IoT solutions to be seamlessly integrated into a variety of industries to promote technical advancement and economic prosperity.

Overall, by offering precise guidelines and standards for technology development, implementation, and regulation, the international standards in IoT promote innovation, facilitate access to global markets, and increase stakeholder confidence. Adoption of these standards by organizations, governments, and participants in the industry encourages sustainability, consistency, and dependability in IoT deployments, which spurs growth and opens up revolutionary applications in the fields of smart cities, healthcare, transportation, agriculture, and other areas. The confluence of these attributes renders the IoT a potent and adaptable technology, finding use in a wide range of sectors such as smart homes, healthcare, transportation, agriculture, and industrial automation. When taken as a whole, these features allow IoT to provide increased productivity, better judgment, lower costs, and better user experiences. International and national standards applied to a variety of real-world domains are being briefed in subsequent Sections 16.4–16.8.

16.4 IoT Standards in Healthcare

The healthcare sector can benefit greatly from IoT technologies, which include automated data collecting, sensing, and monitoring of objects, people, and patients, as well as identity and authentication of persons [24, 25]. Tracking is the capacity to identify a moving person or item. To improve hospital operations, patient flow is tracked in this case. Identification and authentication include accurately identifying hospitalized infants to avoid mismatching, keeping thorough and current electronic medical records, and identifying patients to reduce hazardous events. In this way, sensors are applicable to both inpatient and outpatient care. Some standards are required to create uniformity to carry out the aforementioned task. Thus, some of the requirements listed below enable the use of IoT in healthcare.

Through wearable technology, smart medical equipment, and remote monitoring, IoT revolutionizes patient care in the healthcare industry. Smartwatches and fitness trackers are examples of wearables that continuously monitor vital signs including blood pressure, heart rate, and glucose levels [25]. They also send real-time data to healthcare specialists. This makes it possible to identify any health problems early and create tailored treatment programs. Smart pills monitor patient compliance using built-in sensors. Hospitals use IoT-based systems to control patient flow, monitor equipment, and keep critical care units running smoothly. In general, IoT and blockchain technology adoption in the medical industry improves patient outcomes, lowers hospital readmission rates, and boosts the effectiveness of medical care delivery [26].

16.4.1 IEEE Standards

16.4.1.1 IEEE 11073

The interoperability of devices used for personal health, such as scales, blood pressure monitors, and blood glucose meters, is covered by a series of standards called IEEE 11073 Personal Health Device (PHD) standards [27, 28]. Although these standards are based on earlier IEEE 11073 standards' work, they differ by featuring a simpler device model and giving preference to consumer devices over hospital-grade equipment. PHD standards address the requirement for an openly stated, including computers, cell phones, health appliances, and health gateways. Primary aim is to bring real-time plug-and-play interoperability feature for medicals, emergency, and wellness devices. The dialogue between the managers and agents is depicted in Figure 16.1. This standard family is intended to facilitate smooth communication between medical equipment, guaranteeing that they comprehend and engage with one another in a productive manner [28].

Medical equipment follow a systematic approach and are organized as systems to interact and communicate effectively. The system consists of managers and agents. Managers keep copies of agent data, respond to updates, and have the ability to control the agents remotely. Agents are linked to medical devices and deliver data. Reversible roles between managers and agents enable adaptable hybrid systems. With a flexible implementation not specified by the standard, the agent application process interfaces between proprietary protocols and ISO/IEEE standards. Managed medical objects (MMOs) are arranged hierarchically in a domain information model (DIM) by the

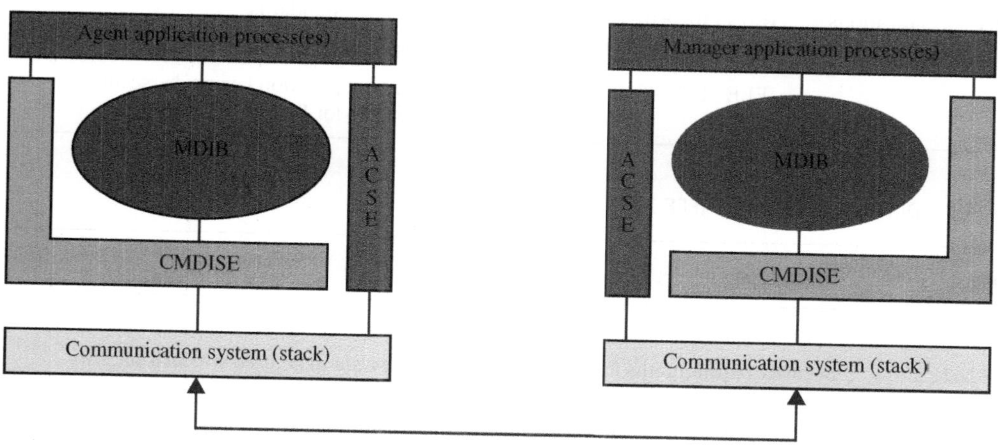

Figure 16.1 Agents and managers in ISO/IEEE 11073.

medical data information base (MDIB), where the implementation is not standardized but the layout is. Without sending MMOs, the Association Service Control Element – regulated by ISO/IEC 15953 and 15954 – manages the creation and dissolution of communication associations. Dynamic data interchange services (Create, Update, Delete) between agent–manager systems are defined by the common medical device information service element for MMOs, enabling sophisticated operations through comprehensive reports. Finally, ASN.1 and medical device encoding rules (MDER) are used by the presentation layer to encode object data [29]. A variety of standards for different categories that deal with medical devices under IEEE 11073 are listed in Table 16.1.

16.4.1.2 IEEE 2621

IEEE 2621 series provides manufacturers with guidelines to document the security of medical devices, enabling the safe use of these devices to control CDDs in conjunction with consumer mobile devices such as smartphones [30]. This set of standards will enable safe communication between wireless diabetes devices, such as insulin pumps, continuous glucose monitoring, and automated insulin dosing systems. The IEEE 2621 series comprises the various substandards as given in Table 16.2.

16.4.1.3 IEEE 2933

This particular standard creates a framework for IoT data and device interoperability based on the TIPPSS principles (trust, identity, privacy, protection, safety, and security) used for clinical purposes. This covers wearable clinical IoT, as well as compatibility with other clinical IoT devices, in-hospital devices, electronic health records (EHR), electronic medical records (EMR), and linked healthcare systems [31–33].

Table 16.1 Standards of IEEE 11073 for medical devices.

Category	Standard	Description
Nomenclature	IEEE 11073-10101	Gives nomenclature for personal devices used.
Modeling and generic protocols	IEEE 11073-20601	Personal Health Devices exchange protocol.
	IEEE 11073-10206	Object-oriented abstract information model for Personal Health Devices.
Device specializations	IEEE 11073-104xx	Device specializations standards.
Security	IEEE 11073-10401	Vulnerability assessment for Personal Health Devices.
	IEEE 11073-40102	Security baseline of application layer cybersecurity mitigation techniques for PHDs.

Table 16.2 Different standards of IEEE 2621.

Standard	Description
IEEE 2621.1	Gives a framework for device security evaluation program.
IEEE 2621.2	Security requirements for connected diabetes devices to be used within a security evaluation program.
IEEE 2621.3	Provides information on mobile phones in diabetes management, within a security evaluation program.

16.4.2 Other Standards

16.4.2.1 P2650

Trained audiologists and/or clinicians use specialized audiometric equipment, such as audiometers, Oto-acoustic emissions (OAE), and auditory brainstem response (ABR) to screen for and diagnose hearing impairment [34]. These devices are costly to buy and run and require a specific anechoic chamber. As a result, in emerging economies, this restricts their three A's, namely availability, accessibility, and affordability. By creating guidelines that permit the prescreening of hearing-impaired individuals on current mobile platforms, this project aims to address these three A's. This enhances existing mobile platforms along with linked portable/wearable devices to screen for the hearing impaired.

16.4.2.2 Health Level Seven International (HL7)

Healthcare systems frequently employ a variety of apps each with a unique set of features and programming languages while hospitals depend on intricate bespoke systems and colleges might use software created especially for medical research. General practitioners usually use premade practice management software these organizations must efficiently exchange patient data HL7s [35]. An HL7 message comprises multiple segments, each on a separate line and separated by a carriage return character (\r, hexadecimal 0D). Segments contain one or more fields, separated by pipe (|) characters, and subfields within these fields are separated by caret (^) characters. The HL7 framework includes an object type definition (OTD) library [19], which provides prefabricated message structures [36]. This allows healthcare providers to design interfaces that adhere to HL7 standards and customize messages by adding segments and optional fields. Each HL7 message must include a message type in its header, indicating the nature of the transmitted message [20]. Different sections and categories pertaining to HL7 are listed in Table 16.3.

16.4.2.3 DICOM

Digital imaging and communications in medicine (DICOM) is an international standard for image processing and medical data, digital imaging, and communication of data [37]. It tells the transmission of medical picture formats that adhere to the quality criteria and data required for clinical application. Nearly every cardiology, radiology, and radiotherapy device (X-ray, computed tomography (CT), magnetic resonance imaging (MRI), ultrasound, etc.) uses DICOM [38]. It is a messaging protocol that is used extensively in the healthcare industry worldwide, powering hundreds of thousands of medical imaging devices. Currently, billions of DICOM images are accessible and used in clinical settings. A DICOM file consists of the header, preamble, prefix, and picture pixel intensity data [39]. The patient's details, the study's acquisition parameters, image dimensions, matrix sizes, and color spaces are all included in the header as shown in Figure 16.2.

The attribute "7FE0" divides the pixel intensity data from the preamble. Use the data from header 6 to decode the series of "1s" and "0's" that make up the pixel intensity data into a picture. When DICOM and the Health Level 7 (HL7) standards are used, picture archiving and communication system (PAS), hospital information system (HIS), and radiology information system (RIS) may all share textual data more readily.

16.5 IoT Standards in Agriculture and Food Industry

IoT in agriculture, sometimes known as "smart farming", uses GPS, drones, and sensors to maximize a range of farming tasks. Farmers can access real-time data from sensors that track

Table 16.3 Different sections of HL7.

Section	Category	Description
Section 1	Primary standards	Used for system integrations, and compliance.
Section 1a	Clinical document architecture (CDA)	Gives document architecture for clinical products.
Section 1b	EHR – electronic health records	Electronic health records managing functional models.
Section 1c	FHIR	Modern health information exchange resources.
Section 1d	Version 2 (V2)	Widely used standards for messaging in healthcare.
Section 1e	Version 3 (V3)	HL7's reference information model (RIM).
Section 1f	Arden Syntax	Formalism for representing procedural clinical knowledge to facilitate sharing among personnel and institutions.
Section 1g	CCOW	Specification for application integration of HL7 clinical context management at the point of use.
Section 1h	Cross-paradigm/domain analysis models	Analysis models that span multiple paradigms based on logical level.
Section 2	Clinical and administrative domains	Primary standards for messaging and documentation.
Section 3	Implementation guides	Guides and support documents for implementing existing standards, serving as supplemental material.
Section 4	Rules and references	Information for software development.
Other classifications	Master grid	All HL7 standards can be located by ANSI/ISO/HITSP approval and search variables in Master Grid.

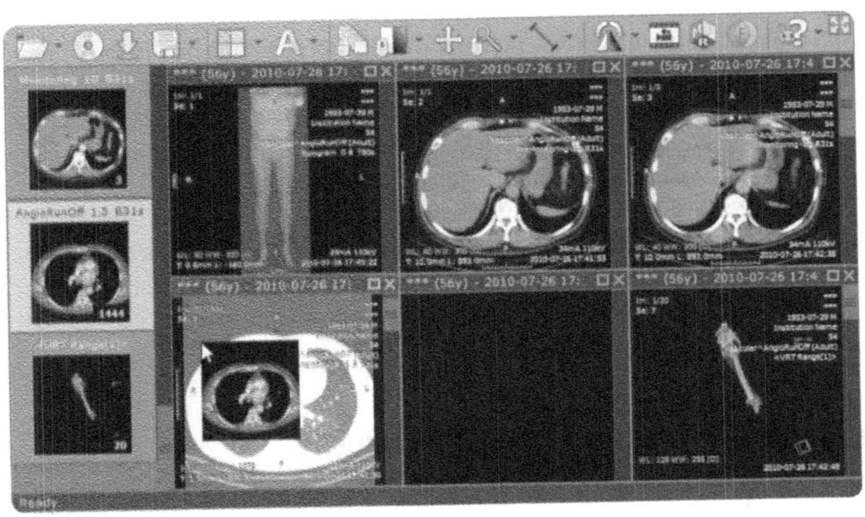

Figure 16.2 DICOM-based details of an image.

crop health, weather patterns, and soil moisture. Drones provide airborne imagery for field study, which aids in detecting problems such as nutrient deficits or pest infestations [40]. Precision planting, watering, and fertilizing are made possible by GPS-enabled equipment, which minimizes resource waste. Water efficiency is increased via automated irrigation systems, which modify water usage based on soil and meteorological data. All things considered, IoT facilitates data-driven decision-making, which raises agricultural yield, lowers expenses, and supports sustainable farming methods. International and national standards pertaining to the agriculture and food industry are presented below.

16.5.1 IEEE Standards

16.5.1.1 IEEE P2796

The IEEE Approved Draft Standard for the Internet of Food (IoF) Framework is IEEE 2796-20241 [41]. For IoF system applications, this standard offers an architectural framework that addresses data trust, scalability, and interoperability in the food chain industry. It supports a wide range of domain applications, including smart home intelligence and agri-logistics as well as smart farming. Further improving IoF solutions are related standards such as IEEE P2796.1 (for data requirements) and IEEE P2796.2 (for data exchange architecture and interface requirements) [42].

The main objectives of IoF systems are interoperability, scalability, and data trust. Interoperability ensures that various components, such as data platforms, actuators, and sensors, can communicate seamlessly through standardized protocols such as CoAP and MQTT. This facilitates cross-domain data transmission, benefiting applications in consumer services, supply chain management, and smart farming [43]. Scalability refers to the system's ability to grow and adapt to changing requirements, making it suitable for diverse scales, from small farms to large agri-logistics networks. This involves considering dynamic scalability, resource allocation, and load balancing during design. Data trust is crucial for decision-making in the food chain, requiring reliable data through mechanisms such as data provenance, secure data transmission with encryption and authentication, and privacy-preserving techniques such as anonymization and differential privacy to ensure data accuracy, security, and privacy. Its related standards are given in Table 16.4.

16.5.2 ISO Standards

16.5.2.1 ISOBUS (ISO11783)

ISO 11783 was developed to provide an open networked framework for onboard electronic systems [44]. Additionally, it standardizes the protocol and format for data exchange between actuators, sensors, control components, portions of the tractor, and implements placed on it, as well as information storage and display devices. This enables communication between electronic control units

Table 16.4 Standards of IEEE P2796.

Standard	Title	Objectives
IEEE P2796.1	Data Requirements for IoF	Ensures scalable, flexible, and interoperable network solutions for IoT applications.
IEEE P2796.2	Data Exchange Architecture for IoF	Optimizes supply chains and produces dependable IoF solutions. Ensures interoperability of devices, systems, and applications.

Table 16.5 Different parts of ISO 11783.

ISO 11783 Part	Description	ISO 11783 Part	Description
Part 1	General standards	Part 8	Power train messages
Part 2	Physical layer	Part 9	Tractor ECU
Part 3	Data link layer	Part 10	Task controller
Part 4	Network layer	Part 11	Mobile data element dictionary
Part 5	Network management	Part 12	Diagnostics services
Part 6	Virtual terminal	Part 13	File server
Part 7	Implement messages	Part 14	Sequence control

(ECUs), which are required to connect the tractor and implement networks. There are 14 sections in the ISO 11783, for agriculture and forestry-serial control as mentioned in Table 16.5.

XML files are used in ISOBUS to standardize data transfer, ensuring interoperability and simplicity of integration across many systems. This makes it possible for the tractor, implements, and farm management software to communicate data seamlessly. Plug-and-play functionality is made possible using ISOBUS, which means that the tractor's ultrasonic testing (UT) will automatically detect and set up the required controls when an implement is attached to it. This guarantees appropriate communication and streamlines the setup procedure [31]. Comprehensive task documenting and record-keeping are made possible by ISOBUS. Application rates, coverage areas, and operating hours are just a few examples of the data that are logged and can be utilized for reporting and analysis to help with decision-making and regulatory compliance.

16.5.2.2 ISO 28258

The digital interchange of soil-related data is outlined in ISO 28258:2013. It seeks to enable any source, holder, or consumer of soil data to locate and transmit data in a clear and straightforward manner [45]. It also seeks to promote sharing accurate, precisely defined, and specified soil-related data between persons and organizations using digital platforms. It includes feature definitions, several parameter standards, and encoding guidelines that facilitate retrievable and consistent data interchange. Additionally, by extending previous international standards, it permits the explicit geo-referencing of soil data, making it easier to utilize soil data in geographic information systems (GIS).

ISO 28258 identifies several interesting features related to soil studies as presented in Figure 16.3 [46]. A site represents the immediate surroundings of a soil study, including topography and land usage. A plot is a specific area where soil research is conducted, often leading to the description of a soil profile and the collection of soil samples for physio-chemical analysis. Plots can be divided into trial pits, surfaces, and boreholes. A profile is an arranged collection of soil horizons or layers that constitute the soil bed at a specific point, used for soil classification. A profile element refers to an element of a soil profile with defined upper and lower depths, and is divided into layers, which are arbitrary and heterogeneous sections, and horizons, which are pedo-genetically homogeneous segments [47]. Lastly, a soil specimen is a homogenized sample of soil taken at a particular depth, primarily for physio-chemical analyses.

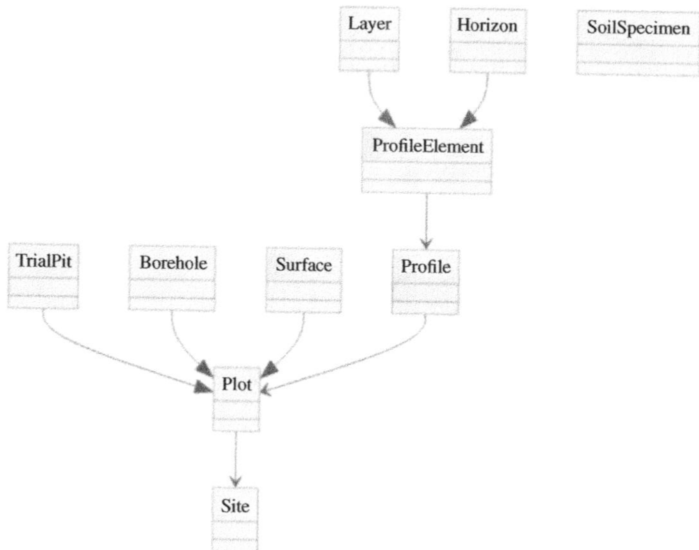

Figure 16.3 Layout of ISO 28258.

Table 16.6 Different standards of ISO 22000 series.

ISO 22000	Requirements for food safety management systems
ISO 22001	Guidelines for the food and drinks industry on the application of the standard
ISO/TS 22002	Food manufacturing, catering, farming, food packing, transport and storage, feed, and the manufacture of animal feed are among the prerequisite courses for food safety
ISO/TS 22003	Conditions for organizations that audit and certify food safety management systems
ISO/TS 22004	Food safety management systems – guidelines on the application of the standard
ISO 22005	In the feed and food chain, traceability
ISO 22006	Quality management systems

16.5.2.3 ISO 22000

The International Organization for Standardization (ISO) developed the outcome-focused ISO 22000 food safety management system, which provides guidelines for any company in the food industry to improve overall food safety performance [48, 49]. These standards aim to ensure food safety throughout the worldwide food supply chain. Besides emphasizing traceability in the feed and food chain, the standards offer general guidance for ensuring food safety. Among the standards in the ISO 22000 family are given in Table 16.6 [50].

16.5.3 Other Standards

16.5.3.1 AgGateway ADAPT Standard

This is an industry standard that is used to improve farm data interoperability. This is a comprehensive framework created especially for the agricultural sector to help with data integration and interoperability between various hardware and software tools used in farming and agriculture. The main function of this standard is that it enhances agricultural data management by promoting data

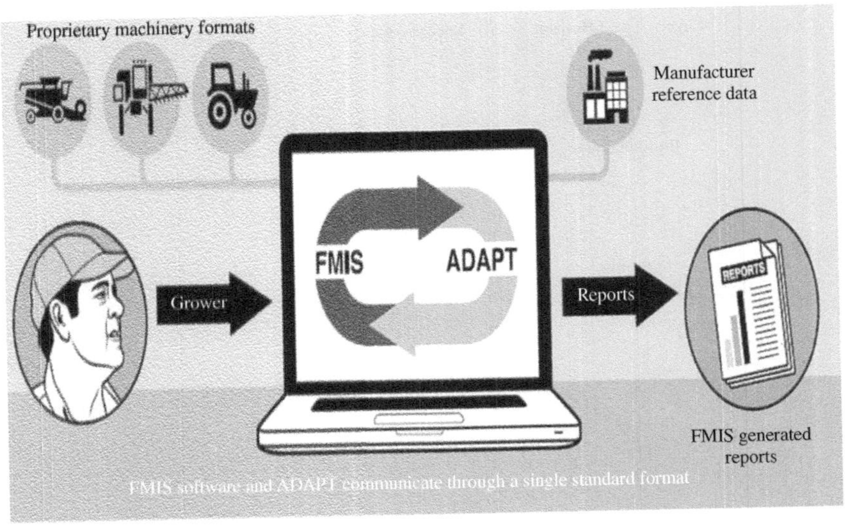

Figure 16.4 FMIS – farm management information system.

interoperability, mapping, and translation [51]. It establishes common data formats, protocols, and application programming interface (API) for smooth communication between diverse agricultural systems, enabling effective data management across platforms. It provides guidelines for integrating data from various sources such as machinery sensors and weather stations. Agricultural data application programming toolkit (ADAPT) encourages plug-and-play integration, allowing easy adoption of new tools without extensive customization. The standard ensures data security and privacy throughout the data lifecycle, protecting sensitive agricultural information. Developed by AgGateway, ADAPT fosters collaboration among agricultural stakeholders and offers comprehensive implementation guidance, including technical assistance, training materials, and testing tools to support the deployment of ADAPT-compliant systems as shown in Figure 16.4.

The AgGateway ADAPT standard encourages data sharing and interoperability among various farming software systems and equipment to improve and expedite agricultural operations. ADAPT makes it easier for technology to integrate seamlessly by defining standard data formats, communication protocols, and APIs. This enables farmers to effectively manage and use agricultural data from multiple sources. Agricultural decision-making skills, operational effectiveness, and sustainability are all enhanced by this standardization. In addition, it guarantees privacy and data security, which promotes confidence and adherence to legal obligations. The overall goals of ADAPT are to facilitate the digital transformation of the agricultural business, encourage innovation, and modernize farming techniques.

16.5.3.2 GlobalGAP

The Integrated Farm Assurance Standard (IFA), also referred to as the Global GAP Certificate, addresses various key issues, such as health, safety and welfare of workers, and food safety etc. The foundation of GlobalGAP is the IFA standard, which provides detailed specifications for various types of agricultural production. It is divided into multiple scopes: the Crops Base covers the production of ornamentals, fruits, vegetables, and flowers, focusing on crop protection, soil management, and postharvest handling. It also covers other facets of the food supply chain and production, such as compound feed manufacturing and chain of custody. GlobalGAP mandates the following: food safety, traceability, the environment, worker health and safety, animal welfare, and crop management [52].

The livestock base addresses the production of dairy, poultry, pigs, and beef, emphasizing animal health, shelter, feed, and well-being. The aquaculture base involves the rearing of fish and seafood, concentrating on feed, animal health, water quality, and environmental sustainability. The chain of custody (CoC) standard ensures the traceability of GlobalGAP-certified products from farm to customer, maintaining the integrity of certification by keeping certified products separate from non-certified ones throughout the supply chain. The global risk assessment on social practice (GRASP) module, an optional addition to the IFA standard, evaluates farm social practices related to worker welfare, health, and safety. Key GRASP features include worker representation, ensuring employees can voice opinions and participate in decision-making; occupational health and safety, implementing policies to prevent workplace accidents and health issues; and fair treatment, ensuring equitable pay, benefits, and working conditions for all agricultural laborers.

16.6 IoT Standards in Smart Home and Industrial Automation

Consumer IoT applications improve everyday life's connectedness, efficiency, and convenience. Automation and remote control are made possible by smart home appliances including security systems, lighting, and thermostats, which enhance comfort and safety [53]. Fitness trackers and smartwatches are examples of wearable technology that tracks health measurements and activity levels and provides real-time data and individualized insights. Smart washing machines and refrigerators are examples of connected appliances that simplify home tasks via remote control and maintenance notifications. Features such as entertainment, maintenance monitoring, and real-time navigation are available in IoT-enabled cars. Together, these apps improve lifestyle, safety, and health, and allow technology to be seamlessly incorporated into daily tasks [54, 55].

16.6.1 Z-Wave

Z-Wave is a popular technology for wireless communication in home automation that makes it easier to build networks of smart homes. It makes communication between devices possible for monitoring and control. Z-Wave minimizes interference with other wireless technologies such as Wi-Fi and Bluetooth by operating in the sub-1 GHz frequency range. Z-Wave networks consist of various controllers, sensors, and actuators that enhance smart home automation. Primary Controllers, such as Aeotec Z-Wave hubs, Hubitat, and SmartThings, act as the nerve center, overseeing and coordinating communication among all units [56, 57]. Secondary controllers, such as mobile apps and wall-mounted keypads, provide device control without managing the network. Sensors include Motion Sensors such as Aeotec MultiSensor and Fibaro Motion Sensor, which detect movement and control lights or send notifications. Key advantages of various other standards of Z-Wave have been listed in Table 16.7.

16.6.2 Matter (Formerly Project CHIP)

Matter is an open-source networking protocol intended for IoT and smart home devices [58, 59]. Apart from always providing local control as an option, it strives for increased security and interoperability with multiple manufacturers. The Connectivity Standards Alliance created Matter, formerly known as Project Connected Home over IP (CHIP), as a single smart home standard to guarantee interoperability across gadgets made by various manufacturers [60]. A number of important

Table 16.7 Different types of Z-wave.

Standard	Key features
Z-Wave (Original)	Foundational protocol, reliable communication, operated at 908.42 MHz (US)
Z-Wave Plus (500 Series)	Increased range, battery life, OTA updates, and better network management
Z-Wave Plus V2 (700 Series)	Further range extension, improved security (S2), lower power consumption, and better sensor technology
Z-Wave Long Range (LR)	Up to 1-mile range, supports over 2000 nodes, backward compatible
Z-Wave 800 Series	Improved performance, range, and battery life, enhanced security

Table 16.8 IEEE and ISO-based standards used in Matter.

Standard	Description
IEEE 802.15.4	Physical and MAC layer standard for low-power, low-data-rate wireless communication, are foundational for IoT devices such as those supported by Matter [61, 62].
IPv6 (ISO/IEC 2460)	Internet Protocol version 6, offering a vast address space and supporting direct communication and integration within IP-based networks, including Matter [63].
CoAP (RFC 7252)	Constrained Application Protocol, enabling RESTful communication between devices, crucial for interoperability and device control in Matter ecosystems [64].
DTLS (RFC 6347)	Datagram Transport Layer Security, providing encryption, authentication, and integrity for secure data exchange between Matter devices and services [65].
CBOR (RFC 7049)	Concise binary object representation, optimizing data encoding for efficient transmission and storage of information in Matter devices and networks [66].

standards are used by Matter (previously Project CHIP) to guarantee functioning and compatibility in home automation. Table 16.8 lists the matter-related IEEE and ISO-based standards.

These standards provide strong communication, interoperability, and security in smart home automation environments, and serve as the foundation for Matter's technical requirements. They make it possible for gadgets made by many manufacturers to function harmoniously, improving user experience and increasing the potential of networked smart devices.

16.6.3 Thread

Using a low-power mesh networking structure, Thread is a wireless networking protocol tailored for IoT and smart home applications [67, 68]. It enables devices to communicate either directly or via intermediary nodes. Thread has several important advantages for home automation. First, its strong, self-healing network allows dependable connectivity and efficient communication even in difficult settings. Its energy efficiency maximizes power consumption, prolonging sensor battery life and improving system performance as a whole. Strong encryption and secure device joining procedures safeguard data and privacy, making security a top priority. Because of its scalability, Thread is perfect for bigger smart home setups, supporting networks with hundreds of devices

Table 16.9 IEEE and ISO standards used in Thread.

Standard	Description
IEEE 802.15.4	Physical and MAC layer standard for low-power, low-data-rate wireless communication, forming the basis of Thread's network communication.
IPv6 (ISO/IEC 2460)	Internet Protocol version 6, providing a large address space and enabling unique IP addresses for each device in the Thread network, facilitating direct communication and integration with other IP-based networks.
6LoWPAN (RFC 6282)	Adapts IPv6 packets for transmission over IEEE 802.15.4 networks, optimizing bandwidth usage and enabling efficient communication in low-power wireless environments.
DTLS (RFC 6347)	Datagram Transport Layer Security, providing encryption, authentication, and data integrity for secure communication between Thread devices.
CoAP (RFC 7252)	Constrained Application Protocol, facilitating RESTful communication between devices in the Thread network, essential for smart home automation applications.

as they grow. To improve device compatibility, it guarantees interoperability by integrating with a variety of smart home ecosystems and platforms. It also improves user experience by streamlining deployment through automated setup and simple installation. Table 16.9 summarizes the IEEE and ISO-based standards and their roles in Thread for home automation:

16.6.4 OPC Unified Architecture (OPC UA)

OPC UA offers a standardized communication platform that guarantees smooth interoperability among machines, systems, and devices, it is essential to industrial automation. The main purpose is to provide safe and dependable data and information interchange across industrial automation systems, irrespective of the platform, operating system, or manufacturer [69, 70].

The OPC foundation created and maintains the open, vendor-neutral OPC UA, which is widely accepted. Its worldwide applicability and compatibility are guaranteed by the endorsement of international standards organizations such as the International Electrotechnical Commission (IEC) and the ISO as mentioned in Table 16.10. The architecture, communication protocols, and information modeling of OPC UA are covered in full in ISO 62541, which is essential for a general understanding of the protocol and its application in industrial automation systems. ISO 62542 focuses specifically on device models within OPC UA, tailored for process automation applications, offering detailed specifications for modeling and communication with devices commonly used in such scenarios [71]. IEEE 21451 (Parts 8–11) defines transducer electronic data sheet (TEDS) formats for sensors, specifying how sensor characteristics and capabilities can be represented and communicated using OPC UA, thus ensuring interoperability and compatibility in sensor networks.

These standards provide a structured framework for implementing OPC UA, enhancing compatibility, interoperability, and reliability across diverse industrial automation environments. By aligning with international standards, OPC UA facilitates broader acceptance and seamless integration with other standardized systems and technologies, enabling manufacturers and developers to create compliant products and solutions that meet global performance, security, and efficiency criteria. This standardization reinforces OPC UA's role as a leading communication protocol in industrial automation.

Table 16.10 IEEE and ISO standards used in OPC.

Standard	Title and description	Focus area
ISO 62541	Industrial automation systems and integration – OPC Unified Architecture (OPC UA) – Part 1: Overview and concepts [72].	Provides an overview and foundational concepts of OPC UA.
ISO 62542	Industrial automation systems and integration – OPC Unified Architecture (OPC UA) – Part 100: Device models for process automation [73].	Specifies device models for use in process automation.
IEEE 21451-8 to 11	Standard for Transducer Electronic Data Sheet (TEDS) Formats for Sensors – Parts 8-11.	Defines TEDS formats for sensors and their integration with OPC UA.

16.6.5 PROFINET

Process field network (PROFINET) is a widely used industrial ethernet standard for real-time communication in industrial automation. It integrates field-level devices such as sensors and actuators with higher-level systems such as programmable logic controllers (PLCs) [74, 75]. This protocol provides a robust and flexible way to exchange data, enabling efficient control and monitoring of industrial processes. PROFINET supports real-time data exchange, ensuring timely responses for critical processes that require precise timing and synchronization. Its scalability ranges from simple device-level networks to complex architectures encompassing entire production facilities, making it suitable for a wide range of applications. Leveraging Ethernet technology, PROFINET benefits from high bandwidth, widespread availability, and familiarity within IT environments. Table 16.11 summarizes the ISO, IEEE, and other standards relevant to PROFINET in industrial automation.

Table 16.11 Standards used in PROFINET.

Standard	Description	Relevance to PROFINET
ISO 16484-5	Standard for communication between building automation and control systems (BACS) using PROFINET.	Specifies PROFINET for building automation integration.
IEEE 802.3	Defines ethernet network specifications, including physical and data link layers.	Utilizes ethernet for communication, ensuring compatibility.
IEC 61158	Defines fieldbus communication protocols, including PROFINET communication over Ethernet (Part 6–10).	Basis for PROFINET communication standards
IEC 61784-2-3	Specifies industrial communication networks and systems, including PROFINET (Part 2–3) [76].	Defines PROFINET within the context of industrial automation [77].
Profibus and PROFINET International (PI) Specifications	PI provides technical specifications and guidelines for PROFINET.	Details protocols, device profiles, and integration aspects.

Table 16.12 Comparison of standards of ISA/IEC 62443.

Standard	Description
ISA/IEC 62443-1-1	Introduction and overview
ISA/IEC 62443-2-1	Lifecycle model for IACS
ISA/IEC 62443-3-2	Requirements for system security and its levels
ISA/IEC 62443-4-1	Product development requirements
ISA/IEC 62443-2-4	Technical security requirements for IACS components
ISO/IEC 27001	Information security management system (ISMS)
ISO/IEC 27002	Information control security codes
ISO/IEC 27019	Guidelines for the process of information security management specifically for IACS (supplements ISO/IEC 27001 and ISO/IEC 27002)
ISO/IEC 27032	Guidelines for cybersecurity
IEEE 1686-2019	Intelligent electronic devices (IEDs) cybersecurity standards
IEEE 802.1X	Port-based network access control
IEEE 802.11i	Robust security network (RSN) enhancements for wireless LANs
IEEE 802.16e	Wireless MAN air interface for fixed wireless access systems

16.6.6 ISA/IEC 62443 Standards

A complete set of standards called ISA/IEC 62443 was created specifically for industrial automation and control systems (IACS) cybersecurity. These standards have a multilayered approach, covering the network, system, and component levels as well as other layers of the industrial control system architecture. They stress how crucial it is to carry out customized risk assessments to properly detect and reduce cybersecurity vulnerabilities. A security lifecycle model with stages for evaluation, design, implementation, operation, and maintenance is defined by ISA/IEC 62443 [76]. It allows companies to match their cybersecurity measures to the criticality of their operations by classifying security requirements into several security levels (SL1 to SL4). To improve control and security of industrial networks, the standards encourage the adoption of the zone and conduit model, which consists of logically grouping assets with comparable cybersecurity requirements (zones) and secure communication connections between them (conduits). A comparison of the ISA/IEC 62443 standards with ISO and IEEE standards related to industrial cybersecurity is presented in Table 16.12.

16.7 IoT Standards for Disaster Management

16.7.1 IEEE 1512 Standards

IEEE 1512 standard gives the details of the exchange of vital data about public safety. The message sets given use Abstract Syntax Notation One (ASN.1 or ASN) syntax and are compliant with the National Intelligent Transportation Systems Architecture [77, 78]. The transportation-related events, through common incident management message sets, are transmitted. Three other companion volumes, which define incident management message sets for data sharing relating to transportation management, hazardous materials, and cargo, among other things, are also members of

Table 16.13 Family of standards under IEEE 1512.

Standard	Title and description	Focus area
IEEE Std 1512-2006	Common incident management message for emergency management centers	Basic information exchanged for any incident, such as incident description.
IEEE Std 1512.1-2006	Common traffic incident management message for emergency management centers	Messages related to traffic flow, control equipment, and repair are defined.
IEEE Std 1512.2-2004	Public safety incident management message for emergency management centers	Warning information, situation awareness, and interagency asset management are supported.
IEEE Std 1512.3-2006	Hazardous material incident management message for emergency management centers	Incidents related to commercial vehicles and homeland security.
Companion volumes	Each defines messages and includes a data dictionary defining the data elements in each message	Companion volumes detailing specific message sets and data elements for traffic, public safety, and hazardous material incidents.

that family. The 1512 Family of Standards will be the collective name for that set of standards. Supporting effective communication for the interagency management of transportation-related events in real time is the aim of that set of standards. These occurrences consist of crises, mishaps, scheduled lane closures, unique events, and man-made or natural calamities. Table 16.13 summarizes the IEEE 1512 family of standards for incident management.

16.7.2 ISO 22320 Standards

ISO 22320 is an international standard developed by the ISO to provide a framework for emergency management and incident response [79]. This standard is designed to help organizations manage and respond to emergencies effectively and efficiently, ensuring that they are prepared to handle a wide range of incidents, from natural disasters to technological and human-made emergencies. ISO 22320 is part of the ISO 22300 series of standards, which focuses on societal security and business continuity management. The primary objective of ISO 22320 is to enhance the ability of organizations to manage emergency situations by providing a clear set of principles and requirements. This includes establishing a common framework for command and control, ensuring effective communication and information management, and promoting coordination and cooperation among all parties involved in the response. By adopting ISO 22320, organizations can improve their overall preparedness, response, and recovery efforts, ultimately minimizing the impact of emergencies on people, property, and the environment.

One of the key aspects of ISO 22320 is its focus on command and control. The standard outlines the roles and responsibilities of those involved in managing an incident, ensuring a clear chain of command and well-defined decision-making processes. This helps to ensure that resources are used effectively and that actions are coordinated and aligned with the overall objectives of the response effort [80].

By adopting ISO 22320, organizations can benefit from a comprehensive and systematic approach to emergency management. The standard helps to ensure that they are prepared to respond to a wide range of incidents and that their response efforts are well-coordinated and effective. This not only helps to protect people, property, and the environment but also enhances the overall resilience of the organization and the community as a whole.

16.7.3 Other Standards

16.7.3.1 Common Alerting Protocol (CAP)

Common alerting protocol (CAP) is an international standard format used for emergency altering and people warning during disasters or emergencies during earthquakes, tsunamis, and volcanoes. When there are widespread emergencies, overlapping jurisdictions can lead to different authorities sending out alerts as the situation develops. When an emergency develops, all alerting authorities utilize CAP to guarantee that important information is communicated coherently in public messaging and private contacts between authorities and affiliates [81]. The trust that people have in alerting systems is increased when the same notice is sent across several platforms, increasing reach and impact. In comparison to alternative approaches, CAP expedites and simplifies the warning issuance procedure. Conventional alerting techniques, such as phone conversations, emails, and online postings, consumes enormous time and take focus away from creating precise and useful notifications. A single message may be quickly distributed across several alerting techniques with CAP, which makes it the least error-prone and fastest means to get information to individuals in danger. Utilizing the globally recognized alerting format, CAP guarantees the unambiguous transmission of crucial information and prevents misunderstandings. CAP is a public warning and all-hazard emergency alert format that is easy to use and adaptable to different information and communication technologies (ICT). Many at-risk people, including those who are blind, deaf, cognitively disabled, or do not understand the language used in the alerts, are frequently not sufficiently served by current public alerting. By utilizing CAP's data features and automated translation, these problems can be lessened while maintaining universal access to warnings [82].

16.8 IoT Standards in Cybersecurity and Data Science Domain

16.8.1 IEEE 802.1X Standards

An IEEE standard for port-based network access control (PNAC) is IEEE 802.1X. It belongs to the class of networking protocols known as IEEE 802.1 [83]. It gives details about the devices looking to connect to a local area network (LAN) or wireless local area network (WLAN). The standard specifically tackles an attack method known as Hardware Addition, in which a hacker assumes the identity of a visitor, client, or employee and sneaks a hacking device into the establishment, which they then plug into the network to get complete access. A computer connected to Walmart's network in 2005 compromised thousands of their servers, providing a prominent illustration of the problem.

IEEE 802.1X specifies how the Extensible Authentication Protocol (EAP), sometimes referred to as "EAP over LAN" or "EAPOL," is encapsulated over wired IEEE 802 networks and 802.11 wireless networks [84]. EAPOL was expanded to include additional IEEE 802 LAN technologies, such as IEEE 802.11 wireless, in 802.1X-2004. Originally designed for IEEE 802.3 Ethernet, IEEE 802.5 Token Ring, and fiber distributed data interface (FDDI) (ANSI X3T9.5/X3T12 and ISO 9314), it was first published in 802.1X-2001. The extensible authentication protocol over LAN (EAPOL) was also modified for use with IEEE 802.1AE ("MACsec") and IEEE 802.1AR (Secure Device Identity, DevID) in 802.1X-2010 to support service identification and optional point-to-point encryption over the internal LAN segment. The structure of IEEE 802.1X is presented in Figure 16.5.

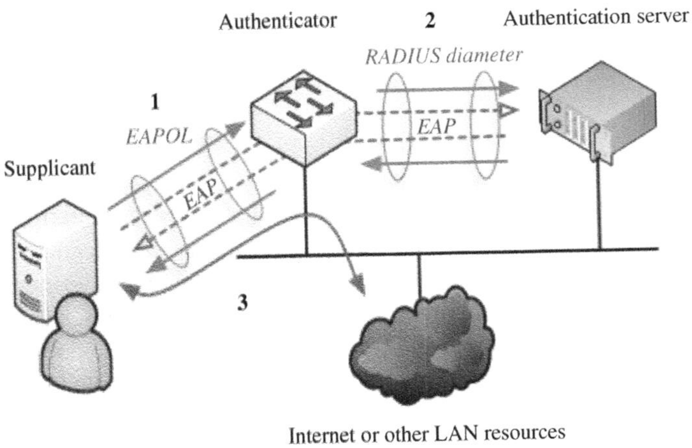

Figure 16.5 Structure of IEEE 802.1X.

16.8.2 ISO Standards

16.8.2.1 ISO/IEC 20547

A set of guidelines known as ISO/IEC 20547 was created to offer a big data reference architecture. This extensive architecture guarantees efficient administration and usage of massive volumes of data by catering to the diverse requirements of big data applications. The principal objective of ISO/IEC 20547 is to establish a uniform methodology that may be implemented in many sectors to augment interoperability, scalability, and performance within big data settings [85]. Table 16.14 shows different standards of ISO/IEC 20547.

Table 16.14 Different standards of ISO/IEC 20547.

Standard	Description	Focus area
ISO/IEC 20547-1:2020	Establishes a framework and application process for big data reference architecture.	Sets foundational principles and processes for implementing big data architectures, ensuring seamless interaction of different components.
ISO/IEC 20547-2:2018	Identifies a wide range of use cases from different industries and sectors to extract essential requirements.	Helps stakeholders understand practical applications of big data and requirements for effective big data solutions.
ISO/IEC 20547-3:2020	Provides the reference architecture for big data and defining components and relationships within the architecture.	Focuses on structural aspects of big data systems to ensure efficient and effective support for various big data operations.
ISO/IEC 20547-4:2020	Addresses security and privacy aspects of big data reference architecture and measures to protect big data systems and their data.	Ensures that big data systems are secure and compliant with privacy regulations, protecting sensitive information from unauthorized access and breaches.
ISO/IEC 20547-5:2020	Identifies gaps where new standards are needed, guiding the development big data ecosystem.	Helps organizations navigate the landscape of big data standards, ensuring the adoption of the most relevant and up-to-date practices.

16.8.2.2 ISO/IEC 19941

A standard called ISO/IEC 19941:2017 aims to improve cloud computing interoperability and portability [86]. It offers a framework to guarantee that various cloud solutions and services may cooperate and that moving data, apps, and services between cloud environments can happen with as little interruption as possible. This standard covers various cloud service categories, including infrastructure as a service (Iaas), platform as a service (PaaS), and software as a service (SaaS), and emphasizes the importance of standardizing APIs and data models to facilitate effective communication and data transfer across platforms [87]. By promoting best practices for designing, implementing, and managing cloud services, ISO/IEC 19941 helps organizations avoid vendor lock-in, optimize costs, and achieve better integration of their IT environments. This flexibility enables organizations to leverage multiple cloud providers, enhance their IT resource management, and prepare for future technological advancements. Related standards, such as ISO/IEC 17788 [88], ISO/IEC 17789 [89], ISO/IEC 19086-1 [90], and ISO/IEC 27017 [91], complement ISO/IEC 19941 [92] provide additional guidelines and frameworks for cloud computing.

16.9 Research Scope for Future Work

Even though IoT has numerous benefits, it also has some drawbacks and difficulties that must be resolved if it is to reach its full potential. It includes security concerns, privacy, energy consumption, data management and analysis, and complexity issues. Because IoT devices frequently have low processing and memory capacities, it might be challenging to put strong security measures in place. Hackers may target the vast amount of data that IoT devices transmit, which could result in data breaches and privacy problems. It is difficult to make sure that only authorized users and devices can access the IoT network [4]. IoT devices' extensive data collection might raise privacy issues, particularly when it comes to the handling of sensitive personal data. The possibility of using IoT devices for surveillance raises moral and legal concerns. A lot of IoT devices run on batteries, and regular data transmission can quickly drain them, requiring replacements or recharging regularly. For IoT devices to last for a longer time and be sustainable, they must function well and use little electricity. Further, IoT devices generate enormous amounts of data, which can be overwhelming and challenging to store, manage, and analyze efficiently. Real-time data processing calls for a large amount of computational power, which can be expensive and difficult to execute. Beyond all these issues, the IoT system installation and maintenance can be difficult, requiring specific expertise and abilities. The intricacy of an interconnected IoT system can make it difficult to locate and fix problems.

16.10 Conclusion

Development and standardization of IoT devices and policies are the need of the hour. In this chapter, the international and national standards that are to be followed for trending domains have been meticulously reviewed and detailed. Reduction of processing time, easier process automation, automated care and procedure audits, and better medical inventory management are the main goals of automatic data collection and transfer. Sensor devices provide patient-centered functions, especially those concerning patient diagnosis, by providing real-time data on health markers. Application domains include monitoring patient adherence to prescribed pharmaceutical regimens, tracking patient well-being alerts, and other telemedicine solutions. Further, the standards applicable to agricultural, food industry, smart-home, and industrial automation have also been reviewed thoroughly and presented in this chapter. This chapter also reviewed IEEE/IEC and ISO standards that are useful for standards cyber-security of data science perspective. An extensive review of various standards presented in the chapter would be a handheld material for the researchers and industrialists working in healthcare, agriculture, smart systems and industrial automation, cyber security, disaster management, and data analytics domains. Furthermore, the possible research scopes in these fields along with the problems in the existing complex ones at the global level are also being highlighted.

References

1 Asghari, P., Rahmani, A.M., and Javadi, H.H.S. (2019). Internet of Things applications: a systematic review. *Computer Networks* 148: 241–261.

2 Gokhale, P., Bhat, O., and Bhat, S. (2018). Introduction to IOT. *International Advanced Research Journal in Science, Engineering and Technology* 5 (1): 41–44.

3 Madakam, S., Ramaswamy, R., and Tripathi, S. (2015). Internet of Things (IoT): a literature review. *Journal of Computer and Communications* 3 (5): 164–173.

4 Stoyanova, M., Nikoloudakis, Y., Panagiotakis, S. et al. (2020). A survey on the Internet of Things (IoT) forensics: challenges, approaches, and open issues. *IEEE Communications Surveys & Tutorials* 22 (2): 1191–1221.

5 George, A.S. and George, A.H. (2020). Industrial revolution 5.0: the transformation of the modern manufacturing process to enable man and machine to work hand in hand. *Journal of Seybold Report* 15 (9): 214, 234.

6 Janiesch, C., Koschmider, A., Mecella, M. et al. (2020). The Internet of Things meets business process management: a manifesto. *IEEE Systems, Man, and Cybernetics Magazine* 6 (4): 34–44.

7 Hassan, R., Qamar, F., Hasan, M.K. et al. (2020). Internet of Things and its applications: a comprehensive survey. *Symmetry* 12 (10): 1674.

8 Whitmore, A., Agarwal, A., and Da Xu, L. (2015). The Internet of Things—A survey of topics and trends. *Information Systems Frontiers* 17: 261–274.

9 Pramanik, P.K.D., Pal, S., and Choudhury, P. (2018). Beyond automation: the cognitive IoT. Artificial intelligence brings sense to the Internet of Things. In: *Cognitive Computing for Big Data Systems Over IoT: Frameworks, Tools and Applications* (ed. A.K. Sangaiah, A. Thangavelu, and V.M. Sundaram), 1–37. Springer.

10 Tomazzoli, C., Scannapieco, S., and Cristani, M. (2023). Internet of Things and artificial intelligence enable energy efficiency. *Journal of Ambient Intelligence and Humanized Computing* 14 (5): 4933–4954.

11 Al-Turjman, F. and Abujubbeh, M. (2019). IoT-enabled smart grid via SM: An overview. *Future Generation Computer Systems* 96: 579–590.

12 Al-Ali, A.R. and Aburukba, R. (2015). Role of Internet of Things in the smart grid technology. *Journal of Computer and Communications* 3 (5): 229–233.

13 Ghasempour, A. (2019). Internet of Things in smart grid: architecture, applications, services, key technologies, and challenges. *Inventions* 4 (1): 22.

14 Syed, A.S., Sierra-Sosa, D., Kumar, A., and Elmaghraby, A. (2021). IoT in smart cities: a survey of technologies, practices and challenges. *Smart Cities* 4 (2): 429–475.

15 Jan, B., Farman, H., Khan, M. et al. (2019). Designing a smart transportation system: an Internet of Things and big data approach. *IEEE Wireless Communications* 26 (4): 73–79.

16 Niture, D.V., Dhakane, V., Jawalkar, P., and Bamnote, A. (2021). Smart transportation system using IOT. *International Journal of Engineering and Advanced Technology* 10 (5): 434–438.

17 Darmois, E., Elloumi, O., Guillemin, P., and Moretto, P. (2022). IoT standards–state-of-the-art analysis. In: *Digitising the Industry Internet of Things Connecting the Physical, Digital and Virtual Worlds* (ed. O. Vermesan and P. Friess), 237–263. River Publishers.

18 Pal, A., Rath, H.K., Shailendra, S., and Bhattacharyya, A. (2018). IoT standardization: the road ahead. In: *Internet of Things-Technology, Applications and Standardization* (ed. J. Sen), 53–74. IntechOpen.

19 Kassab, W.A. and Darabkh, K.A. (2020). A–Z survey of Internet of Things: architectures, protocols, applications, recent advances, future directions and recommendations. *Journal of Network and Computer Applications* 163: 102663.

20 Cirne, A., Sousa, P.R., Resende, J.S., and Antunes, L. (2022). IoT security certifications: challenges and potential approaches. *Computers & Security* 116: 102669.

21 Rosenbaum, S. (2011). The Patient Protection and Affordable Care Act: implications for public health policy and practice. *Public Health Reports* 126 (1): 130–135.

22 Chaturvedi, S. and Mohanty, S.K. (2010). Assessing the market openness effects of regulation in India: an overview of emerging trends and policy issues. *Journal of World Trade* 44 (5): 985.

23 Kedia, M., Sekhani, R., and Katiyar, T. (2020). *The Role of Standards in Diffusion of Emerging Technologies Internet of Things (IoT)*, No. 20-r-04. New Delhi, India: Indian Council for Research on International Economic Relations (ICRIER).

24 Behmanesh, A., Sayfouri, N., and Sadoughi, F. (2020). Technological features of Internet of Things in medicine: a systematic mapping study. *Wireless Communications and Mobile Computing* 2020 (1): 9238614.

25 Lin, Q. and Zhao, Q. (2021). IoT applications in healthcare. In: *Internet of Things: Cases and Studies* (ed. F.P.G. Márquez and B. Lev), 115–133. Springer.

26 Cerchione, R., Centobelli, P., Riccio, E. et al. (2023). Blockchain's coming to hospital to digitalize healthcare services: designing a distributed electronic health record ecosystem. *Technovation* 120: 102480.

27 Yao, J. and Warren, S. (2005). Applying the ISO/IEEE 11073 standards to wearable home health monitoring systems. *Journal of Clinical Monitoring and Computing* 19: 427–436.

28 Badawi, H.F., Laamarti, F., and El Saddik, A. (2018). ISO/IEEE 11073 personal health device (X73-PHD) standards compliant systems: a systematic literature review. *IEEE Access* 7: 3062–3073.

29 Andersen, B., Kasparick, M., Ulrich, H. et al. (2018). Connecting the clinical IT infrastructure to a service-oriented architecture of medical devices. *Biomedical Engineering/Biomedizinische Technik* 63 (1): 57–68.

30 Klonoff, A.N., Lee, W.A., Xu, N.Y. et al. (2023). Six digital health technologies that will transform diabetes. *Journal of Diabetes Science and Technology* 17 (1): 239–249.

31 Liao, Y. and Yeh, S. (2018). Predictability in human mobility based on geographical-boundary-free and long-time social media data. *2018 21st International Conference on Intelligent Transportation Systems (ITSC)*. pp. 2068–2073. IEEE.

32 Taimoor, N. and Rehman, S. (2021). Reliable and resilient AI and IoT-based personalised healthcare services: a survey. *IEEE Access* 10: 535–563.

33 Jayawardena, A., Waller, B., Edwards, B. et al. (2019). Portable audiometric screening platforms used in low-resource settings: a review. *The Journal of Laryngology & Otology* 133 (2): 74–79.

34 Arthi, S. and Sreenivas, T. V. (2020). Binaural spatial audiometry screening using android mobile device audio i/o facility. *2020 National Conference on Communications (NCC)*. pp. 1–6. IEEE.

35 Viangteeravat, T., Anyanwu, M.N., Nagisetty, V.R. et al. (2011). Clinical data integration of distributed data sources using Health Level Seven (HL7) v3-RIM mapping. *Journal of Clinical Bioinformatics* 1: 1–10.

36 Chang, H. and Kwak, Y.S. (2001). Introduction of the communication standard health level seven and HL7 Korea. *Journal of Korean Society of Medical Informatics* 7 (2): 161–172.

37 Chen, P. (2012). Study on medical image processing technologies based on DICOM. *Journal of Computers* 7 (10): 2354–2361.

38 Aggarwal, B. and Sahu, A. (2023). Information technology in radiology. In: *Textbook of Radiology and Imaging, Volume 1-E-Book* (ed. David Sutton and Bharat Aggarwal), 100. Elsevier.

39 Baraskar, T.N. and Mankar, V.R. (2019). The DICOM image compression and patient data integration using run length and Huffman encoder. In: *Coding Theory* (ed. S. Radhakrishnan and M. Sarfraz), 193. IntechOpen.

40 Farooq, M.S., Riaz, S., Abid, A. et al. (2019). A survey on the role of IoT in agriculture for the implementation of smart farming. *IEEE Access* 7: 156237–156271.

41 Leone, L. (2017). Beyond connectivity: the internet of food architecture between ethics and the EU citizenry. *Journal of Agricultural and Environmental Ethics* 30 (3): 423–438.

42 Manuel, A., Guerreiro, C., Ribeiro, J. et al. (2018). P2796 percutaneous treatment of severe mitral regurgitation with mitraclip device: potential role of NT-proBNP in prognosis assessment. *European Heart Journal* 39 (suppl_1): ehy565-P2796.

43 Senoo, E.E.K., Akansah, E., Mendonça, I., and Aritsugi, M. (2023). Monitoring and control framework for IoT, implemented for smart agriculture. *Sensors* 23 (5): 2714.

44 Suomi, P. and Oksanen, T. (2015). Automatic working depth control for seed drill using ISO 11783 remote control messages. *Computers and Electronics in Agriculture* 116: 30–35.

45 Schulz, S., Eberhardt, E., and Reznik, T. (2017). Information model for digital exchange of soil-related data-potential modifications on ISO 28258. *EGU General Assembly Conference Abstracts*. p. 7242.

46 Panagos, P., Van Liedekerke, M., Borrelli, P. et al. (2022). European Soil Data Centre 2.0: soil data and knowledge in support of the EU policies. *European Journal of Soil Science* 73 (6): e13315.

47 Kempen, B., Yigini, Y., Viatkin, K. et al. (2019). The Global Soil Information System (GloSIS)–concept and design. *Geophysical Research Abstracts*. Vol. 21, p. 1.

48 Escanciano, C. and Santos-Vijande, M.L. (2014). Reasons and constraints to implementing an ISO 22000 food safety management system: evidence from Spain. *Food Control* 40: 50–57.

49 Agus, P., Ratna Setyowati, P., Arman, H.A. et al. (2020). The effect of implementation integrated management system ISO 9001, ISO 14001, ISO 22000 and ISO 45001 on Indonesian food industries performance. *Test Engineering and Management* 82 (20): 14054–14069.

50 Gil, L., Ruiz, P., Escrivá, L. et al. (2017). A decade of Food Safety Management System based on ISO 22000: a global overview. *Revista de Toxicología* 34 (2): 84–93.

51 Craker, B., Danford, D. D., Ferreyra, R. A. et al. (2018). ADAPT: a Rosetta Stone for agricultural data. *Proceeding of the 14th International Conference on Precision Agriculture*. Vol. 13, p. 2020.

52 Steidle, M. and Herrmann, G.A. (2019). Group certification: market access for smallholder agriculture. In: *Sustainable Global Value Chains* (ed. M. Schmidt, D. Giovannucci, D. Palekhov, and B. Hansmann), 639–656. Cham: Springer International Publishing.

53 Korneeva, E., Olinder, N., and Strielkowski, W. (2021). Consumer attitudes to the smart home technologies and the Internet of Things (IoT). *Energies* 14 (23): 7913.

54 Babayigit, B. and Abubaker, M. (2023). Industrial Internet of Things: a review of improvements over traditional scada systems for industrial automation. *IEEE Systems Journal* 18 (1): 120–133.

55 Nkuba, C.K., Kim, S., Dietrich, S., and Lee, H. (2021). Riding the IoT wave with VFuzz: discovering security flaws in smart homes. *IEEE Access* 10: 1775–1789.

56 Franco, A., Crisostomi, E., and Hammoud, M. (2023). Monitoring of public buildings via energy-efficient Z-wave wireless sensors. *Journal of Physics: Conference Series* 2648 (1): 012033. IOP Publishing.

57 Elhadi, S., Chhiba, L., Sael, N., and Marzak, A. (2022). Operating modeling of medium range protocols IoT. *International Conference on Digital Technologies and Applications*. pp. 284–293. Cham: Springer International Publishing.

58 Mota, A., Serôdio, C., and Valente, A. (2024). Matter protocol integration using Espressif's solutions to achieve smart home interoperability. *Electronics* 13 (11): 2217.

59 Belli, D., Barsocchi, P., and Palumbo, F. (2023). Connectivity Standards Alliance Matter: state of the art and opportunities. *Internet of Things* 25: 101005.

60 Ortega i Blasi, A. (2022). Evaluating Thread protocol in the framework of Matter. Master's thesis. Universitat Politècnica de Catalunya.

61 Alkama, L. and Bouallouche-Medjkoune, L. (2021). IEEE 802.15.4 historical revolution versions: a survey. *Computing* 103 (1): 99–131.

62 Chen, F., Wang, N., German, R., and Dressler, F. (2010). Simulation study of IEEE 802.15.4 LR-WPAN for industrial applications. *Wireless Communications and Mobile Computing* 10 (5): 609–621.

63 Minoli, D. (2013). *Building the Internet of Things with IPv6 and MIPv6: The Evolving World of M2M Communications*. Wiley.

64 Iglesias-Urkia, M., Orive, A., Urbieta, A., and Casado-Mansilla, D. (2019). Analysis of CoAP implementations for industrial Internet of Things: a survey. *Journal of Ambient Intelligence and Humanized Computing* 10 (7): 2505–2518.

65 Tuexen, M., Stewart, R., Jesup, R., and Loreto, S. (2017). *RFC 8261: Datagram Transport Layer Security (DTLS)*. Encapsulation of SCTP Packets.

66 Rix, T., Detken, K. O., and Jahnke, M. (2016). Transformation between XML and CBOR for network load reduction. *2016 3rd International Symposium on Wireless Systems within the Conferences on Intelligent Data Acquisition and Advanced Computing Systems (IDAACS-SWS)*. pp. 106–111. IEEE.

67 Cilfone, A., Davoli, L., Belli, L., and Ferrari, G. (2019). Wireless mesh networking: an IoT-oriented perspective survey on relevant technologies. *Future Internet* 11 (4): 99.

68 Khattak, S.B.A., Nasralla, M.M., Farman, H., and Choudhury, N. (2023). Performance evaluation of an IEEE 802.15.4-based thread network for efficient Internet of Things communications in smart cities. *Applied Sciences* 13 (13): 7745.

69 Lu, Y. and Asghar, M.R. (2020). Semantic communications between distributed cyber-physical systems towards collaborative automation for smart manufacturing. *Journal of Manufacturing Systems* 55: 348–359.

70 Dahlmanns, M., Lohmöller, J., Fink, I. B. et al. (2020). Easing the conscience with OPC UA: an internet-wide study on insecure deployments. *Proceedings of the ACM Internet Measurement Conference*. pp. 101–110.

71 Ivanova, T.A., Batchkova, I.A., and Belev, Y.A. (2019). Information modeling of intelligent and secure cyber-physical production systems using OPC UA. *Machines. Technologies. Materials.* 13 (12): 542–545.

72 Nazarenko, A.A., Sarraipa, J., Camarinha-Matos, L.M. et al. (2021). Analysis of relevant standards for industrial systems to support zero defects manufacturing process. *Journal of Industrial Information Integration* 23: 100214.

73 Pitsonyane, K.S. and Mnjama, N. (2022). Records management in an ISO certified environment: a case study of Botho University in Botswana. *ESARBICA Journal: Journal of the Eastern and Southern Africa Regional Branch of the International Council on Archives* 41: 101–120.

74 Pereira, C.E., Diedrich, C., and Neumann, P. (2023). Communication protocols for automation. In: *Springer Handbook of Automation* (ed. S.Y. Nof), 535–560. Cham: Springer International Publishing.

75 Silva, M., Pereira, F., Soares, F. et al. (2015). An overview of industrial communication networks. In: *New Trends in Mechanism and Machine Science: From Fundamentals to Industrial Applications* (ed. P. Flores and F. Viadero), 933–940. Springer.

76 Cusimano, J. (2022). Industrial control system risk assessment standards and leading practices in the chemical industry. *Process Safety Progress* 41 (4): 665–669.

77 McLean, C., Lee, Y. T., Jain, S. et al. (2011). Modeling and simulation of critical infrastructure systems for homeland security applications. Tech. Rep. NISTIR, 7785. US Nat. Inst. Standard Technol., Gaithersburg, MD, USA.

78 Williams, B. (2008). *Intelligent Transport Systems Standards.* Artech House.

79 To, N.T. and Kato, T. (2018). Characteristics and development of policy and institutional structures of emergency response in Vietnam. *International Journal of Disaster Risk Reduction* 31: 729–741.

80 Tanimura, K. and Yoshikawa, K. (2014). Crisis management system for largescale disasters. *Hitachi Review* 63 (5): 35–39.

81 Potter, J., Aylward, D., Hatfield, D., and Dwarkanath, S. (2004). Architecture and principles of a modern integrated emergency medical information network. *Advanced Emergency Nursing Journal* 26 (2): 103–109.

82 Dalela, P. K., Saldhi, A., Bhave, P., and Tyagi, V. (2020). Common alerting protocol compliant emergency warning and alert system for legacy broadcasting networks. *2020 IEEE Conference on Cognitive and Computational Aspects of Situation Management (CogSIMA).* pp. 1–5. IEEE.

83 Luo, F., Wang, B., Fang, Z. et al. (2021). Security analysis of the TSN backbone architecture and anomaly detection system design based on IEEE 802.1 Qci. *Security and Communication Networks* 2021 (1): 6902138.

84 Marques, N., Zúquete, A., and Barraca, J.P. (2020). EAP-SH: an EAP authentication protocol to integrate captive portals in the 802.1 X security architecture. *Wireless Personal Communications* 113 (4): 1891–1915.

85 Walshe, R. (2021). The road to big data standardisation. In: *The Elements of Big Data Value* (ed. Edward Curry, Andreas Metzger, Sonja Zillner, Jean-Christophe Pazzaglia, and Ana García Robles), 333–354. Springer.

86 Vale, K.M.A.C. and de Alencar, F.M.R. (2020). Challenges, patterns and sustainability indicators for cloud computing. *Brazilian Journal of Development* 6 (8): 57031–57053.

87 Mohammed, C.M. and Zeebaree, S.R. (2021). Sufficient comparison among cloud computing services: IaaS, PaaS, and SaaS: a review. *International Journal of Science and Business* 5 (2): 17–30.

88 Quilachamin, W.G., Alonso, I.A., and Herrera-Tapia, J. (2018). Overview of service and deployment models offered by cloud computing, based on International Standard ISO/IEC 17788. *International Journal of Advanced Computer Science and Applications* 9 (11): 218.

89 Antonopoulos, N. and Gillam, L. (2010). *Cloud Computing*, vol. 51, No. 7. London: Springer.

90 Terfas, H., Suryn, W., Roy, J., and Eftekhar, S. M. (2018). Extending ISO/IEC 19086 cloud computing SLA standards to support cloud service users with the SLA negotiation process. *SQM XXVI, 127.*

91 Kamaruddin, N.A., Mohamed, I., Jarno, A.D., and Daud, M. (2020). Cloud security pre-assessment model for cloud service provider based on ISO/IEC 27017: 2015 additional control. *Revolution* 2 (5): 1–17.

92 Coallier, F. (2022). IoT standardization strategies in ISO/IEC JTC 1/SC 41. *2022 IEEE 8th World Forum on Internet of Things (WF-IoT)*. pp. 1–6. IEEE.

17

Challenges and Future Directions

Burhan Khan[1], Aabid A. Mir[3], Nur F.L.M. Rosely[4], and Khang W. Goh[2,5]

[1]*Department of Computer System and Technology, Faculty of Computer Science and Information Technology, Universiti Malaya, Kuala Lumpur, Malaysia*
[2]*School of Engineering, Shinawatra University, Pathum Thani, Thailand*
[3]*Malaysian Institute of Information Technology, Universiti Kuala Lumpur, Kuala Lumpur, Malaysia*
[4]*School of Computer Science, Faculty of Innovation and Technology, Taylor's University, Kuala Lumpur, Malaysia*
[5]*Faculty of Data Science and Information Technology, INTI International University, Nilai, Malaysia*

17.1 Introduction

Innovations such as the use of the Internet of Things (IoT) in smart grid technology are paving the way for a new era in energy systems, where the way power has been traditionally generated, distributed, and managed in the past is being redefined. This integration of IoT into smart grids makes the system intelligent, efficient, reliable, sustainable, and capable of surviving the rising requirements of the 21st century. This introduction provides insight into the background and importance of IoT in smart grids, along with the evolution and state of the art of this technology, its paradigm-changing effects, and the need to address the emerging challenges of the developments in the future.

17.1.1 Context and Significance of IoT in Smart Grids

A smart grid is a transformation of traditional energy networks, which have been largely centralized, unidirectional, and mechanically controlled. Advanced smart grids can be defined as a technology-driven strategy in which intelligent solutions enable a self-balancing two-way power flow and information exchange, thereby empowering them to self-manage and respond to new requirements of electrical end users for the digital age [1]. This infrastructure is further enhanced upon integration with IoT technologies, which enable improved energy management, advanced fault detection, and more precise balancing of power loads from renewable sources [2].

The grid is populated with IoT devices, including smart meters, sensors, and automated controllers, which measure and control energy from multiple sources. These data decide how to better use electric power among generations, distributions, and consumption, which are also collected by many of those devices. IoT data can also be used to improve the grid's reliability by predicting when system failures are likely to occur and also helps to model an energy-stressed grid with greater detail using real-time information and dynamic pricing that promotes consumer energy efficiency [3].

IoT for Smart Grid: Revolutionizing Electrical Engineering, First Edition.
Edited by Rahiman Zahira, Palanisamy Sivaraman, Chenniappan Sharmeela, and Sanjeevikumar Padmanaban.

17.1.2 Evolution and Current Landscape of Smart Grid Technology

The evolution of smart grid technology started as an endeavor to enable an electric power distribution system with information technology (IT) capabilities, aiming to address some reliability and efficiency challenges associated with the grid. This has evolved to include various technologies, such as automatic control, novel energy generation and storage forms, and advanced metering infrastructure (AMI) [4]. The necessity for such innovations cannot be overstated, as utilities look to tackle the demands of aging infrastructure, growing peak demand, and diminishing customer tolerance for blackout while upholding high levels of resilience and demand for low-carbon energy, which threaten to strangle utility business models. Figure 17.1 illustrates the evolution from traditional to modern smart grids, highlighting the role of IoT integration in facilitating this transition.

In 2021, we witnessed the rapid global expansion of smart grids, primarily in response to pressure to improve grid management and decarbonize imposed by government environmental targets. Europe, North America, and some parts of Asia have taken the first steps in implementing renewable energy sources on a scale beyond control thus far [5].

17.1.3 The Transformative Impact of IoT Integration on Energy Systems

The IoT's influence on the world's energy systems, such as smart grids, prospects significant transparency secrets from their keystones. The first benefit is the improvement in grid resilience. The real-time monitoring and rapid response capabilities of IoT devices can enhance the reliability of the energy supply by reducing the potential impact of power outages and system failures [6].

Furthermore, IoT allows the incorporation of variable and distributed renewable energy sources such as solar and wind. The complications leading to these phenomena arise from managing a more complex system. IoT solutions are part of this complexity, assisting in ensuring that the energy produced is effectively distributed or conserved, depending on the requirements of the initial wave [7]. Table 17.1 summarizes the key impacts of the IoT on smart grids.

Table 17.1 illustrates the revolutionary energy management and distribution era by integrating the IoT into smart grid systems. The well-planned integration of IoT strengthens grid reliability through real-time monitoring and automated control systems and caters to potential deviations in the energy flow to ensure consistency. The infrastructure is designed to collect data from multiple devices over the entire grid, from driving advanced demand–response programs to providing on-site energy-efficiency programs using new IoT sensors and smart meter hardware. This type of data-driven control lowers operational expenditure and increases the lifespan of the grid by being able to predict and prevent breakdowns and collapses before they occur.

Furthermore, IoT technologies are critical in leveraging intermittent renewable energy sources, including wind and solar energy, predicting energy outputs, and managing fluctuations to stabilize the grid. Moreover, IoT can also serve as a tool in the hands of consumers with information on consumption flowing through the network, helping consumers make informed decisions, increasing awareness about energy, and even encouraging efficiency and sustainability behaviors, which, in turn, increases the efficiency and sustainability of the grid.

Furthermore, IoT integration into smart grids solves the central scalability and standardization issue, as it scales systems to work efficiently with escrowing loads as systems grow in scope. These enhanced measures provide additional cybersecurity security that avoids potential attacks on sensitive data and grid integrity that can occur in the digitally integrated future. As IoT grows and scales, it will provide the grid with greater intelligence and redefine the interaction between consumers and their energy environments, making smart grids an essential enabler of modern infrastructure.

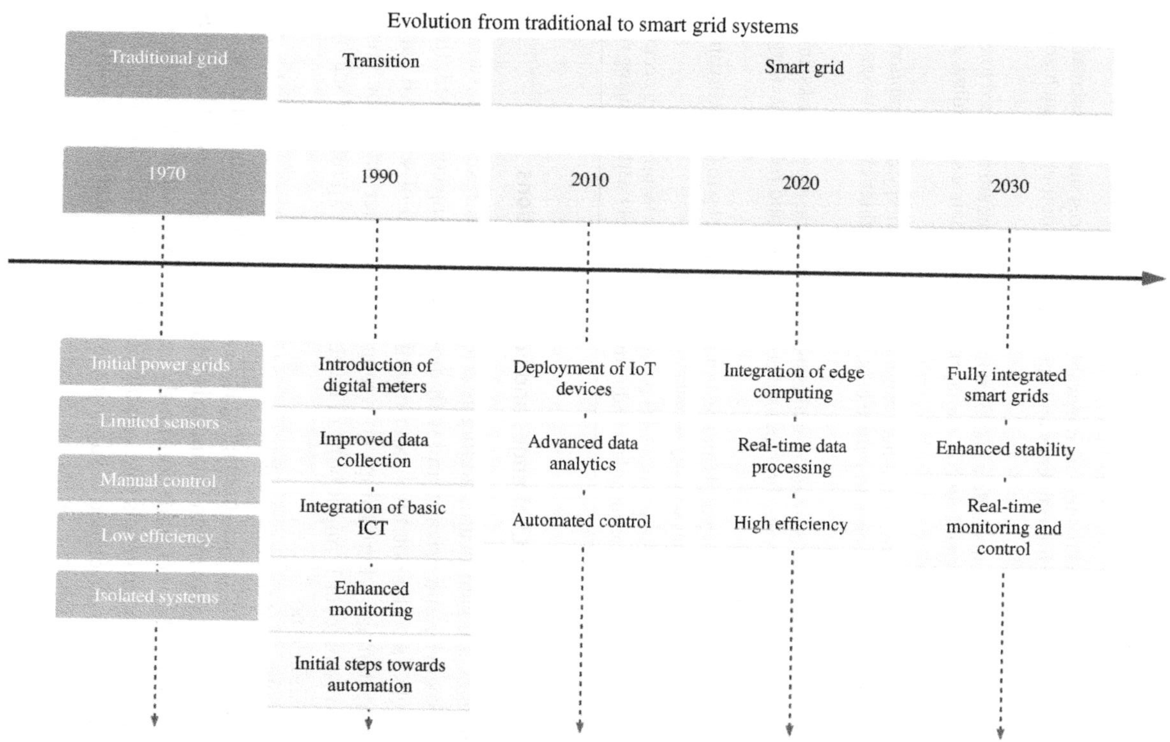

Figure 17.1 Schematic representation of the evolution from traditional to smart grid systems, emphasizing the role of IoT technologies.

Table 17.1 Overview of the key transformative impacts of IoT integration into smart grid systems.

Impact category	Description
Enhanced grid reliability	IoT devices facilitate real-time monitoring and automated control systems that predict and react to system disturbances, significantly improving grid reliability
Efficient energy management	Smart meters and sensors allow for more precise energy usage tracking and management, enabling demand–response systems that adjust energy flow based on real-time demand
Renewable energy integration	IoT technologies help integrate renewable energy sources such as solar and wind power by managing their intermittency and optimizing their output to match grid demand
Advanced-data analytics	The massive amounts of data collected from IoT devices are analyzed to enhance decision-making regarding grid management, maintenance scheduling, and load distribution
Consumer engagement and control	With IoT applications, consumers can monitor their energy consumption in real time, make informed decisions to reduce bills, and manage their energy usage more actively
Operational efficiency	IoT applications reduce operational costs by automating routine tasks, improving maintenance operations through predictive analytics, and reducing energy losses across the grid
Cybersecurity enhancements	Advanced security protocols and real-time monitoring by IoT devices can help identify and mitigate potential security threats to the grid
Regulatory compliance	IoT enables better compliance with energy regulations by providing accurate data reporting and facilitating quicker adjustments to comply with changing regulations

17.1.4 Importance of Addressing Challenges and Future Directions

Although the advantages are significant, adding IoT to smart grids results in a few challenges. The top concerns are security and privacy, and with more things being connected, there are more avenues for cyberattacks [8]. The potential exposure of a great deal of sensitive data that might cause more significant damage where things go wrong is yet another flag to be raised. Personal consumer information (whatever it might be) can be accessed and exploited if security measures differ from what they should be.

Another challenge is the need for more scalability and standardization of IoT technologies. As the number of connected devices increases, managing the system's performance and the compatibility challenges between various platforms and standards for connected devices [9] becomes complex.

These challenges need to be met with constantly evolving cybersecurity, strong data privacy standards, and interoperability protocols that can adapt to potential rapid changes in technology. Moreover, regulations must be developed to support the technology and keep current with its state to ensure that smart grids can operate safely and effectively.

The rest of this chapter details these challenges, discusses the state of the art and new trends, and outlines what should be done for future developments of IoT-based smart grid technologies.

17.2 Security and Privacy Concerns in Transactive Systems

The primary objective is the optimal energy distribution in an entirely peer-to-peer (P2P) scenario targeting classical economic theory mechanisms (market mechanisms) in real life, for consumption and production, for consumers and producers, particularly in IoT-enabled smart grids, the so-called transactive systems. Such systems also face immense security and privacy challenges because their interconnected nature must be carefully understood. The following subsections 17.2.1 and 17.2.2 will address the security and privacy concerns in the transactive system.

17.2.1 Data Security Challenges

Implementing IoT devices in smart grids has caused unprecedented problems regarding data security and communication networks between devices and market participants. Ensuring data integrity, confidentiality, and availability over diverse meshes of devices and systems is a significant challenge. Further elaboration is shown below.

17.2.1.1 Vulnerabilities Specific to IoT-Enabled Smart Grids

IoT-enabled smart grids depend highly on functional interconnected devices and smart data-driven operations. Key vulnerabilities include the following:

- **Attack Surface Expansion:** Introducing IoT devices to smart grids significantly increases the attack surface. Weak authentication protocols and unencrypted communication channels allow attackers to compromise devices such as smart meters and connected home appliances [10]. One example is a compromised smart meter that could hijack energy, leading to illegal energy theft or incorrect billing.
- **Data Integrity Risks:** In transactive energy systems, energy data are utilized in crucial market-ing decisions. Playing with this data can lead to incorrect pricing and financial loss. False data injection attacks (FDIA) can disrupt the integrity [8]. The result of attacks in FDIA may mean that incorrect data informed by smart meters or sensors is being provided to the system operators, and this incorrect data may lead to wrong market transactions.
- **Distributed Denial of Service (DDoS):** Botnets have the potential to invoke many compromised IoT devices to launch DDoS attacks on the smart grid infrastructure, disrupting grid stability and availability [11]. These attacks may result in significant financial losses and disruptions in critical infrastructure, such as hospitals and public transportation.
- **Inadequate Firmware Updates:** Most IoT devices in smart grids do not have secure and automatic firmware update procedures, making them vulnerable to known vulnerabilities [12]. A case in point is the Mirai botnet, which uses unpatched firmware-based vulnerabilities in IoT devices to carry out large-scale DDoS attacks.
- **Lack of Device Authentication:** Rogue devices can enter the smart grid network without strong authentication. This gap allows malicious devices to modify data, interrupt services, or pry into legitimate devices [13].

17.2.1.2 Case Studies of Cybersecurity Incidents

This section is designed to present several case studies of cybersecurity incidents. Cybersecurity incidents highlight the urgent need for new security paradigms in transactive systems:

- **Colonial Pipeline Ransomware Attack (2021):** In 2021, Colonial Pipeline experienced a ransomware attack that required the company to cease pipeline operations, resulting in the unavailability of fuel across the US East Coast. The source of entry was a secret password exposed to a successful attack.

- **Texas Winter Storm Power Outage (2021):** Indirectly, the 2021 Texas Winter Storm power outage focused on how a modern-day power distribution network is vulnerable and could be exploited in the utility sector, as none of it is a cybersecurity appliance; thus, the much broader outage showed us that the grid needs to be much more secure and resilient [14].
- **Oldsmar Water Treatment Plant Attack (2021):** In February 2021, a hacker tried to poison Oldsmar, Florida's water supply, by tampering with the level of sodium hydroxide at the water treatment plant. Commensurate with cybersecurity vulnerability, the attempted attack was a sober reminder that critical infrastructure should be in safe hands [15].
- **Ukraine Power Grid Attack (2015):** The Ukraine power grid attack is still considered one of the foundational case studies for smart grid security. Defenders of Democracy, the Sandworm group, named after the series of complex and well-thought attacks they reportedly conducted, used spear-phishing emails to access the IT network. Subsequently, they instigated supervisory control and data acquisition (SCADA) systems with remote access tools (RAT), causing a blackout of approximately 230,000 people [16]. During the incident, weak password policies and ineffective network segmentation were targeted, highlighting the need for a holistic cybersecurity approach.
- **Stuxnet Worm (2010):** The virus, believed to be aimed at nuclear establishments in Iran, was a highly sophisticated worm that could exploit zero-day vulnerabilities in industrial control systems (ICS). It propagates through USB drives and exploits the vulnerability of Siemens programmable logic controllers (PLCs) to blow up centrifuges. At inception, Stuxnet was developed to target nuclear facilities; nonetheless, because smart grids share a similar ICS dependence on critical infrastructure as nuclear plants, many of the methods developed for Stuxnet could be easily converted [17].

These events highlight the critical vulnerabilities in modern infrastructure, stressing the need for robust cybersecurity. It covers issues from ransomware and power outages to attempted water supply poisoning and sophisticated cyberattacks on control systems. These underscore the urgency for comprehensive and resilient security measures across all sectors.

17.2.2 Privacy Issues

Transactive systems improve the availability of energy distribution, but they come at the cost of serious privacy concerns as customers consume granular data. In addition, privacy becomes more vulnerable owing to the interconnectedness of devices and systems.

17.2.2.1 Risks Associated with Consumer Data Collection and Analysis

Specific risks from this extensive data collection relate to potential impacts on consumers and possible effective tools to mitigate each of these risks.

- **Granular Energy Consumption Data:** The long-term potential of smart meters to gather microdata on consumer energy usage is possible. For instance, they can identify when some appliances are off or on, enabling general predictions regarding the target household's user routines, presence, and lifestyle [18]. This level of insight can also reveal consumer behavior and presents a privacy issue when exposed to parties who should not have access.
- **Third-Party Data Sharing:** Data consumed by an individual is shared with third parties and is often necessary for data analytics, which increases the probability of unauthorized access or misuse. For example, sharing data with home appliance manufacturers to run targeted ads might violate privacy if permission has not been explicitly granted [13].

- **Behavioral Profiling:** This practice connects energy consumption to other data sources to develop exacting consumer behavior profiles, which can facilitate unjust, discriminatory practices [19]. Additionally, insurance companies could use it to offer lower premiums, whereas employers could use it to monitor their employees.
- **Identity Theft:** If hackers can hack into smart meter databases through poor data protection, they can access a list of personal information, opening up an opportunity for identity theft. For example, a hacked database can provide a potential adversary with physical addresses and usage patterns, as well as credit card billing [8].
- **Smart Home Device Integration:** This connects the smart grid with the smart home, making the consumer's data liable to privacy concerns. For example, smart thermostats and security cameras generate sensitive data, which, if compromised, could enable the tracking of consumer behavior using the data [10].
- **IoT Device Misconfiguration:** Misconfigured IoT devices, including smart meters or home energy management systems, may accidentally reveal private consumer data for unauthorized access [12].

17.2.2.2 Regulations and Best Practices for Data Privacy

The expanding value of data privacy has directed the development of several regulatory frameworks worldwide. The General Data Protection Regulation (GDPR) in the EU, the California Consumer Privacy Act (CCPA) in the United States, the Personal Data Protection Act (PDPA) in Singapore, and Brazil's General Data Protection Law (LGPD) intend to protect consumer data through consent, access, rectification, and deletion rights. Additionally, best practices for data privacy in transactive systems include data anonymization, purpose limitation, strong access controls, encryption, data retention policies, and consumer education. These measures collectively ensure the protection of personal data and uphold the privacy rights of individuals. More elaboration is as follows:

- **GDPR:** It places many stipulations on information nonpublication, user consent, access to and correction of knowledge, and right to erasure enforcement beginning 25 May 2018 [20]. Key principles include:
- **Consent:** Individuals must consent to collect and use their data in the first instance.
- **Access and Rectification:** Consumers should have the right to access and correct their data.
- **Data Minimization:** Data should be adequate, relevant, and limited to what is necessary in relation to the purposes for which they are processed.
- **Right to be Forgotten:** Users can ask for their data to be removed.
- **CCPA:** It provides California residents with the right to knowledge about the personal data that companies collect and the ability to request that data be deleted. Key provisions include:
- **Right to Know:** Consumers can demand to know what personal data is being collected.
- **Right to Delete:** Consumers also have the right to delete their information.
- **Right to Opt-out:** The consumer is given the right to opt out of having their data sold [21].
- **PDPA – Singapore:** PDPA is an omnibus data protection regulation encompassing data protection, consent, and data breach notification requirements. Key provisions include:
- **Data Protection:** Organizations must safeguard personal data from unauthorized access.
- **Consent:** Data collection and processing require each user's permission.
- **Data Breach Notification:** Organizations should inform affected individuals and authorities if their system is breached [22].
- **LGPD:** The LGPD is the Brazilian General Data Protection Law that stipulates the requirement for organizations to obtain informed consent to process data, protect data, and give data subjects rights to access and delete their data [23].

- **Best Practices for Data Privacy in Transactive Systems:**
- **Data Anonymization:** Apply privacy protection schemes to ensure customer anonymity [24], for example, aggregate energy consumption to protect identities.
- **Purpose Limitation:** Data should be collected and processed only for specific purposes. It clearly articulates the reason for the collection and avoids repurposing the data without explicit consent.
- **Access Control:** Implement strong access controls to prevent unauthorized access to consumer information. Limit access based on the user and allow only approved users to create and implement tags. Restrict tag creation and usage based on user roles using role-based access control (RBAC) and attribute-based access control (ABAC) to restrict access to necessary stakeholders.
- **Encryption:** Encrypt all data at rest and in transit to protect data from unauthorized access. Employ robust encryption algorithms such as AES-256.
- **Data Retention Policies:** Use data retention policies to prevent unnecessary data retention. Periodically purge or de-identify old data that are no longer required.
- **Consumer Education:** Educate consumers about their data privacy rights and how to manage their data. This indicates the method for opting out of data collection or requesting data deletion.

Wrapping these up, the global highlighting on data privacy has directed robust regulatory frameworks such as the GDPR, CCPA, PDPA, and LGPD, all focusing on consent, access, rectification, and deletion rights to protect consumer data. Employing best practices such as data anonymization, purpose limitation, strong access controls, encryption, data retention policies, and consumer education is critical to ensure the privacy and security of personal data in transactive systems. These measures together safeguard individuals' privacy rights and promote trust in data handling practices.

17.3 Scalability and Standardization Issues

Scalability and standardization remain two main challenges in IoT-supported smart grids and are critical in providing seamless implementation of transactive smart grid solutions. As the number of interconnected devices grows exponentially, ensuring efficient data exchange management and communication becomes increasingly complex. In some cases, it becomes even more of a mammoth task. This scalability issue is due to the massive data traffic created by millions of IoT devices, which causes traffic congestion, data latency, and data storage issues. In addition, devices from various manufacturers use varying proprietary protocols, exacerbating the complexity and presenting interoperability problems. However, standardization is essential for maintaining consistent protocols, communication standards, and security. The provision of uniform data formats contributes to improved security and privacy landscapes. The disadvantage of this is the unavailability of global standards, and this fragmentation in protocols and security measures makes it more challenging to combine multi-factor authentication (MFA) with other systems, which adds to the complexity of the scale. In transactive systems, scalability and standardization may present crucial concerns, particularly in managing the interoperation and efficiency embedded in the IoT adaptation of smart grids.

17.3.1 Scalability Challenges

The ability to scale is critical in managing the increasing number of deployed devices and the massive volume of data generated in IoT-enabled smart grids. Problematic areas include technological and operational challenges. Also, performance at scale is almost impossible to maintain even with guarded consideration, regardless of the hardware, storage, or network level. The following section 17.3.1.1 address the scalability challenges in further detail.

17.3.1.1 Technological and Operational Barriers to Scaling IoT Infrastructures

Scaling IoT-enabled smart grids involves major technological and operational hurdles. Technological challenges include network congestion, data storage and processing limits, and device variety. Operations challenges include system management, power consumption, and data quality assurance. The main reason is to create seamless and efficient communication among the millions of connected devices, which is the first barrier to overcome.

- **Technological Barriers:**
- **Network Congestion:** It occurs when many IoT devices normally communicate with a smart grid. Growing data traffic results in packet loss and latency issues that cause a decrease in the overall network performance [25].
- **Data Storage and Processing:** When discussing IoT, it produces a large amount of data that must be stored and processed. However, most traditional cloud-based architectures fail to satisfy latency and data management at scale [26].
- **Device Heterogeneity:** Differences in the capabilities of IoT devices lead to IoT interoperability issues, such as different communication protocols or data formats [27].
- **Operational Barriers:**
- **System Management:** As the network grows, managing millions of devices, performing firmware updates, and guaranteeing comprehensive security for each device becomes infeasible [28].
- **Energy Consumption:** IoT devices should consume little energy, particularly in intelligent geological deployments. Large-scale devices increase energy consumption, which decreases network sustainability [29].
- **Data Quality:** With the increase in the volume of data, there is a critical need to maintain good data quality. In addition, inaccurate or incomplete data [30] affects decision-making processes in smart grids [30].

17.3.2 Solutions for Maintaining System Performance at Scale

Innovative ways to maintain system performance in line with the scale of IoT-enabled smart grids can be the way forward. Some major strategies include edge computing, data aggregation, load balancing, network virtualization, and blockchain-based access control [31]. Alleviating network congestion reduces data retention, energy efficiency, and seamless communication between homogeneous/heterogeneous device solutions.

- **Edge Computing:** Edge computing involves moving data processing closer to the data source, thereby improving latency, data fidelity, and network congestion.
- **Data Aggregation:** Aggregating data at intermediate nodes helps reduce the data transmission overhead and improve scalability [12].
- **Load Balancing:** Evenly distributes network traffic across all nodes to prevent overloading and improve system performance.
- **Network Virtualization:** This approach deploys network functions without changing or adding hardware, enhancing network capacity to manage better and allocate resources among IoT devices [32].

Table 17.2 Scalability challenges and solutions.

Challenge	Solution
Network congestion	Edge computing
Data storage and processing	Data aggregation
Device heterogeneity	Network virtualization
System management	Blockchain-based access control
Energy consumption	Energy harvesting
Data quality	Load balancing

- **Energy Harvesting:** Energy harvesting techniques enable IoT devices to work with renewable sources, contributing to efficient energy consumption [26].
- **Blockchain-Based Access Control:** Applying blockchain for decentralization and data integrity helps scale transactive systems (Table 17.2) [33].

17.3.3 Standardization and Interoperability

Given the millions of devices running in IoT-enabled smart grids, standardization and dependence are two important elements that facilitate smooth data exchange and communication among all data generation devices. The proprietary nature of IoT devices and their different protocols causes interoperability issues. Standards work to implement common protocols and standards to enable all devices from different vendors and systems to communicate with one another effectively.

They also have security and privacy standards for encryption, access control, and some (limited) data privacy implementations. These standards are crucial in transactive systems that process, analyze, and share consumers' data among several parties.

17.3.3.1 The Need for Uniform Protocols and Standards

Smart grids require uniform protocols and standards to achieve interoperability and flawless connectivity between IoT devices. Real-time application programming interfaces (APIs) facilitate information exchange between agencies and are instrumental in mitigating issues with interoperability, data integrity, and data privacy. Standardization helps make proper decisions, involves no discrepancy in data formats, and ensures the same level of security across myriad devices.

- **Interoperability Issues:** IoT devices use proprietary protocols, and often, each device must be from the same manufacturer to operate together, which limits integration and complexity. Standardization guarantees that devices communicate with each other [27].
- **Security and Privacy Concerns:** A standardized protocol enables a consistent and ubiquitous form of security and privacy practices on devices, reducing security vulnerabilities [29].
- **Data Quality and Accuracy:** Using APIs/standard data formats can enhance data quality and enable the accuracy required for decision-making in a transactive system [30].

17.3.3.2 Efforts Toward Global Standardization

Standardization efforts at a global level aim to establish comprehensive guidelines and protocols for the integration and interoperability of various IoT devices to work in a coordinated manner. We are obliged to use IoT devices in daily life for many applications. Institute of Electrical and Electronics

Table 17.3 Key standards for smart grid interoperability.

Standard	Organization	Description
IEEE 2030.5 (SEP)	IEEE	Smart energy profile for secure communication
IEEE 802.15.4	IEEE	Low-power wireless communication protocols
IEC 61850	IEC	Communication networks for power utility automation
IEC 62351	IEC	Security standards for communication networks
ISO/IEC 27001	ISO/IEC	Information security management framework
NIST Smart Grid Framework	NIST	Guidelines for smart grid interoperability

Engineers (IEEE), International Electrotechnical Commission (IEC), International Organization for Standardization (ISO), and National Institute of Standards and Technology (NIST) have defined standards (e.g., IEEE 2030.5, IEC 61850, and ISO/IEC 27001) for better security, interoperability, and information management in smart grids.

- **IEEE Standards Association:**
- **IEEE 2030.5 (Smart Energy Profile):** This provides a standard for integrating smart grid applications and establishing a secure and interoperable communication link between utilities and IoT devices [34].
- **IEEE 802.15.4:** Low-power wireless communication protocol for IoT devices commonly found in ZigBee networks [35].
- **IEC:**
- **IEC 61850:** Communication networks and systems for power utility automation (data model, interoperability) [36].
- **IEC 62351:** Provides securities specifications for communication networks in a smart grid context [37].
- **ISO:**
- **ISO/IEC 27001:** Provides information security management in IoT-enabled smart grids [38].
- **NIST:**
- **NIST Framework for Smart Grid Interoperability:** Provides guidelines and standards for interoperability in smart grids (Table 17.3) [39].

17.4 Emerging Trends in Transactive IoT

Owing to the technologies and market models of the transactive IoT, the energy industry is becoming more secure, transparent, and efficient. Blockchain and artificial intelligence (AI) have helped predict maintenance or load in a specific area, indications shared in a decentralized energy market, etc. [40]. Market dynamics are also changing, influencing adoption rates, market drivers, and growth barriers in the transactive IoT.

17.4.1 Technological Innovations

Blockchain is at the forefront of technological development, enabling transactive IoT systems to process energy transactions securely, transparently, and efficiently. Blockchain can also roll out tamperproof, decentralized energy trading platforms and AI and machine learning (ML) technologies to optimize predictive maintenance, load management, and energy forecasting. Smart inverters make prosumers active in helping maintain the grid, which increases reliability and reduces the work of maintaining the grid and changing market dynamics.

17.4.1.1 Blockchain for Secure and Transparent Transactions

Transactive IoT will be possible through blockchain's significant benefits and features (decentralized, secure, and transparent energy transactions). Blockchain uses smart contracts and an immutable ledger, which, in return, wipes out intermediaries and minimizes transaction costs, enabling consumers to engage in the energy market directly. It also helps with data integrity and improves security in energy trading.

- **Decentralized Energy Trading:** Blockchain makes it easier for consumers to participate in P2P energy trading, buying and selling excess power directly. This decentralized model minimizes transaction costs and puts consumers at the center of the energy markets [41].
- **Smart Contracts:** Smart contracts are self-executing contracts with the terms of the agreement between buyer and seller being directly written into lines of code, in which the codes and the agreements contained therein exist across a distributed, decentralized blockchain network, allowing the contracts to both be self-enforcing and more efficient by reducing the role of intermediation [42].
- **Data Integrity and Security:** Blockchain using its immutable ledger form volunteer impracticable with unwanted data washed off and tamper sage stopped, and the involved party from changing any information in energy transactions. In addition, the decentralized nature of blockchain makes security processes more robust, mitigating the risk of single-point failures [33].
- **Energy Credit Management:** Blockchain can play a role in the accurate and real-time tracking of energy credits in renewable energy markets. It provides transparency and accountability regarding energy production and use [43].

17.4.1.2 Advances in AI and Machine Learning for Predictive Maintenance and Load Management

AI models incorporate historical and real-time data to achieve these predictions. These improvements help lower the downtime, stabilize the grid, increase efficiency, and reduce energy consumption in the aggregate.

- **Predictive Maintenance:** ML algorithms ingest data from historical systems and real-time data powered through sensors and IoT devices to predict when equipment will likely fail and schedule maintenance well in advance. Thus, downtime and maintenance costs can be minimized [44].
- **Load Management:** Complex AI models optimize load management, forecast energy demand, and adjust supply in real time. This helps increase grid stability and eliminates the need to use more expensive peak load power plants [45].
- **Energy Forecasting:** This application is a quintessential example of how AI-based energy forecasting models are used to predict renewable energy generation and consumer demand with high fidelity, thus ensuring efficient energy trading and load balancing [46].
- **Fraud Detection:** ML algorithms can be employed in fraud detection to capture abnormalities in energy consumption patterns and identify potential energy theft or fraud in transactive IoT networks (Table 17.4) [47].

Table 17.4 Technological innovations in transactive IoT.

Technology	Applications
Blockchain	Decentralized energy trading, smart contracts, and data integrity
Artificial intelligence	Predictive maintenance, load management, and energy forecasting
Machine learning	Anomaly detection and fraud detection

17.4.2 Market Trends

The greater uptake of transactive IoT owing to mandates and government support, a rise in consumer demand for clean energy solutions, and demands for improved energy efficiency and grid reliability to fuel the rapid growth of the market. However, interoperability, security, and high implementation costs remain obstacles to growth. Global adoption rates, market drivers, and barriers ensure that those looking to take advantage of emerging opportunities in the transactive IoT landscape are equipped with critical background information.

17.4.2.1 Adoption Rates, Market Drivers, and Growth Barriers

The fragmented industry also presents several challenges; once these are addressed, adoption rates will remain uneven. However, as per regional trends, while the global adoption of transactive IoT continues to improve at a global level steadily, the transition to adoption varies between regions, and the rates of smart grid deployment between regions differ greatly. Key market drivers, including sustainability targets, regulatory backing, and consumer appetite, are fueling uptakes, with profound challenges such as interoperability challenges, security fears, and cost implications inhibiting progress.

- **Adoption Rates:**
- **Global Adoption:**
  The global IoT energy market is growing at a compound annual growth rate (CAGR) of 16% and is expected to increase by the end of 2027 [48].
- **Regional Trends:**
  - **North America:** Leading the smart grid deployment and P2P energy trading with regulatory support [49].
  - **Europe:** The adoption rates are high due to regulatory standards on pollution control and consumer interest in green energy production [50].
  - **Asia-Pacific:** Smart city projects expansion, booming energy consumption, and growth [51].
- **Market Drivers:**
- **Sustainability Goals:** The rising shift toward renewable energy and sustainability goals propel the demand for transactive IoT solutions across energy markets [52].
- **Regulatory Support:** Governments worldwide encourage smart grid deployment, P2P energy trading, and Industrial Internet of Things (IIoT) edge management systems [49].
- **Consumer Demand:** Consumers are more specific and demand better control, information, and environmentally friendly energy services, encouraging P2P trading and smart meters [50].
- **Growth Barriers:**
- **Interoperability Issues:** The use of proprietary protocols and the absence of a unified standard make it difficult for IoT devices to communicate well with each other [27].

Table 17.5 Market drivers and growth barriers in transactive IoT.

Market drivers	Growth barriers
Sustainability goals	Interoperability issues
Regulatory support	Security and privacy concerns
Consumer demand	High implementation costs

- **Security and Privacy Concerns:** The transactive IoT involves designing a decentralized, interconnected system with a concealed security/privacy risk [13].
- **High Implementation Costs:** It costs a lot to make IoT devices and retrofit smart grids or similar devices on a small scale (Table 17.5) [48].

17.5 Future Developments in Transactive IoT

The transactive IoT is evolving further with the advent of new technologies and integration in future energy systems. These advancements are expected to define the future of energy distribution, smart cities, and sustainable development.

17.5.1 Next-Generation IoT Technologies

Next-generation IoT technologies such as 5G, quantum computing, and edge computing would make a great revolution in how more efficient transactions are performed in IoT by providing ultralow latency, enhanced processing capabilities, and much more efficient data management.

17.5.1.1 Potential Impacts of 5G, Quantum Computing, and Edge Computing

- **5G Networks:** The relatively recent introduction of 5G networks will offer ultralow latency and high bandwidth, which can be used for real-time energy trading and demand–response for transactive IoT systems. 5G, on the other hand, will enable massive machine-type communication (mMTC) billions of IoT devices to communicate with each other seamlessly and exchange data [19].
- **Quantum Computing:** Real-life data processing applications in transactive IoT involve solving complex optimization problems in real time. Quantum computing is a technology ready to change how data can be processed through machines using quantum bits (qubits versus existing bits). These practices will improve load balancing, predictive maintenance, and grid optimization. Quantum cryptography will also benefit from better security and privacy in the transaction of energy.
- **Edge Computing:** Here, data processing is farmed out to IoT devices or edge nodes, which can ease the central infrastructure's data processing requirements while reducing latency and network traffic load – real-time data processing at the network edge.
- **Artificial Intelligence (AI):** AI will remain a centerpiece for processing data from IoT devices, predicting maintenance, forecasting load, and detecting fraud. Advanced AI models will also facilitate decentralized energy trading and transactive IoT network optimization (Table 17.6) [45].

Table 17.6 Impacts of next-generation IoT technologies on transactive IoT.

Technology	Potential impacts
5G networks	Real-time energy trading and massive IoT device connectivity
Quantum computing	Real-time optimization, enhanced security, and grid optimization
Edge computing	Reduced latency, predictive maintenance, and real-time load management
Artificial intelligence	Load forecasting, fraud detection, and decentralized energy trading

17.5.2 Future Energy Systems

The future of both energy systems will depend heavily on the growth of renewable energy sources such as wind and solar energy and the development of smart cities. These systems will use the transactive IoT to distribute energy while optimally promoting sustainability.

17.5.2.1 Integration with Renewable Energy Sources

Grids incorporate more renewable energy production from sources such as solar and wind, which supply energy intermittently, so it makes sense for the various bits to serve as a system rather than independent players.

- **Distributed Energy Resources (DERs):** Transactive IoT will provide a secure channel for prosumers created by DERs, such as rooftop solar and wind turbines, to sell their surplus energy back to the grid through P2P trading.
- **Energy Storage Systems:** Energy storage systems, such as batteries and pumped hydro, are of utmost importance to cope with the intermittency of renewable energy sources. The transactive IoT will also enable the functionality of stored energy as a "tradeable".
- **Microgrids:** Microgrids are small local energy systems operating in isolation or concert with an overarching electrical grid. The transactive IoT will be used for demand–response, energy trading, and grid optimization (Table 17.7).

17.5.2.2 Smart Cities and IoT: Expanding the Ecosystem

The IoT ecosystem is increasing with smart cities incorporating energy management systems, smart grids, and sustainable infrastructure.

- **Smart Grids:** Transactive IoT will also operate smart grids for real-time energy management, predictive maintenance, and decentralized trading. For better billing and energy theft detection and AMI [13].
- **Sustainable Infrastructure:** Green buildings and electric vehicle (EV) charging stations will deploy transactive IoT for energy management and optimization.

Table 17.7 Key components of future renewable energy systems.

Component	Description
Distributed energy resources	Rooftop solar panels and wind turbines
Energy storage systems	Batteries pumped hydro
Microgrids	Local energy systems and demand–response

Table 17.8 Expanding the smart city ecosystem with transactive IoT.

Smart city component	Role of transactive IoT
Smart grids	Real-time energy management and decentralized trading
Sustainable infrastructure	Energy management and optimization
Urban mobility	Energy trading and efficient EV charging

- **Urban Mobility:** Using Ubiquitous Smart Grid Urban Solutions, urban mobility solutions (e.g., electric buses or autonomous vehicles) will have smart grid-powered flexible charging and load management capabilities. Transactive IoT allows EVs to trade energy with the grid, which is enabled through transactive IoT (Table 17.8).

17.6 Policy, Regulation, and Ethical Considerations

The rapid proliferation of the transactive IoT mandates the development of sound policy frameworks, regulations, and ethical guidelines that can guarantee secure, fair, and sustainable future energy markets. This study explores the policy, regulatory settings, and ethical considerations shaping transactive IoT.

17.6.1 Regulatory Landscape

They must develop and implement comprehensive policy frameworks governing transactive IoT and smart grid activities.

17.6.1.1 Energy Market Regulation

- **European Energy Union Strategy:** The EU Energy Union Strategy is intended to establish a single European energy market. The single Euro market is an example of such integration that facilitates online trading and aids in promoting the integration of renewable energy, decentralized markets, and smart grid deployment. Thus, the objectives foster customer empowerment and cross-border energy trading to enable a more integrated and flexible EU energy system [53].
- **Federal Energy Regulatory Commission (FERC):** FERC oversees wholesale electricity markets in the United States, where P2P energy trading and demand–response programs may fall under their jurisdiction. These markets are organized around regional wholesale balancing authorities, where the FERC Order 841/841-A requires wholesale market participation and a level playing field for energy storage [54].
- **Office of Gas and Electricity Markets (OFGEM):** This is the UK's regulator of electricity markets and comes under the energy sector. This can foster the use of smart meters, demand-side responses, and decentralized energy trading. The regulator supports developing a smart, low-carbon energy system per the UK government's net-zero goals [55].
- **Smart Grid Regulation in India:** IoT-enabled transactive systems are envisaged as part of India's smart grid vision. Regulatory Framework: The Central Electricity Regulatory Commission (CERC) elaborates a regulatory framework for demand–response and decentralized energy markets [56].

17.6.1.2 Cybersecurity Standards

- **ISO/IEC 27001:** International standard providing a model for establishing, implementing, operating, monitoring, reviewing, maintaining, and improving an information security management system. It assists organizations in creating, implementing, maintaining, and improving information security management systems [38].
- **NIST Cybersecurity Framework:** Offer guidelines for protecting smart grids and IoT-enabled energy systems from cyber threats developed by the US NIST. This consists of better risk control strategies and developing a cybersecurity posture for businesses [39].
- **IEC 62351:** This standard, created by the IEC, covers the communication networks used in smart grids, focusing on cybersecurity. It offers security communication protocols and access control and ensures data integrity [37].

17.6.2 Ethical Considerations

An ethical framework must steer the design and implementation of transactive IoT systems.

17.6.2.1 Energy Equity

Clean and affordable energy should be accessible to all consumers [52]. The inequality in energy access is too serious problem for us to allow it to worsen, so appropriate measures must be enforced to prevent this from happening with the transactive IoT. An example of this is the ability of grid-connected microgrids to ensure that some of the energy reaches underserved rural area outlets to those areas.

17.6.2.2 Consumer Consent

Users must provide informed consent for data collection and use, ensuring transparency in how their data is handled. Establishing clear and understandable data policies is essential for building consumer trust. Additionally, embedding privacy into the design of systems helps to secure all consumer data and safeguard user rights.

17.6.2.3 Sustainability

Broader sustainability motives for Active IoT include using more renewable energy sources and reducing environmental impacts [52]. It involves cutting carbon emissions and backing energy efficiency to help build a circular economy.

17.6.2.4 Algorithmic Fairness

The input lists are linked to the articles you liked and want to share that you believe reflect well-argued cases based on facts about the development and deployment of energy pricing, load management, and predictive maintenance algorithms that need to be transparent and nondiscriminatory. They cannot be based on those that discriminate between certain consumer groups or promote unfair pricing practices [57].

17.6.2.5 Cybersecurity and Data Integrity

Energy transaction data are important, and they must be secure and correct. Cybersecurity actions are to be executed in an environment where data artistic taste for end users and an evil-natured market fan of energy [13].

17.7 Conclusion

In this chapter, we present a detailed review of the hurdles and problem statements of transactive IoT systems in the future. This would allow for unleashing the transformational role of IoT in smart grid systems, provided that security and privacy beliefs, scalability concerns, and ethical problems are targets for stakeholders. This could still be achieved by welcoming new technologies such as blockchain and AI, combined with strong regulations, making the perfect formula for an energy future that is safe, fair, and with a lasting foundation. In the long run, all these collective endeavors will leverage IoT's massive transformative power to bring a new outlook to smart grid technology, helping the whole technology and infrastructure a step forward into the modern age.

17.7.1 Summary of Key Points

A system in which the smart grid is integrated with IoT technologies means that this is the path to the fourth industrial revolution in energy management and distribution. However, this introduces a wide range of security, privacy, scalability, and standardization problems. This chapter selects a complete review of these fears and discusses the imminent patterns and future of the transactive IoT framework.

17.7.1.1 Security and Privacy Concerns in Transactive Systems

Transactive systems provide market mechanisms for energy distribution in almost real time; however, they are essentially interlinked and introduce critical issues in terms of security and privacy. Some notable data security challenges are attacking surface expansion, data integrity risks, and DDoS attacks. Case studies of cybersecurity incidents such as the Colonial Pipeline and Ukraine power grid hacks are just a few examples of the many factors demonstrating the urgent need for better security practices. Granular energy consumption data and sharing data with third parties only worsen privacy risks associated with collecting consumer data.

17.7.1.2 Scalability and Standardization Issues

Network congestion, data storage, and device heterogeneity are major obstacles to the scale integration of IoT-based smart grids. Performance may be affected by technological and operational barriers in the system. Solutions such as edge computing, data aggregation, load balancing, network virtualization, and blockchain-based access control can help achieve scalability. Standards such as IEEE 2030.5 and ISO/IEC 27001 are important for tackling these interoperability problems and ensuring that security regulations are enforced every time.

17.7.1.3 Emerging Trends in Transactive IoT

Blockchain, AI, and ML help to ensure that transactions are secure, maintenance is performed promptly, and the load is controlled. Blockchain enables decentralized energy trading and smart contracts, while AI optimizes energy distribution through effective load management. Many businesses have adopted circular business models, driven by market trends such as sustainability goals, regulatory measures, and consumer demand.

17.7.1.4 Future Developments in Transactive IoT

Transactive IoT is poised to change on a large scale with next-generation technologies such as 5G, quantum computing, and edge computing, which enable ultralow latency, enhanced processing, and data management efficiencies. As future-proof energy systems, higher integration with renewable energies, microgrids, and smart city ecosystems is expected to be realized.

17.7.1.5 Policy, Regulation, and Ethical Considerations

A strong foundation for policy frameworks and ethical guidelines is necessary. All three include consumer consent, data access, and rectification and thus closely echo the motivations behind GDPR, CCPA, and LGPD: energy equity, consumer consent, algorithmic fairness, and cybersecurity.

17.7.2 Call to Action for Stakeholders

Recognizing the benefits of transactive IoT systems requires aligning stakeholders across the energy ecosystem and the wider technology and policy industries to help solve these challenges and unlock the potential of these systems.

17.7.2.1 Industry Leaders

Leverage technological advancements such as blockchain or AI to make energy transactions secure, transparent, and more effective. Invest more in research and development (R&D) for improved cybersecurity capabilities and scalable solutions.

17.7.2.2 Academia and Researchers

Undertake interdisciplinary research to design novel security paradigms, large-scale IoT architectures, and ethical algorithms. Post results help to increase the level of information and develop a virtual best practice library.

17.7.2.3 Regulatory Bodies and Policymakers

Develop an overarching policy that balances technological advancement with consumer rights and privacy – implementing well-defined protocols and cybersecurity recommendations to secure an unbiased and fun energy market.

17.7.3 Recommendations for Industry, Academia, and Policymakers

17.7.3.1 Industry

- **Adopt Blockchain Technology:** Establishment of decentralized energy trading platforms that help empower prosumers and reduce transaction costs.
- **Enhance Cybersecurity Measures:** Form cybersecurity teams to guard against new threats and apply for ISO/IEC 27001 and NIST.
- **Invest in Predictive Maintenance:** AI models trained in the cloud can be run to make predictions on the local devices/edge locations where data originate and immediately allow corresponding decisions.

17.7.3.2 Academia

- **Develop Ethical Algorithms:** Create transparent and nondiscriminatory algorithms for energy pricing and load management.
- **Research Scalable IoT Architectures:** Create edge computing frameworks for reduced network congestion and storage bottlenecks.
- **Promote Collaborative Research:** Celebrate and encourage interdisciplinary and industry/regulatory collaborations.

17.7.3.3 Policymakers

- **Implement Uniform Data Privacy Regulations:** Data privacy regulations should be harmonized across regions to guarantee identical consumer protection.
- **Establish Global Standardization Efforts:** Work with IEEE, IEC, NIST, and other standards organizations to standardize globally and create common protocols.
- **Encourage Innovation through Regulatory Sandboxes:** Permit companies to test new business models and technology under certain conditions.

References

1 Gonzalez-Longatt, F. and Torres, J.L.R. (ed.) (2018). *Advanced Smart Grid Functionalities Based on Powerfactory*, 19–48. Cham, Switzerland: Springer International Publishing.

2 Zhang, Q., Cheng, L., and Boutaba, R. (2017). Cloud computing: state-of-the-art and research challenges. *Journal of Internet Services and Applications* 1 (1): 7–18.

3 Kabalci, Y. (2016). A survey on smart metering and smart grid communication. *Renewable and Sustainable Energy Reviews* 57: 302–318.

4 Martins, J.F., Pronto, A.G., Delgado-Gomes, V., and Sanduleac, M. (2019). Smart meters and advanced metering infrastructure. In: *Pathways to a Smarter Power System* (ed. A. Taşcıkaraoğlu and O. Erdinç), 89–114. Academic Press.

5 International Energy Agency (2021). *World Energy Outlook 2021*. Paris, France: IEA.

6 Saleem, Y., Crespi, N., Rehmani, M.H., and Copeland, R. (2019). Internet of things-aided smart grid: technologies, architectures, applications, prototypes, and future research directions. *IEEE Access* 7: 62962–63003.

7 Sarker, E., Halder, P., Seyedmahmoudian, M. et al. (2021). Progress on the demand side management in smart grid and optimization approaches. *International Journal of Energy Research* 45 (1): 36–64.

8 Gunduz, M.Z. and Das, R. (2020). Cyber-security on smart grid: threats and potential solutions. *Computer Networks* 169: 107094.

9 Stankovic, J.A. (2014). Research directions for the Internet of Things. *IEEE Internet of Things Journal* 1 (1): 3–9.

10 Guan, Z., Lu, X., Yang, W. et al. (2021). Achieving efficient and privacy-preserving energy trading based on blockchain and ABE in smart grid. *Journal of Parallel and Distributed Computing* 147: 34–45.

11 Shah, Z., Ullah, I., Li, H. et al. (2022). Blockchain based solutions to mitigate distributed denial of service (DDoS) attacks in the Internet of Things (IoT): a survey. *Sensors* 22 (3): 1094.

12 Zhang, Y., Wang, G., and Wang, Q. (2018). IoT-enabled real-time data-driven energy management in a microgrid. *IEEE Internet of Things Journal* 5 (6): 4377–4388.

13 Leszczyna, R. (2018). Cybersecurity and privacy in standards for smart grids–A comprehensive survey. *Computer Standards & Interfaces* 56: 62–73.

14 Walsh, C. (2022). The Texas storm that proved the need for grid investments. https://www.nrdc.org/bio/christy-walsh/texas-storm-proved-need-grid-investments (accessed 29 October 2024).

15 Greenberg, A. (2021). A hacker tried to poison a Florida City's water supply. Wired. https://www.wired.com/story/oldsmar-florida-water-utility-hack/ (accessed 29 October 2024).

16 Lee, R.M., Assante, M.J., and Conway, T. (2016). Analysis of the cyber-attack on the Ukrainian power grid. *Electricity Information Sharing and Analysis Center (E-ISAC)* 388 (1–29): 3.

17 Bakić, B., Milić, M., Antović, I. et al. (2021). 10 years since Stuxnet: what have we learned from this mysterious computer software worm? *2021 25th International Conference on Information Technology (IT)*. pp. 1–4. IEEE.

18 Armoogum, S. and Bassoo, V. (2019). Privacy of energy consumption data of a household in a smart grid. In: *Smart Power Distribution Systems* (ed. Qiang Yang, Ting Yang, and Wei Li), 163–177. Academic Press.

19 Salam, T., Rehman, W.U., and Tao, X. (2019). Data aggregation in massive machine type communication: Challenges and solutions. *IEEE Access* 7: 41921–41946.

20 Hansen, J., Wilson, P., Verhoeven, E. et al. (2021). *Assessment of the EU Member States' Rules on Health Data in the Light of GDPR*. Luxembourg: European Union.

21 State of California Department of Justice (2024). *California Consumer Privacy Act (CCPA)*. State of California - Department of Justice - Office of the Attorney General. https://oag.ca.gov/privacy/ccpa.

22 Baskaran, H., Yussof, S., Rahim, F.A., and Bakar, A.A. (2020). Blockchain and the personal data protection act 2010 (PDPA) in Malaysia. *2020 8th International Conference on Information Technology and Multimedia (ICIMU)*. pp. 189–193. IEEE.

23 Usercentrics (2024). *LGPD: An Overview of Brazil's General Data Protection Law*. Consent Management Platform (CMP) Usercentrics. https://usercentrics.com/knowledge-hub/brazil-lgpd-general-data-protection-law-overview/.

24 Wang, J., Wu, L., Zeadally, S. et al. (2021). Privacy-preserving data aggregation against malicious data mining attack for IoT-enabled smart grid. *ACM Transactions on Sensor Networks (TOSN)* 17 (3): 1–25.

25 Ketu, S. and Mishra, P.K. (2022). A contemporary survey on IoT-based smart cities: architecture, applications, and open issues. *Wireless Personal Communications* 125 (3): 2319–2367.

26 Liu, Y., Peng, M., Shou, G. et al. (2020). Toward edge intelligence: multiaccess edge computing for 5G and Internet of Things. *IEEE Internet of Things Journal* 7 (8): 6722–6747.

27 Astropekakis, K., Drakakis, E., Grammatikakis, K., and Goumopoulos, C. (2022). A survey of IoT software platforms. In: *Advances in Computing, Informatics, Networking and Cybersecurity: A Book Honoring Professor Mohammad S. Obaidat's Significant Scientific Contributions* (ed. P. Nicopolitidis, S. Misra, L.T. Yang, et al.), 299–326. Cham: Springer International Publishing.

28 Bittencourt, L., Immich, R., Sakellariou, R. et al. (2018). The internet of things, fog and cloud continuum: integration and challenges. *Internet of Things* 3: 134–155.

29 Ponnusamy, V. and Sharma, B. (2021). Investigation on IoT intrusion detection in wireless environment. *2021 International Conference on Computer & Information Sciences (ICCOINS)*. pp. 7–13. IEEE.

30 Abir, S.A.A., Anwar, A., Choi, J., and Kayes, A.S.M. (2021). IoT-enabled smart energy grid: applications and challenges. *IEEE Access* 9: 50961–50981.

31 Al Hwaitat, A.K., Almaiah, M.A., Ali, A. et al. (2023). A new blockchain-based authentication framework for secure IoT networks. *Electronics* 12 (17): 3618.

32 Lin, G., Dong, M., Ota, K. et al. (2019). Security function virtualization based moving target defense of SDN-enabled smart grid. *ICC 2019-2019 IEEE International Conference on Communications (ICC)*. pp. 1–6. IEEE.

33 Srivastava, S., Chaurasia, B.K., and Singh, D. (2023). Blockchain-based IoT security solutions. In: *Distributed Computing to Blockchain*, 327–339. Academic Press.

34 *IEEE Standard for Smart Energy Profile Application Protocol* (IEEE Std 2030.5-2018 (Revision of IEEE Std 2030.5-2013)). (2018). https://doi.org/10.1109/IEEESTD.2018.8608044

35 Dutta, D. (2019). IEEE 802.15.4 as the MAC protocol for Internet of Things (IoT) applications for achieving QoS and energy efficiency. *Advances in Communication, Cloud, and Big Data: Proceedings of 2nd National Conference on CCB 2016*. pp. 127–132. Singapore: Springer.

36 Kumar, S., Abu-Siada, A., Das, N., and Islam, S. (2023). Review of the legacy and future of IEC 61850 protocols encompassing substation automation system. *Electronics* 12 (15): 3345.

37 Hussain, S.S., Ustun, T.S., and Kalam, A. (2019). A review of IEC 62351 security mechanisms for IEC 61850 message exchanges. *IEEE Transactions on Industrial Informatics* 16 (9): 5643–5654.

38 Kitsios, F., Chatzidimitriou, E., and Kamariotou, M. (2023). The ISO/IEC 27001 information security management standard: how to extract value from data in the IT sector. *Sustainability* 15 (7): 5828.

39 Gopstein, A., Nguyen, C., O'Fallon, C. et al. (2021). *NIST Framework and Roadmap for Smart Grid Interoperability Standards, Release 4.0*. Gaithersburg, MD: Department of Commerce. National Institute of Standards and Technology.

40 Lu, X. (2024). Application of artificial intelligence algorithms in precision marketing with flow data analysis models. *2024 5th International Conference on Mobile Computing and Sustainable Informatics (ICMCSI)*. pp. 378–383. IEEE.

41 Kouhizadeh, M., Saberi, S., and Sarkis, J. (2021). Blockchain technology and the sustainable supply chain: theoretically exploring adoption barriers. *International Journal of Production Economics* 231: 107831.

42 Munsing, E., Mather, J., and Moura, S. (2017). Blockchain for decentralized optimization of demand response. *Energy Research & Social Science* 39: 321–331.

43 Sikorski, J.J., Haughton, J., and Kraft, M. (2017). Blockchain technology in the chemical industry: machine-to-machine electricity market. *Applied Energy* 195: 234–246.

44 Wang, X., Wang, H., Bhandari, B., and Cheng, L. (2024). AI-empowered methods for smart energy consumption: a review of load forecasting, anomaly detection and demand response. *International Journal of Precision Engineering and Manufacturing-Green Technology* 11 (3): 963–993.

45 Han, W., Yuan, Y., Wang, Y., and Wang, Z. (2018). Load prediction model based on machine learning algorithm. *International Journal of Smart Grid and Clean Energy* 7 (1): 31–37.

46 Raza, S., Wallgren, L., and Voigt, T. (2019). AI-based energy load forecasting for smart grids. *Ad Hoc Networks* 11 (8): 2661–2674.

47 Aldegheishem, A., Anwar, M., Javaid, N. et al. (2021). Towards sustainable energy efficiency with intelligent electricity theft detection in smart grids emphasising enhanced neural networks. *IEEE Access* 9: 25036–25061.

48 Statista. (2024). Global IoT enterprise market worldwide 2019-2027. https://www.statista.com/statistics/1369033/global-iot-market-size/#:~:text=The%20global%20market%20for%20enterprise,around%20238%20billion%20by%202023 (accessed 29 October 2024).

49 Brown, M.A. and Zhou, S. (2019). Smart-grid policies: an international review. In: *Advances in Energy Systems: The Large-scale Renewable Energy Integration Challenge* (ed. P.D. Lund), 127–147. Hoboken, NJ: Wiley.

50 Sospiro, P., Amarnath, L., Di Nardo, V. et al. (2021). Smart grid in China, EU, and the US: state of implementation. *Energies* 14 (18): 5637.

51 Thakur, A. (2023). Smart cities: mapping the market analysis for the future. In: *Perspectives on the Transition Toward Green and Climate Neutral Economies in Asia* (ed. P. Ordóñez de Pablos, M.N. Almunawar, and M. Anshari), 143–158. IGI Global.

52 Ball, C.S. and Degischer, D. (2024). IoT implementation for energy system sustainability: the role of actors and related challenges. *Utilities Policy* 90: 101769.

53 Leal-Arcas, R., Grasso, C., and Ríos, J.A. (2016). The creation of the European Energy Union. In: *Energy Security, Trade and the EU*, 381–448. Edward Elgar Publishing.

54 Konidena, R. (2019). FERC Order 841 levels the playing field for energy storage. *MRS Energy & Sustainability* 6: E5.

55 Ofgem. (2024). *Welcome to Ofgem*. https://www.ofgem.gov.uk/ (accessed 29 October 2024).

56 Rose, A., Wayner, C., Koebrich, S. et al. (2020). *Policy and Regulatory Environment for Utility-Scale Energy Storage: India*, No. NREL/TP-6A20-78101. Golden, CO: National Renewable Energy Lab (NREL).

57 Rahwan, I., Cebrian, M., Obradovich, N. et al. (2019). Machine behavior. *Nature* 568 (7753): 477–486.

Index

IoT for Smart Grid: Revolutionizing Electrical Engineering, First Edition.
Edited by Rahiman Zahira, Palanisamy Sivaraman, Chenniappan Sharmeela, and Sanjeevikumar Padmanaban.

Printed and bound by CPI Group (UK) Ltd, Croydon, CR0 4YY
17/09/2026
14964335-0006